Innovations in Engineered Porous Materials for Energy Generation and Storage Applications

Innovations in Engineered Porous Materials for Energy Generation and Storage Applications

Editors

Ranjusha Rajagopalan

Institute of Superconducting and Electronics Materials
University of Wollongong
Innovation Campus, Squires Way
North Wollongong, NSW
Australia

Avinash Balakrishnan

Suzlon Energy Limited
Material Technology Lab
Paddhar, Bachau Road, Kukama
Bhuj, Kutch, Gujarat
India

CRC Press
Taylor & Francis Group
Boca Raton London New York

CRC Press is an imprint of the
Taylor & Francis Group, an **informa** business

A SCIENCE PUBLISHERS BOOK

CRC Press
Taylor & Francis Group
6000 Broken Sound Parkway NW, Suite 300
Boca Raton, FL 33487-2742

First issued in paperback 2020

© 2018 by Taylor & Francis Group, LLC
CRC Press is an imprint of Taylor & Francis Group, an Informa business

No claim to original U.S. Government works

ISBN-13: 978-1-138-73902-4 (hbk)
ISBN-13: 978-0-367-78130-9 (pbk)

Library of Congress Cataloging-in-Publication Data

Names: Rajagopalan, Ranjusha, editor. | Balakrishnan, Avinash, editor.
Title: Innovations in engineered porous materials for energy generation and
storage applications / editors, Ranjusha Rajagopalan (Institute of
Superconducting and Electronics Materials, University of Wollongong,
Innovation Campus, Squires Way, North Wollongong, NSW, Australia), Avinash
Balakrishnan (Suzlon Energy Limited, Material Technology Lab, Paddhar,
Bachau Road, Kukama, Bhuj, Kutch, Gujarat, India).
Description: Boca Raton, FL : CRC Press, 2018. | "A science publishers book."
| Includes bibliographical references and index.
Identifiers: LCCN 2018001641 | ISBN 9781138739024 (hardback)
Subjects: LCSH: Energy storage. | Electrodes. | Porous materials.
Classification: LCC TK2980 .I56 2018 | DDC 621.31/260284--dc23
LC record available at https://lccn.loc.gov/2018001641

Visit the Taylor & Francis Web site at
http://www.taylorandfrancis.com

and the CRC Press Web site at
http://www.crcpress.com

Preface

The field of renewable energy generation and storage sectors has seen an upsurge in research and development activities and has made significant and rapid strides in device development. We have foreseen a renewed interest in this emerging field (specifically the field of porous based materials) by both the student population and scientists and engineers. This book originated from Dr. Balakrishnan and Dr. Rajagopalan's sustained research and substantial research background in the area of porous energy materials and their application to energy generation and storage devices. This book intends to cater to a broad base of seniors and graduate students having varied backgrounds such as physics, electrical and computer engineering, chemistry, mechanical engineering, materials science, nanotechnology and even to a reasonably well-educated layman interested in porous based materials for variety applications. Given the present unavailability of a "mature" textbook having suitable breadth of coverage (although basic books and plethora of journal articles are available with the added difficulty of referring to multiple sources), we have carefully designed the book layout and contents with contributions from well-established experts in their respective fields. This book is aimed at, graduate and postgraduate students/researchers in the aforementioned disciplines.

The book consists of 13 well-rounded chapters arranged in a logical and distilled fashion. Each chapter is intended to provide an overview with examples chosen primarily for their educational purpose. The readers are encouraged to expand on the topics discussed in the book by reading the exhaustive references provided towards the end of each chapter. The chapters have also been written in a manner that fits the background of different science and engineering fields. Therefore, the subjects have been given a primarily qualitative structure and in some cases providing detailed quantitative analysis. Based on our own experience, the complete set of topics contained in this book can be covered in a single semester and prepare the student for a research program in the advancing field of porous materials, apart from equipping the student for mastering the subject.

In order to augment the research topics and help the reader grasp the fundamental nuances of the subject each chapter caters several simple, well-illustrated equations and schematic diagrams. The progression of chapters is designed in such a way that the basic theory and techniques are introduced early on, leading to the evolution of the field of porous materials in the areas of energy storage and generation. The readers will find this logical evolution highly appealing as it introduces a didactic element to the reading of the textbook apart from grasping the essentials of an important subject. Wherever possible, color versions of the figures are incorporated, and they can also be made accessible through online prints.

We, the editors (Avinash Balakrishnan and Ranjusha Rajagopalan) express our thanks to the dedicated scientists who have written the individual chapters. Their enthusiasm in writing the chapters of high quality and delivering on time after incorporating the review comments, made the release of the textbook a simplified task for us. We would also like to thank the editorial team (CRC Press) for encouraging us to begin this project and guiding it to its completion. Thanks for their excellent attention to detail and for their constant review of the project progress. In addition, we express our thanks to our colleague Ms. Shaymaa Al-Rubaye, and Professors from Institute superconducting and electronics materials (ISEM), University of Wollongong (UOW) (Distinguished Professor Hua Kun Liu and Director Professor Shi Xue Dou). Our sincere thanks, to Suzlon Energy Limited team members (Mr. Hitesh Nanda, Mr. Thanu Subramoniam, Dr. Sachin Bramhe, Mr. Vinayak Sabane, Mr. Deepu Surendran, Mr. Harinath P.N.V., Mr. Alok Singh, Mr. Nagaprakash M.B., Mr. Rishikesh Karande) for their immense support. The completion of this book would not have been possible without support from the funding agency, ARENA Smart Sodium Storage System program, under which Dr. Ranjusha Rajagopalan is working at ISEM, UOW.

Ranjusha Rajagopalan
Associate Research Fellow
University of Wollongong
Wollongong, Australia

Avinash Balakrishnan
Manager, Suzlon Blade Technology
Materials Laboratory
Suzlon Energy Limited, Bhuj, India

Contents

POROUS MATERIALS IN ENERGY GENERATION

NEW PERSPECTIVES AND TRENDS

POROUS MATERIALS IN ENERGY STORAGE

1

Exploration for Porous Architecture in Electrode Materials for Enhancing Energy and Power Storage Capacity for Application in Electro-chemical Energy Storage

Malay Jana[1] and *Subrata Ray*[2,*]

1. Introduction

Electrical Energy Storage (EES) technology enables us to convert one form of energy, mainly electrical energy, to another form of energy, store it and convert it back when it is to be used. Presently, the plants generating electrical energy are located remotely from users and the energy is distributed through grids. EES is considered a critical technology to help power grid operations and load balancing as it helps in (i) meeting peak load demand, (ii) managing the time variation of energy, and (iii) improving the power quality and reliability.

Emission of greenhouse gases primarily from power plants and vehicles during generation of energy by burning fossil fuels is leading progressively to global warming, which is melting the polar icecaps and threatens to submerge shoreline countries apart from the adverse climatic change and catastrophes. In addition, the polluting gases like SO_x and NO_x, and the solid particles generated during burning of fossil fuels, particularly in vehicles, expose the living species to a host of lungs related diseases adversely affecting the quality of life. On top of these hazards, fossil fuels are also resource limited and for the sustenance of civilisation, there is a need to reduce our dependence on them as source of energy. One may, therefore, produce more clean energy from sources like hydroelectric and nuclear power plants, which are free from greenhouse gases as well as polluting gases and at the same time reduce our dependence on energy from resource limited fossil fuels. But there are safety issues for nuclear power and hydroelectric power based on large dams, which require construction of huge man-made water reservoirs that may trigger earthquakes and other disasters. Therefore, it is imperative to exploit commercially renewable energy from solar, wind and other sources in order to sustain our civilisation and preserve the quality of our life.

[1] School of Materials Science and Engineering, Oklahoma State University, Tulsa, OK 74106, United States.
[2] School of Engineering, Indian Institute of Technology Mandi, Mandi 175001, Himachal Pradesh, India.
* Corresponding author: surayfmt@gmail.com

Responding to these requirements, the energy basket is already a mixed bag of renewable and non-renewable energy sources as indicated in the data on world energy consumption from various sources in 1999 in quads (1 quad = 10^{15} British thermal unit = 2.9×10^{11} kWh)—petroleum: 149.7, natural gas: 87.3, coal: 84.9, nuclear: 25.2, hydro, geothermal, solar wind and other renewable: 29.9, out of the total energy production of 377.1 quads (Energy Information Administration Office of Energy Market and End Use 1999). Thus, it is imperative to integrate more renewable energy in the vehicles and in the grid.

Renewable energy does not provide a steady source of energy and suffers from the problem of intermittent generation of electricity when there is intervention of cloud in solar energy or fall of wind velocity in wind energy, etc. There is also a mismatch between the time of generation (ex: day for solar energy) and use (mostly night for domestic use) requiring energy storage for time shifting to match generation and demand. Integration of renewable energy to the grid without storage will enhance the mismatch of supply and demand posing a problem for energy management of the grid. If the intermittent renewable energy is 15–20 per cent of the overall energy consumption, the grid operators are able to absorb its effect on grid stability (European Commission 2013). But, when the demand is high and the contribution of intermittent energy exceeds 20–25 per cent (US Energy Information Administration 2014), EES is required for alleviating the effect of intermittence of renewable power generation on grid stability and performance.

Apart from integrating more renewable energy in the grid, there should be efficient energy management by minimising wastage of energy through better technology and recovering as much of energy, which may go waste. The vehicles are always decelerating either to reduce speed or stop altogether by braking to dissipate energy and also, while the dock cranes are lowering the crate (Whittingham 2008). If we could provide appropriate storage technologies one could recover and store these energies in suitable capacitors or batteries.

Even with conventional energy, grid faces a problem in matching supply and demand, which varies during the hours of the day as shown in Fig. 1(a). The generation, if responds to such variation, requires to run plants away from the optimum conditions of operation increasing not only the fuel consumption per unit production of electricity but also, the wear and tear of the components of the power plant. It is possible to run the plants for a minimum base load and to store energy when the demand is lower than the base load and use the stored energy for meeting the peak demand as explained schematically in Fig. 1(a). Apart from daily variation in demand for energy there is also seasonal variation as shown in Fig. 1(b, c) typically for India. Thus, energy storage is a key technology for the grid management even with conventional sources of energy.

One also observes ramping load and the energy storage technology to be used, must be able to pump energy responding to it. There are also small fluctuations in load as the energy use changes continuously amongst individual users and the EES technology should be such as to provide power quickly to compensate for voltage and frequency stability.

Luo et al. (Luo et al. 2015) summarises the following functions of EES systems in power network operation and load balancing: (i) helping to meet peak load demand, (ii) management of time varying energy, (iii) alleviating intermittence of renewable energy generation, (iv) improving power quality/reliability, (v) meeting remote and mobile energy needs of vehicles, (vi) supporting realisation of smart grids, (vii) helping the management of distributed and standby power generation and (viii) reducing electrical energy imports during peak demand periods.

2. Present Status of EES Technologies

The current technologies for EES may be broadly classified on the basis of storage mechanisms of energies as mechanical, chemical, electrical, electro-chemical and thermal as shown in Fig. 2. Hydrogen and synthetic natural gas could be used as energy carriers and electrical energy to be stored may be used for electrolysis of water to produce hydrogen for storage, which could be used to generate

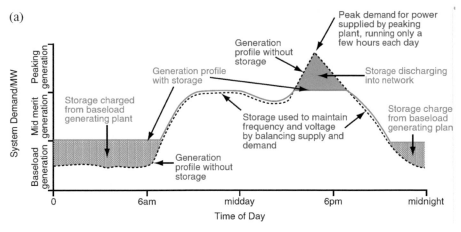

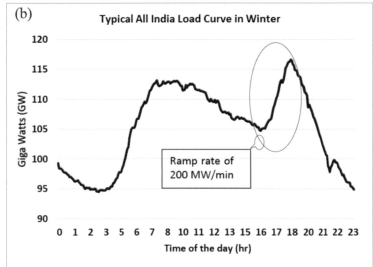

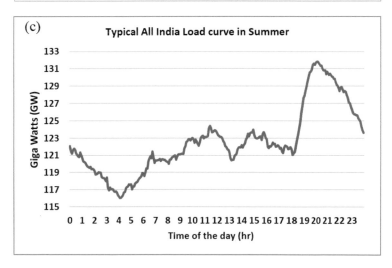

Fig. 1. (a) Schematic variation of load curve during the hr of a day (Whittingham 2008) and typical all India load curve for (b) winter and (c) summer (Power System Operation Corporation 2016).

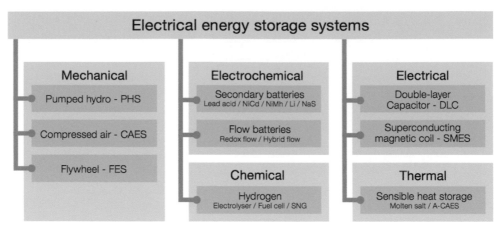

Fig. 2. Different types of Electrical Energy Storage (EES) systems (IEC 2011).

electricity in fuel cell by oxidising hydrogen as and when required. The combined electrolysis and fuel cell may be classified as electro-chemical storage. Many would not classify thermal storage under EES as electrical energy is not input to such systems. But thermal storage may be used to buffer renewable energy and could be used when required.

Apart from the need of EES in the context of management of power in a grid, there is requirement of stored energy to run numerous mobile devices and different applications require different specifications like power capacity and response time, as summarised in the second column of Table 1, where the functions are mentioned in the first column (Luo et al. 2015). The last column of the table lists the EES technologies, which meet the specifications and their status for a given application. The response time, which varies depending on the application area, as mentioned in Table 1 is the time it takes for a system to provide energy at its full rated power. Those technologies, proven for an application and those showing promise are also listed in the last column.

Amongst different EES technologies, pumped hydro accounts for 127,000 MW of worldwide storage capacity and the capacity for compressed air storage is only 440 MW. Sodium sulphur battery has become commercially viable and it is used in 200 installations across the world with total capacity of 315 MW. In spite of the importance of EES, the use of energy storage is only for about 2.5 per cent of power delivered in US while in Europe and Japan it is for 10 per cent and 15 per cent of power respectively, significantly more due to favourable policies (Dunn et al. 2011). The mobile (transportation) and stationary EES technologies have different cost and capacity parameters for commercial viability of electrochemical storage technology and the present challenge is to meet them through the development of better and cheaper materials. US department of energy (DOE) and automobile industries set the goal for development of batteries in vehicles to enable a midsized sedan to cover 300 mile range: energy density of 300 Wh/L and 250 Wh/kg at a cost of $125/kWh. For stationary application in grid the target cost is still lower at $100/kWh to achieve 20 per cent penetration of wind energy in the grid in US by 2030. Currently, Li-ion battery is too costly (exceeding $700/kWh) for mobile application in electric vehicles. For stationery storage, the cost of Li-ion battery is about $3000/kW for power applications and $500/kWh for energy applications. So, a significant cost reduction by a factor of 3 to 5 is required for its commercial viability for energy storage applications (Liu et al. 2013).

Li-ion batteries are attractive for mobile energy storage applications due to their high energy/power density but the other emerging batteries of high capacity based on lithium such as Li-S and Li-air batteries are yet to overcome their poor cycle life and high cost (Bruce et al. 2012). Response time is also very important. Batteries take considerably longer time to charge compared to that for filling liquid fuel in cars while capacitors can be charged very fast—in s or min. But supercapacitors

Table 1. The status of EES technologies in different areas of application (Luo et al. 2015).

Application Area	Application Characteristics and Specifications	Experienced and Promising Energy Storage Options
Power quality	~ < 1 MW, response time (~ ms, < 1/4 cycle), discharge duration (ms to s)	Experienced: flywheels, batteries, SMES, capacitors, supercapacitors; Promising: flow batteries
Ride-through capability (bridging power)	~ 100 kW–10 MW, response time (up to ~ 1 s), discharge duration (s to min and even hr)	Experienced: batteries and flow batteries; Promising: fuel cells, flywheels and supercapacitors
Energy management	Large (> 100 MW), medium/small (~ 1–100 MW), response time (min), discharge duration (hr–d)	Experienced: Large (PHS, CAES, TES); small (batteries, flow batteries, TES); Promising: flywheels, fuel cells
More specific applications		
Integration renewable smoothing intermittent	Up to ~ 20 MW, response time (normally up to 1 s, < 1 cycle), discharge duration (min to hr)	Experienced: flywheels, batteries and supercapacitors; Promising: flow batteries, SMES and fuel cells
Integration renewable for back-up	~ 100 kW–40 MW, response time (s to min), discharge duration (up to days)	Experienced: batteries and flow batteries; Promising: PHS, CAES, solar fuels and fuel cells
Emergency back-up power	Up to ~ 1 MW, response time (ms to min), discharge duration (up to ~ 24 hr)	Experienced: batteries, flywheels, flow batteries; Promising: small-scale CAES and fuel cells
Telecommunications back-up	Up to a few of kW, response time (ms), discharge duration (min to hr)	Experienced: batteries; Promising: fuel cells, supercapacitors and flywheels
Ramping and load following	MW level (up to hundreds of MW), response time (up to ~ 1 s), duration (min to a few hr)	Experienced: batteries, flow batteries and SMES; Promising: fuel cells
Time shifting	~ 1–100 MW and even more, response time (min), discharge duration (~ 3–12 hr)	Experienced: PHS, CAES and batteries; Promising: flow batteries, solar fuels, fuel cells and TES
Peak shaving	~ 100 kW–100 MW and even more, response time (min), discharge duration (< 10 hr)	Experienced: PHS, CAES and batteries; Promising: flow batteries, solar fuels, fuel cells and TES
Load levelling	MW level (up to several hundreds of MW), response time (min), discharge duration (> 12 hr)	Experienced: PHS, CAES and batteries; Promising: flow batteries, fuel cells and TES
Seasonal energy storage	Energy management, 30–500 MW, quite long-term storage discharge duration (up to wk), response time (min)	Promising: PHS, TES and fuel cells; Possible: large-scale CAES and solar fuels
Low voltage ride-through	Normally lower than 10 MW, response time (~ ms), discharge duration (up to min)	Experienced: Flywheels, batteries; Promising: flow batteries, SMES and supercapacitors
Transmission and distribution stab.	Up to 100 MW, response time (~ ms, < 1/4 cycle), discharge duration (ms to s)	Experienced: batteries and SMES; Promising: flow batteries, flywheels and supercapacitors
Black-start	Up to ~ 40 MW, response time (~ min), discharge duration (s to hr)	Experienced: small-scale CAES, batteries, flow batteries; Promising: fuel cells and TES
Voltage regulation and control	Up to a few of MW, response time (ms), discharge duration (up to min)	Experienced: batteries and flow batteries; Promising: SMES, flywheels and supercapacitors
Grid/network fluctuation suppression	Up to MW level, response time (ms), duration (up to ~ min)	Experienced: batteries, flywheels, flow batteries, SMES, capacitors and supercapacitors
Spinning reserve	Up to MW level, response time (up to a few s), discharge duration (30 min to a few hr)	Experienced: batteries; Promising: small-scale CAES, flywheels, flow batteries, SMES and fuel cells
Transportation applications	Up to ~ 50 kW, response time (ms–s), discharge duration (s to hr)	Experienced: batteries, fuel cells and supercapacitors; Promising: flywheels, liquid air storage and solar fuels

Table 1 contd....

...Table 1 contd.

Application Area	Application Characteristics and Specifications	Experienced and Promising Energy Storage Options
End-user electricity service reliability	~ up to 1 MW, response time (ms, < 1/4 cycle), storage time at rated capacity (0.08–5 hr)	Experienced: batteries; Promising: flow batteries, flywheels, SMES and supercapacitors
Motor starting	Up to ~ 1 MW, response time (ms–s), discharge duration (s to min)	Experienced: batteries and supercapacitors; Promising: flywheels, SMES, flow batteries and fuel cells
Uninterruptible power supply	Up to ~ 5 MW, response time (normally up to s), discharge duration (~ 10 min to 2 hr)	Experienced: Flywheels, supercapacitors, batteries; Promising: SMES, small CAES, fuel cells, flow batteries
Transmission upgrade deferral	~ 10–100 + MW, response time (~ min), storage time at rated capacity (1–6 hr)	Experienced: PHS and batteries; Promising: CAES, flow batteries, TES and fuel cells
Standing reserve	Around 1–100 MW, response time (< 10 min), storage time at rated capacity (~ 1–5 hr)	Experienced: batteries; Promising: CAES, flow batteries, PHS and fuel cells

can store less energy than that can be stored in a battery by 1–2 orders of magnitude. Thus, the requirements of more energy with low response time could be achieved in future in a hybrid of batteries and supercapacitors involving both the capacitive mechanism as well as Faradaic ionic transport as in a battery. Such a combination of electro-chemical and capacitive (Electrical) storage technology will retain the advantages of pollution free operation, high round trip efficiency, long cycle life and low maintenance, apart from flexible power characteristics (Lukatskaya et al. 2016).

2.1 Electrochemical and Capacitive Storage Technology

In the storage technology under this category, energy is stored either in a battery through electrochemical mechanism or in a capacitor through capacitive mechanism. Supercapacitors operate on two storage mechanisms: (i) double layer capacitance and (ii) pseudo-capacitance. Electric double layer capacitance is due to the reversible adsorption of ions at the interface of an electrode and electrolyte to provide for electrostatic storage of electrical energy. Electrochemical storage of energy in pseudo-capacitance involves chemical reaction like redox reaction resulting in continuous change in oxidation state or intercalation and change in oxidation state on the electrode surface. A supercapacitor may have both the mechanisms of storage depending on the design and composition of the electrode. Depending on the dominant mechanism of storage, supercapacitors may be classified into three types—Electrical Double Layer Capacitor (EDLC), pseudo-capacitor and hybrid capacitor, where a combination of both the storage mechanisms of electrical double layer and pseudo capacitance are equally prominent. There is often confusion when one tries to distinguish batteries and pseudo-capacitors although there are proposed guidelines to distinguish them (Simon et al. 2014, Brousse et al. 2015). The batteries may involve phase transition as revealed by distinct peaks and plateaus in cyclic voltammograms (CV) but for supercapacitors, there is continuous highly reversible change in oxidation state during charge/discharge resulting in: (i) broadened peaks in CV due to intercalation and little separation between the peaks during charge/discharge or (ii) perfectly rectangular CV due to redox reaction (Simon and Gogotsi 2008, Conway 1999). Further, the intrinsic kinetics are also different as the battery is characterized by electrode process involving semi-infinite diffusion indicated by $i \sim v^{0.5}$ where i is the current in mA and v is the voltage sweep, while for supercapacitor there is linear sweep rate as $i \sim v$. Phase change in a battery electrode material is often accompanied by strain, threatening dimensional stability and limiting cycle life. Typical cyclic voltammetry and galvanostatic profiles for different electrochemical and capacitive energy storage mechanisms are shown in Fig. 3.

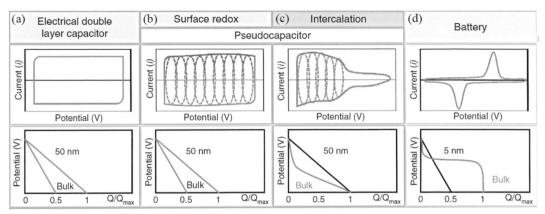

Fig. 3. Typical cyclic voltammetry and galvanostatic profiles (showing influence of surface area through size) of (a) EDLC, (b) Pseudo-capacitor based on redox reaction, (c) Pseudo-capacitor based on intercalation and (d) battery involving intercalation and phase change; i~current and v~sweep rate (Lukatskaya et al. 2016).

The double layer based capacitance is characterised by nearly rectangular voltammograms in cyclic voltammetry (CV) since there is instantaneous charge separation as soon as external electrical field is applied. The galvanostatic charge-discharge profiles are also linear as in Fig. 3(a), observed in high specific surface area materials like porous carbon derived from carbide or activated carbon, graphene, carbon onions and nanotubes. Both pseudo-capacitance and battery involve Faradaic chemical reaction. The CV as shown in Fig. 3(b), reveals pseudo capacitance due to continuous highly reversible change in oxidation state observed in compounds of transition metals with specific structures like those of RuO_2, birnessite MnO_2, 2D Ti_3C_2. However, galvanostatic profile is linear. But pseudo-capacitance involving intercalation shows significantly broadened peaks in CV as in Fig. 3(c) but galvanostatic profile is linear, as observed in compounds of transition metals with large channelled structure like T-Nb_2O_5. Battery also involves change in oxidation state by intercalation but the phase change results in distinct peaks as shown in Fig. 3(d) as observed in $LiCoO_2$, $LiFePO_4$ and Si. Often there is non-martensitic phase transformation involving nucleation and growth, limited by diffusion kinetics. When there are large pathways for movement of ions in the structure of a material, the kinetics is expected to improve. In the following section, the electrochemical and capacitive storage devices are described.

It is apparent from the mechanisms of electrochemical and capacitive storage that the extent of storage will depend on the electrode-electrolyte interaction, which takes place on the surface of the electrodes at sites favourable for absorption, redox reaction or intercalation, as applicable in a given circumstance. The pores also provide paths for faster diffusion of electro-active species, resulting in faster response. Nanostructures and porous structures are extremely useful in electrodes as they have relatively much higher specific surface area offering more active sites for intercalation or redox reaction, thereby increasing specific energy and power density. However, the size of the pores should be large enough to allow ions of active species to access the surface area inside. These structures also have the added advantage of accommodating the strain resulting from volume change that often accompanies intercalation and de-intercalation.

Apart from electrodes and electrolytes, any of these storage devices has passive components like separators to prevent short circuit between the electrodes, current collectors and casings. Thus, a small storage device weighs 5–10 times the weight of active storage materials in the electrodes, thereby lowering the energy density of the device. There could be efforts to reduce the passive components to enhance energy density. There are three useful directions for this purpose: (i) development of improved materials architecture to achieve high energy density by use of thicker electrodes, (ii) development of electrode materials (or composite materials) with good conductivity eliminating the need for current collectors and (iii) use of solid or gel electrolyte eliminating the need for separators.

2.1.1 Battery Energy Storage (BES) System

The rechargeable batteries are widely used as BES systems for domestic and industrial purposes. The schematic of the system is shown in Fig. 4, where cells are combined to give battery system.

Each cell has an anode, a cathode and electrolyte between them. The electrolyte could be solid, liquid or viscous. The cell converts electrical energy to chemical energy for storage during charging and the stored chemical energy is converted back to electrical energy during discharging for the use of the energy for different purposes describes earlier. The chemical reactions taking place at the anodes and cathodes during charging and discharging of cells in different types of batteries used in BES systems are shown in Table 2 along with the voltage obtained in each unit cell combined.

In lead acid batteries, the anode is lead, the cathode is PbO_2 and the electrolyte is H_2SO_4. Apart from having low capital cost of 50–600 \$/kWh, these batteries have relatively high cycle efficiencies of ~ 63–90 per cent, fast response times and small daily self-discharge rates of < 0.3 per cent (Chen et al. 2009, Beaudin et al. 2010, Hadjipaschalis et al. 2009, Kondoh et al. 2000). But the limitations are low cycling life of ~ 2000, energy density of 50–90 Wh/l and specific energy of 25–50 Wh/kg (Chen et al. 2009, Farret and Simoes 2006, Baker 2008). They also perform poorly at low temperature and so, require thermal management facility adding to the cost. The thrust of research in lead acid battery is to develop materials for extending cycle life and depth of discharge.

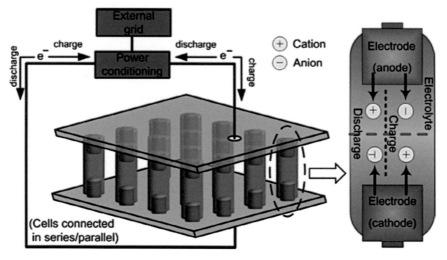

Fig. 4. Schematic diagram showing a combination of electro-chemical cells into BES system connected to grid (Luo et al. 2015).

Table 2. The chemical reactions at the anode and cathode of different batteries and the resulting cell voltage (Luo et al. 2015).

Battery Type	Chemical Reactions at Anodes and Cathodes	Unit Voltage
Lead-acid	$Pb + SO_4^{2-} \leftrightarrow PbSO_4 + 2e^-$ $PbO_2 + SO_4^{2-} + 4H^+ + 2e^- \leftrightarrow PbO_4 + 2H_2O$	2.0 V
Lithium-ion	$C + nLi^+ + ne^- \leftrightarrow Li_nC$ $LiXXO_2 \leftrightarrow Li_{1-n}XXO_2 + nLi^+ + ne^-$	3.7 V
Sodium-sulphur	$2Na \leftrightarrow 2Na^+ + 2e^-$ $\chi S + 2e^- \leftrightarrow \chi S^{2-}$	~ 2.08 V
Nickel-cadmium	$Cd + 2OH^- \leftrightarrow Cd(OH)_2 + 2e^-$ $2NiOOH + 2H_2O + 2e^- \leftrightarrow 2Ni(OH)_2 + 2OH^-$	1.0–1.3 V
Nickel-metal hydride	$H_2O + e^- \leftrightarrow 1/2H_2 + OH^-$ $Ni(OH)_2 + OH^- \leftrightarrow NiOOH + H_2O + e^-$	1.0–1.3 V
Sodium nickel chloride	$2Na \leftrightarrow 2Na^+ + 2e^-$ $NiCl_2 + 2e^- \leftrightarrow Ni + 2Cl^-$	~ 2.58 V

In Li-ion batteries, the anode is graphitic carbon, the cathode is a lithium metal oxide like $LiCoO_2$ or $LiMO_2$ (M – metal), and the electrolyte is $LiClO_4$ or $LiPF_6$ dissolved in non-aqueous organic liquid (Diaz-Gonzalez et al. 2012). It has response time of ms, high cycle efficiency of ~ 97 per cent and relatively high energy density of ~ 1500–10,000 Wh/l and specific energy of 150–200 Wh/kg (Chen et al. 2009, UKDTI 2004, IEC 2011, Hadjipaschalis et al. 2009). These batteries suffer from the depth of discharge (DOD) in a cycle, which affects battery life and the on-board computer necessary to manage its operation adds to the cost. The thrust in research for these batteries is to increase battery power and to develop materials for anode, cathode and the electrolyte to increase specific energy of the cell.

Applied Energy Services (AES) energy storage in US has commercially employed 8 MW/ 2 MWh BES system based on Li-ion battery in New York for frequency regulation since 2010 and enhanced the power to 16 MW in 2011 (Taylor et al. 2012, USDOE). AES also employed 32 MW/8 MWh Li-ion battery system in 2011 for 98 MW wind generation plant in Laurel Mountain (USDOE, Subburaj et al. 2014). The cost effectiveness of Li-ion battery system is under assessment in EES trial of European lithium battery in UK employing 6 MW/10 MWh battery system (Tweed 2013). For integrating renewable energy to the grid, Toshiba plans to install 40 MW/20 MWh Li-ion battery system in Tohuku (Daly 2014). Li-ion battery systems are now increasingly applied in mobile power sources for electric vehicle (EV) and hybrid electric vehicle (HEV) in capacities up to 50 kW and 15–20 kW respectively (Intrator et al. 2011).

Sodium-sulphur batteries has electrodes of molten sodium and molten sulphur, and the electrolyte of β-alumina. To ensure that the electrodes are molten, a temperature of 574–624 K has to be maintained although there is high reactivity (Taylor et al. 2012). These batteries have high energy densities of 150–300 Wh/l, higher rated capacity up to 244.8 MWh and high pulse power capability along with almost no daily self-discharge (Diaz-Gonzalez et al. 2012, IEC 2011, Kawakami et al. 2010). But the problems are high annual operating cost of $80/kW/year and thermal system necessary to maintain the temperature (Luo et al. 2015). This battery system has high potential and it is already employed for EES in various locations as given in Table 3.

There are a number of other cells used in battery systems for energy storage like Ni-Cd, Ni-MH (metal hydride). Ni-Cd based batteries have limited EES applications but by replacing Cd by a hydrogen absorbing alloy leads to a moderate specific energy of ~70–100 Wh/kg and a relatively high energy density of ~ 170–420 Wh/l. Ni-MH batteries have cycle life, even more than Li-ion batteries, reduced 'memory effect' compared to Ni-Cd, and environment friendly (Zhu et al. 2013, Fetecenko et al. 2007). Ni-MH batteries find applications in mobile power sources in portable products, EVs, HEVs and industrial UPS devices. But these batteries are sensitive to deep cycling affecting its performance and have high self-discharge, losing ~ 5–20 per cent of its capacity within a day. A battery similar to Na-S battery, operating at ~ 523–623 K, called ZEBRA battery, has been developed based on Na-nickel chloride, which has specific energy of ~ 94–120 Wh/kg, energy density of ~ 150 Wh/l and specific power of 150–170 W/kg. It is maintenance free and has good pulse power capability, very little self-discharge and high cycle life. Rolls Royce has used this battery to replace lead acid battery

Table 3. Details of commercial exploitation of Na-S battery system (Luo et al. 2015).

Name/Locations	Rated Power/Capacity	Application Area
Kawasaki EES test facility, Japan	0.05 MW	The first large-scale, proof principle, operated in 1992
Long Island Bus's BES System, New York, US	1 MW/7 MW h	Refueling the fixed route vehicles
Rokkasho Wind Farm ES project, Japan	34 MW/244.8 MW h	Wind power fluctuation mitigation
Saint Andre, La reunion, France	1 MW	Wind power on an island
Graciosa Island, Younicos, Germany	3 MW/18 MW h	Wind and solar power EES for islands, commissioning 2013
Abu Dhabi Island, UAE	40 MW	Load levelling

in its EV. GE-Durathon has introduced this battery based UPS in the market. FIAMM Energy Storage Solutions have produced such batteries named Sonick batteries and marketing it for energy storage.

2.1.2 Flow Battery Energy Storage (FBES) System

Flow batteries store energies by reduction-oxidation reactions in the electrolytes contained in the flow compartments around the electrodes separated by ion selective membrane. During charging, the electrolyte at the anode, called anolyte, is oxidised while that at the cathode is reduced converting the electrical energy fed through the electrodes into chemical energy. The opposite takes place during discharging to convert stored chemical energy into electrical energy by the reduction of anolyte and oxidation of catholyte. Vanadium based redox reaction is used in the Vanadium Redox Battery System (VRBS), which is the most mature technology for energy storage using flow batteries. VRBS uses redox couples V^{2+}/V^{3+} and V^{4+}/V^{5+} respectively at the anode and the cathode in the cell separated by ion selective membrane, which only allows H^+ to pass through it, as shown in Fig. 5. The electrode reactions are: $V^{3+} + e^- \leftrightarrow V^{2+}$ at the anode and $V^{4+} \leftrightarrow V^{5+} + e^-$ at the cathode during charging and discharging. The cell voltage is $\sim$ 1.4 V. VRB systems have responses faster than 0.001 s, efficiencies up to $\sim$ 85 per cent and life exceeding 10,000–16,000 cycles (Gonzalez et al. 2004). They can provide continuous power as discharge duration time exceeds 24 hr. The challenges in this system are high operating cost, low electrolyte stability and solubility, resulting in low energy density.

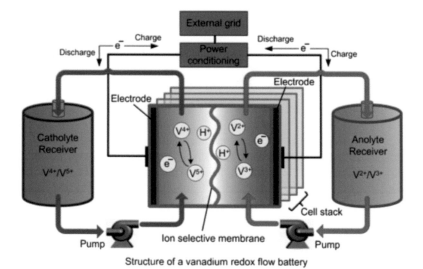

Structure of a vanadium redox flow battery

Fig. 5. The schematic diagram showing Vanadium Redox Battery (VRB) System connected to grid (Luo et al. 2015).

VRB's are mainly employed for stationary storage and UPS for improving load levelling, integrating renewable energy to grid and power security. Some selected storage facilities using VRB systems are given in Table 4.

2.1.3 Capacitive Energy Storage (CES) System

The supercapacitor, called electrical double layer capacitor (EDLC), consists of two conductor electrodes, an electrolyte and a membrane separator. The energy is stored in the double layer between the electrolyte and the conductor electrodes as shown in Fig. 6 (Diaz-Gonzalvez et al. 2012). The figure schematically shows an EES based on double layer supercapacitor connected to the grid.

Table 4. Selected instances of application of VRB systems for storage (Luo et al. 2015).

Name/Locations	Power/Capacity	Application Area
Edison VRB EES facility, Italy	5 kW, 25 kW h	Telecommunications back-up application
Wind power EES facility King Island, Australia	200 kW, 800 kW h	Integrated wind power, foil fuel energy with EES
Wind Farm EES project, Ireland	2 MW, 12 MW h	Wind power fluctuation mitigation, grid integration
VRB EES facility installed by SEI, Japan	1.5 MW, 3 MW h	Power quality application
VRB facility by PacifiCorp, Utah, US	250 kW, 2 MW h	Peak power, voltage support, load shifting
VRB EES system build by SEI, Japan	500 kW, 5 MW h	Peak shaving, voltage support

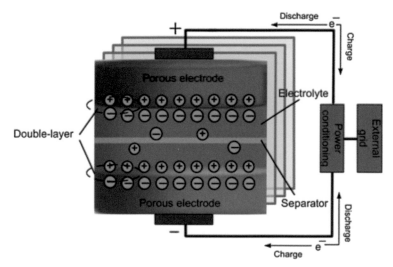

Fig. 6. Electrical Double Layer Capacitor (EDLC) with two conductor electrodes, electrolyte and a membrane separator (Luo et al. 2015).

And it has energy and power densities between those of batteries and traditional capacitors. The supercapacitors have high cycle efficiency of 84–97 per cent and long cycle life exceeding 10^5 cycles but the capital cost is more than \$6,000/kWh and daily self-discharge rate is high at ~ 5–40 per cent (Chen et al. 2009, Smith et al. 2008). Thus, supercapacitors are suited for short term storage typically in power quality, hold-up or bridging power to equipment, solenoid or valve actuation, etc. but not for large scale or long-term storage. The thrust of research in this area is for developing electrode materials with higher energy density and low cost so as to arrive at capacitive storage of durability of ~ 10^6 cycles and specific power of 10 kW/kg (Conway 1999).

Batteries and supercapacitors are combined to develop fast response systems like Ecoult UltraBattery smart systems. Xtreme Power super dry battery and Axion lead carbon batteries are other advanced systems based on lead acid batteries (Rastler 2010, Ultrabattery by Ecoult).

2.1.4 Porosity and Critical Issues for Electrode Materials

There are several critical issues in active materials limiting the performance of supercapacitors and batteries: (i) change in microstructure with cycling, (ii) volume changes on intercalation and deintercalation, (iii) phase changes during cycling and (iv) formation of insulating phase.

During cycling, there may be change in shape size and distribution of phases in the electrode materials affecting the connectivity of the phases. In non-nanostructured electrode of Co_3O_4, there is serious agglomeration and cracking during cycling leading to capacity fading (Li et al. 2014a).

The intercalation and deintercalation of active species in an electrode material create stress as a result of change in volume, which may cause loss of adhesion between particles and between the particles and current collector leading to capacity fading during cycling (Jana and Singh 2017). Silicon, when intercalated by lithium to $Li_{4.4}Si$ changes its molar volume about four times and so, silicon electrode pulverises leading to severe capacity fading and poor rate performance (Chan et al. 2008). However, when silicon, applied to the anode of Li-ion battery, has demonstrated very high initial capacity of 4200 mAhg^{-1} to form an alloy of $Li_{22}Si_5$ (Axel et al. 1966) and a low discharge potential of 0.22 V with respect to lithium metal. At room temperature, the capacity is 3600 mAhg^{-1} to form $Li_{15}Si_4$ (Obrovac and Christensen 2004) but it is not possible to achieve this capacity due to low diffusion rate of lithium in silicon. Porous structure in silicon is of interest to accommodate this large volume change while maintaining integrity. The stress distribution in porous silicon, as given in Fig. 7, clearly shows that the maximum stress on lithiation decreases with increasing size of pores (Ge et al. 2012). The porous structure also results in relatively large surface area and the pores are expected to increase the access of electrolyte inside, reducing the diffusion length of lithium ion for transport from electrolyte to silicon allowing charge/discharge at high current rates overcoming the limitation of small diffusion rate of lithium in bulk silicon.

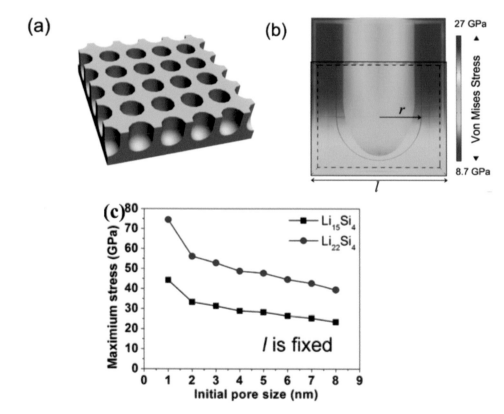

Fig. 7. (a) Porous structure of silicon, (b) Von Mises stress distribution in one unit cell containing a pore (pore-to-pore distance l = 12 nm) and (c) the variation of maximum stress with size of the pore (Ge et al. 2012).

Sometimes, there is phase change in the electrode materials during cycling as in $LiMnO_2$, which changes from cubic to tetragonal structure on cycling leading to severe capacity fading (Shao-Horn et al. 1999). There is often formation of an electronically insulating layer on the electrode surface blocking the passage of electroactive species to the active sites in the electrode materials and it leads to capacity fading. The insulating layer forms by decomposition of the electrolyte, thereby increasing the impedance and also consuming recyclable electroactive ion (Agubra and Fergus 2013).

The accessibility of electrolyte to a large surface area of the electrode is required to enhance the extent and the rate of absorption/intercalation, which are the mechanisms of storing energy in electro-chemical and capacitive storage. The occupation of intercalation sites inside the electrode by the electroactive species is limited by diffusion distance, which is small and even when one increases the loading of active material in an electrode, the material accessible for storage of energy is still limited by diffusion distance. Thus, to increase the energy capacity one needs to increase the access of active material in the electrode to electroactive species by making the electrode material highly porous with large specific area such that electrolyte percolates extensively through open channels.

3. Architecture of Porous Materials

There are two approaches to increase specific surface area of a material—firstly, by decreasing the size of the particle to nanometer range and secondly, by incorporating porosity in the material. Decreasing size will make proportionately more material accessible to electroactive species and also, the diffusion distance will be a function of size at the nano-level. In the first generation of electrode the researchers explored monolithic homogeneous nanomaterials such as nanoparticles (0D), nanowires and nanotubes (1D), layered materials (2D) and mesoporous structures (3D). The composite structures of different dimensions have also been conceived and fabricated as shown in Fig. 8. The core-shell structure is another interesting 0D composite structure developed particularly for materials like Si or Ge where there is high volume change on intercalation/deintercalation. The active material may breathe inside the conducting carbon and protect its electrochemical performance. A pomegranate 3D structure also helps to provide conduction path and to accommodate strain from high volume change of active material.

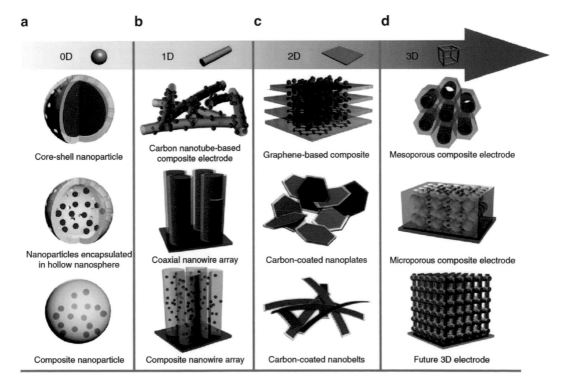

Fig. 8. Schematic of heterogeneous nanostructures of (a) 0D, (b) 1D, (c) 2D and (d) 3D (Lukatskaya et al. 2016, Liu et al. 2011).

If the nano-sized material becomes porous with open channels for percolation of electrolyte inside and the electroactive species will access more material for adsorption/intercalation. Nanoporous materials may help us to attain the target performance, if its cost is within what a given application could afford.

In a crystalline porous material, pores may be integral part of the crystalline arrangement of the structure or may occur between the crystallites. There may be pores in non-crystalline materials in more open atomic arrangement or in pores inside particles. Porous materials may be classified on the basis of the size of the pores or porous channels, D, as microporous (D < 2 nm), mesoporous (2 nm > D > 50 nm) or macroporous (D > 50 nm). When pores are integral part of the structure, these pores are uniform and permanent in the sense that they do not collapse during post synthesis processing. The porous materials may also be looked from the pseudo-dimensionality of their basic form—porous spherical particles—hollow or filled of 0D, rods or tubes of 1D, planar sheets of 2D and blocks of truly 3D. These materials of basic form and size are aggregated to make a 3D electrode. There is also issue of the dimension of pores—isolated pores of 0D, unidirectional parallel channels (1D), bi-directional parallel channels or channels directed in all three directions, 2D planar channels and interpenetrating channels and solid phase. Isolated pores do not allow percolation of electrolyte inside.

Nanoporous materials have pores of size between 1 nm to 100 nm. Mesoporous structure is more suitable for quick transport of electrolyte while microporous structure is more suitable for ion adsorption (Frackowiak and Beguin 2001). Thus, porous materials should be balanced in pore size distribution for optimum performance. Apart from electrochemical and capacitive storage, porous materials are also important in chemical energy storage like fuel cells. Both capacitive and fuel cell require storage of electroactive species or hydrogen by adsorption. The porous structure is required both for access of electrolyte as well for adsorption. Theoretical studies on hydrogen adsorption in porous materials have confirmed that presence of micropores influence the extent of hydrogen adsorption. Grand Canonical Monte Carlo simulation has shown that the optimum pore size is below 1 nm (Rzepka et al. 1998).

3.1 Molecular Design for Pore Space

The design at the molecular level of inorganic structure involves generally layered solids, which is doped by atoms or molecules of different sizes so as to create instability in the layered structure leading to the evolution of different structures with porosity. The design of inorganic-organic combination gives flexibility in controlling the distances between inorganic units by bonding it with organic molecules of suitable size so as to result in the desired crystalline structure with integrated porosity. In inorganic-polymer combination, monomer may be so chosen as to bond with inorganic unit and the degree of polymerisation may also be controlled to result in the desired distance between the inorganic units. Depending on the bond strength of the covalent bond, the resulting framework may be crystalline or non-crystalline with the desired amount of porosity. Since the nature of the material to be used for energy storage is inorganic primarily, so one has to arrive at porous structure either by suitable modification of inorganic structure to induce transformation to more open structure incorporating pore space or by combining it with organic molecules/polymers of different sizes. All these combinations are the results of different types of chemical bonds or physical trapping by surrounding one by the other, and are obtained chemically. There is immense opportunity to tailor new materials following these routes. These combinations of inorganic-organic or inorganic-polymer hybrid materials incorporating pore space may be used as such or the organic/polymer component could be evaporated, decomposed or carbonized as the case may be, in order to create a porous inorganic material or a porous composite containing what remained after burning—mostly carbon.

4. Synthesis of Porous Materials

The macropores occur in metals, oxides and composites during powder processing or liquid metallurgy and these are considered defects as those have adverse effects on particularly mechanical properties. But our primary task now is to develop porous materials with micropores and mesopores. A large variety of chemical methods have been employed to develop these porous materials in different shapes and sizes including hydrothermal, solvothermal and micro-emulsion processes. Electro-spinning has been employed to make a large variety of porous nanotubes/nanofibers/nanowires. In order to create pores or channels, heat treatment is often the last step in the synthesis of nanoporous material, for expelling gaseous or volatile component by decomposition, carbonisation or even oxidation, depending on temperature and the atmosphere.

There has been a long-standing challenge of designing and synthesising crystalline solid-state materials from molecular building blocks. The concept involves use of secondary building units (SBU) to direct assembly of ordered frameworks. This approach is called reticular synthesis, which has resulted in tailored materials with predetermined structure, composition and properties including highly porous frameworks with exceptionally large surface area. The aim is to develop a network of organic or inorganic-organic units. The organic unit may be a monomer capable of polymerisation resulting in covalently bonded framework or molecules capable of forming supramolecular network bonded by coordinative bond. The resulting material could be crystalline or polymeric, accommodating large pore space, depending on the nature of bond. It is also possible to design porous structure from inorganic materials with a given structure by molecular intervention like doping so as to evolve to a more open structure accommodating porous channels.

Traditionally, mould is used to shape materials and similarly, mesoporous structure has also been shaped using templates. The routes of synthesis of porous materials are numerous and it is not our intention to cover the entire spectrum. Our effort is to capture the logic of the routes to arrive at different shapes of porous materials of nano-size, which may have potential for application in electrodes for electrochemical and capacitive energy storage.

4.1 Synthesis by Inorganic Molecular Design

This design presently involves layered solids, where inorganic species of different sizes are inserted between the layers to destabilise the structure so as to evolve to a more open crystalline structure accommodating space as pores and porous channels. The size of the inserted species may control the size of channels. Development of processes to arrive at such architecture starting from a suitable precursor layered material is challenging as it will be illustrated in the context of natural and synthetic octahedral molecular sieve (OMS) of manganese oxide.

The different allotropes of manganese oxides have different size and geometry of porous channels—one dimensional channels (1D), two-dimensional layer channels (2D) and three dimensional interconnected channels (3D) as shown in Fig. 9. The different sizes of channels result in variation in specific capacitance, ionic conductivity and specific surface area as shown in the same figure. It has been observed that increase in channel size and connectivity results in improved electrochemical performance.

The control of tunnel size in manganese oxide has been carried out by hydrothermal synthesis from planar manganese oxide precursor under controlled pH. In the mixed valent manganese framework of (+2, +3, +4) or (+3, +4), a small number of guest cations are required for charge balance in layered or tunnel structured manganese oxides. These guest cations between the layers act as structure directors, causing instability to transform the layers to a structure having different tunnel sizes. The guest cations under aqueous or hydrothermal condition are generally hydrated. The size of the hydrated cation depends on the state of hydration and the state of hydration could be varied to control the size of hydrated cation to obtain different tunnel size.

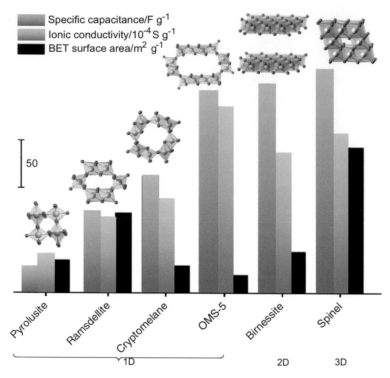

Fig. 9. Different allotropes of manganese oxide with different size and dimensionality of channels leading to different specific capacitance, ionic conductivity and specific surface area (Ghodbane et al. 2009, Lukatskaya et al. 2016).

Sodium ion may easily be hydrated in aqueous environment and the bond strength between Na^+ and water increases with pH. Thus, hydrothermal treatment of sodium birnessite at different pH will result in tunnelled structure of different tunnel sizes as shown in Fig. 10. Under hydrothermal condition, sodium ion will have more hydration at higher pH and so the size of the hydrated ion will be larger resulting in larger size of tunnel in the OMS structure of MnO_2. The hydration bond strength increases with pH and the bond lengths of Na^+-O bond are 2.47 and 2.3 Å respectively under neutral and basic condition at room temperature. Thus, at pH = 1.0, one gets tunnel size of 1×1 (OMS-7) and it increases to tunnel size of 2×3 (OMS-6) at pH = 7.0 and tunnel size of 2×4 (OMS-5) at pH = 13 (Shen et al. 2005).

The tunnels in the crystallographic structure in different allotropic phases of MnO_2 provide interconnected pathways of different size and geometry for ion transport. Increase in channel size and connectivity improves electrochemical performance as there is a strong correlation between capacitance and ionic conductivity, observed in different allotropes of MnO_2. Thus, the example of MnO_2 illustrates the importance of developing appropriate porous structure in the architecture of inorganic materials through innovative processing and one may develop new inorganic porous materials to explore their potential for application in electrodes of electrochemical and capacitive storage. There are a number of materials, which have the right crystal structure with interconnected channels of different dimensions: (i) 2D-layered structure (transition metal oxides, carbides, and dichalcogenides) and (ii) 3D-interconnected pores (T-Nb_2O_5, spinel MnO_2, etc.). One may look for processing methods to evolve into new structures with the required size of the channels as it has been done for layered MnO_2 structure.

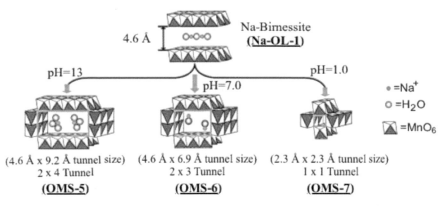

Fig. 10. Development of different sizes of tunnel structure in octahedral molecular sieve (OMS) in MnO_2 by hydrothermal processing at different pH (Shen et al. 2005).

4.2 Synthesis by Inorganic-organic Molecular Design

There are two classes or inorganic-organic hybrid materials—firstly those where the inorganic component traps organic component or vice versa and secondly, those where there is bonding between the inorganic and organic components. In the latter type, the bonding could be strong covalent or ionic between the inorganic and organic unit. If the bonds are strong and irreversible, these could not be broken easily and reformed as required during the growth of crystalline arrangement under ambient condition. On the other hand, if the bonds are relatively weak and reversible, it is possible for the bonds to break and reform as required during the growth of crystals. Thus, one may design such material combination as to result in relatively weak and reversible bond so as to result in crystalline inorganic-organic combination. But strong irreversible bond will result in a non-crystalline combination. There are three distinct lines of development of nanoporous inorganic-organic materials—Metal Organic Framework (MOF) linked by coordinative bond, Supramolecular network linked by non-covalent bonds and Covalent Organic Framework (COF).

In the third alternative, there is combination atoms of carbon and other lighter atoms, which are covalently bonded in an open network. Bonds in typical MOF and supramolecular networks are reversible and weak enough to permit growth of crystals with porosity integral to the structure at ambient temperature. But covalently bonded frameworks have irreversible strong bonds, which generally do not permit growth of covalently bonded large crystalline arrangement similar to those in diamond, graphite or graphene. The attempts to grow large crystals involving polymers have at best resulted in microcrystalline powders.

4.2.1 Covalent Organic Framework (COF)

Covalently linked organic network may result either by reversible or irreversible reactions. If the organic building blocks are rigid and sterically demanding, mostly based on aromatic subunits, the networks form generally by coupling reactions, which are kinetically controlled and leads to formation of irreversible covalent bond. The features of such network are permanent porosity, physical and chemical stability but it lacks long range order or crystallinity. The pore size distribution is generally narrow.

If the network forms by reversible reactions, this class of materials is called covalent organic frameworks (COFs), which is composed of covalent building blocks made of boron, carbon, oxygen, hydrogen, and also, sometimes, nitrogen or silicon, stitched together by organic subunits. Most

COFs are synthesised by condensation reaction, particularly by boronic acid condensation forming boronic anhydrite or boronic acid condensation with catechol (El-Kaderi et al. 2007). Two crystalline porous structures COF-1 and COF-5 could be prepared by condensation reactions based on molecular dehydration reaction as shown in Fig. 11 (Zhu and Ren 2015). For COF-1, three boronic acid molecules comes together to form a planar six-membered B_3O_3 (boroxine) ring with the elimination of three water molecules, and a honeycomb-like structure is expected to form using 1,4-benzenediboronic acid (BDBA) as monomers. For COF-5, an analogous condensation reaction is employed, which forms borate ester. First, the dehydration reaction between boronic acid and diol generates a five-membered borate ester ring (BO_2C_2). Then, it is found that the entire coplanar extends to a sheet structure. In the condensation reactions leading to formation of borate anhydride and borate ester, the result is formation of reversible bond, which could be formed, broken, and reformed to finally obtain a stable state as a result of dynamic covalent chemistry. This is dynamic covalent chemistry (DCC) and the reaction is thermodynamically controlled and not kinetically but it allows an 'error checking' or 'proof-reading' process to adjust itself to reduce its structural defects and form a stable state. Therefore, crystalline COFs would finally form.

Apart from these condensation reactions accompanied by molecular dehydration, there are other condensation reactions of aldehyde and amine or aldehyde and hydrazide where imine (Uribe-Romo et al. 2009) or hydrozone (Uribe-Romo et al. 2011) forms. Two reaction mechanism not based on condensation reaction, are (i) trimerization of dicyano compounds to give covalent triazine frameworks (CTFs), however, to generate reversibility the reaction needs to be carried out under much harsher

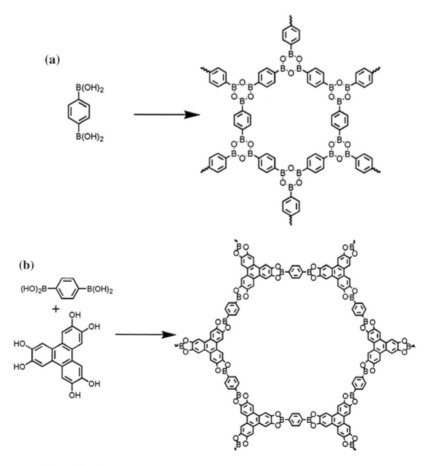

Fig. 11. Formation of (a) COF-1 and (b) COF-5 by condensation reaction (Zhu and Ren 2015).

conditions (Kuhn et al. 2008), and (ii) dimerisation of nitroso compound to azodioxides (Beaudoin et al. 2013). The synthesis of the latter will be discussed in some details as these compounds show excellent crystallinity and even enabled the first single crystal COF structures. However, due to the low stability, no permanent porosity could be achieved and upon removal of the solvent molecules the crystallinity was lost.

Depending on the selection of building blocks, the COFs may form 2D or 3D networks. Planar building blocks are the constituents of 2D COFs, whereas for the formation of 3D COFs, typically tetragonal building blocks are involved.

Beaudoin et al. (Beaudoin et al. 2013) in an effort to form crystalline COF, followed a strategy of using covalent bonds of lower strength ($\sim$ 20–30 kcal mol^{-1}) and succeeded in generating four tetrahedrally oriented nitroso groups monomers to induce spontaneous formation of diamondoid network, NPN-1, NPN-2 and NPN-3 grown respectively from 3:2 (vol/vol) mesitylene/ethanol or 3:2 (vol/vol) benzene/ethanol, 3:2 (vol/vol) methanol/ethanol and 4:4:1 (vol/vol) mesitylene/ethanol, tetrahydrofuran as shown in Fig. 12.

Apart from covalently bonded porous organic network discussed so far, it is possible to synthesise covalently bonded inorganic-organic network through sol-gel processing by following one of the three approaches. In the first approach, the inorganic component like metal alkoxide, $[M(RO)_n]$, forms compound with an organic group or polymer chain, Y, linking two (x = 2) or more (x > 2) metal alkoxide units like $[(RO)_nM]_xY$, which may retain the structure of Y in the final product. The organic groups like saturated or unsaturated hydrocarbon chains or polyaryls, have different lengths and may be substituted by the inorganic $Si(OR)_3$ group at both ends or the inorganic group may be grafted in the polymer (Sanchez and Ribot 1994, Loy and Shea 1995, Judeinstein and Sanchez 1996). The second approach involves functionalisation of inorganic building block and formation of inorganic structure, followed by crosslinking of the organic functions through polymerisation (Ribot and Sanchez 1999, Kickelbick and Schubert 2001). In the third approach a bifunctional molecular precursor

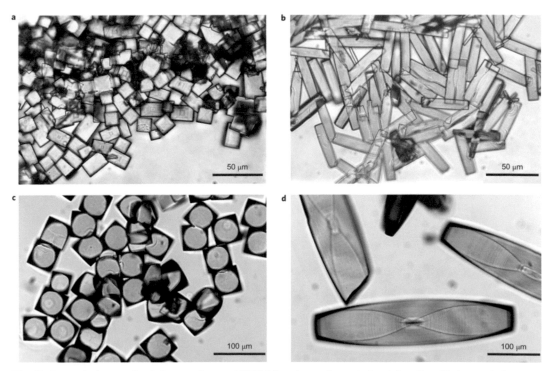

Fig. 12. Large single crystals of nitroso polymer of NPN-1 in a (grown in mesitylene/ethanol) and b (grown in benzene/ethanol), NPN-2 in c (grown in 3:2 (vol/vol) mesitylene/methanol) and NPN-3 in d (grown in 4:4:1 (vol/vol) mesitylene/ethanol/tetrahydrofuran) (Beaudoin et al. 2013).

$(RO)_nM$—X-A is formed with the inorganic group $(RO)_nM$ such that the organic functionality A is capable of polymerisation or crosslinking. In this approach inorganic group is formed *in situ*, usually with no defined nanostructure, as distinct from the second approach. The precursor is usually reacted with water and then sol-gel processing resulting in inorganic network followed by polymerisation or crosslinking of the organic function creating an extended organic frame work (Schubert et al. 1995). For inorganic-organic hybrid polymers, the third approach is often used and the majority of materials are based on polysiloxane backbone. But, the first two approaches with well-defined structures of the inorganic and organic building blocks are more suitable for developing nano-scale structures. There are two variations of the approaches described—the grafting of inorganic group $Si(OR)_3$ to the organic polymer as a variation of the first approach and as a variation of the second approach, incorporation of nanometer to micrometer size inorganic particles as fillers in organic polymer with strong surface interaction between them.

When one wants to use the inorganic materials properties associated with smaller size like nano-size, it is better to use a well-defined cluster as inorganic building blocks with well-defined stoichiometry, size and shape. The polyhedral oligomeric silsesquioxanes (POSS) cluster like $[RSiO_{3/2}]_n$ (POSS), or spherosilicates, $[ROSiO_{3/2}]_n$ have already been investigated in some detail as a constituent of inorganic-organic hybrid materials. Representative examples are shown in Fig. 13. The most frequently used cluster is the cubic octamer, $R_8Si_8O_{12}$ or $(RO)_8Si_8O_{12}$ silicate cage. The groups R can be used for crosslinking or polymerisation reactions by which the silicate cages are incorporated into hybrid polymers.

There are efforts to develop inorganic clusters based on transition metal oxides for developing inorganic-polymer network but it is challenging to find suitable organically modified transition metal oxide clusters (OMTOC) as the transition-metal equivalents to the POSS (Schubert 2001).

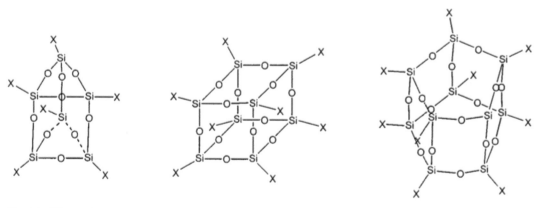

Fig. 13. Silicate clusters or cages for the development of inorganic-organic (polymer) hybrid where X = R or RO is polymerisable organic group (Schubert 2001).

4.2.2 *Metal-Organic Framework (MOF)*

In metal organic framework (MOF), there are two constituents—metal ion or clusters called secondary building unit (SBU) and organic linkers. In SBU, the metal ions are stitched into polyhedral or polygonal cluster involving different metal ions and multidentate (chelating or bridging) functional groups like carboxylates, etc. The following organic carboxylates have been used in many MOFs and the abbreviations for the different carboxylates are—ADC for acetylenedicarboxylate, BTC for benzenetricarboxylate, NDC for 2,6-naphthalenedicarboxylate, BTB for benzenetribenzoate, MTB for methanetetrabenzoate, ATC for adamantanetetracarboxylate and ATB for adamantanetetrabenzoate.

Direct joining of polyhedral or polygonal nodes and linkers extend their geometry into nets of definite topology and structure. The topologies of the structures of MOF-31 to MOF-39 are described in Fig. 14 along with the organic and inorganic SBU with their corresponding linkers. (Kim et al. 2001). MOF-31 ($Zn(ADC)_2 \bullet (HTEA)_2$), MOF-32 ($Cd(ATC) \bullet [Cd(H_2O)_6](H2O)_5$), MOF-33 ($Zn_2(ATB)(H_2O) \bullet (H_2O)_3(DMF)_3$) have tetrahedral SBUs linked into diamond networks. Less symmetric tetrahedral SBU results in MOF-35 ($Zn_2(ATC) \bullet (C_2H_5OH)_2(H_2O)_2$), which has a network similar to Ga in $CaGa_2O_4$.

MOF-34 ($Ni_2(ATC)(H_2O)_4 \bullet (H_2O)_4$) has the structure akin to Al network in SrA_{12}. Square and tetrahedral SBUs in MOF-36 ($Zn_2(MTB)(H_2O)_2 \bullet (DMF)_6(H_2O)_5$) are linked into PtS network. The octahedral SBUs in MOF-37 ($Zn_2(NDC)_3 \bullet [(HTEA)(DEF)(ClBz)]_2$) forms a network linking

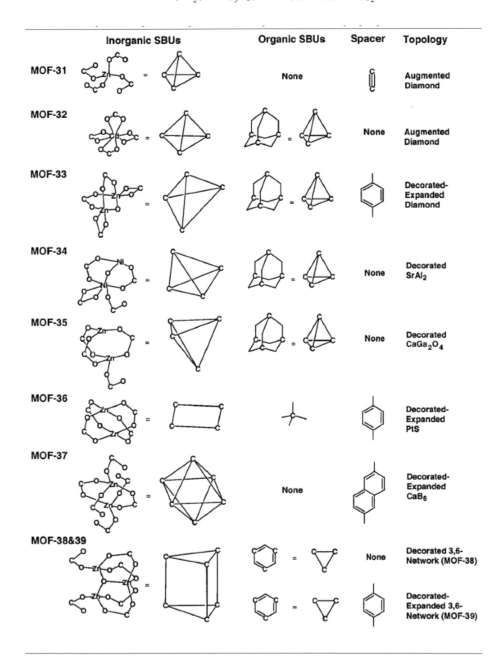

	Inorganic SBUs	Organic SBUs	Spacer	Topology
MOF-31		None		Augmented Diamond
MOF-32			None	Augmented Diamond
MOF-33				Decorated-Expanded Diamond
MOF-34			None	Decorated $SrAl_2$
MOF-35			None	Decorated $CaGa_2O_4$
MOF-36				Decorated-Expanded PtS
MOF-37		None		Decorated-Expanded CaB_6
MOF-38&39			None	Decorated 3,6-Network (MOF-38)
				Decorated-Expanded 3,6-Network (MOF-39)

Fig. 14. Inorganic and organic secondary building units (SBU) involved in MOF-31 to MOF-39 along with linkers and the resulting topology (Kim et al. 2001).

octahedral shapes, similar to B in CaB_6. The structures arising from linking of triangular and trigonal prismatic SBUs are found in MOF-38 ($Zn_3O(BTC)_2 \bullet (HTEA)_2$) and MOF-39 ($Zn_3O(HBTB)_2(H_2O) \bullet (DMF)_{0.5}(H_2O)_3$). One may observe that the points of extension in MOF are the number of connections between one metal cluster and the other metal clusters and the points could be from 3 to 12, 24 and infinity, as shown in Fig. 15.

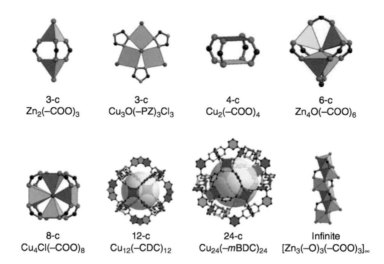

Fig. 15. Selected SBU involving M-N and M-O clusters of Zn and Cu having different connectivity; n in n-c indicates connectivity of nodes; colour codes—black for carbon, red for oxygen, green for nitrogen, purple for chlorine, blue polyhedra for zinc or copper and yellow ball for free space; PZ stands for pyrazolate, CDC or cdc for 9H-carbazole-3,6-dicarboxylate and mBDC for 1,3-benzenedicarboxylate (Schoedel and Yaghi 2016).

To describe net topologies, one commonly uses three letter codes provided by the Reticular Chemistry Structure Resources (RCSR) database. For example, the SBU, $Zn_4O(CH_3COO)_6$ having 4-c or four connections when linked by BDC or bdc (written in both capital or small in literature) results in MOF-5 with pcu net as shown in Fig. 16. It is observed in this figure that the structural net of MOF-5 could be retained but the arm of the cubic net could be enhanced by changing the linker to terphenyl dicarboxylate.

One may vary the size and the nature of structure without changing topology of the net and the MOFs belonging to such a family are called isoreticular (IR) family indicated by adding IR before MOF. A large number of MOFs have been designed with high porosity and large pore openings and one example of isoreticular (IR) family is the 16 members of cubic MOFs of pcu net from IRMOF—1 to 16 where the parent IRMOF-1, is MOF-5, which has the smallest structure and is made of $Zn_4O(fumarate)_3$. The largest member of this IR family is IRMOF-16, where the edge is doubled and the volume increases eightfold, and it is made of $Zn_4O(tpdc)_3$; $tpdc^{2-}$ is terphenyl-4,4"dicarboxylate as shown in Fig. 16.

The synthesis of MOF is carried out by conventional and unconventional methods. The conventional synthesis is often carried out by solvothermal method by heating a mixture of organic linker and metal salt in a solvent (DMF or DEF, etc.) usually containing formamide functionality. The materials produced is thermally unstable or reactive to the solvent and thus, breaking the bonds or exposing metal sites for binding guest species accessing into the pores of MOF. The unconventional synthesis is normally by mechano-chemical method where organic linker and metal salt are ground together in agate mortar and pestle or ball mill without solvent. The reaction is mechano-chemically initiated to result in hydrated MOF, which is gently heated to dehydrate MOF and expose metal sites to bind guest species.

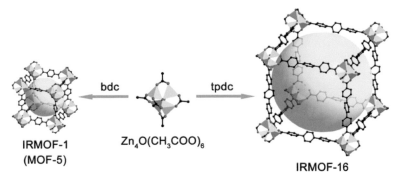

Fig. 16. SBU of $Zn_4O(CH_3COO)_6$ linked by 1,3-benzenedicarboxylate (bdc) resulting IRMOF-1 or MOF-5 while the same SBU linked by terphenyl-4,4"dicarboxylate (tpdc) resulting in bigger cubic net of IRMOF-16 (Lu et al. 2014, Zhao et al. 2016).

Post synthetic modification of MOFs may be carried out by reacting the links with organic units or metal organic complexes in order to enhance reactivity within pores. Multivariate MOFs (MV-MOF) have been created by incorporating multiple organic functionalities within a single framework to create further complexities. Multivariate mixed-metal oxides with high surface area have been prepared using MOF-74 as the precursor to obtain MOF-74-NiCo and on annealing it results in porous $Ni_xCo_{3-x}O_4$ as shown in Fig. 17 (Chen et al. 2015). Transition metals, alkaline earth metals, p-block metals, actinides and even mixed metals have been used to develop MOF and the organic linker could be divalent or multivalent organic carboxylates. Together they form 3D structures

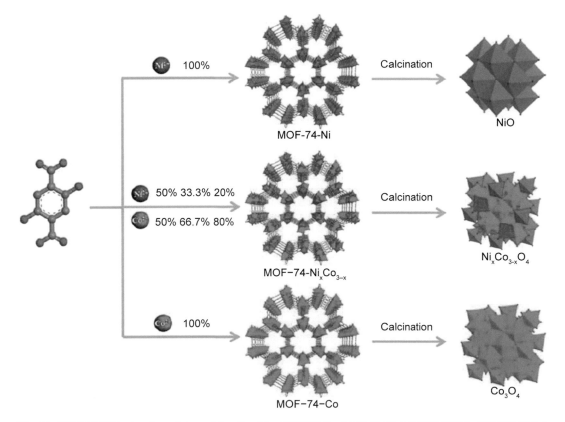

Fig. 17. MTV-MOF-74, showing mixture of functionalities MOF-74-Co, MOF-74-Ni, MOF-74-NiCo1, MOF-74-NiCo2 and MOF-74-NiCo4 decorating the interior of crystals to provide an environment capable of highly selective binding and on calcinations resulting in oxides and mixed oxides (Chen et al. 2015).

with definite pore size (as large as 9.8 nm) distribution and high surface area in the range of 1000 to 10,000 m^2/g by various combinations of metal and organic linkers. The variety of the geometry of constituents, size and functionality has led to more than 20,000 reported MOFs so far.

Due to controlled micro- and meso-porous structures, MOFs are considered materials for high potential for application in energy storage. In one study, copper hexacyanoferrate nanoparticles have been prepared as active electrode material with MOF structure and it is found to be capable of accommodating high strain during the redox reaction (Wessells et al. 2011). These materials are inexpensive and environment friendly. In addition, there have been materials derived from MOF by one step calcination process to produce porous nanostructures of carbon, metal and metal-carbon composites for application in electrode materials.

4.3 Synthesis by Direct Chemical Method

In these methods, nanomaterials form in the liquid phase often as precipitate by using either hydrothermal, solvothermal or micro-emulsion methods. These methods provide well known routes for the synthesis of nanomaterials using aqueous solvent, inorganic-organic hybrid solvent or pure organic solvents. Pores in the inter-crystalline space may be created by evaporation/expulsion of a gaseous or volatile constituent of a compound by annealing leading to porous materials.

Wang et al. (Wang et al. 2013) precipitated micron size crystals of $MnCO_3$ from a solution of $MnSO_4$, H_2O in water and ethanol by using NH_4HCO_3 solution in water and ethanol as precipitant. These crystals of $MnCO_3$ were then heated at 600°C for 10 hr in air to obtain porous Mn_2O_3, which was mixed with LiOH in molar ratio of 1:1.05, ground together and annealed at 750°C for 10 hr. The reaction of molten LiOH with porous Mn_2O_3 to result in porous spheres of $LiMn_2O_4$ is shown schematically in Fig. 18.

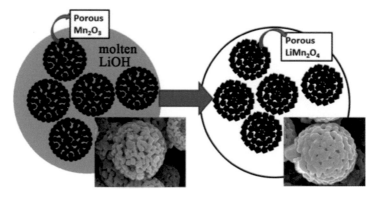

Fig. 18. The schematic of the processing of porous sphere of MnO_2 to porous spheres of $LiMn_2O_4$ along with SEM images (Wang et al. 2013).

Yellow powders of microcubes of $MnCO_3$ were prepared by Wu et al. (Wu et al. 2012) by hydrothermal reaction of $KMnO_4$ with sucrose, as shown in Fig. 19(a), with the aim to synthesize hollow porous (HP) microcubes of $LiMn_2O_4$. The powders were decomposed at 290°C for 2 hr for partial conversion of $MnCO_3$ to black MnO_2 phase and the resulting powder was dispersed in diluted HCl to preferentially dissolve $MnCO_3$ because of its faster dissolution, leaving behind hollow porous (HP) MnO_2 shown in Fig. 19(b). The amount of pore could be controlled by controlling decomposition time, pH of the acid and the dissolution time in it. HP-MnO_2 was lithiated by LiI in acetonitrile to get HP-$LiMn_2O_4$ shown in Fig. 19(c).

Another processing route is the microemulsion method, which is an energy saving route and at the same time easy for synthesising nanomaterials (Ganguli et al. 2010). During the reaction, nucleation takes place during collisions among reverse micelles, which contains the reactants. A

Fig. 19. SEM micrographs of (a) microcubes of $MnCO_3$, (b) hollow porous Mn_2O_3 and (c) hollow porous $LiMn_2O_4$ (Wu et al. 2012).

major advantage of this method is its ability to control the morphology and pore size by the variation of reactant concentrations, temperature, water to surfactant ratios, and aging time (Mai et al. 2014).

To form reverse micelles, one uses surfactant and the commonly used surfactant is cetyltrimethyl ammonium bromide (CTAB). Xu et al. (Xu et al. 2009) assembled nanoparticles to make porous Co_3O_4 nanorods with a length of 3–5 μm and diameter of ≈ 200 nm. The synthesis route is schematically shown in Fig. 20(a). Reverse micelles containing Co^{2+} coming in contact with $C_2O_4^{2-}$ resulted in the formation of precursor CoC_2O_4 nuclei. After absorption of CTAB surfactant molecules onto the surface of the CoC_2O_4 nuclei, nanorods form by direct growth. Annealing of the CoC_2O_4 nanorods transformed them to porous Co_3O_4 nanorods due to the release of CO_2 and the microstructure of these rods are also shown in Fig. 20. Similarly, one could synthesise other binary metal oxides.

4.4 Synthesis by Electro-spinning

Jayraman et al. (Jayraman et al. 2013) synthesised interconnected network of one dimensional fibers with diameter around 600–700 nm by electro-spinning, as shown in Fig. 21(a). Homogeneous sol-gel was prepared by stirring 1.5 g of polyvinylpyrrolidone (PVP, M_w:130000) in 15 ml ethanol for an hour, then mixing in it simultaneously 0.3 g of lithium nitrate and 2.33 g of manganese acetate dehydrate and stirred again. Then 1 ml of acetic acid was added to the solution under vigorous stirring. The resulting sol-gel solution was subjected to electro-spinning under 22 kV at a flow rate of 0.7 ml h^{-1} using a syringe pump. The spun fiber was preheated to 500°C for 1 hr before calcining at 800°C for 5 hr in air resulting in highly porous tubular structure of 3–10 μm length, about 500 nm diameter and about 65–85 nm thickness. Decomposition of polymer and organic moieties present in the spun fiber results in aggregated nano-sized $LiMn_2O_4$ particles over the surface of the fibers, as shown in Fig. 21(b).

Niu et al. (Niu et al. 2015) has introduced a novel gradient approach to electro-spinning to make porous tubes. The precursor solution, which is to be spun, contains a mixture of three polymers—low, medium and high molecular weight polyvinyl alcohol (PVA). At high voltage, the mixture separates

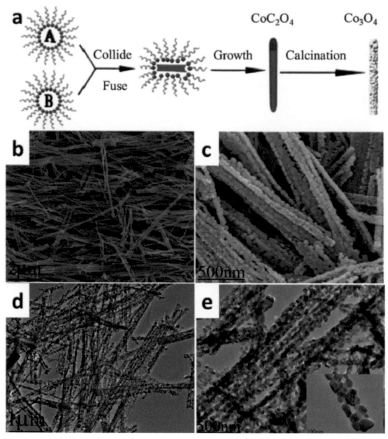

Fig. 20. (a) Schematic processing route of porous Co_3O_4 nanorods (Xu et al. 2009); (b, c) SEM micrographs and (d, e) TEM micrographs of the $ZnCo_2O_4$ (Du et al. 2011).

out in three layers, the inner one is of low molecular weight, then the medium molecular weight and the outer layer of high molecular weight but the inorganic materials are homogeneously distributed in the three layers. When the spun tube is heated, the pyrolysis temperature of low molecular weight layer is arrived first and with increasing temperature, the layer shrinks towards medium molecular weight layer along with inorganic material resulting in a tube, whose inner diameter increases further with the pyrolysis of medium molecular weight layer and finally a mesoporous tube of nanoparticles results after the pyrolysis of the high molecular weight layer. This approach is shown schematically in Fig. 22(a) and the micrographs of the materials produced are also shown in Fig. 22. Through this approach, mesoporous tubes of metal oxides like CuO, Co_3O_4, SnO_2 and MnO_2, mixed metal oxides like $LiMn_2O_4$, $LiCoO_2$, $NiCo_2O_4$, LiV_3O_8, $Na_{0.7}Fe_{0.7}Mn_{0.3}O_2$, and $LiNi_{1/3}Co_{1/3}Mn_{1/3}O_2$, and, Phosphates like $A_3V_2(PO_4)_3$ where, A = Li, Na were fabricated. A variation of heat treatment involving preheating at 300°C in air decomposes all the three layers and there is material movement towards the outer periphery but the inorganic materials in it are not carried outward. During high temperature annealing, carbonisation results in pea like carbon nanotubes with inorganic materials growing into spheres along the centre as shown schematically in Fig. 22(b).

Pea like nanotubes of Co, $LiCoO_2$, $Na_{0.7}Fe_{0.7}Mn_{0.3}O_2$ and $Li_3V_2(PO_4)_3$ (shown in Fig. 22(h)) have been grown by similar gradient electro-spinning approach. Heating rates are important in the evolution of morphology and Peng et al. (Peng et al. 2015) have grown porous nanowires, nanotubes and tube-in-tube morphology of $CoMn_2O_4$ at different heating rates, following the same approach but

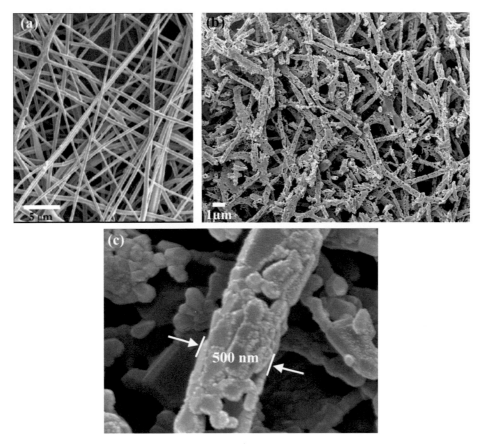

Fig. 21. (a) The polymer based fibers prepared by electro-spinning, (b, c) hollow electro-spun $LiMn_2O_4$ nanofibers (Jayaraman et al. 2013).

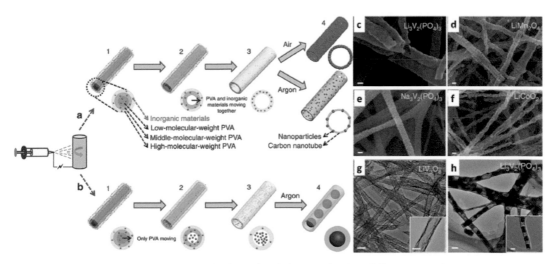

Fig. 22. (a) Schematic of the gradient electro-spinning and evolution towards tube at different stages of pyrolysis, (b) evolution towards pea like nanotubes, (c–g) SEM and TEM images of nanotubes, and (h) TEM image of pea like nanotubes; scale bar of SEM images—200 nm and of inset TEM images—100 nm (Niu et al. 2015, Wei et al. 2017).

using different set of polymers. They also succeeded in growing porous tube in tube morphology of other mixed metal oxides like $NiCo_2O_4$, $CoFe_2O_4$, $NiMn_2O_4$ and $ZnMn_2O_4$.

Bubble nanorod morphology of carbon-Fe_2O_3 has been developed by Ji et al. (Ji et al. 2014) by combining electro-spinning with Kirkendall effect as shown schematically in Fig. 23(a to d). The precursor nanofiber composed of Fe(acac)$_3$ (iron acetylacetonate) and polyacrylonitrile (PAN) has been prepared by electro-spinning. On annealing the precursor nanofiber in a reducing atmosphere, one obtains FeO_x—carbon nanofiber due to carbonisation of PAN and decomposition of Fe(acac)$_3$. In reducing atmosphere, FeO_x was reduced further by carbon to Fe. Then the annealing is carried out in oxidising atmosphere to develop Fe-Fe_2O_3 core-shell structure taking advantage of Kirkendall effect and finally, it leads to carbon nanorod with hollow nanosphere of Fe_2O_3 embedded in it as shown in the microstructure in Fig. 23(e–g).

Electro-spinning has been used also to make porous heterogeneous nanofibers/nanowires of oxides like NiO/ZnO (Qiao et al. 2013), TiO_2/ZnO (Wang et al. 2010), GeO_2/SnO_2 (Lei et al. 2015), CuO/SnO_2 (Zhao et al. 2012).

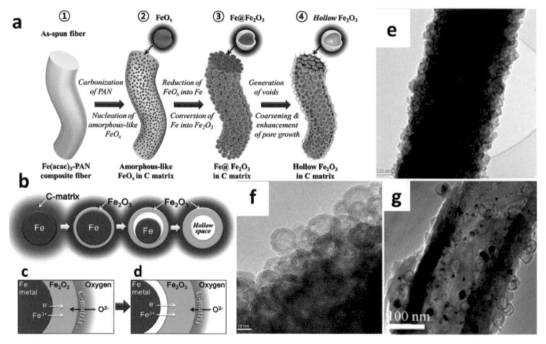

Fig. 23. (a–d) Schematics of the formation of bubble-nanorod structure of carbon-Fe_2O_3, (e, f) TEM images of this bubble-nanorod structure, ((a–f) Cho et al. 2015), (g) pure carbon-based bubble-nanorod structure (Chen et al. 2012, Wei et al. 2017).

4.5 Synthesis by Electro-chemical/Electroless Etching

Etching of Ag selectively from Ag-Au alloy nanowire is a standard method of preparing porous alloy nanowires. One may develop porosity in a metal by galvanic replacement by more noble metal. It may be illustrated in the context of bulk silicon, which may be converted into porous silicon by electroless or electrochemical etching and there could be control over the porosity content and the pore size. Electroless etching of silicon when carried out in a solution of $AgNO_3$ in HF, the reaction proceeds as,

$$4Ag^+ + 4e^- \rightarrow 4Ag$$
$$Si + 6F^- \rightarrow (SiF_6)^{2-} + 4e-$$

Silicon donates electron to reduce Ag^+ to Ag and in the process, gets oxidised to be etched away by F^-. Since the redox potential of Ag^+/Ag electrode is below the valence band of silicon, the etching takes place preferentially near the dopant site in p-type of silicon leaving pores on the surface.

The nanoparticles of silicon could be doped by boron and etched to obtain porous nanoparticles, as shown schematically in Fig. 24. The size of the pore could be controlled by varying the extent of doping as indicated in Fig. 25. In general, metal-assisted electroless etching takes place in an etchant solution containing HF and metal salts, such as $AgNO_3$, $KAuCl_4$, K_2PtCl_6.

Electroless etching of silicon wafers using silver nanoparticles results in silicon nanowires, as shown schematically in Fig. 26, and the porous silicon nanowires bonded by alginate binder results in superior performance in the anode of lithium ion battery.

Electrochemical etching of silicon wafer in HF at a constant current density also results in macroporous silicon as shown schematically in Fig. 27. The control of porosity and depth of porosity could be done by current density and concentration of HF (Turner 1958). The layer of porous silicon may be lifted off by suddenly changing current density. When combined with pyrolized polyacrilonitrile (PPAN) it gives a capacity of 1600 mAhg^{-1} (Thakur et al. 2012a) and if coated by gold, it results in a still higher capacity of 2000 mAhg^{-1} (Thakur et al. 2012b). Mesoporous silicon has been prepared by electrochemical etching of heavily boron doped silicon wafer in aqueous 48 per cent HF:ethanol (3:1 v/v) electrolyte at current densities of ~ 225 mA cm^{-2} for 358 s. The free-standing film of mesoporous silicon was then removed in aqueous HF:ethanol (1:30 v/v) at current densities of ~ 10 mA cm^{-2} for 750 s, washed and sonicated in ethanol to get ~ 40 μm particles and dried. The particles are coated with carbon in a muffle furnace, evacuated to < 1 mtorr, heated to 600°C before introducing precursor gas of argon:acetylene (9:1) and heated further to 690°C for 30 min (Li et al. 2014b).

Fig. 24. Schematic of the process to convert bulk silicon to porous silicon by electroless etching (Ge et al. 2013a).

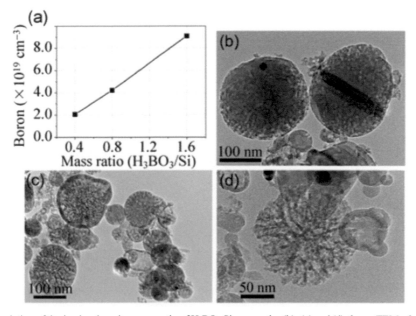

Fig. 25. (a) variation of doping by changing mass ratio of H_3BO_3:Si mass ratio, (b), (c) and (d) shows TEM of porous silicon doped by H_3BO_3:Si mass ratios of 2:5, 4:5 and 8:5 respectively (Ge et al. 2013a).

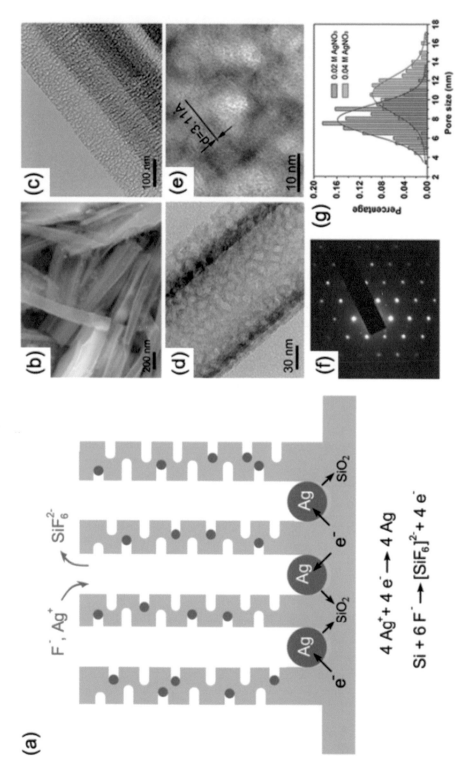

Fig. 26. (a) Schematic of mechanism of nanowire formation (Ge et al. 2013b), (b) SEM, (c), (d), (e) TEM images, (f) diffraction pattern and (g) pore size distribution for etchant containing 0.02 and 0.04 M AgNO$_3$ (Ge et al. 2012).

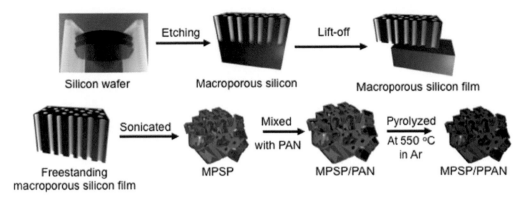

Fig. 27. The schematic of electrochemical etching to make macroporous silicon (MPS) and the composite of its particle (MPSP) with pyrolyzed polyacrilonitrile (PPAN) (Thakur et al. 2012c).

4.6 Synthesis by Using Template

This approach requires a template, which may leave behind a porous structure. The first task is to find suitable templates, which may be removed to leave behind porosity and it started in 1997, when Velev first used latex sphere to synthesize macroporous silica (Velev et al. 1997). The template of latex spheres arranged in cubic array was called 'colloidal crystals'. Close packed spheres of polystyrene, silica and PMMA have been used to form colloidal crystals to be used as a template for porous materials and the size of spheres determines the pore size left behind in the synthesised porous material.

Microporous molecular sieves in the family of zeolites have regular array of uniform sized channels but the channels in mesoporous sieves developed earlier were irregular and distributed in size. Mesoporous mono-disperse silica nanoparticles have been developed in 1992 by Mobil researchers, who introduced a new family of mesoporous materials (2–10 nm), M41S, formed by liquid crystal mechanism by aggregation of organic template molecules (Kresge et al.1992, Beck et al. 1992). These materials possess a well-defined pore structure and the most popular materials in these families are MCM (Mobil Composition of Matter)-41 and MCM-48, which respectively has hexagonal arrangement of unidirectional pores and 3D cubic pore system belonging to space group Ia3d. MCM-48 has bi-continuous structure centred on the gyroid minimal surface that divides the pore space into two nonintersecting sub-volumes (Schumacher et al. 2000, Vartuli et al. 1996). There is also MCM-50, which has lamellar pores. The pore structures in MCM-50, MCM-41 and MCM-48 are shown in Fig. 28.

Mesoporous spherical amorphous particles of silica with larger pore size (4.6 to 30 nm) in hexagonal array was prepared by University of California, Santa Barbara and this material is called SBA (Santa Barbara Amorphous)-15. The designation of different families of mesoporous material is given in Table 5 and pore structure of some selected materials in different families is given in Table 6.

SBA-15 has been used by Gu et al. (Gu et al. 2015) as a template to prepare mesoporous peapod-like Co_3O_4 over carbon nanotube arrays, which have high surface areas (up to 750 m^2 g^{-1}) and large pore sizes. Many other porous nanowires, such as mesoporous CoN, CrN (Shi et al. 2008), porous Ni (Li et al. 2008), mesoporous $CuCo_2O_4$ (Pendashteh et al. 2015) and mesoporous Si over carbon core-shell (Kim and Cho 2008a) have been prepared by using SBA-15 as template for their synthesis.

Carbon nanostructures could be synthesised using a porous template of silica and by insertion of a carbon precursor into the pores of the template by means of a gas or liquid route, followed by carbonisation (Ryoo et al. 1999, Vix-Guterl et al. 2003). The silica could then be removed from SiO_2/C combination by acid leaching. Three precursors used were propylene, sucrose and pitch. The synthesised carbons have been designated by CXY where X and Y correspond to the type of carbon precursor and template used, respectively as given in Table 7.

Fig. 28. Pore structure in mesoporous materials—MCM-50, MCM-41 and MCM-48 (Ying et al. 1999).

Table 5. Designation of families of mesoporous materials.

Designation	Full Name
MSU	Michigan State University
SBA	Santa Barbara Amorphous
MCM	Mobil Composition of Matter
HMS	Hollow Mesoporous Silica
OMS	Ordered Mesoporous Silica
TUD	Technische Universiteit Delft
MCF	Meso Cellular Form
FSM	Folded Sheet Mesoporous

Table 6. Structural characteristics of pores in selected types of mesoporous materials.

Types of Mesoporous Material	Pore Structure
MCM-41	Hexagonal, 1D
MCM-48	Cubic Bi-continuous, 3D
SBA-1	Cubic, 3D
SBA-3	Hexagonal, 2D
SBA-15	Hexagonal, 1D
SBA-16	Cage like arrangement
MSU	Hexagonal, 2D
HMS	Hexagonal

Table 7. Porous carbon materials synthesised using different templates.

Template	Carbon Source		
	Propylene (Pr)	Sucrose (S)	Pitch (P)
MCM-48	CPr48	CS48	CP48
SBA-15	CPr15	CS15	CP15
MSU-1	CPrU	CSU	CPU

Carbon nanotubes (CNTs) both single walled and multiwalled could be synthesised (Ray et al. 2016) in the desired diameter and these tubes have good conductivity. CNTs have been used as template for the synthesis of porous 1D nanomaterials. Zhu et al. (Zhu et al. 2012) prepared CNT-Ni_3S_2 hybrid nanostructures via a multistep synthesis route with CNT as template, shown in Fig. 29(a).

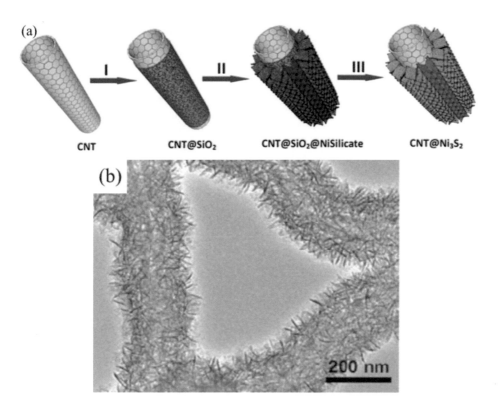

Fig. 29. (a) Schematic of multi-step synthesis of Ni_3S_2 over CNT and (b) its TEM (Zhu et al. 2012).

A silica was coated on CNT by hydrolysing tetraethyl orthosilicate. Nickel silicate nanosheets were then grown *in situ* on the silica layer to form $Ni_3Si_2O_5(OH)_4$ precursor over SiO_2 on CNT. The final Ni_3S_2 on CNT was obtained as shown in Fig. 29(b), by the conversion of $Ni_3Si_2O_5(OH)_4$ to Ni_3S_2 and removal of the silica layer. Other similar porous metal oxide/CNT nanostructures like Co_3O_4/CNT (Du et al. 2007a), CuO/CNT, (Hoa et al. 2010) SnO_2/CNT (Du et al. 2009, Jia et al. 2009) and mesoporous carbon/CNT (Qian et al. 2011) have been synthesised using CNT as the template. Metal oxide nanotubes have been made after removal of CNT for $LiMn_2O_4$ (Tang et al. 2013), TiO_2 (Li et al. 2015), and In_2O_3 (Du et al. 2007b).

Silica used as template could also be reduced directly to silicon but carbothermic reduction of silicate takes place at temperature exceeding 2000°C, above melting point of silicon. Magnesium is able to reduce silica at 650°C. Following the processing route of magnesiothermic reduction given in Fig. 30(a), monodisperse silica microspheres in Fig. 30(b) was converted to porous silicon shown in Fig. 30(c).

Silica spheres with solid core and mesoporous shell was produced (Xiao et al. 2015) using simultaneous hydrolysation and condensation. A mixture of absolute ethanol, deionised water and aqueous ammonia was heated to 30°C and tetraethyl orthosilicate was added rapidly under stirring. After 1 hr a mixture of tetraethyl orthosilicate and n-octadecyltrimethoxysilane was added drop by drop for 20 minutes and the solution was kept at ambient temperature for 12 hr. The resulting white powder was centrifuged and calcined for 6 hr at 550°C for the removal of organic species. These silica spheres were reduced following the processing route given in Fig. 31(a) to produce hierarchically porous silicon nanosphere (hp-SiNS) whose microstructures are shown in Figs. 31(b, c).

There has been use of small molecular organic acid like citric acid to synthesise mesoporous microspheres of composite $LiFePO_4$/C by spray drying at 200°C a precursor solution of Li_2CO_3, $Fe(NO_3)_2.9H_2O$, $NH_4H_2PO_4$ and citric acid in the stoichiometric ratio of $n_{Li}:n_{Fe}:n_p:n_{acid}::1:1:1:x$ using a carrier gas of air (Yu et al. 2009a). The proportion of acid could be increased by increasing x to

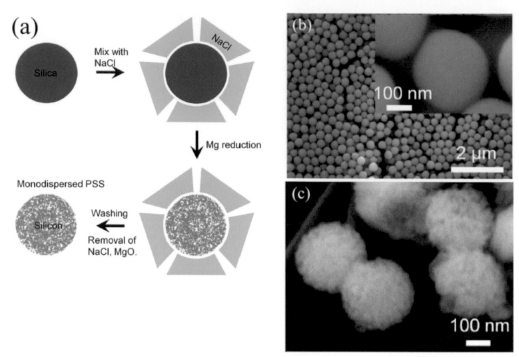

Fig. 30. (a) Schematic of magnesiothermic reduction of silica; SEM micrographs of (b) monodisperse silica microspheres and (c) porous silicon (Wang et al. 2015).

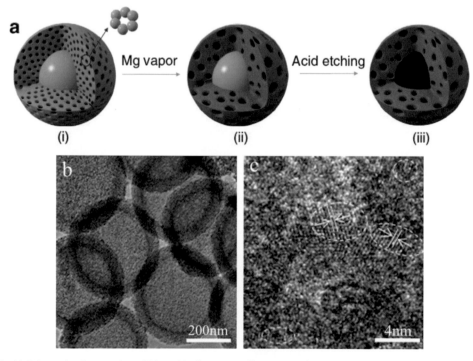

Fig. 31. (a) Schematic of processing of hierarchically porous silicon nanosphere (hp-SiNS), (b) its TEM and (c) HRTEM micrograph (Xiao et al. 2015).

obtain higher porosity. The sprayed powder is then heated in air limited box furnace for 5 hr and calcined at 700°C for 12 hr in argon atmosphere. The process is schematically shown in Fig. 32(a) and the SEM micrographs of the resulting sphere are shown in Fig. 32(b, c).

A template directed growth of silicon nanotube (SiNT) has been reported by (Tesfaye et al. 2015) where sacrificial ZnO nanowire template was used. Seeds of ZnO nanocrystals were formed on stainless steel substrate and annealed at 300°C. Thereafter, ZnO nanowires were grown over the seeds by hydrothermal process at 95°C using 0.02 M $Zn(NO_3)_2$ and 0.02 M hexamethylenetetramine as precursors. Silicon was deposited on nanowires of ZnO by exposing them to SiH_4 (0.5 per cent in He) at 530°C. The nanowire template was removed by subjecting it to NH_4Cl vapor at 450°C leaving behind packed array of SiNT as shown in Fig. 33.

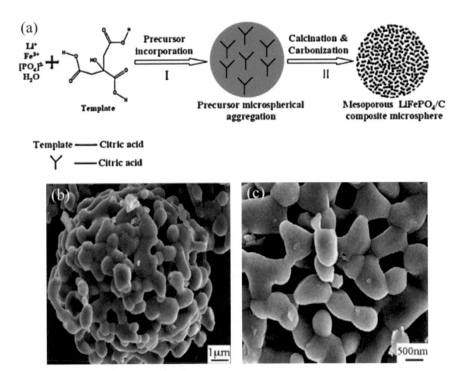

Fig. 32. (a) Schematic of spray drying process using molecular template of citric acid and (b, c) SEM micrographs of mesoporous composite microsphere of $LiFePO_4/C$ (Yu et al. 2009a).

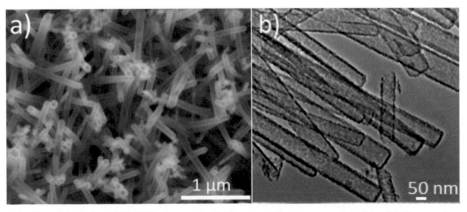

Fig. 33. (a) SEM micrograph and (b) TEM micrograph of silicon nanotube grown over template of ZnO nanowires, removed to obtain tubes (Tesfaye et al. 2015).

Liu et al. (Liu et al. 2010b) used a biological template of crab shell ground after annealing in air to remove organics, to impregnate it with a precursor solution followed by carbonisation to get mesoporous carbon nanofibers. Bamboo like graphitic carbon nanofiber have been made by growing uniform distribution of the template of silica nanoparticles in carbon nanofibers and these templates of silica particles have been etched away by HF to obtain porous bamboo like structure of porous carbon nanofiber (Sun et al. 2015). Similarly, one may use metal nanocrystals of chosen size as prefilled templates for getting porous nanofibers.

Doherty et al. (Doherty et al. 2009) prepared colloidal crystals from PMMA spheres of sizes 100, 140 and 270 nm as shown in Fig. 34 and $LiFePO_4$ solution was added drop by drop over the templates, dried in air and heated to the calcining temperatures of 320, 500, 600, 700 and 800°C for several hours in a reducing atmosphere. It has been observed that smaller bead diameter in the colloidal crystal template results in higher surface area but there is decrease in surface area due to increased temperature of calcinations attributed to higher growth of $LiFePO_4$ crystals.

An opal structure may be made by pre-sintering silica microspheres and then silicon may be deposited in the interspaces by chemical vapour deposition (CVD) using SiH_4 as the precursor gas or by filling the space with precursor gel containing silicon, which on annealing will solidify to result in rigid porous silicon. Silica may be removed by treating with HF and an inverse opal structure of silicon is left behind as shown in Fig. 35.

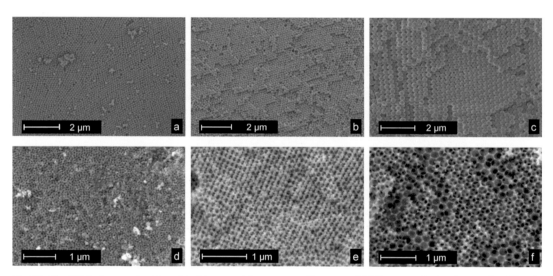

Fig. 34. The SEM micrographs of template (above) and the corresponding porous LiFeO₄ prepared using the template (directly below) of sizes (a, d) 100 nm, (b, e) 140 nm and (c, f) 270 nm, calcined at 500°C (Doherty et al. 2009).

Similarly, nickel has been electro-deposited on silica opal structure to arrive at inverse opal structure of nickel, on which silicon was deposited by CVD (Zhang and Braun 2012). Also, carbon inverse opal structure has been obtained and silicon deposited on it (Esmanski and Ozin 2009). Porous nickel has been prepared by selective leaching of copper from Cu-Ni alloy and silicon is then deposited in these pores by CVD process (Gowda et al. 2012). Silicon gel was mixed with silica nanospheres and annealed at high temperature to obtain micro sized porous silicon powder after treating with HF as shown in Fig. 35(c, d).

It may be noted that to prevent reaction between nanoparticles of silicon and silica there should be protective coating of n-butyl group in gel.

Ordered mesoporous spinel of Li-Mn-O was made by infiltrating aqueous solution of $Mn(NO_3)_2$ into 3D pore structure of ordered mesoporous silica, followed by removal of silica, which was then heated in a reducing atmosphere to form mesoporous structure of spinel Mn_3O_4 from Mn_2O_3. Mesoporous spinel of $LiMn_2O_4$ was made by reacting Mn_3O_4 with LiOH (Jiao et al. 2008).

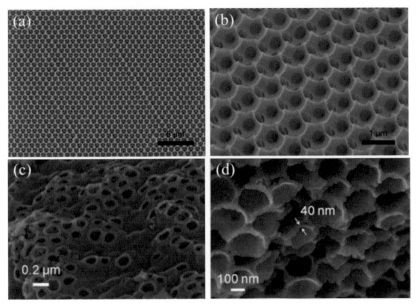

Fig. 35. SEM images of (a, b) inverse opal structure of porous silicon and (c, d) 3D porous silicon (Esamanski and Ozin 2009, Kim et al. 2008b).

The use of template to generate porosity may be used to generate further porosity in porous materials like MOF or COF and it is expected that these routes may now be explored more extensively.

5. Electro-chemical Performance of Porous Materials

Initially, efforts were directed to increase specific surface area of an electrode by using nanoparticles for various compositions in the electrode. It demonstrated good potential for storage with high rate capability. In Li-ion battery, use of nanoparticles of Si, Ge, $LiCoO_2$, etc. resulted in substantial improvements in power performance and cycling behaviour attributed respectively to shorter diffusion paths necessary for the access of electroactive species and accommodation of strain change accompanying intercalation and deintercalation (Jason and Feng 2013, Okubo and Honma 2013, Chan et al. 2008).

5.1 Electrochemical Capacitors

In capacitors, there is only charge storage and the electrode material is not involved in any reaction. Thus, there is no transformation involved and limiting issues like accompanying strain or diffusion in the bulk electrode are not relevant. Thus, the capacitors have long life of about 10^5 cycles and higher rate capabilities compared to the batteries and high cycle efficiency between 84–99 per cent but also, high daily discharge rate between 5–40 per cent. The energy delivered by the capacitor is not at constant potential but at potential decreasing finally to zero. The energy stored in an electrode of a capacitor uses only surface area and so, the extent of storage of energy is ordinarily limited from 0.1 Wh/kg to 1 Wh/kg, compared to 100 Wh/kg in Li-batteries (Whittingham 2008). Therefore, significant scope for the application of porous materials exists to increase energy density in a capacitor.

EDLC type of supercapacitors generally uses carbon of high surface area as electrode and sulphuric acid or acetonitrile as the electrolyte but the energy stored is still less than 10 Wh/kg and cell voltage is limited by the decomposition potential of the electrolyte, which is < 1 V for aqueous electrolyte and around 3 V for non-aqueous electrolyte. The supercapacitors are more suited for delivery of energy over short period of say, 10^{-2} to 10^{-3} s, and for large number of cycles at low cost.

In a study, electrochemical capacitors were synthesized with porous nanocarbon structures templated by mesoporous silica and also, by MCM-48, SBA-15 and MSU-1 having cubic, hexagonal and 3D wormhole porous structure respectively. The pores were filled either with propylene (Pr), sucrose (S) or pitch (P) and then were carbonised. Silica was removed by acid leaching to get the porous carbon structure. The designation of the resulting carbon is as described in the section on template based synthesis. The electrolytes were either aqueous 1 mol/l of H_2SO_4 or nonaqueous 1.4 mol/l or 1 mol/l of tetraethyl ammonium tetrafluroroborate—$TEABF_4$ in CH_3CN. It was observed that (i) the carbon prepared with MCM-48 and SBA-15 has highly ordered periodic structure and that from the template MSU-1 has wormhole porous structure lacking in symmetry, (ii) carbon materials have significant total surface area from 630 to 2200 m^2g^{-1} and (iii) the pores consisted of narrow micropores (ultramicropores), secondary micropores (supermicropores, 0.7 nm < diameter < 2 nm) and mesopores (2 nm < diameter < 50 nm). Sucrose resulted in higher surface area because of additional microporosity during its carbonisation.

Irrespective of the type of nanostructured carbon, voltammetry curves at 2 mV/s and 10 mV/s show near rectangular shape in both the electrolytes, characteristic of an ideal capacitor (Vix-Guterl et al. 2004). For a given carbon precursor, the total surface area, volume of micropores and the capacitance of carbon are less for SBA-15 template compared to those for MCM-48, which had wall thickness of ~ 1.5 nm compared to 4.5 nm in SBA-15, resulting in relatively larger pores with template SBA-15. It was observed that details of pore structure symmetry are not very important for the performance of supercapacitor but the amounts of micropores, mesopores and their connectivity are important. The super microporosity that results from the crosslinking of wormholes, are expected to give better access of electrolyte and consequently, better charging of the electrical double layer but the capacitance of CS48 and CSU were found comparable. However, the frequency dependence of capacitance in organic medium and charge storage efficiency was found to be better in CS48 than those in CSU (Vix-Guterl et al. 2005).

Pseudo-capacitors, on the other hand, uses both—the surface as well as the bulk, for energy storage and so, could store energies in excess of 10 Wh/kg. But it faces material challenges similar to batteries as it uses the same materials like transition metal oxides like vanadium and manganese oxides for chemical storage of energy. Some of these crystalline oxides say vanadium oxide, loses its crystallinity on reacting with lithium as lithium and vanadium ions get randomised in the lattice and show a typical sloping discharge profile of a capacitor. The design of electrode for this purpose should start with electrode-electrolyte combination of high conductivity and charge transfer but at reasonable cost. The pores in the electrode material should be of such size so that they can deliver and remove ions at a very high rate. An oxide, which has generated considerable interest, is ruthenium oxide, which has taken a significant jump in stored energy to around 200 Wh/kg.

Vanadium oxide has been employed as electrode material in batteries like silver vanadium oxide used in pacemaker and this oxide has been synthesised into a broad range of morphologies. Wang et al. (Wang et al. 2006) have explored the nanovanadium oxide of different morphologies, whose performance has been measured by Ragone plots of energy density in Wh/kg and power density in W/kg. The former gives energy, the capacity for doing work while the latter gives the rate of doing work. The Ragone plot for different morphologies of nanovanadium oxide is given in Fig. 36.

Microcapacitors using carbon onions of 6–7 nm diameter demonstrated superior power density at charging rate as high as 10 Vs^{-1} (Pech et al. 2010). Nanostructures of 1D offer most of the advantages of 0D structures but in addition, there is advantage of its high aspect ratio for making free-standing and flexible electrode without the necessity of addition of any binder (Chan et al. 2008, Please put Simon and Gogotsi 2013, Seng et al. 2011). Wei et al. (Wei et al. 2017) have summarised performance of 1D nanomaterials in supercapacitors in a table, which is reproduced in Table 8. However, for references to individual studies reported in the table, one may refer to Table 3 in his review. The investigators

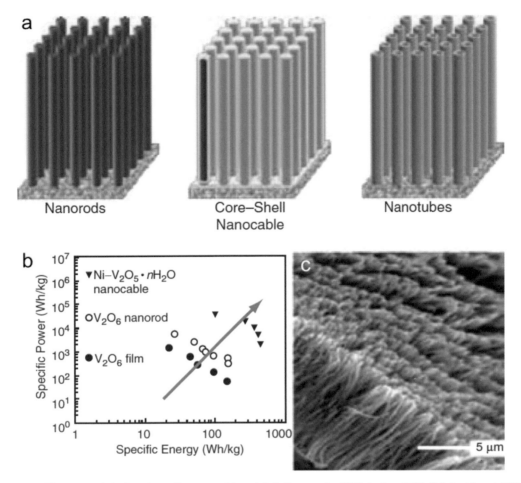

Fig. 36. Different morphologies of vanadium pentoxide and their Ragone plot (Whittingham 2008, Takahashi et al. 2005, Takahashi et al. 2004).

may take clue from this table to identify the potential materials for exploring further the impact of various architectures on electrochemical performance.

Nanomaterials with templated 3D mesoporosity are an important class of EES electrode materials. The pore structure is well balanced as mesopores are generated by templates and also, the micropores occur between the nanocrystals, resulting in superior capacity and rate performance compared to the non-templated nanocrystalline materials (Brezesinski et al. 2010a). Several pseudocapacitive metal oxides like Nb_2O_5, CeO_2, etc., have demonstrated superior electro-chemical behaviour (Brezesinski et al. 2010b, Rauda et al. 2013).

Hybrid electrochemical capacitors (EC) consisting of two electrodes—a porous-carbon double-layer electrode and a pseudocapacitive electrode, using different charge storage mechanisms provide superior performance, particularly at high charge/discharge rates. Amatucci et al. (Amatucci et al. 2001) have shown that these hybrid ECs have superior power density than Li-ion batteries and also, a higher energy density than EDLCs. Using activated carbon (AC) at both electrodes, there are commercial Li-ion hybrid ECs available with energy densities of about 20–30 Whkg^{-1}, about 3–5 times higher than that of conventional EDLCs. This high energy density has been attributed to added redox capacitance and a wider voltage window (Naoi et al. 2013). The 'Aquion battery' has a spinel-MnO_2/AC (activated carbon) electrodes with sodium sulphate solution as electrolyte and its energy density is ~ 20 Whkg^{-1} showing no degradation even after 10,000 cycles. In the next generation of

Table 8. Electrochemical performance of some 1D nanomaterials in supercapacitors (Wei et al. 2017).

Electrode Material	Initial Specific Capacitance	Electrolyte	Potential Range	Energy Density/ Power Density	Test Method
Mesoporous carbon nanofiber webs (E)	262 F g^{-1} at 0.2 A g^{-1}	1 M H$_2$SO$_4$	-0.2–0.5 V	–	Three electrode cell
Porous carbon fibers (E)	197 F g^{-1} at 5 mV s^{-1}	1 M H$_2$SO$_4$	0–1 V	–	Three electrode cell
Porous carbon nanofiber networks (E)	302 F g^{-1} at 0.2 A g^{-1}	6 M KOH	0–1 V	5.2 Wh kg^{-1}/10 kW kg^{-1}	Two electrode cell
Porous carbon nanofibers (E)	–	0.5 M H$_2$SO$_4$	0–1.2 V	3.22 Wh kg^{-1}/0.6 kW kg^{-1}	Two electrode cell
Bamboo-like porous carbon nanofibers (E)	2.1 F cm^{-3} at 33 mA cm^{-3}	H$_3$PO$_4$/PVA gel	0–0.9 V	2.37 Wh kg^{-1}/61.3 kW kg^{-1}	Two electrode cell
KOH activated CNTs (CD)	53.6 F g^{-1} at 50 mA g^{-1}	7 M KOH	0–1 V	–	Two electrode cell
N-doped porous carbon Nanofibers (T)	202 F g^{-1} at 1 A g^{-1}	6 M KOH	-1–0 V	–	Three electrode cell
N, P-co-doped carbon nanofibers (T)	204.9 F g^{-1} at 1 A g^{-1}	2 M H$_2$SO$_4$	0–1 V	1.86 Wh kg^{-1}/26.61 kW kg^{-1}	Two electrode cell
PEDOT-nanotube (E)	132 F g^{-1} at 5 mA cm^{-2}	1 M LiClO$_4$	0–1.2 V	–	Two electrode cell
Hydrous RuO$_2$ nanotubes (T)	861 F g^{-1} at 0.5 A g^{-1}	1 M H$_2$SO$_4$	0–1 V	–	Three electrode cell
Hydrous RuO$_2$ nanotube arrays (T)	1300 F g^{-1} at 0.1 mV s^{-1}	1 M H$_2$SO$_4$	0–1 V	7.5 Wh kg^{-1}/4320 kW kg^{-1}	Three electrode cell
MnO$_2$ nanotube (E)	320 F g^{-1} at 20 mV s^{-1}	1 M Na$_2$SO$_4$	0–1 V	–	Three electrode cell
Mesoporous MnO$_2$ nanotubes (T)	365 F g^{-1} at 0.25 A g^{-1}	1 M Na$_2$SO$_4$	-0.2–0.8 V	–	Three electrode cell
Porous Co$_3$O$_4$ nanowires (L)	260 F g^{-1} at 2 A g^{-1}	2 M KOH	0.1–0.6 V	–	Three electrode cell
Mesoporous Co(OH)$_2$ nanowires (T)	993 F g^{-1} at 1 A g^{-1}	1 M KOH	-0.1–0.5 V	–	Three electrode cell
Mesoporous NiO nanotubes (L)	405 F g^{-1} at 0.5 A g^{-1}	6 M KOH	0–0.5 V	–	Three electrode cell
H-TiO$_2$ nanotube arrays (CE)	3.24 mF cm^{-2} at 100 mV s^{-1}	0.5 M Na$_2$SO$_4$	0–0.8 V	–	Three electrode cell
Porous NiCo$_2$O$_4$ nanowires (L)	743 F g^{-1} at 1 A g^{-1}	1 M KOH	-0.05–0.45 V	–	Three electrode cell
Mesoporous NiCo$_2$O$_4$ nanowire arrays (L)	1283 F g^{-1} at 1 A g^{-1}	6 M KOH	0–0.4 V	–	Three electrode cell
MnO$_2$/Mn/MnO$_2$ sandwich-structured nanotube arrays (T)	955 F g^{-1} at 1.5 A g^{-1}	1 M Na$_2$SO$_4$	0–0.8 V	52 Wh kg^{-1}/15 kW kg^{-1}	Three electrode cell
MnO$_2$/CNT composites (L)	201 F g^{-1} at 1 A g^{-1}	1 M Na$_2$SO$_4$	0–0.9 V	–	Three electrode cell
Mesoporous Co$_3$O$_4$ nanowire arrays (L)	1160 F g^{-1} at 2 A g^{-1}	6 M KOH	0–0.55 V	–	Three electrode cell
Porous CoO@PPy nanowire arrays (L)	–	3 M NaOH	0–1.8 V	43.5 Wh kg^{-1}/87.5 W kg^{-1}	Two electrode cell

Table 8 contd.....

Table 8 contd. ...

Electrode Material	Initial Specific Capacitance	Electrolyte	Potential Range	Energy Density/ Power Density	Test Method
Porous VN nanowires-graphene composites (L)	73 F g^{-1} at 0.1 A g^{-1}	1 M LiPF$_6$ in EC/DEC (1:1 v/v)	0–4 V	162 Wh kg^{-1}/200 W kg^{-1}	Two electrode cell
Porous TiNb$_2$O$_7$ nanotubes (T)	–	1 M LiPF$_6$ in EC/DMC (1:1 v/v)	0–3 V	≈ 34.5 Wh kg^{-1}/7.5 kW kg^{-1}	Two electrode cell
Intertwined CNT/ V$_2$O$_5$ nanowires (L)	–	1 M LiClO$_4$ in PC	0.1–2.7 V	40 Wh kg^{-1}/210 W kg^{-1}	Two electrode cell
Porous CNT/V$_2$O$_5$ nanowire composites (L)	≈ 35 F g^{-1} at 0.5 mA	1 M NaClO$_4$ in PC	0–2.8 V	38 Wh kg^{-1}/140 W kg^{-1}	Two electrode cell

*(Synthesis methods indicated within brackets: E = Electrospinning method, L = liquid phase method, T = template-assisted method, CE = chemical etching method, and CD = chemical deposit method).

lead-acid batteries, hybrids have been employed by replacing one of the lead plates with an activated carbon electrode resulting in the improvement of power density by a factor of two and increased cycle life by a factor of three (Lam and Louey 2006). The hybrid ECs provide a prospective line of exploration where one may try different material architecture in electrodes like porous 1D, 2D and 3D and even, try composite electrodes for getting higher energy storage, improve rate performance and better cyclability to reduce overall cost for energy storage.

5.2 Battery Systems

Li-ion batteries is already being used extensively in portable electronics and it has received considerable attention for increasing its energy density. But, the resource for lithium is also limited so it is worthwhile to look for batteries with other electroactive species. It will be interesting to start with these emerging alternate battery systems, to focus on breakthrough necessary in these batteries for commercial viability as energy storage systems.

Sodium ion based battery has emerged as an alternative class of energy storage devices with the advantage of natural abundance and low cost of sodium (Kundu et al. 2015, Pan et al. 2013). But Na$^+$ ion is 1.5 times larger and 3.3 times heavier than Li$^+$ ion and so, the diffusion of Na$^+$ ion is inherently slower than that of Li$^+$ ion (Pan et al. 2013). Thus, it is necessary to develop a suitable host with large interstitial space to accommodate Na$^+$ ions and allow faster diffusion. Highly porous host structure may give extensive access to electrolyte, reducing the diffusion length and accommodating volume change more than that for lithium for comparable performances.

Sodium–sulphur batteries using β-alumina as electrolyte have become commercial and are operating at around 300°C with liquid sodium, which raises safety concerns. The temperature could be reduced if suitable sulphur solvents and electrolyte could be found. However, sodium-sulphur batteries have relatively high energy density of 150–300 Whl^{-1}, almost no daily self-discharge, higher rated capacity of batteries up to 244.8 MWh and high pulse power capability. There is no toxic material in the battery and so, recyclability is up to 99 per cent (Liu et al. 2010a). But the operating cost is high at 80$/kW/year and a heating system is required to maintain temperature. Sumitomo Electric Industries and Kyoto University has reported to have developed a low temperature sodium bearing material, which melts at 330 K and it is claimed that the battery developed using this material has achieved a capacity of 290 Whl^{-1} (Sumitomo Electric 2011).

If a battery could have pure lithium as anode overcoming dendritic plating and consequent short circuiting, and a chalcogen as cathode, the volumetric capacity could be improved significantly. Zinc-air cells has been exploited commercially in hearing aids and the energy density in lithium-air cells could be doubled over that of present day Li-ion batteries. Lithium-air cell forming Li_2O_2 has demonstrated reversibility but there is large difference between charging and discharging voltages, respectively 4.2 V and 2.6 V, resulting in low efficiency.

Elemental sulphur is abundantly available and Li-S batteries have a high theoretical capacity of 1,675 mAhg^{-1} and specific energy of 2600 Whkg^{-1} apart from being low cost. But Li-S batteries have several problems. The elemental sulphur is an electrical insulator (5×10^{-30} S cm^{-1} at 25°C) and during discharge, there are reaction products, which lower both the electrochemical activity and the utilisation of sulphur. The cyclability is also poor due to high solubility of the intermediate product, lithium polysulphide Li_2S_x ($2 \leq x \leq 8$), formed during both charging and discharging (Ji et al. 2011, Mikhaylik and Akridge 2004). Dissolved polysulfides diffuse to the lithium anode to be reduced to short-chain polysulfides, which may move back to the cathode for its reoxidation to long chain polysulfides. This parasitic shuttle reaction results in low coulombic efficiency. Some of the soluble polysulfides are reduced to insoluble Li_2S_2 and/or Li_2S, to deposit on the anode surface and eventually forming a thick layer during repeated cycling. During discharge similar deposits may form on the cathode surface. The formation of insoluble sulphide on both the electrodes results in progressive capacity degradation due to the loss of active material, lower accessibility of the electrodes leading to degradation of electrode and increased cell impedance. Porous encapsulation could provide a solution to these problems. Ordered mesoporous carbon has been used to encapsulate the sulphur species within its nanochannels and then this composite was coated with a thin-layer of polymer and the material shows an initial reversible capacity of 1320 mAhg^{-1} and improved cyclability (Ji et al. 2009). A high initial reversible capacity of 1071 mAh g^{-1} at a rate of 0.5 C was obtained in porous hollow carbon spheres with a mesoporous shell structure composed of sulphur and 91 per cent of capacity was retained after 100 cycles (Jayaprakash et al. 2011). Wu et al. (Wu et al. 2011) made a sulphur/polythiophene composite with a core/shell structure and it shows a high initial discharge capacity of 1119 mAg^{-1} and 69.5 per cent of the capacity could be retained after 80 discharging/charging cycles. Xiao et al. (Xiao et al. 2012) synthesised self-assembled polyaniline hollow nanowires (PANI-HNW) for encapsulation of sulphur. The polymer forms a 3D, cross-linked, structurally-stable sulphur-polyaniline (SPANI) polymer backbone through in situ vulcanisation for creating interconnected inter- and/or intra-chain disulfide bonds. This SPANI-HNW/S composite electrode could retain a discharge capacity of 837 mAhg^{-1} after 100 cycles at a rate of 0.1 C. At high discharge rate of 1 C, the electrode still manifests a very stable cycling performance up to 500 cycles. Porous conductive polymers like polythiophene and polypyrrole have been used for physical absorption to hold sulphur and so, one may explore conducting MOF for application in Li-S batteries.

Li-air battery is being explored as a rechargeable battery due to its future potential. It transforms the chemical energy in lithium (anode) and oxygen (cathode) into electrical energy during discharge. Presently, it has low power density, poor cyclability and low energy efficiency. The other important challenges are dendrite formation in lithium anode, its incompatibility with electrolyte and air, instability of electrolyte in oxygen rich electrochemical conditions, conductivity and evaporation of non-aqueous electrolytes, etc. These challenges are discussed in details in the review by Shao et al. (Shao et al. 2013). There is considerable scope to explore porous materials in this battery.

The Vanadium redox batteries are marked by relatively high efficiency of 85 per cent, response faster than 0.001 s and the ability to operate for cycles exceeding 10,000–16,000 cycles (Gonzalez et al. 2004). The power of the flow batteries depends on the size of the electrodes considering the number of stacks while the energy storage capacity depends on the concentration and the amount of electrolyte. There are still some technical challenges like low electrolyte stability and solubility, low energy density and high operating cost (Leung et al. 2012). These batteries are now commercially available and the advantage of stacked bipolar plates allows the battery pack to provide high power in a small volume. When all the electroactive species are not dissolved as in flow batteries, these are called hybrid flow batteries and Zinc-Bromine (ZnBR) battery belongs to this category. In this battery, two aqueous electrolytes containing electroactive components of Zn^+ and Br^- respectively are stored in two external tanks and during charging/discharging, these two electrolytes flow through the cell having carbon-plastic composite electrode stacked in compartments where electrochemical reaction takes place (Chen et al. 2009). Polysulfide-bromine (PSB) flow battery uses sodium bromide and sodium polysulfide as electrolytes, which are soluble, abundantly available and cost effective (Weber et al. 2011). However, bromine and sodium sulfate produced during chemical reactions in the cell may create some environmental issues. The technology is still in demonstration stage and application of porous electrode may increase the rate capability of such batteries.

In commercial Li-ion battery, the cathode is made of $LiCoO_2$, and the anode is made of graphitic carbon, which, when fully intercalated, becomes LiC_6 corresponding to a charge capacity of 372 mAhg^{-1}. The electrolyte is a non-aqueous organic liquid, which has dissolved lithium salt, generally $LiPF_6$ or $LiClO_4$. For one electroactive lithium ion, there are six carbon atoms in the anode and the gravimetric energy density that could be practically attained is around 350 mAhg^{-1}. In order to create high energy density at low cost in both the electrodes there is search for alternate materials with porous architecture, which will boost energy density and improve rate capability as well. Thus, porous electrode materials and their architecture have received considerable attention. Carbon with different pore sizes as microporous (pore size < 2 nm), mesoporous (2 nm > pore size > 50 nm), and macroporous (pore size > 50 nm) are promising anode materials.

Porous carbon has been employed in the anode of Li-ion battery and it shows improved electrochemical performance due to additional intercalation sites and easy accessibility of electrolyte. Yang et al. (Yang et al. 2013) used α-cyclodextrin as a carbon precursor in presence of block copolymer Pluronic F127 for hydrothermal reaction, followed by pyrolysis to result in hollow carbon nanospheres of diameter 200–400 nm with surface area of 400 m^{2}g^{-1} and this material in anode of Li-ion battery demonstrated a specific charge capacity of 450 mAhg^{-1}.

Zhou et al. (Zhou et al. 2003) prepared ordered mesoporous carbon structure having BET surface area of 1030 m^{2}g^{-1} for anode and observed a gravimetric energy density of 1100 mAhg^{-1} at a current rate of 100 mAg^{-1}. The capacity was stable for the first 20 cycles with a low fading rate of 0.075 per cent per cycle. The high capacity and stable performance have been attributed to increased surface area and mesoporosity. Hu et al. (Hu et al. 2007) made hierarchically porous carbon with a 3D network of mesopores and micropores, and it demonstrated after 40 cycles a stable capacity of ~ 500 mAhg^{-1}, almost twice that of conventional graphite based anode. The anode also showed very high rate capability. These improvements have been attributed to the significant presence of connected porosity providing easy access to electrolyte and increased sites for intercalation. Yu et al. (Yu et al. 2009b) incorporated tin particles in porous multichannel carbon microtubes by electrospinning and the composite anode has resulted in a stable capacity of ~ 300 mAhg^{-1} even at a high rate of 10C (1C = 3.6 Ag^{-1}). Rahman et al. (Rahman et al. 2015) has summarised the electrochemical performance of porous carbon anode in a table which is reproduced here in Table 9 but for individual work, one may see his article.

Silicon may hold lithium till $Li_{22}Si_5$ resulting in a high specific capacity of 4200 mAhg^{-1} and is marked by low discharge potential but it has been overlooked because of low conductivity and fast capacity degradation due to strain associated with intercalation and deintercalation as mentioned earlier. However, porous nanostructure of silicon may significantly improve performance by accommodating strain and giving better access to electrolyte.

Hierarchically porous silicon nanospheres (hp-SiNSs) when employed in the anode of Li-ion battery shows reversible specific capacity of 1850 mAhg^{-1} at 0.1C (1C = 3.6 Ag^{-1}). Therefore, it is apparent that most of the silicon in the electrode is active due to accessibility by Li$^+$ ions. Following first two cycles at C/20, a capacity exceeding 1800 mAhg^{-1} could be maintained even after 200 cycles. The Coulombic efficiency for this morphology is 52 per cent at the first cycle but it increases to above 99 per cent after twentieth cycle. In the first few cycles, a stable SEI is formed and the Coulombic efficiency increases to above 99 per cent. When compared to commercial Si nanoparticles under similar conditions, there is faster decay of capacity in the first 100 cycles from 2000 mAhg^{-1} to < 1000 mAhg^{-1}. The enhanced cyclability of the hp-SiNSs may be attributed to its unique porous morphology. The discharge capacities are 1850, 1430, 1125, 920 and 700 mAhg^{-1} respectively at

Table 9. Electrochemical performance of porous carbon (Rahman et al. 2015).

Anode Types	Specific Capacity (Mah G^{-1})	Remarks
Microporous carbon (2–6 μm)	Reversible capacity: 536 Irreversible capacity: 2547 (at first cycle)	The high irreversible capacity was caused by the high surface area of anode which led to formation of SEI High capacity fade was observed in the first few cycles
Microporous spherical carbon (6.5 μm)	Reversible capacity: 430 Irreversible capacity: 590 (at first cycle)	The high irreversible capacity is due to the formation of SEI on outer surface and within micropores of anode material Surface modification is applied to reduce the pore size, and solved this problem
Mesoporous CMK-3 (3.9 nm)	Reversible capacity: 1100 Irreversible capacity: 3100 (at first cycle)	The high irreversible capacity loss of 2000 mAh g^{-1} is due to the surface containing [H] and [O], SEI formation, and corrosion like reactions of Li$_x$C$_6$ Average 1.5 mAh g^{-1} capacity fades during first 5 cycles
Macroporous carbon (220 nm)	High discharge capacity of 320 mAh g^{-1} after 60th cycles with 98 per cent capacity retention	The enhanced cyclic performances due to the relatively good graphitic crystallinity of carbon GMC1000
Macroporous carbon (285 nm)	Discharge capacity of 260 mAh g^{-1} with a capacity retention of 83 per cent after 30 cycles	Large irreversible capacity was observed due the formation of SEI and inherent nature of hard carbon

the rates of C/10, C/5, C/2, C and 2C. The cyclability was improved to 600 cycles at a rate of C/2 when the loading decreases from 1 to 0.5 mAhcm^{-2} (Xiao et al. 2015). The electrodes made of hp-SiNSs achieve a moderate improvement in volumetric capacity of ~ 760 mAhcm^{-3} compared to ~ 620 mAhcm^{-3} observed in graphitic anodes (Magasinski et al. 2010). Thus, morphology of hp-SiNSs holds a significant promise as anode material in Li-ion battery.

Wang et al. synthesised monodisperse silicon nanospheres by hydrolysis followed by magnesiothermic reduction as described earlier. When the material was mixed with CNT for enhancing conductivity and employed in anode of Li-ion battery it demonstrated a reversible capacity of 1500 mAhg^{-1} at a rate of C/2 even after 500 cycles (Wang et al. 2015). When silicon nanoparticles were inside carbon based rigid spheres with interconnected channels for fast access of electrolyte, the anode made of this material demonstrated a capacity of ~ 1400 mAhg^{-1} for 100 cycles without any appreciable decay and the volumetric capacity is 1270 mAhcm^{-3} at the rate C/2, which is much higher than graphite based electrode (Magasinki et al. 2010). Chen et al. (Chen et al. 2012) synthesised nanocomposite of hollow porous silicon and silver nanoparticles by magnesiothermic reduction of hollow porous silica nanoparticles and this composite electrode demonstrated high specific reversible capacity of 3762 mAhg^{-1}, good cycling stability with over 93 per cent capacity retention and rate performance of over 2000 mAhg^{-1} at 4000 mAg^{-1}. Tesfaye et al. (Tesfaye et al. 2015) prepared silicon nanotubes with thin porous walls using sacrificial template of zinc oxide nanowires. These porous nanotubes remained stable at multiple C-rates. The second discharge has shown a capacity of 3095 mAhg^{-1} at a rate of C/20, with a Coulombic efficiency of 63 per cent. After 30 cycles it attained a stable capacity of 1670 mAhg^{-1}. Kim et al. (Kim et al. 2016) synthesised meso-porous Si-coated carbon nanotube (CNT) composite powders by sol□gel method followed by magnesiothermic reduction and the resulting composite powder when employed in the anode of Li-ion battery demonstrated a reversible capacity of 1640 mAhg^{-1} at the rate of 200 mAg^{-1}. Li et al. (Li et al. 2014b) synthesised large (> 20 nm) mesoporous silicon sponge by anodisation method and this material demonstrated a capacity ~ 750 mAhg^{-1} based on total electrode weight with > 80 per cent capacity retention over 1000 cycles. The irreversible capacity loss in the first cycle is < 5 per cent.

A porous carbon scaffold based silicon anode shows some capacity fading during operation for long cycles although it was claimed to have accommodated the strain of intercalation/deintercalation (Guo et al. 2010). A large number of porous silicon-carbon anodes have been synthesised and Rahman et al. has summarised the electrochemical performance of these porous carbon-silicon anodes in a table which is reproduced in Table 10 but for individual work one may refer to his article (Rahman et al. 2015).

Table 10. Electrochemical performance of porous silicon/carbon (Rahman et al. 2015).

Anode Types	Specific Capacity (mAh g^{-1})	Remarks
Ordered porous Si@C nanorods	Reversible specific capacity of 850 mAh g^{-1} and 627 mAh g^{-1} after 1st and 220th cycles, respectively	Good rate capability, intimate contact of the silicon and carbon which ensure accessibility of active materials Decomposed the electrolyte and formed SEI layer
Hollow core–shell structured Si/C	Reversible specific capacity of 1370 mAh g^{-1} after 1st cycle and retain 98 per cent capacity retention after 100 cycles	Huge initial irreversible capacity is due to the formation of a SEI layer and the decomposition of the electrolyte
Interconnected Porous Si/C	Reversible specific capacity of 3573.1 mAh g^{-1} and 996.5 mAh g^{-1} after 1st and 100th cycle, respectively	High initial irreversible capacity is due to the formation of SEI and irreversible insertion of Li in Si particles
Si/porous carbon nanofiber	Reversible specific capacity of 2071 mAh g^{-1} and 870 mAh g^{-1} after 1st and 100th cycle, respectively	Stable SEI layers and uniform distribution of Si enhanced the electrochemical performances The rapid capacity loss is mainly attributed to the agglomeration of Si on the surface of nanofibers
Three-dimensional porous Si@C	The first discharge and charge capacities are 2240 and 1860 mAh g^{-1}, indicating an initial coulombic efficiency of 83 per cent High reversible discharge capacity of over 1700 mAh g^{-1} after 70 cycles at 0.2 C	The porous structures allowed the volume expansion inward and outward at the same time, and provided more lithiation sites, while the carbon layer could stabilise the porous structure and the SEI film during the lithiation/delithiation process
Si/porous-Carbon composite with buffering voids	Stable reversible capacity of 980 mAh g^{-1} after 80 cycles with small capacity fade of 0.17 per cent/cycle and high rate capability (721 mAh g^{-1} at 2000 mA g^{-1})	The capacity degraded quickly due to the cracking and pulverisation of Si electrode over cycling

Metal organic framework (MOF) has been employed in the anode of Li-ion battery. Cubic MOF-177 (Zn_4O (1,3,5-benzenetribenzoate)$_2$) when used as anode, demonstrated the first charge and discharge capacities of 400 mAhg^{-1} and 105 mAhg^{-1} respectively and the involved electrode reactions are (Li et al. 2006):

$$Zn_4O(BTB)_2 . (DEF)_m (H2O)_n + e^- + Li^+ \rightarrow Zn + Li_2O$$
$$DEF + H_2O + e^- + Li^+ \rightarrow Li_2(DEF) + LiOH$$
$$Zn + Li^+ + e^- \rightarrow LiZn$$

There are large numbers of MOF derived materials employed in Li-ion battery and their electrochemical performance are summarised in a table in the review by Zhao et al. given in Table 11 (Zhao et al. 2016). For details of these studies one may refer to his article.

The commercial cathode in Li-ion battery is $LiCoO_2$, which has capacity of 140 mAhg^{-1} but there is fast capacity fading at higher cycling rates. The capacity increases to 185 mAhg^{-1} for nanotube morphology, which is higher than that observed in nanoparticles or solid nanowires (Li et al. 2005). Spinel $LiMn_2O_4$ has been explored as a high-power cathode but it is marked by sluggish Li$^+$ ion movement. Wang et al. (Wang et al. 2013) prepared the porous sphere of $LiMn_2O_4$ and when it was used in cathode, it demonstrated stable high-rate capability with discharge capacity of 83 mAhg^{-1} at a rate of 20 C and a good capacity retention after cycling. The excellent high rate capability for porous $LiMn_2O_4$ spheres has been attributed to porous morphology and tightly compacted nanocrystallites forming three-dimensional channels which are highly suitable for lithium ion diffusion. It has been

Table 11. Electrochemical performance of MOF derived nanostructures (Zhao et al. 2016).

Materials	Original MOFs	Current Density	Cycle Numbers	Capacities (mAh g^{-1})
N doped graphene particle analogs	ZIF-8	100 mA g^{-1}	50	2132
Co_3O_4 nanoparticles	$Co_3(NDC)_3(DMF)_4$	100 mA g^{-1}	100	730
Mesoporous Co_3O_4	Co(bdc)(DMF)	200 mA g^{-1}	60	913
Co-CP-A	Co(bdc)(DMF)	100 mA g^{-1}	50	536
Co-CP-B	Co(bdc)(DMF)	100 mA g^{-1}	50	651
Co-CP-C	Co(bdc)(DMF)	100 mA g^{-1}	50	391
Co_3O_4 nanoplates	Co(HO-BDC)(bbb)	500 mA g^{-1}	500	852
p-Co_3O_4	$Co_3(BDC)_3(DMF)_4$	0.1 C	100	300
r-Co_3O_4	Co(BDC)(DMSO)	0.1 C	100	800
Co_3O_4 hollow dodecahedra	ZIF-67	100 mA g^{-1}	100	780
Co_3O_4 hexagonal nanorings	Co-NTCDA	100 mA g^{-1}	30	1370
α-Fe_2O_3	$Fe_3O(H_2O)_2Cl(BDC)_3 \cdot nH_2O$	0.2 C	50	911
Fe_2O_3 microboxes	Prussian blue	200 mA g^{-1}	30	950
Porous Fe_2O_3 nanocubes	Prussian blue	200 mA g^{-1}	50	800
CuO hollow octahedra	$Cu_3(btc)_2$	100 mA g^{-1}	100	470
ZnO@C	MOF-5	75 mA g^{-1}	50	1200
Co doped ZnO@C	Co doped MOF-5	100 mA g^{-1}	50	725
$CoFe_2O_4$ nanocubes	$Co[Fe(CN)_6]_{0.667}$	1 C	200	1115
$Mn_{1.8}Fe_{1.2}O_4$ nanocubes	$Mn_3[Fe(CN)_6]_2 \cdot nH_2O$	60 mA g^{-1}	200	827
$ZnxCo_{3x}O_4$ hollow polyhedra	Zn-Co-ZIF	100 mA g^{-1}	50	990
$NiFe_2O_4$@TiO_2 nanorod	Fe_2Ni-MIL-88	100 mA g^{-1}	100	1034
ZnO@ZnO quantum dots/C	ZIF-8	500 mA g^{-1}	100	699
ZnO/$ZnFe_2O_4$/C	ZIF-8	500 mA g^{-1}	100	1390

described earlier that Wu et al. (Wu et al. 2012) prepared HP-$LiMn_2O_4$ microcubes, which demonstrated its ability to maintain a reversible capacity of above 100 mAhg^{-1} even after 200 cycles, with constantly changing current density when solid-state $LiMn_2O_4$ sample shows a capacity of only 76 mAhg^{-1}, which fades faster. Hollow nanofibers of $LiMn_2O_4$ prepared by electro-spinning demonstrated an initial discharge capacity of ~ 120 mAhg^{-1} at a current density of 15 mAg^{-1} with a coulombic efficiency of ~ 90 per cent (Jayraman et al. 2013) while Liu et al. observed a reversible capacity of only 80 mAhg^{-1} and a capacity retention of ~ 85 per cent after five cycles for $LiMn_2O_4$ nanofibers (Liu et al. 2012). Jiao et al. (Jiao et al. 2008) prepared a mesoporous lithium manganese oxide spinel having composition $Li_{1.12}Mn_{1.88}O_4$, to obtain superior rate capability with 50 per cent higher specific capacity compared to that in corresponding bulk material at a rate of 30C, 3000 mAg^{-1} under ambient temperature.

Multi-electron systems in both cathode and anode materials offer an opportunity for doubling the theoretical gravimetric energy density compared with current systems based on intercalation. However, to enable multi-electron systems to operate effectively it is essential to apply 3D nanostructuring of electrodes. Polyanionic compounds like $Na_3V_2(PO_4)_3$ is considered to hold promise as a cathode for Na-ion batteries. By electro-spinning, Liu et al. (Liu et al. 2014) encapsulated nanoparticles of $Na_3V_2(PO_4)_3$ in interconnected porous carbon nanofibers, which is conducting, and the electrode demonstrated high capacity and a good rate performance. Because of high conductivity of 1D carbon nanostructure, these materials are commonly added in battery electrodes to decrease resistance and thereby, reducing ohmic losses (Landi et al. 2009). Jiang et al. (Jiang et al. 2015) incorporated the same nanoparticles after carbon coating, into mesoporous carbon with highly ordered 1D channel and the material demonstrated a rate capability of 78 mAhg^{-1} at 30C and a life of 2000 cycles at 5C.

Layered materials capable of intercalation are attractive for application in electrodes. Graphene (Zhu et al. 2011), some transition metal oxides (Ghodbane et al. 2009), MoS$_2$ (Acerce et al. 2015), Ti$_2$C (MXenes morphology) (Come et al. 2012) are some good examples. 2D morphology is very attractive because these are a few atoms thick and offer a large number of active sites. Often, in these materials thermodynamics is favourable for spontaneous intercalation. The structure also has the advantage of accommodating large ions by solvation. Therefore, this morphology is very suitable for constructing flexible electrode (Feng et al. 2015) and its performance is superior to the corresponding bulk material. Metallic 2D materials like 1T transition metal dichalcogenides MX$_2$ where transition metal M (Mo, W, etc.) is sandwiched between two layers of X (S, Se, Te) have great potential because of their conductivity and ability to support redox reactions.

Wei et al. (Wei et al. 2017) have reviewed the electrochemical performance of a number of 1D nanostructures in the electrodes of Li-ion battery (LIB), sodium ion battery (SIB), Li-S battery and Li-O$_2$ battery and the details of performance are summarised by them as given in Table 12. For references to these individual results, the reader may see Table 2 in this review.

Although nanostructuring results in benefits as anticipated but there are often issues to be faced. The transition metal oxides, which are primarily insulators or wide gap semiconductors, there is need to increase conductivity by doping, through partial reduction, by creating good contact between

Table 12. Electro-chemical performance of some 1D nanomaterials in batteries (Wei et al. 2017).

Materials	LIBs				
	Highest Reversible Capacity		Cycling Performance		
	Capacity (mAh g^{-1})	Specific Current (mA g^{-1})	Capacity (mAh g^{-1})	Cycle	Specific Current (mA g^{-1})
LiCoO$_2$ nanotubes (T)	185	10	168	100	10
LiNi$_{0.8}$Co$_{0.2}$O$_2$ nanotubes (T)	205	10	145	100	10
LiMn$_2$O$_4$ porous nanotubes (T)	110	495	120	1200	500
LiNi$_{0.5}$Mn$_{1.5}$O$_4$ porous nanorods (L)	140	147	113	500	735
Li$_3$V$_2$(PO$_4$)$_3$/C mesoporous nanowires (L)	128	133	96	3000	665
Li$_3$V$_2$(PO$_4$)$_3$ mesoporous nanotubes (E)	131	133	86	9500	1330
N-doped porous carbon nanofibers (E)	632	1000	625	300	1000
Hierarchical tubular structures constructed from ultrathin TiO$_2$(B) nanosheets (T)	216	335	160	400	1675
Elongated bending TiO$_2$-based nanotubes (L)	267	33.5	114	10000	8375
Double-walled Si nanotubes (CD)	1780	200	874	6000	24000
Porous Si/SiO$_x$ nanowires (CE)	1936	6000	1503	560	600
Bubble-nanorod-structured Fe$_2$O$_3$/C nanofibers (E)	913	500	812	300	1000
3D hierarchical tubular CuO/CuO core/shell heterostructure arrays (CD)	1364	100	1140	1000	1000

Table 12 contd....

Table 12 contd. ...

Materials	SIBs				
	Highest Reversible Capacity		Cycling Performance		
	Capacity (mAh g^{-1})	Specific Current (mA g^{-1})	Capacity (mAh g^{-1})	Cycle	Specific Current (mA g^{-1})
Na$_3$V$_2$(PO$_4$)$_3$ nanoparticles in 1D carbon nanofibers (E)	77	228	–	–	–
Na$_3$V$_2$(PO$_4$)$_3$/nanopaticles in 1D channel mesoporous carbon (T)	115	57	78	5000	228
Na$_{0.7}$Fe$_{0.7}$Mn$_{0.3}$O$_2$ mesoporous nanotubes (E)	109	100	74	5000	500
Hollow carbon nanowires (L)	251	100	251	400	50
Nitrogen doped porous carbon fibers (L)	310	50	23	100	50
Sn-doped TiO$_2$ nanotubes (L)	252	50	257	50	50
Spider web-like Na$_2$T$_{i3}$O$_7$ nanotubes (L)	425	50	107	500	500
SnSb nanoparticles in the Porous carbon nanofiber (E)	392	50	110	200	10000
NiCo$_2$O$_4$ hierarchical porous nanowire array on the carbon fiber (L)	761	50	542	50	50
Porous CuO Arrays (L)	674	20	290	460	200
MoS$_2$ nanoparticles in carbon nanofibers (E)	381.7	100	283.9	600	100
Li-S Battery					
S in hollow carbon nanofiber (T)	≈ 1000	335	730	150	837.5
S in hollow carbon nanofibers filled with MnO$_2$ nanosheets (T)	1161	83.75	~ 700	300	837.5
S in the three-dimensional carbon nanofibers coated with ethylenediamine-functionalised reduced graphene oxide (T)	1314	167.5	950	200	837.5
Li-O$_2$ Battery					
Materials	Voltage Plateau (vs Li$^+$/Li)	Achieved Capacity		Cycling performance	
		Capacity (mAh g^{-1})	Voltage (V)	Cycles	Capacity Limit
La$_{0.5}$Sr$_{0.5}$CoO$_{2.91}$ hierarchical mesoporous nanowires (L)	2.7 V	11509	2	–	–
La$_{0.75}$Sr$_{0.25}$MnO$_3$ porous nanotubes (E)	2.8 V	9000–11000	2.4	124	1000
Co$_3$O$_4$ hierarchical porous nanowires (L)	≈ 2.65V	11160.8	2.4	73	1000

*(Synthesis methods indicated within brackets: E = Electro-spinning method, L = liquid phase method, T = template-assisted method, CE = chemical etching method, and CD = chemical deposit method).

particles and by addition of binders. There are efforts to overcome the lack of conductivity by creating composite of active material with conducting materials like carbon black, carbon nanotubes or graphene. One may also create a structure of active material over a backbone of conducting material. There are also synergistic benefits for electrochemical performance and mechanical properties, which is important to maintain mechanical integrity, important to fight degradation of performance (Huang et al. 2016, Jana et al. 2014). When one increases thickness of electrodes by stacking 2D layers, often the electrolyte access inside the stacked layers to redox sites may get obstructed if not done carefully. If the density of material in the electrode is too low then there is danger of not meeting the volumetric power density.

Concluding Remarks

Electrical Energy Storage (EES) is a critical technology for increased use of renewable energies to reduce our dependence on fossil fuels as it provides solution for problems arising out of integration of renewables to the grid, ensuring quality power with stable voltage and frequency. The mobile power in electric or hybrid vehicles from EES is also very important to avoid using fossil fuels. The technology is considered critical because it will enable the civilisation to overcome the challenges of environment pollution and global warming. Amongst various EES technologies available, electrochemical storage like batteries or supercapacitors has a significant place but the batteries have limited rate capabilities while supercapacitors have energy density 1–2 orders of magnitude less than that of battery. For development of batteries for vehicles, automobile industries have set the goal of energy density of 300 Wh/L and 250 Wh/kg at a cost of $125/kWh. For stationary application in grid the target cost is still lower at $100/kWh. Thus, there is need to enhance the energy density of supercapacitors and batteries and also, the rate capability of batteries while reducing the cost significantly by a factor of three to five. To achieve these objectives, extensive research is going on for finding better electrode materials for the cathode and anode.

The energy is stored in the electrode by absorption, redox reaction or intercalation of the electroactive species from the electrolyte at the favourable surface sites. To increase energy density in a supercapacitor by absorption or surface reaction, one needs to enhance the surface area of the electrode in contact with electrolyte. For batteries, in addition to surface, the access of the active material inside the electrode is necessary to increase its participation in storage of electroactive species. The access inside is by diffusion of the electroactive species. The time taken for diffusion will affect the rate performance during charge/discharge. The diffusion distance together with the size of the individual entity like particle or grain of the active material in the electrode, decides the rate performance of the electrode. The entry of electroactive species into the active material may results in problems including strain, phase change and formation of electronically insulating layer on the surface, degrading the electrochemical performance. The surface area increases when size of individual entity of the active material reduces and so, active materials of nano-size are expected to result in better energy density because of large surface area and better rate performance due to higher diffusion coefficient resulting from relatively open structure. These are possible provided the adverse factors are not aggravated to counter these advantages. The surface area could be further enhanced if the material becomes porous and thus, nanoporous active materials in the electrode are investigated extensively. The nanoporous materials are also able to accommodate the strain on entry and exit of electroactive species due to lower stress developing in the electrode compared to bulk material and thus, help to maintain the integrity of the electrode and counter capacity fading due to it. But the amount of pore should be such that the pores do not reduce strength to a level so as to adversely affect mechanical integrity.

The routes of synthesis of nanoporous materials are varied. The methods could be purely inorganic, inorganic-organic or organic. Doping has been carried out to place dopant between layers in layered compounds to drive the structure towards instability in order to transform it to an open structure with tunnels of pores in it.

Secondary building unit (SBU) of metal or organic cluster may be stitched together with organic linker to create a material with tailored distances between the clusters. Monomers may be linked to clusters for subsequent polymerisation creating a crosslinked structure. The chemists have created different classes of solids called porous organic framework (POF) including metal organic framework

(MOF), covalent organic framework (COF) or covalent organic polymers (COP). There are also chemical routes like hydrothermal methods involving precipitation and heating, sol-gel method and emulsion methods. Sometimes the last step in synthesis is calcinations to expel a gaseous component under controlled environment, leaving behind a porous product. The methods are also aimed to create certain dimensionality of the porous material, particle (0D), rod and wire (1D) or blocks (3D), and in it, the desired size and architecture of porosity has to be incorporated to make the porous nanomaterial. The pores could be isolated ones (0D) and porous channels (1D), either disordered or ordered. The channels could be ordered parallel to one direction, two directions or three directions. The channels could be an interconnected network. Else, the pores and channels could be random. Isolated pores are not very useful to enhance accessibility of electrolyte inside active material and the focus should be on porous channels. A particle could be porous or hollow inside with porous shell. There could be composite core-shell assembly with porous shell or core or both. There are synthesis methods to create such particles or hollow cubes. There could be tubes or hollow wires synthesised directly or by sacrificing a template of nanowire of another material. Electro-spinning is a popular method for making porous rods, tubes or nanowires of a material with or without nanoparticles of another material embedded in the wall. Porous template could be used to make a block of a material with ordered or somewhat random porous channels. Thus, there are large variety of methods to synthesise porous materials of different dimensionality and different architecture of pores.

There are studies on electrochemical performance of a variety of nanoporous materials with different dimensionality and various porous architectures. There are some bulk materials, which are known for high energy density when employed in electrodes like silicon, although it has problem of disintegration under strain leading to capacity degradation. Nanoporous structures of silicon have the ability to accommodate strain without disintegration and also, improve cyclability and energy density significantly. Similar studies are available for other well-known anode and cathode materials when having nanoporous structure. It is still not clear as to which materials dimensionality and porous structure results in the best electrochemical performance. The porous channels are important for the flow of electrolyte inside the active material to give more access of electrolyte for intercalation or absorption on the surface. It is generally claimed that mesoporous channels are suitable for this purpose. It is also claimed that microporous veins as it exists between nanocrystals, connecting mesoporous channels results in more effective absorption/intercalation inside. The size of microporous channels should be greater than the size of electro-active ion. Enhanced surface area through accessible porosity is important.

But how should the porous channel be distributed in the active material?

It is apparent that well distributed channels with adjacent wall distance less than twice the diffusion distance is desirable for good rate performance. Thus, if microchannels are along the boundary of nanocrystals, the size of the crystals should be less than twice the diffusion distance for quick intercalation/reaction in the active material of the electrode. But if the crystals are porous, larger size may work as well.

But, is it really necessary for the porous channels to be ordered?

A randomly distributed channel may also give good electrochemical performance. There is need to study the desired dimensionality and optimum porous architecture in an active material to arrive at the best electrochemical performance in electrode and one has to find a cheaper route of synthesis of such material to arrive at economic batteries with acceptable performance for various applications in energy storage.

References

Acerce, M., D. Voiry and M. Chhowalla. 2015. Metallic 1T phase MoS_2 nanosheets as supercapacitor electrode materials. Nat. Nanotech. 10: 313–318.

Agubra, V. and J. Fergus. 2013. Lithium ion battery anode aging mechanisms. Materials 6: 1310–1325.

Amatucci, G.G., F. Badway, A. Du Pasquier and T. Zheng. 2001. An asymmetric hybrid nonaqueous energy storage cell. J. Electrochem. Soc. 148: A930–A939.

Axel, H., H. Schafer and A. Weiss. 1966. Zur Kenntnis der Phase $Li_{22}Si_5$. Z. Naturforsch. B 21: 115–117.

Baker, J. 2008. New technology and possible advances in energy storage. Energy Policy 36: 4368–4373.

Beaudin, M., H. Zareipour, A. Schellenberglabe and W. Rosehart. 2010. Energy storage for mitigating the variability of renewable electricity sources: An updated review. Energy Sust. Dev. 14: 302–314.

Beaudoin, D., T. Maris and J.D. Wuest. 2013. Constructing monocrystalline covalent organic networks by polymerization. Nature Chem. 5: 830–835.

Beck, J.S., J.C. Vartuli, W.J. Roth, M.E. Leonowicz, C.T. Kresge, K.D. Schmitt et al. 1992. A new family of mesoporous molecular sieves prepared with liquid crystal templates. J. Am. Chem. Soc. 114: 10834–10843.

Brezesinski, K., J. Wang, J. Haetge, C. Reitz, S.O. Steinmueller, S.H. Tolbert et al. 2010a. Pseudocapacitive contributions to charge storage in highly ordered mesoporous group V transition metal oxides with iso-oriented layered nanocrystalline domains. J. Am. Chem. Soc. 132: 6982–6990.

Brezesinski, T., J. Wang, S.H. Tolbert and B. Dunn. 2010b. Ordered mesoporous α-MoO_3 with iso-oriented nanocrystalline walls for thin-film pseudocapacitors. Nat. Mater. 9: 146–151.

Brousse, T., D. Belanger and J.W. Long. 2015. To be or not to be pseudocapacitive? J. Electrochem. Soc. 162: A5185–A5189.

Bruce, P.G., S.A. Freunberger, L.J. Hardwick and J.-M. Tarascon. 2012. Li-O_2 and Li-S batteries with high energy storage. Nat. Mater. 11: 19–29.

Chan, C.K., H. Peng, G. Liu, K. McIlwrath, X.F. Zhang, R.A. Huggins et al. 2008. High-performance lithium battery anodes using silicon nanowires. Nat. Nanotech. 3: 31–35.

Chen, S., M. Xue, Y. Li, Y. Pan, L. Zhu and S. Qiu. 2015. Rational design and synthesis of $Ni_xCo_{3-x}O_4$ nanoparticles derived from multivariate MOF-74 for supercapacitors. J. Mater. Chem. A 3: 20145–20152.

Chen, H., T.N. Cong, W. Yang, C. Tan, Y. Li and Y. Ding. 2009. Progress in electrical energy storage system: A critical review. Prog. Nat. Sci. 19: 291–312.

Chen, Y., Z. Lu, L. Zhou, Y-W. Mai and H. Huang. 2012. Triple-coaxial electrospun amorphous carbon nanotubes with hollow graphitic carbon nanospheres for high-performance Li ion batteries. Energy Environ. Sci. 5: 7898–7902.

Cho, J.S., Y.J. Hong and Y.C. Kang. 2015. Design and synthesis of bubble-nanorod-structured Fe_2O_3-carbon nanofibers as advanced anode material for Li-ion batteries. ACS Nano. 9: 4026–4035.

Come, J., M. Naguib, P. Rozier, M.W. Barsoum, Y. Gogotsi, P.-L. Taberna et al. 2012. A non-aqueous asymmetric cell with a Ti_2C-based two-dimensional negative electrode. J. Electrochem. Soc. 159: A1368–A1373.

Conway, B.E. 1999. Electrochemical supercapacitors: Scientific fundamentals and technological applications. Springer US.

Daly, J. 2014. Japan looks at recycling vehicle batteries for renewable power. http://oilprice.com/Alternative-Energy/Renewable-Energy/Japan-looks-at-Recycling-Vehicle-Batteries-for-Renewable-Power.html.

Diaz-Gonzalez, F., A. Sumper, O. Gomis-Bellmunt and R. Villafafila-Robles. 2012. A review of energy storage technologies for wind power applications. Renew. Sust. Energy Rev. 16: 2154–2171.

Doherty, C.M., R.A. Caruso, B.M. Smarsly and C.J. Drummond. 2009. Colloidal crystal templating to produce hierarchically porous $LiFePO_4$ electrode materials for high power lithium ion batteries. Chem. Mater. 21: 2895–2903.

Du, N., H. Zhang, B. Chen, J. Wu, X. Ma, Z. Liu et al. 2007a. Porous Co_3O_4 nanotubes derived from $Co_4(CO)_{12}$ clusters on carbon nanotube templates: A highly efficient material for Li-battery applications. Adv. Mater. 19: 4505–4509.

Du, N., H. Zhang, B. Chen, X. Ma, X. Huang, J. Tu et al. 2009. Synthesis of polycrystalline SnO_2 nanotubes on carbon nanotube template for anode material of lithium-ion battery. Mater. Res. Bull. 44: 211–215.

Du, N., H. Zhang, B. Chen, X. Ma, Z. Liu, J. Wu et al. 2007b. Porous indium oxide nanotubes: Layer-by-layer assembly on carbon-nanotube templates and application for room-temperature NH_3 gas sensors. Adv. Mater. 19: 1641–1645.

Du, N., Y. Xu, H. Zhang, J. Yu, C. Zhai and D. Yang. 2011. Porous $ZnCo_2O_4$ nanowires synthesis via sacrificial templates: High-performance anode materials of Li-ion batteries. Inorg. Chem. 50: 3320–3324.

Dunn, B., H. Kamath and J-M. Tarascon. 2011. Electrical energy storage for the grid: A battery of choices. Science 6058: 928–935.

El-Kaderi, H.M., J.R. Hunt, J.L. Mendoza-Cortés, A.P. Côté, R.E. Taylor, M. O'Keeffe et al. 2007. Designed synthesis of 3D covalent organic frameworks. Science 316: 268–272.

Energy Information Administration, Office of Energy Markets and End Use. US Department of Energy, Washington DC. 2000. Annual Energy Review 1999. http://www.eia.doe.gov/aer.

Esmanski, A. and G.A. Ozin. 2009. Silicon inverse-opal-based macroporous materials as negative electrodes for lithium ion batteries. Adv. Funct. Mater. 19: 1999–2010.

European Commission. 2013. The future role and challenges of energy storage. DG ENER Working Paper. http://ec.europa.eu/energy/infrastructure/doc/energy-storage/2013/energy_storage.pdf.

Farret, F.A. and M.G. Simoes. 2006. Integration of alternate sources of energy. John Wiley & Sons Inc.

Feng, F., J. Wu, C. Wu and Y. Xie. 2015. Regulating the electrical behaviors of 2D inorganic nanomaterials for energy applications. Small 11: 654–666.

Fetcenko, M.A., S.R. Ovshinsky, B. Reichman, K. Young, C. Fierro, J. Koch et al. 2007. Recent advances in NiMH battery technology. J. Power Sources 165: 544–551.

Frackowiak, E. and F. Beguin. 2001. Carbon materials for the electrochemical storage of energy in capacitors. Carbon 39: 937–950.

Ganguli, A.K., A. Ganguly and S. Vaidya. 2010. Microemulsion-based synthesis of nanocrystalline materials. Chem. Soc. Rev. 39: 474–485.

Ge, M., J. Rong, X. Fang and C. Zhou. 2012. Porous doped silicon nanowires for lithium ion battery anode with long cycle life. Nano Lett. 12: 2318–2323.

Ge, M., J. Rong, X. Fang, A. Zhang, Y. Lu and C. Zhou. 2013a. Scalable production of porous silicon nanoparticles and their application in lithium-ion battery anode. Nano Res. 6: 174–181.

Ge, M., X. Fang, J. Rong and C. Zhou. 2013b. Review of porous silicon preparation and its application for lithium-ion battery anodes. Nanotechnol. 24: 1–10.

Ghodbane, O., J.-L. Pascal and F. Favier. 2009. Microstructural effects on charge-storage properties in MnO_2-based electrochemical supercapacitors. ACS Appl. Mater. Interfaces 1: 1130–1139.

Gonzalez, A., B. O'Gallachoir, E. McKeogh and K. Lynch. 2004. Study of electricity storage technologies and their potential to address wind energy intermittency in Ireland. Final report. Sustainable Energy Research Group and Rockmount Capital Patterns, Cork. Project - RE/HC/03/001.

Gowda, S.R., V. Pushparaj, S. Herle, G. Girishkumar, J.G. Gordon, H. Gullapalli et al. 2012. Three-dimensionally engineered porous silicon electrodes for Li ion batteries. Nano Lett. 12: 6060–6065.

Gu, D., W. Li, F. Wang, H. Bongard, B. Spliethoff, W. Schmidt et al. 2015. Controllable synthesis of mesoporous peapod-like Co_3O_4@carbon nanotube arrays for high-performance lithium-ion batteries. Angew. Chem. 127: 7166–7170.

Guo, J., X. Chen and C. Wang. 2010. Carbon scaffold structured silicon anodes for lithium-ion batteries. J. Mater. Chem. 20: 5035–5040.

Hadjipaschalis, I., A. Poullikkas and V. Efthimiou. 2009. Overview of current and future Energy storage technologies for electric power applications. Renew. Sust. Energy Rev. 13: 1513–1522.

Hoa, N.D., N.V. Quy, H. Jung, D. Kim, H. Kim and S.K. Hong. 2010. Synthesis of porous CuO nanowires and its application to hydrogen detection. Sens. Actuators B. 146: 266–272.

Hu, Y.-S., P. Adelhelm, B.M. Smarsly, S. Hore, M. Antonietti and J. Maier. 2007. Synthesis of hierarchically porous carbon monoliths with highly ordered microstructure and their application in rechargeable lihtium batteries with high-rate capability. Adv. Func. Mater. 17: 1873–1878.

Huang, P., C. Lethien, S. Pinaud, K. Brousse, R. Laloo, V. Turq et al. 2016. On-chip and freestanding elastic carbon films for micro-supercapacitors. Science 351: 691–695.

International Electrotechnical Commission (IEC). 2011. Electrical energy storage white paper. http://www.iec.ch/whitepaper/pdf/ iecWP-energystorage-LR-en.pdf.

Intrator, J., E. Elkind, A. Abele, S. Weissman, M. Sawchuk and E. Bartlett. 2011. 2020 Strategic analysis of energy storage in California. Center for Law, Energy & The Environment Report. Publication no. CEC-500-2011-047.

Jana, M., A. Sil and S. Ray. 2014. Morphology of carbon nanostructures and their electrochemical performance for lithium ion battery. J. Phys. Chem. Solids 75: 60–67.

Jana, M and N. Singh, Raj. 2017. A study of evolution of residual stress in single crystal silicon electrode using Raman spectroscopy. Appl. Phys. Lett.111: 1–5.

Jason, G. and W. Feng. 2013. Nanoscale anodes of silicon and germanium for lithium batteries. pp. 69–90. *In*: R. Yazami (ed.). Nanomaterials for Lithium-Ion Batteries: Fundamentals and Applications. Pan Stanford Publishing.

Jayaprakash, N., J. Shen, S.S. Moganty, A. Corona and L.A. Archer. 2011. Porous hollow carbon@sulfur composites for high-power lithium–sulfur batteries. Angew. Chem. 50: 5904–5908.

Jayaraman, S., V. Aravindan, P.S. Kumar, W.C. Ling, S. Ramakrishna and S. Madhavi. 2013. Synthesis of porous $LiMn_2O_4$ hollow nanofibers by electrospinning with extraordinary lithium storage properties. Chemical Comm. 49: 6677–6679.

Ji, L., M. Gu, Y. Shao, X. Li, M.H. Engelhard, B.W. Arey et al. 2014. Controlling SEI formation on SnSb-porous carbon nanofibers for improved Na ion storage. Adv. Mater. 26: 2901–2908.

Ji, L., M. Rao, H. Zheng, L. Zhang, Y. Li, W. Duan et al. 2011. Graphene oxide as a sulfur immobilizer in high performance lithium/sulfur cells. J. Am. Chem. Soc. 133: 18522–18525.

Ji, X.L., K.T. Lee and L.F. Nazar. 2009. A highly ordered nanostructured carbon–sulphur cathode for lithium–sulphur batteries. Nat. Mater. 8: 500–506.

Jia, Y., L. He, Z. Guo, X. Chen, F. Meng, T. Luo et al. 2009. Preparation of porous tin oxide nanotubes using carbon nanotubes as templates and their gas-sensing properties. J. Phys. Chem. C. 113: 9581–9587.

Jiang, Y., Z. Yang, W. Li, L. Zeng, F. Pan, M. Wang et al. 2015. Nanoconfined carbon-coated $Na_3V_2(PO_4)_3$ particles in mesoporous carbon enabling ultralong cycle life for sodium-ion batteries. Adv. Energy Mater. 5: 1–8.

Jiao, F., J. Bao, A.H. Hill and P.G. Bruce. 2008. Synthesis of ordered mesoporous Li–Mn–O spinel as a positive electrode for rechargeable lithium batteries. Angew. Chem. 47: 9711–9716.

Judeinstein, P. and C. Sanchez. 1996. Hybrid organic–inorganic materials: a land of multidisciplinarity. J. Mater. Chem. 6: 511–525.

Kawakami, N., Y. Iijima, M. Fukuhara, M. Bando, Y. Sakanaka, K. Ogawa et al. 2010. Development and field experiences of stabilization system using 34 MW NaS batteries for a 51 MW wind farm. IEEE Int. Symp. Ind. Electron. 2371–2376.

Kickelbick, G. and U. Schubert. 2001. Inorganic clusters in organic polymers and the use of polyfunctional inorganic compounds as polymerization initiators. Monatsh. Chem. 132: 13–30.

Kim, H. and J. Cho. 2008a. Superior lithium electroactive mesoporous Si@carbon core–shell nanowires for lithium battery anode material. Nano Lett. 8: 3688.

Kim, H., B. Han, J. Choo and J. Cho. 2008b. Three-dimensional porous silicon particles for use in high-performance lithium secondary batteries. Angew. Chem. 47: 10151–10154.

Kim, J., B. Chen, T.M. Reineke, H. Li, M. Eddaoudi, D.B. Moler et al. 2001. Assembly of metal-organic frameworks from large organic and inorganic secondary building units: New examples and simplifying principles for complex structures. J. Am. Chem. Soc. 123: 8239–8247.

Kim, W-S., J. Choi and S-H. Hong. 2016. Meso-porous silicon-coated carbon nanotube as an anode for lithium-ion battery. Nano Res. 9: 2174–2181.

Kondoh, J., I. Ishii, H. Yamaguchi, A. Murata, K. Otani, K. Sakuta et al. 2000. Electrical Energy storage systems for energy networks. Energy Conversion Management 41: 1863–1874.

Kresge, C.T., M.E. Leonowicz, W.J. Roth, J.C. Vartuli and J.S. Beck. 1992. Ordered mesoporous molecular sieves synthesized by a liquid-crystal template mechanism. Nature. 359: 710–712.

Kuhn, P., M. Antonietti and A. Thomas. 2008. Porous, covalent triazine-based frameworks prepared by ionothermal synthesis. Angew. Chem. 47: 3450–3453.

Kundu, D., E. Talaie, V. Duffort and L.F. Nazar. 2015. The emerging chemistry of sodium ion batteries for electrochemical energy storage. Angew. Chem. 54: 3431–3448.

Lam, L.T. and R. Louey. 2006. Development of ultra-battery for hybrid-electric vehicle applications. J. Power Sources 158: 1140–1148.

Landi, B.J., M.J. Ganter, C.D. Cress, R.A. DiLeo and R.P. Raffaelle. 2009. Carbon nanotubes for lithium ion batteries. Energy Environ. Sci. 2: 638–654.

Lei, D., B. Qu, H.T. Lin and T. Wang. 2015. Facile approach to prepare porous GeO_2/SnO_2 nanofibers via a single spinneret electrospinning technique as anodes for lithium-ion batteries. Ceram. Int. 41: 10308–10313.

Leung, P., X. Li, C. Ponce de León, L. Berlouis, C.T.J. Low and F.C. Walsh. 2012. Progress in redox flow batteries, remaining challenges and their applications in energy storage. RSC Adv. 2: 10125–10156.

Li, H., H. Lin, S. Xie, W. Dai, M. Qiao, Y. Lu et al. 2008. Ordered mesoporous Ni nanowires with enhanced hydrogenation activity prepared by electroless plating on functionalized SBA-15. Chem. Mater. 20: 3936–3943.

Li, H., Q. Zhou, Y. Gao, X. Gui, L. Yang, M. Du et al. 2015. Templated synthesis of TiO_2 nanotube macrostructures and their photocatalytic properties. Nano Res. 8: 900–906.

Li, L., G. Zhou, X-Y. Shan, S. Pei, F. Li and H-M. Cheng. 2014a. Co_3O_4 mesoporous nanostructures@graphene membrane as an integrated anode for long-life lithium-ion batteries. J. Power Sources 255: 52–58.

Li, X., F. Cheng, B. Guo and J. Chen. 2005. Template-synthesized $LiCoO_2$, $LiMn_2O_4$, and $LiNi_{0.8}Co_{0.2}O_2$ nanotubes as the cathode materials of lithium ion batteries. J. Phys. Chem. B 109: 14017–14024.

Li, X., F. Cheng, S. Zhang and J. Chen. 2006. Shape-controlled synthesis and lithium-storage study of metal-organic frameworks $Zn_4O(1,3,5-benzenetribenzoate)_2$. J. Power Sources 160: 542–547.

Li, X., M. Gu, S. Hu, R. Kennard, P. Yan, X. Chen et al. 2014b. Mesoporous silicon sponge as an anti-pulverization structure for high-performance lithium-ion battery anodes. Nat. Commun. 5: 1–7.

Liu, C., F. Li, L-P. Ma and H-M. Cheng. 2010a. Advanced materials for energy storage. Adv. Mater. 22: E28–62.

Liu, H.J., X.M. Wang, W.J. Cui, Y.Q. Dou, D.Y. Zhao and Y.Y. Xia. 2010b. Highly ordered mesoporous carbon nanofiber arrays from a crab shell biological template and its application in supercapacitors and fuel cells. J. Mater. Chem. 20: 4223–4230.

Liu, J., J.-G. Zhang, Z. Yang, J.P. Lemmon, C. Imhoff, G.L. Graff et al. 2013. Materials science and materials chemistry for large scale electrochemical storage: From transportation to electrical grid. Adv. Funct. Mater. 23: 929–946.

Liu, J., K. Tang, K. Song, P.A. van Aken, Y. Yu and J. Maier. 2014. Electrospun $Na_3V_2(PO_4)_3$/C nanofibers as stable cathode materials for sodium-ion batteries. Nanoscale 6: 5081–5086.

Liu, R., J. Duaya and S.B. Lee. 2011. Heterogeneous nanostructured electrode materials for electrochemical energy storage. Chem. Commun. 47: 1384–1404.

Liu, S., Y.Z. Long, H.D. Zhang, B. Sun, C.C. Tang, H.L. Li et al. 2012. Preparation and electrochemical properties of $LiMn_2O_4$ nanofibers via electrospinning for lithium ion batteries. Adv. Mater. Res. 562: 799–802.

Loy, D.A. and K.J. Shea. 1995. Bridged polysilsesquioxanes. Highly porous hybrid organic-inorganic materials. Chem. Rev. 95: 1431–1442.

Lu, W., Z. Wei, Z.-Y. Gu, T.-F. Liu, J. Park, J. Park et al. 2014. Tuning the structure and function of metal-organic frameworks via linker design. Chem. Soc. Rev. 43: 5561–5593.

Lukatskaya, M.R., B. Dunn and Y. Gogotsi. 2016. Multidimensional materials and device architectures for future hybrid energy storage. Nat. Commun. 7: 1–13.

Luo, X., W. Jihong, M. Dooner and J. Clarke. 2015. Overview of current development in electrical energy storage technologies and the application potential in power system operation. Appl. Energy 137: 511–536.

Magasinski, A., P. Dixon, B. Hertzberg, A. Kvit, J. Ayala and G. Yushin. 2010. High-performance lithium-ion anodes using a hierarchical bottom-up approach. Nat. Mater. 9: 353–358.

Mai, L., X. Tian, X. Xu, L. Chang and L. Xu. 2014. Nanowire electrodes for electrochemical energy storage devices. Chem. Rev. 114: 11828–11862.

Mikhaylik, Y.V. and J.R. Akridge. 2004. Polysulfide shuttle study in the Li/S battery system. J. Electrochem. Soc. 151: A1969–1976.

Naoi, K., W. Naoi, S. Aoyagi, J. Miyamoto and T. Kamino. 2013. New generation 'nanohybrid supercapacitor'. Acc. Chem. Res. 46: 1075–1083.

Niu, C., J. Meng, X. Wang, C. Han, M. Yan, K. Zhao et al. 2015. General synthesis of complex nanotubes by gradient electrospinning and controlled pyrolysis. Nature Commun. 6: 1–9.

Obrovac, M.N. and L. Christensen. 2004. Structural changes in silicon anodes during lithium insertion/extraction. Electrochem. Solid State Lett. 7: A93–A96.

Okubo, M. and I. Honma 2013. High-rate Li-ion intercalation in nanocrystalline cathode materials for high-power Li-ion batteries. pp. 227–251. *In*: Yazami, R. (ed.). Nanomaterials for Lithium-Ion Batteries: Fundamentals and Applications. Pan Stanford Publishing, Singapore.

Pan, H., Y.-S. Hu and L. Chen. 2013. Room-temperature stationary sodium-ion batteries for large-scale electric energy storage. Energy Environ. Sci. 6: 2338–2360.

Pech, D., M. Brunet, H. Durou, P. Huang, V. Mochalin, Y. Gogotsi et al. 2010. Ultrahigh-power micrometre-sized supercapacitors based on onion-like carbon. Nat. Nanotech. 5: 651–654.

Pendashteh, A., S.E. Moosavifard, M.S. Rahmanifar, Y. Wang, M.F. El-Kady, R.B. Kaner et al. 2015. Highly ordered mesoporous $CuCo_2O_4$ nanowires, a promising solution for high-performance supercapacitors. Chem. Mater. 27: 3919–3926.

Peng, S., L. Li, Y. Hu, M. Srinivasan, F. Cheng, J. Chen et al. 2015. Fabrication of spinel one-dimensional architectures by single-spinneret electrospinning for energy storage applications. ACS Nano. 9: 1945–1954.

Power System Operation Corporation Ltd. 2016. Flexibility requirement in Indian power system. Jan, 2016. https://posoco.in/download/flexibility_requirement_in_indian_power_system/?wpdmdl=711.

Qian, X., Y. Lv, W. Li, Y. Xia and D. Zhao. 2011. Multiwall carbon nanotube@mesoporous carbon with core-shell configuration: A well-designed composite-structure toward electrochemical capacitor application. J. Mater. Chem. 21: 13025–13031.

Qiao, L., X. Wang, X. Sun, X. Li, Y. Zheng and D. He. 2013. Single electrospun porous NiO–ZnO hybrid nanofibers as anode materials for advanced lithium-ion batteries. Nanoscale. 5: 3037–3042.

Rahman Md. A., Y.C. Wong, G. Song and C. Wen. 2015. A review on porous negative electrodes for high performance lithium-ion batteries. J. Porous Mater. 22: 1313–1343.

Rastler, D. 2010. Electricity energy storage options: A white paper primer on applications, costs, and options. Electric Power Research Institute (EPRI).

Rauda, I.E., V. Augustyn, B. Dunn and S.H. Tolbert. 2013. Enhancing pseudocapacitive charge storage in polymer templated mesoporous materials. Acc. Chem. Res. 46: 1113–1124.

Ray, S., M. Jana and A. Sil. 2016. Nanostructure and atomic arrangement. pp. 123–146. *In*: Aliofkhazraei, M., N. Ali, W.I. Milne, C.S. Ozkan, S. Mitura and J.L. Gervasoni (eds.). Graphene Science Handbook. CRC Press, Taylor & Francis, FL, USA.

Ribot, F. and C. Sanchez. 1999. Organically functionalized metallic oxo-clusters: Structurally well-defined nanobuilding blocks for the design of hybrid organic-inorganic materials. Comments Inorg. Chem. 20: 327–371.

Ryoo, R., S.H. Joo and S. Jun. 1999. Synthesis of highly ordered carbon molecular sieves via template-mediated structural transformation. J. Phys Chem. B. 103: 7743–7746.

Rzepka, M., P. Lamp and M.A. De la Casa-Lillo. 1998. Physisorption of hydrogen on microporous carbon and carbon nanotubes, J. Phys. Chem. B. 102: 10894–10898.

Sanchez, C. and F. Ribot. 1994. Design of hybrid organic-inorganic materials synthesized via sol-gel chemistry. New J. Chem. 18: 1007–1047.

Schoedel, A. and O.M. Yaghi. 2016. Reticular Chemistry of Metal-Organic Frameworks Composed of Copper and Zinc Metal Oxide Secondary Building Units as Nodes. pp. 43–72. *In*: Kaskel, S. (ed.). The Chemistry of Metal–Organic Frameworks: Synthesis, Characterization, and Applications. Wiley-VCH Verlag GmbH & Co. KGaA, Weinheim, Germany.

Schubert, U. 2001. Polymers reinforced by covalently bonded inorganic clusters. Chem. Mater. 13: 3487–3494.

Schubert, U., N. Huesing and A. Lorenz. 1995. Hybrid inorganic-organic materials by sol-gel processing of organofunctional metal alkoxides. Chem. Mater. 7: 2010–2027.

Schumacher, K., P.I. Ravikovitch, A.D. Chesne, A.V. Neimark and K.K. Unger. 2000. Characterization of MCM-48 Materials. Langmuir. 16: 4648–4654.

Seng, K.H., J. Liu, Z.P. Guo, Z.X. Chen, D. Jia and H.K. Liu. 2011. Free-standing V_2O_5 electrode for flexible lithium ion batteries. Electrochem. Commun. 13: 383–386.

Shao-Horn, Y., S.A. Hackney, A.R. Armstrong, P.G. Bruce, R. Gitzendanner, C.S. Johnson et al. 1999. Structural characterization of layered $LiMnO_2$ electrodes by electron diffraction and lattice imaging. J. Electrochem. Soc. 146: 2404–2412.

Shao, Y., F. Ding, J. Xiao, J. Zhang, W. Xu, S. Park et al. 2013. Making Li-air batteries rechargeable: Material challenges. Adv. Funct. Mater. 23: 987–1004.

Shen, X.-F., Y.-S. Ding, J. Liu, J. Cai, K. Laubernds, R.P. Zerger et al. 2005. Control of nanometer-scale tunnel sizes of porous manganese oxide octahedral molecular sieve nanomaterials. Adv. Mater. 17: 805–809.

Shi, Y., Y. Wan, R. Zhang and D. Zhao. 2008. Synthesis of self-supported ordered mesoporous cobalt and chromium nitrides. Adv. Funct. Mater. 18: 2436–2443.

Simon, P. and Y. Gogotsi. 2008. Materials for electrochemical capacitors. Nat. Mater. 7: 845–854.

Simon, P. and Y. Gogotsi. 2013. Capacitive energy storage in nanostructured carbon–electrolyte systems. Acc. Chem. Res. 46: 1094–1103.

Simon, P., Y. Gogotsi, and B. Dunn. 2014. Where do batteries end and supercapacitors begin? Science 343: 1210–1211.

Smith, S.C., P.K. Sen and B. Kroposki. 2008. Advancement of energy storage devices and applications in electrical power system. IEEE Power Energy Soc. Gen. Meet. – Conversion Deliv. Electrical Energy in 21st Century. 1–8.

Subburaj, A.S., P. Kondur, S.B. Bayne, M.G. Giesselmann and M.A. Harral. 2014. Analysis and review of grid connected battery in wind applications. Sixth Annual IEEE Green Technol. Conf. 1–6.

Sumitomo Electric annual report 2011 (year-end 31st March 2011). Sumitomo Electric ingenious dynamic.

Sun, Y., R.B. Sills, X. Hu, Z.W. Seh, X. Xiao, H. Xu et al. 2015. A bamboo-inspired nanostructure design for flexible, foldable, and twistable energy storage devices. Nano Lett. 15: 3899–3906.

Takahashi, K., S.J. Limmer, Y. Wang and G. Cao. 2004. Synthesis and electrochemical properties of single-crystal V_2O_5 nanorod arrays by template-based electrodeposition. J. Phys. Chem. B. 108: 9795–9800.

Takahashi, K., Y. Wang and G. Cao. 2005. Ni−V_2O_5·nH_2O core−shell nanocable arrays for enhanced electrochemical intercalation. J. Phys. Chem. B 109: 48–51.

Tang, W., Y. Hou, F. Wang, L. Liu, Y. Wu and K. Zhu. 2013. $LiMn_2O_4$ nanotube as cathode material of second-level charge capability for aqueous rechargeable batteries. Nano Lett. 13: 2036–2040.

Taylor, P., R. Bolton, D. Stone, X-P. Zhang, C. Martin, P. Upham et al. 2012. Pathways for energy storage in UK, Technical report, March, 2012. Centre for low carbon futures.

Tesfaye, A.T., R. Gonzalez, J.L. Coffer and T. Djenizian. 2015. Porous silicon nanotube arrays as anode material for Li-ion batteries. ACS Appl. Mater. Interfaces. 7: 20495−20498.

Thakur, M., M. Isaacson, S.L. Sinsabaugh, M.S. Wong and S.L. Biswal. 2012b. Gold-coated porous silicon films as anodes for lithium ion batteries. J. Power Sources. 205: 426–432.

Thakur, M., R.B. Pernites, N. Nitta, M. Isaacson, S.L. Sinsabaugh, M.S. Wong et al. 2012a. Freestanding macroporous silicon and pyrolyzed polyacrylonitrile as a composite anode for lithium ion batteries. Chem. Mater. 24: 2998–3003.

Thakur, M., S.L. Sinsabaugh, M.J. Isaacson, M.S. Wong and S.L. Biswal. 2012c. Inexpensive method for producing macroporous silicon particulates (MPSPs) with pyrolyzed polyacrylonitrile for lithium ion batteries. Sci. Rep. 2: 1–7.

Turner, D.R. 1958. Electropolishing silicon in hydrofluoric acid solutions. J. Electrochem. Soc. 105: 402–408.

Tweed, K. 2013. UK launches Europe's largest energy storage trial. IEEE Spectrum. Published on August 1, 2013. http://spectrum.ieee.org/energywise/energy/the-smarter-grid/uk-launches-europes-largest-energy-storage-trial.

U.S. Energy Information Administration. 2014. Today in Energy (Sep. 2014). http://www.eia.gov/ todayinenergy/.

UK Department of Trade and Industry (DTI). 2004. Review of electrical energy storage technologies and systems and of their potential for the UK. DTI report DG/DTI/00055/00/00.

Ultrabattery: The new dimension in lead acid battery technology. www.ecoult.com.

Uribe-Romo, F.J., C.J. Doonan, H. Furukawa, K. Oisaki and O.M. Yaghi. 2011. Crystalline covalent organic frameworks with hydrazone linkages. J. Am. Chem. Soc. 133: 11478–11481.

Uribe-Romo, F.J., J.R. Hunt, H. Furukawa, C. Klöck, M. O'Keeff and O.M. Yaghi. 2009. A crystalline imine-linked 3-D porous covalent organic framework. J. Am. Chem. Soc. 131: 4570–4571.

US department of energy (DOE) global energy storage database. www.energystorageexchange.org/projects.

Vartuli, J.C., C.T. Kresge, W.J. Roth, S.B. McCullen, J.S. Beck, K. D.Schmitt et al. 1996. Designed synthesis of mesoporous molecular sieve systems using surfactant-directing agents (ed. Moser, W.R.) Advanced Catalysts and Nanostructured Materials: Modern Synthesis Methods, Academic Press, New York, pp. 1–19.

Velev, O.D., T.A. Jede, R.F. Lobo and A.M. Lenhoff. 1997. Porous silica via colloidal crystallization. Nature. 389: 447–448.

Vix-Guterl, C., E. Frackowiak, K. Jurewicz, M. Friebe, J. Parmentier and F. Beguin. 2005. Electrochemical energy storage in ordered porous carbon materials. Carbon 43: 1293–1302.

Vix-Guterl, C., S. Saadallah, K. Jurewicz, E. Frackowiak, M. Reda, J. Parmentier et al. 2004. Supercapacitor electrodes from new ordered porous carbon materials obtained by a templating procedure. Mat. Sci. Eng. B 108: 148–55.

Vix-Guterl, C., S. Saadallah, L. Vidal, M. Reda, J. Parmentier and J. Patarin. 2003. Template synthesis of a new ordered carbon structure from pitch. J. Mat. Chem. 13: 2535–2539.

Wang, H.Y., Y. Yang, X. Li, L.J. Li and C. Wang. 2010. Preparation and characterization of porous TiO_2/ZnO composite nanofibers via electrospinning. Chinese Chem. Lett. 21: 1119–1123.

Wang, W., Z. Favors, R. Ionescu, R. Ye, H.H. Bay, M. Ozkan et al. 2015. Monodisperse porous silicon spheres as anode materials for lithium ion batteries. Sci. Rep. 5: 1–6.

Wang, Y., K. Takahashi, K.H. Lee and G.Z. Cao. 2006. Nanostructured vanadium oxide electrodes for enhanced lithium-ion intercalation. Adv. Funct. Mater. 16: 1133–1144.

Wang, Y., X. Shao, H. Xu, M. Xie, S. Deng, H. Wang et al. 2013. Facile synthesis of porous $LiMn_2O_4$ spheres as cathode materials for high-power lithium ion batteries. J. Power Sources 226: 140–148.

Weber, A.Z., M.M. Mench, J.P. Meyers, P.N. Ross, J.T. Gostick and Q. Liu. 2011. Redox flow batteries: A review. J. Appl. Electrochem. 41: 1137–1164.

Wei, Q., F. Xiong, S. Tan, L. Huang, E.H. Lan, B. Dunn et al. 2017. Porous one-dimensional nanomaterials: design, fabrication and applications in electrochemical energy storage. Adv. Mater. 1602300: 1–39.

Wessells, C.D., R.A. Huggins and Y. Cui. 2011. Copper hexacyanoferrate battery electrodes with long cycle life and high power. Nat. Commun. 2: 1–5.

Whittingham, M.S. 2008. Materials challenges facing electrical energy storage. MRS Bulletin 33: 411–419.

Wu, F., J.Z. Chen, R.J. Chen, S.X. Wu, L. Li, S. Chen et al. 2011. Sulfur/polythiophene with a core/shell structure: Synthesis and electrochemical properties of the cathode for rechargeable lithium batteries. J. Phys. Chem. C. 115: 6057–6063.

Wu, Y., Z. Wen, H. Feng and J. Li. 2012. Hollow porous $LiMn_2O_4$ microcubes as rechargeable lithium battery cathode with high electrochemical performance. Small 8: 858–862.

Xiao, L., Y. Cao, J. Xiao, B. Schwenzer, M.H. Engelhard, L.V. Saraf et al. 2012. A soft approach to encapsulate sulfur: Polyaniline nanotubes for lithium-sulfur batteries with long cycle life. Adv. Mater. 24: 1176–1181.

Xiao, Q., M. Gu, H. Yang, B. Li, C. Zhang, Y. Liu et al. 2015. Inward lithium-ion breathing of hierarchically porous silicon anodes. Nat. Commun. 6: 1–8.

Xu, R., J. Wang, Q. Li, G. Sun, E. Wang, S. Li et al. 2009. Porous cobalt oxide (Co_3O_4) nanorods: facile synthesis, optical property and application in lithium-ion batteries. J. Solid State Chem. 182: 3177.

Yang, Z-C., Y. Zhang, J.H. Kong, S.Y. Wong, X. Li and J. Wang. 2013. Hollow carbon nanoparticles of tunable size and wall thickness by hydrothermal treatment of α-Cyclodextrin templated by F127 block copolymers. Chem. Mater. 25: 704–710.

Ying, J.Y., C.P. Mehnert and M.S. Wong. 1999. Synthesis and applications of supramolecular-templated mesoporous materials. Angew. Chem. 38: 56–77.

Yu, F., J. Zhang, Y. Yang and G. Song. 2009a. Preparation and characterization of mesoporous $LiFePO_4$/C microsphere by spray drying assisted template method. J. Power Sources. 189: 794–797.

Yu, Y., L. Gu, C. Zhu, P.A. van Aken and J. Maier. 2009b. Tin nanoparticles encapsulated in porous multichannel carbon microtubes: Preparation by single-nozzle electrospinning and application as anode material for high-performance Li-based batteries. J. Am. Chem. Soc. 131: 15984–15985.

Zhang, H.G. and P.V. Braun. 2012. Three-dimensional metal scaffold supported bicontinuous silicon battery anodes. Nano Lett. 12: 2778–2783.

Zhao, Y., X. He, J. Li, X. Gao and J. Jia. 2012. Porous CuO/SnO_2 composite nanofibers fabricated by electrospinning and their H_2S sensing properties. Sens. Actuators B. 165: 82–87.

Zhao, Y., Z. Song, X. Li, Q. Sun, N. Cheng, S. Lawes et al. 2016. Metal organic frameworks for energy storage and conversion. Energy Storage Mater. 2: 35–62.

Zhou, H., S. Zhu, M. Hibino, I. Honma and M. Ichihara. 2003. Lithium storage in ordered mesoporous carbon (CMK-3) with high reversible specific energy capacity and good cycling performance. Adv. Mater. 15: 2107–2111.

Zhu, G. and H. Ren. 2015. Porous Organic Frameworks: Design, Synthesis and Their Advanced Applications Springer Berlin Heidelberg, Germany.

Zhu, T., H.B. Wu, Y. Wang, R. Xu and X.W.D. Lou. 2012. Formation of 1D hierarchical structures composed of Ni_3S_2 nanosheets on CNTs backbone for supercapacitors and photocatalytic H_2 production. Adv. Energy Mater. 2: 1497–1502.

Zhu, W.H., Y. Zhu, Z. Davis and B.J. Tatarchuk. 2013. Energy efficiency and capacity retention of Ni-MH batteries for storage applications. Appl. Energy 106: 307–313.

Zhu, Y., S. Murali, M.D. Stoller, K.J. Ganesh, W. Cai, P.J. Ferreira et al. 2011. Carbon-based supercapacitors produced by activation of graphene. Science 332: 1537–1541.

2

Graphene-based Porous Materials for Advanced Energy Storage in Supercapacitors

Zhong-Shuai Wu,[1,*] Xiaoyu Shi,[1,2,3] Han Xiao,[1] Jieqiong Qin,[1,4] Sen Wang,[1,4] Yanfeng Dong,[1] Feng Zhou,[1] Shuanghao Zheng,[1,4,2] Feng Su[1,4] and Xinhe Bao[1,2]

1. Introduction

1.1 Definition and Category of Supercapacitors

With the rapid growth of portable electronics and hybrid electric vehicles, energy storage systems have attracted a lot of attention to meet this increasing demand. Lithium ion batteries are the most common electrical energy storage devices used in our modern society due to their huge storage capacity, which can store large amount of energy in a relatively smaller volume and weight, while providing suitable levels of energy density up to $150 \sim 200$ Wh kg^{-1} for different applications. However, as illustrated in a Ragone plot (Fig. 1) for evaluating the most important energy storage systems, lithium ion batteries suffer from low power density and limited cyclability (Simon and Gogotsi 2008). Therefore, the widespread use of batteries are significantly restricted, especially in these energy storage systems where fast charge and discharge is required. On the other hand, conventional electrolytic capacitors show extremely low energy density, which can't meet the energy requirement of certain electronic devices.

Supercapacitors, also known as electrochemical capacitors or ultracapacitors, can provide higher specific energy than electrolytic capacitors, several orders of magnitude higher power density than lithium ion batteries, superior long cycle life ($> 10^5$ times), and fast charge and discharge processes (within several minutes to several seconds) (Zhai et al. 2011). Therefore, supercapacitors could fill the gap between lithium ion batteries and conventional electrolytic capacitors, and become a very promising candidate of next-generation energy storage devices.

According to energy storage mechanism, supercapacitors can be divided into three types, including electrical double layer capacitors (EDLCs), pseudocapacitors and hybrid supercapacitors (Simon and Gogotsi 2008). In principle, EDLCs store energy through fast adsorption and desorption of positive and

[1] Dalian National Laboratory for Clean Energy, Dalian Institute of Chemical Physics, Chinese Academy of Sciences, 457 Zhongshan Road, Dalian 116023, P. R. China.
[2] State Key Laboratory of Catalysis, Dalian Institute of Chemical Physics, Chinese Academy of Sciences, 457 Zhongshan Road, Dalian 116023, P. R. China.
[3] Department of Chemical Physics, University of Science and Technology of China, 96 Jinzhai Road, Hefei 230026, P. R. China.
[4] University of Chinese Academy of Sciences, 19 A Yuquan Road, Shijingshan District, Beijing, 100049, P. R. China.
* Corresponding author: wuzs@dicp.ac.cn

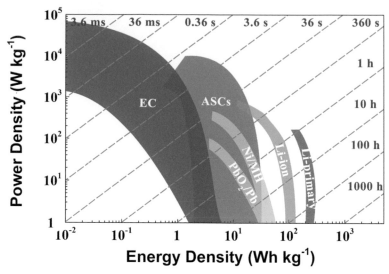

Fig. 1. Ragone plot related to specific power against specific energy for various electrical energy storage devices, including conventional electrolytic capacitors, supercapacitors, and different batteries (reprinted with permission from Zheng et al. 2017. Copyright 2017, Elsevier).

negative charges at the electrode surface. Nanocarbon materials are widely used as EDLC electrodes, including templated carbons, carbon nanotubes (CNTs), graphene, and so on. The key advantage of EDLCs lies in ultralong cycle stability and high power density. Pseudocapacitors use fast and reversible surface and near surface redox reactions for charge storage. In general, pseudocapacitive electrode materials include various transition metal oxides and hydroxides (e.g., MnO_2, RuO_2, $Ni(OH)_2$ and $Co(OH)_2$) and conductive polymers (e.g., polyaniline (PANI), polypyrrole (PPy), and polythiophene). In comparison with EDLCs, pseudocapacitors possess higher specific capacitance and energy density, but its cycling stability is far inferior to EDLCs. The third type is hybrid supercapacitor, consisting of EDLC electrode and pseudocapacitive electrode, which can efficiently combine high power of EDLCs with high energy density of pseudocapacitors (Zhai et al. 2011).

1.2 Advantages of Utilising Graphene for Supercapacitors

The key to develop supercapacitors with high performance is to develop electrode materials. As a kind of newly-developed material, graphene is attracting dramatically increasing interest after Novoselov, Geim, and coworkers exfoliated it from graphite using the mechanical cleavage method in 2004 (Novoselov et al. 2004). Graphene is a single-atom-thick, crystalline carbon film that exhibits various unprecedented properties, such as ultrahigh carrier mobility at room temperature ($\sim 10,000$ cm^2 V^{-1} s^{-1}), quantum hall effect, excellent optical transparency (~ 97.7 per cent), high Young's modulus (~ 1 TPa), and excellent thermal conductivity ($\sim 5,000$ W m^{-1} K^{-1}) (Zhang 2015).

Importantly, graphene possesses unique properties of large theoretical specific surface area (SSA, $2,630$ m^2 g^{-1}), superior electrical conductivity, theoretical specific capacitance of 550 F g^{-1}, and thus was demonstrated to be a highly potential electrode material for supercapacitors. For instance, graphene has ultrathin 2D structure that is beneficial for the fast transport of electrolyte ions along the plane of graphene for in-plane supercapacitors and high-power micro-supercapacitors (Fig. 2) (Yoo et al. 2011). Further, graphene is composed of an atom-thick layer of sp^2 carbon atoms, which are bonded through conjugated π–π interactions, endowing graphene remarkable strength and flexibility. As a consequence, graphene-based thin films hold great promise for developing new types of flexible,

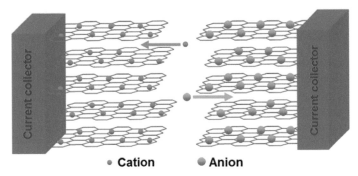

Fig. 2. Schematic of the operating principle of in-plane supercapacitors utilised for the performance evaluation of graphene as electrodes (Adapted with modification from Yoo et al. 2011).

transparent and miniaturised ultrathin supercapacitors, which are very flexible power sources for portable and wearable electronics, such as mobile phones, smart watches and intelligent glasses.

1.3 Pore Role in Supercapacitors

Though graphene has remarkable properties, the performance of supercapacitors with graphene as electrode materials reported is still far from the theoretical value. The main reason for this is the strong aggregation and restacking of graphene sheets (GSs) under Van der Waals force, which greatly reduces the available surface area of graphene and the amount of charge adsorption active sites. To solve the problem, materials' researchers take the porous graphene into account naturally, originating from the advanced experience of porous carbon. It has been reported, in the porous electrodes for supercapacitors, the macropores can act as a bulk buffering reservoir for electrolytes to minimise the diffusion distances to the interior surfaces of the pores, while the mesopores can provide a large accessible surface area for ion transport/charge storage, and micropores can continuously increase charge accommodation. These intriguing features are desirable for potential applications in high-performance supercapacitors (Han et al. 2014).

Combined with graphene and pores together, porous graphene materials have several obvious advantages over graphene and other porous carbon materials. First, high mechanical strength of graphene with large aspect ratio can help enhance the stability of porous frameworks and prevent the shrinkage and collapse of the porous structures. Second, the excellent thermal and chemical stability of graphene can enable these porous materials to withstand harsh conditions. Third, the rapid diffusion of electrolytes in channels, and outstanding electrical conductivity of porous graphene, make it an ideal current collector for fast transportation of charge carriers within the porous frameworks. Fourth, graphene derivatives containing oxygen functional groups such as graphene oxide (GO) and reduced GO (RGO) can serve as appealing substrates for the binding of various organic or inorganic species, which provide numerous new opportunities for constructing various graphene-based porous materials with different complexity. As a result, these uncommon features endow these porous graphene materials to serve as key electrode components in high-performance supercapacitors (Han et al. 2014).

In this chapter, we summarise the state-of-the-art fabrication approaches for two-dimensional and three-dimensional graphene-based porous materials with micro-, meso-, and macro-porous structures and highlight unprecedented importance of porosity in the performance enhancement in supercapacitors including electrical double layer capacitors, pseudocapacitors, asymmetric supercapacitors, lithium (sodium) ion supercapacitors, fiber supercapacitors and planar micro-supercapacitors. We also evaluate advantages and disadvantages of all kinds of supercapacitors and discuss their appropriate application fields respectively. Finally, we discuss the current challenge and future trends of next-generation supercapacitors based on porous graphene in term of both academic and industrial aspects.

2. Porous Graphene for EDLCs

2.1 Fundamental Principle of EDLCs

EDLCs store the charge electrostatically by reversible ions adsorption formed at electrodes and electrolyte interface. The mechanism of charge and discharge for EDLCs is illustrated in Fig. 3, in which the electrolyte anions and cations move towards the positive and negative electrodes, respectively, to generate an electrical double-layer at each of the two electrode–electrolyte interfaces in the charged state (Chen and Dai 2013). When the charges are released, the reverse process takes place within the electrolyte. This storage mechanism allows very fast energy uptake and delivery, as well as high stability of EDLCs during millions of charge/discharge cycles.

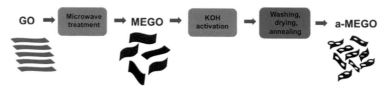

Fig. 3. Schematic showing the microwave exfoliation/reduction of GO and the following chemical activation of MEGO with KOH.

The electrode and electrolyte interface can be acknowledged as a conventional dielectric capacitor, and the capacitance C is defined by equation:

$$C = \varepsilon_r \varepsilon_0 A/d \tag{1}$$

where ε_r is the electrolyte dielectric constant, ε_0 is the dielectric constant of the vacuum ($\varepsilon_0 = 8.854 \times 10^{-12}$ F m^{-1}), d is the effective thickness of the double layer (charge separation distance) and A is the SSA of electrode. A two-electrode supercapacitor cell can be considered as two capacitors in series, in which C_P and C_N represent positive capacitance and negative capacitance, respectively. And the total capacitance (C_T) of the cell can be calculated as following formula:

$$1/C_T = 1/C_P + 1/C_N \tag{2}$$

The thickness of the double layer depends on the concentration of the electrolyte and the size of the ions which is in the order of 5 ~ 10 Å. EDLCs are high power density energy storage devices but with relatively poor energy density at least one order of magnitude lower than battery because of its surface ions adsorption nature. The double layer capacitance is about only 5 ~ 20 µF cm^{-2} depending on the electrolyte used for a smooth electrode (Kotz and Carlen 2000, Simon and Gogotsi 2008). Usually, a high-performance electrode requires high electrical conductivity, ion-accessible surface area, ionic transport rate and electrochemical stability. Thereby, porous and highly conductive electrode materials with high SSA are desired.

2.2 Porous Graphene-based Materials for EDLCs

Graphene has recently received intensive interest as EDLC electrode materials, however, the strong π–π interaction between GSs would introduce the restacking to form graphite-like powder, which would severely decrease the accessible surface area and reduce the ion diffusion rate, resulting in unsatisfactory gravimetric capacitance and relatively low charge/discharge rates. In this context, hierarchical porous graphenes with macro-/meso-, meso-/micro-, macro-/micro-, and macro-/meso-/micropores were proposed for high-power supercapacitors, since macropores can act as a bulk buffering reservoir for electrolytes to minimise the diffusion distance to the interior surfaces of the pores, while mesopores can provide large accessible surface area for fast ion transport and enhanced charge storage, and micropores can continuously increase charge accommodation, respectively.

As a typical example, Müllen's group fabricated 3D graphene aerogels loaded with mesoporous silica composites (GA-SiO$_2$), by the hydrolysis of tetraethoxysilane (TEOS) with graphene aerogel (GA) as the support and cetyltrimethyl ammonium bromide (CTAB) as soft template (Fig. 5a) (Wu et al. 2012a). The SiO$_2$ in GA-SiO$_2$ then serve as sacrificial template to synthesise GA-mesoporous carbon (GA-MC) by using sucrose as the carbon source. Featuring a fully interconnected macroporous graphene framework and mesopores on the carbon walls with a size of 2–3.5 nm, the resulting GA-MC exhibited a SSA of up to 295 m^2 g^{-1} (Fig. 5b). GA-MC as electrode material for supercapacitors manifested an outstanding specific capacitance of 226 F g^{-1} at 1 mV s^{-1}, high rate capability, and excellent cycling stability without capacitance loss after 5,000 cycles (Fig. 5c), which are benefited from the integration of meso- and macroporous structures. It has been recognised that macropores facilitate the penetration and wetting of electrolyte ions to the interior surface and hence improve the rate capability and cycle stability of the electrodes, while mesopores provide a large accessible surface area for ion transport and charge storage, and micropores contribute to the increase of both SSA and capacitance.

Park et al. prepared CO$_2$-activated macroscopic graphene architectures with trimodal pore systems that consist of 3D inter-networked macroporosity arising from self-assembly, mesoporosity arising from the intervoids of nanosheets, and microporosity via CO$_2$ activation (Yun et al. 2014). The micropores in hierarchical structures as synthesised trimodal porous graphene frameworks (tGFs) contributes to significant increase in the surface area of up to 829 m^2 g^{-1} and pore volume to 2,829 cm^3 g^{-1}. A specific capacitance of 278.5 F g^{-1} was obtained for tGFs in 1 M H$_2$SO$_4$ aqueous electrolyte with a three-electrode configuration, which is much better than graphene framework counterpart of 157.7 F g^{-1}. Ruoff and coworkers reported a simple activation with KOH of microwave exfoliated GO (a-MEGO) with high SSA values up to 3,100 m^2 g^{-1}, which is much higher than theoretical SSA of 2,630 m^2 g^{-1}. The two-electrode symmetrical supercapacitor in BMIMBF$_4$ under a working voltage of 3.5 V, showed a high energy density of ~ 75 Wh kg^{-1}, and a power density as high as ~ 250 kW kg^{-1} for packaged cell. Moreover, the porous graphene showed outstanding cycling stability (Zhu et al. 2011).

Although gravimetric capacitance was widely used as the figure-of-merit to evaluate an active material for EDLCs, volumetric performance is a more important parameter in some practical applications with limited space such as portable electronic products. For most porous graphene electrode design, there is usually a trade-off relationship between the gravimetric and volumetric capacitances. The reason is that a highly porous electrode can offer a large specific surface area and favour rapid ion diffusion for high gravimetric capacitance, but usually results in a lower volumetric capacitance due to its relatively low packing density. Moreover, low packing density active materials with abundant empty spaces could be flooded by the electrolyte, which could increase the total mass of the device and lower the energy density of the entire devices. In order to simultaneously achieve high gravimetric and high volumetric capacitances while retaining excellent rate capability,

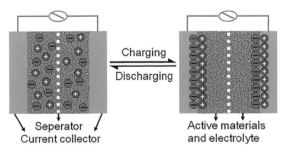

Fig. 4. Charge storage mechanism of EDLCs (Adapted with modification from Chen and Dai 2013).

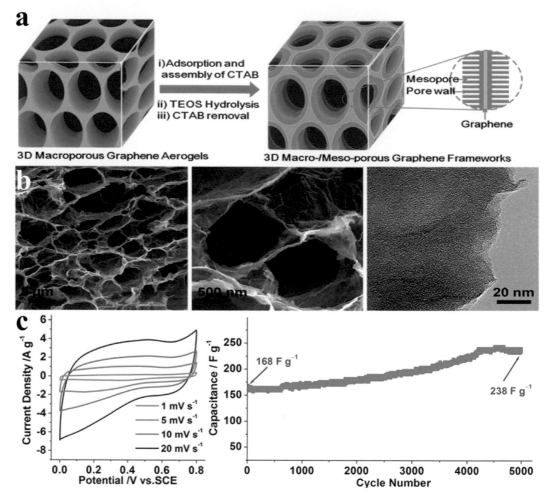

Fig. 5. (a) Fabrication illustration of hierarchical macro- and mesoporous GA-SiO$_2$ frameworks: (i) electrostatic adsorption and assembly of CTAB on the surface of 3D GAs, (ii) TEOS hydrolysis for nucleation and growth of mesoporous silica on the surface of CTA$^+$-adsorped GAs, and (iii) CTAB removal through ethanol washing, drying, and thermal annealing. (b) Morphology and microstructure of GA-MC. (c) Electrochemical performance of GA-MC electrode for supercapacitors (Adapted with modification from Wu et al. 2012a).

Duan et al. prepared a compact holey graphene framework (HGF) by hydrothermal treatment of a GO solution in the presence of H$_2$O$_2$, where H$_2$O$_2$ induced the formation of in-plane holes by oxidation of GO sheets (Fig. 6). The 3D structure containing in-plane holes exhibited a high SSA of 1,560 m^2 g^{-1}. With large ion-accessible surface area, efficient electron and ion transport pathways as well as a high packing density, the holey graphene framework electrode can deliver both high gravimetric and volumetric capacitance (289 F g^{-1} and 205 F cm^{-3}), as well as energy densities (123 Wh kg^{-1} and 87 Wh L^{-1}) (Xu et al. 2014).

The physical presence of holes benefited the performance in electrochemical processes of holey-graphene in general. However, the presence of holes was not the only factor for improving capacitance values. Connell et al. found that the electrochemical capacitive properties for holey graphene are highly correlative with the topology (Lin et al. 2015). They compared the electrochemical performance of holey-graphene obtained by air oxidation of graphene process with different temperature. They found that the increase of both mesopore fractions and oxygen species around the hole edges for the holey

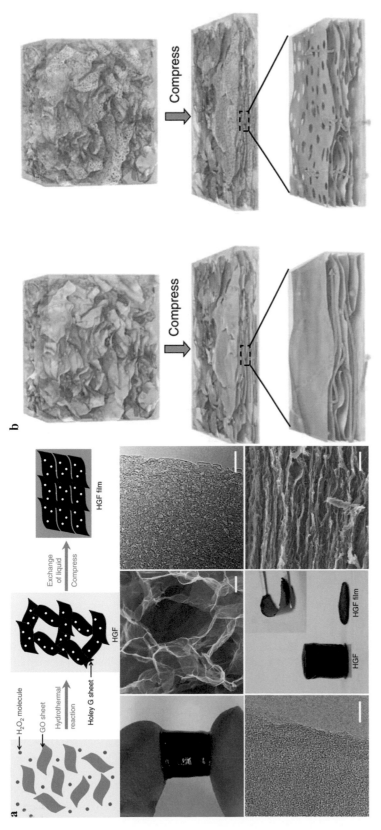

Fig. 6. (a) Preparation and structural characterisation of HGF before and after mechanical compression. (b) Schematic illustration of the HGFs as an ideal material for EDLC electrodes (Adapted with modification from Xu et al. 2014).

graphene samples contributed to the improvement of specific capacitance due to the increased ions adsorption sites. When the heating was carried out in the temperature range where graphitic carbon loss occurs, both the mesopore fraction and the oxygen content continued to increase. However, the capacitive performance of holey-graphene samples reached a maximum, plateaued and then began to exhibit signs of reduction. Apparently, the loss of too much graphitic carbon negatively affected the capacitive performance despite the positive effects from mesopore formation and oxygen doping. Clearly, there is a fine balance of chemical and morphological features that contribute to optimise the capacitive performance of h-Graphene-based electrodes.

3. Porous Graphene Materials for Pseudocapacitors

Pseudocapacitors, as a crucial type of supercapacitors, store energy chemically by rapid redox reactions at the surface or near-surface of electrodes. Compared to EDLCs, pseudocapacitors can afford higher specific capacitance and energy density (Simon and Gogotsi 2008). In principle, pseudocapacitive materials include metal oxides (e.g., RuO_2, MnO_2 and NiO) and conducting polymers (e.g., poly (3,4-ethylenedioxythiophene) (PEDOT), polypyrrole (PPy) and polyaniline (PANi)). Unfortunately, the incident volume change and poor electrical conductivity of metal oxides or conducting polymers result in low power density, poor rate performance and cycling stability for pseudocapacitors.

To address these issues of electrode materials and improve the electrochemical performance of pseudocapacitors, two effective strategies are carried out, including (i) fabrication of graphene-based hybrids of metal oxides or polymers, and (ii) introduction of porous structures to electrode materials. First, graphene is one of the most promising substrate materials due to its unique 2D structure, ultrahigh surface area, and excellent intrinsic physical and chemical properties such as high conductivity and good chemical stability (Sun et al. 2011). Second, porous materials have been recognised as one of most appealing electrode materials because of two- or three-dimensionally interconnected pores, large surface area and controlled porous structure (Han et al. 2014). Based on the above views, the design and construction of graphene-based porous metal oxides or polymers could remarkably improve electrochemical performance of pseudocapacitors.

3.1 Graphene-based Porous Hybrid Nanosheets for Pseudocapacitors

2D graphene-based porous (micro- and meso-porous) hybrid nanosheets are usually obtained by decoration of graphene with porous metal oxides or polymers. Such hybrid nanosheets not only possess the morphological features of 2D graphene and micro- or mesoporous structure, but also exhibit outstanding properties such as high surface area, ultrathin thickness and impressive electrical conductivity. Generally, synthesis methods are categorised into soft-template and hard-template approaches (Han et al. 2014).

Soft-template method is a bottom-up synthesis strategy, in which the aggregates of surfactants or block copolymers serve as templates for the formation of ordered porous structure, and then porous metal oxides or polymers *in situ* grow on graphene nanosheets (Han et al. 2014, Li et al. 2016). In general, the reaction conditions are mild and the structures are easily tuned. For instance, sandwich-like GO-based mesoporous conducting polymers (mPPy@GO or mPANi@GO) nanosheets have been synthesised by using block copolymer polystyrene-b-poly(ethylene oxide) (PS-b-PEO) as the soft template (Fig. 7). The free-standing GO nanosheets with abundant oxygen-containing functional groups serve as a typical 2D surface to enable spherical micelles formation by self-assembly of the PS-b-PEO onto surface. After pyrrole monomers were added and then polymerised, polymer networks around the micelle templates were formed. Finally, mPPy@GO nanosheets were obtained after removing micelle templates by repeated washing. Notably, mPPy@GO nanosheets exhibited 2D free-standing morphology, tailored pore sizes ranging from 5 to 20 nm, adjustable thickness of $35 \sim 45$ nm and enlarged specific surface area of 85 $m^2\,g^{-1}$ (Liu et al. 2015).

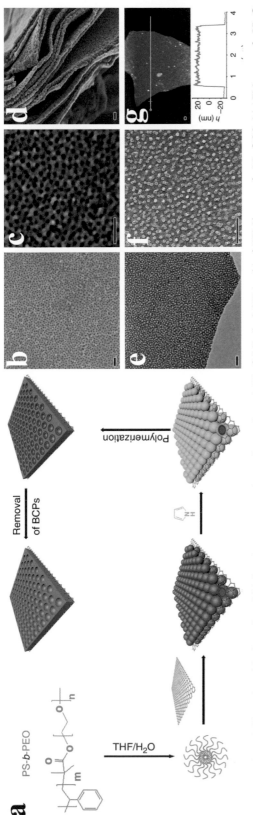

Fig. 7. (a) Schematic illustration of the formation of GO-based mesoporous polypyrole nanosheets (mPPy@GO). (b–d) SEM and (e, f) TEM images of mPPy@GO. (g) AFM survey of mPPy@ GO (Scale bar, 100 nm) (Adapted with modification from Liu et al. 2015).

The hard template method, also called nanocasting approach, uses the preformed templates, such as mesoporous silica, carbon, or aggregates of nanoparticles, to nanocast precursors into the inner pores (Li et al. 2016). After removing the template, the ordered micro- or mesoporous material can be gained. For example, various graphene-based mesoporous hybrid nanosheets can be fabricated using graphene-based mesoporous silica nanosheets as hard template. Typically, with $(NH_4)_2TiF_6$ as the titania precursor, graphene-based mesoporous titania (G-TiO$_2$) nanosheets with a ultrathin thickness, uniform mesopore (~ 2 nm) and high specific surface can be successfully obtained via a simple nanocasting approach (Fig. 8) (Yang et al. 2011a,b).

2D graphene-based porous hybrid nanosheets as electrode materials of pseudocapacitors could remarkably improve electrochemical performance. A remarkable example was recently reported for mPPy@GO nanosheets as electrode in three-electrode system and all-solid-state planar micro-supercapacitors (Liu et al. 2015). Notably, mPPy@GO nanosheets exhibit typical pseudocapacitive behaviour observed in cyclic voltammetry curves and delivered a high specific capacitance of 383 F g^{-1} at 1 mV s^{-1} in three-electrode systems, and a high areal capacitance of 75.5 µF cm^{-2} and rate capability for micro-supercapacitors. In addition, Wong and coworkers (Chen et al. 2015) successfully fabricated graphene-Fe$_2$O$_3$ porous hybrid using hard-template method. The graphene-Fe$_2$O$_3$ porous hybrid with a conductive graphene scaffold and interconnected Fe$_2$O$_3$ porous structure exhibited high specific capacitance of 1,095 Fg^{-1} as anode, outstanding energy density of 98 Wh kg^{-1} and power density of 22,826 W kg^{-1} for pseudocapacitors.

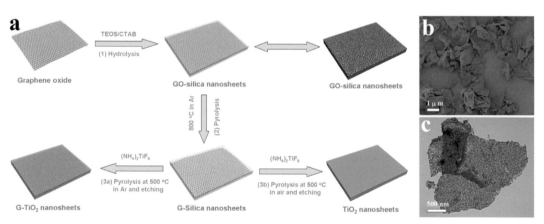

Fig. 8. (a) Schematic illustration of the formation of graphene-based mesoporous titania nanosheets (G-TiO$_2$). (b) SEM and (c) TEM images of G-TiO$_2$ (Adapted with modification from Yang ct al. 2011).

3.2 3D Graphene-based Macroporous Hybrid Frameworks for Pseudocapacitors

3D graphene hybrid frameworks usually refer to macroporous interconnected graphene-based network structure with intriguing properties of graphene nanosheets and intriguing features of individual components. Graphene foams (GFs), as typical 3D macroscopic graphene architecture, possess high surface area, excellent conductivity, low weight density and strong mechanical strength, which are impressive in electronic devices, environmental engineering and biomedical sciences fields (Jiang and Fan 2014). Normally, GFs and its derivatives can be obtained by hydrothermal method, chemical reduction method, template-directed chemical vapor deposition (CVD) method or their combination. GFs incorporated with metal oxides or polymers are intensively developed for high-performance electrodes for pseudocapacitors.

Preparation methods of 3D macroporous graphene-based metal oxides mainly include electrochemical deposition and *in situ* growth of metal oxides on GFs. For example, Chen and coworkers (Dong et al. 2012) synthesised a 3D graphene-Co_3O_4 hybrid using an *in situ* hydrothermal procedure, in which GFs can act as a 3D support to uniformly anchor metal oxides (e.g., Co_3O_4) with well-defined size, shapes, and crystallinity. The resulting 3D hybrid frameworks exhibited improved performance and promising application due to the synergistic cooperation between graphene and metal oxides. 3D macroporous graphene-based polymers are synthesised by chemical polymerisation or electrochemical polymerisation. For instance, Qu and coworkers (Zhao et al. 2013) developed a direct electrochemical polymerisation strategy to form 3D graphene-PPy hydrogel (PPy-G) with high compression tolerance, in which the pyrrole-containing 3D GFs served as the working electrode, Ag/AgCl acted as reference electrode and $NaClO_4$ aqueous solution was used as electrolyte. Interestingly, the PPy-G electrode was durably tolerant to high compressive strain without structural collapse, elastic loss and performance decline for pseudocapacitors.

3D macroporous graphene-based frameworks of metal oxides or polymers can provide a large surface area, high conductivity, robust scaffolds and improved electrochemical performance. As a typical example, Wu and coworkers (Wu et al. 2015) fabricated a graphene-V_2O_5 hybrid aerogel with a low density of 27.7 mg cm^{-3} and high specific surface area of 172 m^2 g^{-1}. The graphene-V_2O_5 hybrid aerogel-based pseudocapacitors exhibited excellent specific capacitance of 486 F g^{-1}, high energy density of 68 Wh kg^{-1} and long cycle life. Other 3D macroporous graphene-based metal oxides, binary metal oxides and conducting polymer also have been reported to demonstrate excellent performance for pseudocapacitors (Jiang and Fan 2014).

3.3 Graphene-based Hierarchically Porous Materials for Pseudocapacitors

2D or 3D graphene-based hierarchically porous architectures with macro/meso/microscale pores are desirable electrode materials for high-performance supercapacitors. The combination of different pores possesses a positive synergic effect, in which macropores serve as a bulk buffering reservoir for electrolytes, mesopores provide a high accessible surface area for ion transport and charge storage, and micropores constantly increase ion and charge accommodation. Unfortunately, the fabrication and pseudocapacitor application of such graphene-based hierarchically porous metal oxides or polymers have been rarely reported. Notably, Wu and coworkers (Wu et al. 2012b) presented 3D graphene-based carbon and metal oxide (Co_3O_4, RuO_2) with hierarchical macropores and mesopores (2–3.5 nm) by combination method of soft and hard templates. The 3D graphene-based RuO_2 exhibited outstanding specific capacitance of 560 F g^{-1} for pseudocapacitors and 1,090 F g^{-1} for RuO_2 active material. However, synthesis of graphene-based hierarchically porous metal oxides or polymers constructed with three kinds of pores (micropores, mesopores, and macropores) still remains a tremendous challenge. It is expected that future challenging directions in this field are mainly centered on graphene-based hierarchically porous metal oxides or polymers with precisely controlled pore morphology including shape, size and wall thickness.

4. Porous Graphene for Asymmetric Supercapacitors

Supercapacitors are of great interest for energy management, being able to deliver a very high power in a short time. However, for broadening the spectrum of their applications, the limited energy density (E) has to be improved. Based on equation of $E = 1/2CV^2$, the enhancement of energy density can be realised through increasing the capacitance (C), and/ or cell voltage (V). As for the operation voltage of supercapacitors, typically, in a symmetric cell, the working voltage is limited to less than 1.23 V in aqueous electrolytes because of the thermodynamic breakdown potential of water molecules, and extended beyond 2.5 V in organic electrolytes or ionic liquid based electrolytes. But organic or ionic liquid based electrolytes suffer from high cost, low ionic conductivity, high toxicity, inflammability and other safety issues, which severely hinder their applications. Another promising approach to achieve higher working voltage for aqueous electrolytes is to design asymmetric supercapacitors (ASCs) using two different electrode materials as the anode and cathode (Fig. 9) (Zheng et al. 2017).

Figure 9 illustrates the principle of ASCs, where two dissimilar materials are assembled together as anode and cathode. The Ragone plot given in Fig. 1 compares the energy and power densities of various energy-storage devices. It is apparent that ASCs deliver significantly higher power density as compared to batteries. Furthermore, the energy densities of asymmetric devices are much higher than that of symmetric supercapacitors, and well comparable to PbO_2/Pb and Ni/MH batteries, suggestive of widespread use of ASCs for next-generation electronics.

Elaborated screening and integrity of different main device components (e.g., electrodes, electrolyte, current collectors, and separator) for designing high-performance ASCs are greatly challenging. Screening of new high-performance nanostructured electrode material is the core part in developing stable high-energy ASCs. So far, great progress has been made and many advanced energy materials have been applied as both negative and positive electrode materials for ASCs. For instance, the negative electrodes reported include nanocarbon materials (carbon nanotubets, and graphene), metal oxides (Fe_2O_3, MoO_3, WO_3) and metal nitrides (TiN, VN, Fe_2N). Among various materials, graphene and its derivatives have shown a great combination of desired properties, which leads to better electrochemical performance in comparison with ACs and CNTs. However, the tendency of GSs to form irreversible agglomerates via van der Waals interactions, resulting in the significant surface area loss and limited ion diffusion and capacitance. Making graphene into porous structures is an effective strategy to prevent the agglomerates of graphene nanosheets and to create graphene materials with high surface area and high specific capacitance.

Zhang et al. (Yan et al. 2012) successfully synthesised porous graphene (Fig. 10(a,b)) using porous MgO sheets as the template by a CVD approach. It should be pointed out that, compared to rGO, porous graphene not only exhibited high specific capacitance but also maintained high capacitance at high current density (Fig. 10c), e.g., 245, 236, 231, 220, and 209 F g^{-1} at different current densities of 1, 2.5, 5, 10, and 25 A g^{-1}, respectively, because of its narrow mesopore distribution and open flat layer with high surface area. Then, the typical ASCs with porous graphene as negative electrode and hierarchical flowerlike high-capacitance $Ni(OH)_2$ anchored on graphene (1,735 F g^{-1} obtained at 1 mV s^{-1}) as positive electrode were assembled. In light of the unique structure and efficient match of these two electrode materials, the as-fabricated ASCs possessed a high voltage region of 0–1.6 V in KOH aqueous electrolyte, displayed high specific capacitance of 218.4 F g^{-1}, maximum energy density of 77.8 Wh kg^{-1} and power density of 15.2 kW kg^{-1}, respectively. Moreover, the $Ni(OH)_2$/graphene//porous graphene ASCs presented excellent cycling stability with capacitance retention of

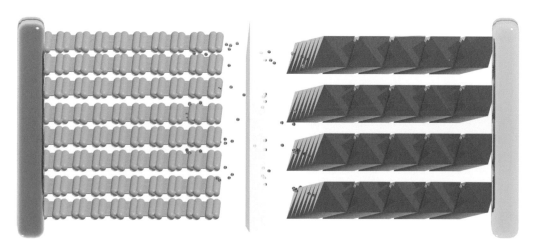

Fig. 9. Schematic of an ASC (Adapted with modification from Zheng et al. 2017).

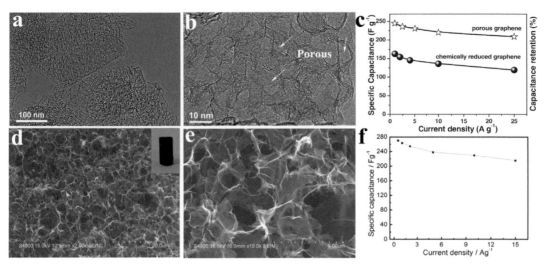

Fig. 10. (a) TEM and (b) high-magnification TEM image of the porous graphene. (c) Specific capacitance of porous graphene and chemically reduced graphene as a function of current density (Adapted with modification from Yan et al. 2012). (d) Low- (inserted with the digital photo of GF) and (e) high-magnification SEM images of GF, (f) specific capacitance at different current densities (Adapted with modification from Tang et al. 2015).

94.3 per cent after 3,000 cycles. In another key example, Tang et al. (Tang et al. 2015) fabricated the novel ASCs based on a highly porous graphene foam (270 F g^{-1}) (Fig. 10(d–e)) negative electrode prepared by mild reduction. The fabricated ASC device was found to work in a voltage window of 1.6 V with an excellent energy density of 34.5 Wh kg^{-1} at power density of 547 W kg^{-1}. Furthermore, it exhibited superior cycling performance with 92.2 per cent capacitance retention after 10,000 cycles. Shen et al. developed a combined chemical foaming, thermal reduction and KOH activation method to produce activated porous graphene (aGNS) with high specific surface area of 1,383 m^2 g^{-1} for negative electrode, and used an interfacial polymerisation method to synthesise sulfonated graphene nanosheet (sGNS)/carboxylated multi-walled CNT/PANI (sGNS/cMWCNT/PANI) composite for positive electrode. The fabricated ASCs could be cycled reversibly at a cell voltage of 1.6 V in 1M H$_2$SO$_4$ aqueous electrolyte, and showed an energy density of 20.5 Wh kg^{-1} at a power density of 25 kW kg^{-1} (Shen et al. 2013). Recently, Thomas et al. developed a functionalised GA by uniformly doping palladium (Pd) nanoparticles on graphene nanosheets (P-GA) followed by a lyophilisation and two-step reduction method. The high surface area (328 m^2 g^{-1}) and low electrical resistivity (50 times lower than one without Pd) of the GA composite generated a high specific capacitance (175.8 F g^{-1} at 5 mV s^{-1}), excellent rate capability (48.3 per cent retention after a tenfold increase of scan rate), and remarkable reversibility. The MnO$_2$//P-GA ASC delivered an average energy density of 13.9 Wh kg^{-1} at a power density of 13.3 kW kg^{-1} (Yu et al. 2015). These impressive performances demonstrate that porous graphene is a good candidate of high conducting active material as negative electrode for ASCs.

Positive electrodes for ASCs such as conductive polymers, transition-metal oxides/hydroxides, generally utilise the materials that exhibit a large amount of pseudocapacitance originating from the Faradaic charge transfer via fast and reversible redox reactions, electrosorption and desorption, or intercalation and deintercalation of the electrolyte with the electrode. So the main drawbacks of positive electrodes include poor electrical conductivity, the fast fading of electrochemical performance and stability, and low power density.

One possible way to overcome this issue is to fabricate the hybrid nanostructures with nanocarbon material such as graphene. 3D porous graphene based frameworks, such as sponges, foams, hydrogels, and aerogels, have been widely explored as the stable yet conductive substrates due to the continuous and highly conductive network providing a highly conductive pathway for electron and ion transportation. Thus, efficient deposition of active materials through convenient and reliable methods, will result in high power density, high mass loading and excellent cycling stability.

For instance (Wang et al. 2014), used a pulsed laser deposition process for the growth of NiO nanoparticles on highly conductive 3D GF (Fig. 11(a–c)), displaying a high specific capacitance of 1,225 F g^{-1}. Using this NiO/GF as positive electrode and hierarchical porous nitrogen-doped CNTs as negative electrode, the assembled ASCs in aqueous KOH solution presented a working voltage of 0.0–1.4 V and energy density of 32 Wh kg^{-1} achieved at power density of 700 W kg^{-1}. Notably, ultrahigh power density of 42 kW kg^{-1} was attained at an extremely short charge-discharge time of 2.8 s. Additionally, Lou's group (Liu et al. 2014) reported a 3D GF/CNT hybrid film with high flexibility and robustness as ideal support for deposition of a large amount of electrochemically active materials per unit area (Fig. 11(d–f)). To demonstrate the concept, MnO$_2$ (Fig. 11g) and PPy have been deposited on the GF/CNT film as positive and negative electrode for lightweight and flexible ASCs in an aqueous electrolyte. The ASCs could work with an output voltage of 1.6 V (Fig. 11(h, i)), deliver high energy/power density (22.8 Wh kg^{-1} at 860 W kg^{-1} and 2.7 kW kg^{-1} at 6.2 Wh kg^{-1}), and exhibited remarkable cycling stability (capacitance retention of 90.2–83.5% after 10,000 cycles). Moreover, the ASCs could retain their electrochemical performance at different bending angles. In

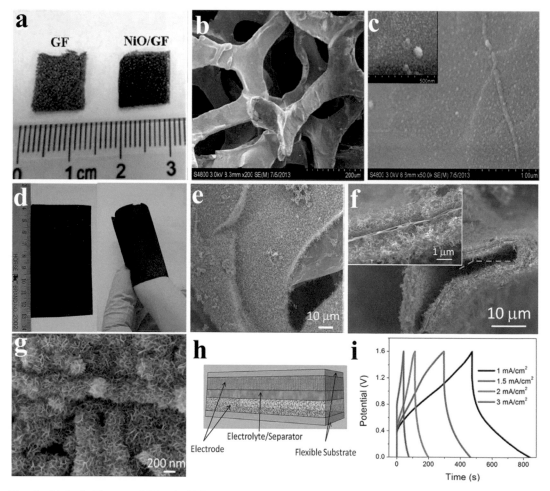

Fig. 11. (a) Optical images of GF and NiO/GF. (b, c) SEM images of (b) GF and (c) NiO/GF (Adapted with modification from Wang et al. 2014). (d) Photograph showing the large size and flexibility of a GF/CNT film. (e, f) Typical SEM images of CNTs in GF/CNT hybrid film. (g) SEM image of MnO$_2$ nanosheets on GF/CNT/MnO$_2$ hybrid film. (h) Schematic of an ASC consisting of the GF/CNT/MnO$_2$ positive electrode, electrolyte-soaked separator and GF/CNT/PPy negative electrode. (i) GCD curves of the ASCs obtained at different current densities (Adapted with modification from Liu et al. 2014).

another example, Zhai et al. synthesised MnO_2 on porous graphene gel/Ni foam (MnO_2/G-gel/NF) as supercapacitor electrode, delivering a large areal capacitance of 3.18 F cm^{-2} at a high mass loading of 13.6 mg cm^{-2} of MnO_2. Moreover, the ASCs based on MnO_2/G-gel/NF as positive electrode and G-gel/NF as negative electrode achieved a remarkable energy density of 0.72 mWh cm^{-3}, excellent cycling stability, with capacitance retention of 98.5 per cent after 10,000 cycles (Zhai et al. 2013).

5. Graphene-based Materials for Hybrid Supercapacitors

5.1 Hybrid Supercapacitors

The continuously surging demand in large-scale energy applications such as electric vehicles further boosts great interest in developing high-performance energy storage devices with both high energy and power densities. Lithium ion batteries (LIBs) and supercapacitors are considered as the predominant power sources, however, LIBs showed limited power density, and supercapacitors delivered low energy density (Fig. 12a) (Aravindan et al. 2014). Therefore, it is highly required to develop the devices for filling the gap between LIBs and supercapacitors. In this regard, lithium ion hybrid supercapacitors (Li-HSCs, or call lithium ion capacitors) consist of a battery-type faradaic high-energy electrode and a capacitive high-power electrode (Ma et al. 2015), which rationally combine the rapid charge/discharge and long cycle life of supercapacitors and high energy-storage capacity of LIBs. Therefore, Li-HSCs can effectively bridge the gap between the LIBs and SCs (Fig. 12b).

Sodium is considered as an ideal alternative to Li for energy storage applications due to its rich abundance (Na), low-cost, high theoretical capacities as well as similar redox potentials ($E_{Na+/Na}$ = –2.7 V) to lithium ($E_{Li+/Li}$ = –3.0 V) (Yabuuchi et al. 2014). Similar to Li-HSCs, sodium ion hybrid supercapacitors (Na-HSCs) employ an insertion-type sodium ion battery (SIB) electrode as anode and high surface area carbonaceous SC electrode as cathode. However, larger ionic radius of Na^+ (1.02 Å) in comparison with Li^+ (0.76 Å) usually results in more sluggish reaction kinetics with lower power/energy density (Mei et al. 2016).

The electrochemical performance of Li-HSCs or Na-HSCs is closely related to chemical compositions and nanostructure design of both cathode and anode. As discussed above, graphene has gained much attention in energy storage and conversion (Geim 2009). In most cases, graphene exhibits superior electrochemical performance to other carbon materials (e.g., activated carbon, graphite and carbon nanotubes). Moreover, GSs could be chemically tailored with defects (holes and heteroatom

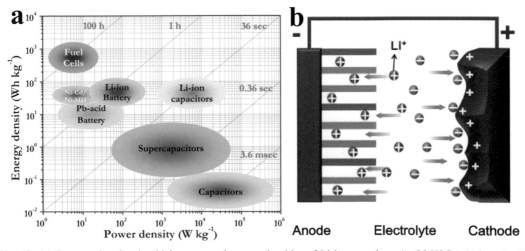

Fig. 12. (a) Ragone plot showing high power and energy densities of Li-ion capacitors (or Li-HSCs) (Adapted with modification from Aravindan et al. 2014). (b) Electrochemical reaction mechanism of Li-HSCs (Adapted with modification from Ma et al. 2015).

doping) and various functional groups, resulting in various graphene derivative materials, such as holy graphene (Wang et al. 2016), nitrogen doped graphene (Wu et al. 2011), GO and rGO (Dreyer et al. 2010). Furthermore, GSs could be assembled into 1D fibers, 2D films and 3D aerogels with novel properties and structures via versatile chemical routes, highlighting the tailored structures and properties for energy storage applications (Huang et al. 2012, Wu et al. 2012c, Xu et al. 2015a,b).

In this section, we overview the recent advances of the state-of-the-art graphene containing electrodes for Li-HSCs and Na-HSCs, and discuss the importance of the pore, doping, assembly, hybridisation of different nanostructure design in improving electrochemical performance, and the major roles of graphene as a superior active material and ultrathin 2D flexible support are highlighted in both Li-HSCs and Na-HSCs.

5.2 Graphene-based Materials for Li-HSCs

5.2.1 Graphene Electrodes

In comparison with graphite, GSs possess high specific surface area and porous nanostructures for ion storage and efficient transport, therefore, graphene electrodes usually exhibit enhanced electrochemical performance. Zhou and coworkers compared the electrochemical performances of pre-lithiated GSs and conventional graphite as anodes for Li-HSCs (Ren et al. 2014). They found that Li-HSCs with pre-lithiated graphene anode showed a specific capacitance of 168.5 F g^{-1} with 74 per cent capacitance retention at 400 mA g^{-1} after 300 cycles, and delivered a maximum power density of 222.2 W kg^{-1} at an energy density of 61.7 Wh kg^{-1}, higher than that of conventional graphite anodes ($\sim$ 113 W kg^{-1}, $\sim$ 20 Wh kg^{-1}) (Ren et al. 2014).

The electrochemical performance of rGO is closely relative to its surface functional groups. As an example, urea-reduced GO could increase the specific energy density by 37.5 per cent compared to hydrazine-reduced GO counterparts as cathodes in Li-HSCs, which resulted from the reversible Li binding with the amide functional groups generated from urea reduction process (Lee et al. 2012).

Porous graphene can work as cathode materials in Li-HSCs. For example, chemically activated graphene could show ultrahigh specific surface area up to 3,100 m^2 g^{-1} with abundant micro-, meso-porous structures for ion storage (Zhu et al. 2011), when it was combined with a graphite anode. The resulting Li-HSCs yielded high specific capacitance of 266 F g^{-1} and energy density of 53.2 Wh kg^{-1}, which was at least five times higher than symmetrical supercapacitors, and greater than lead-acid batteries (Stoller et al. 2012).

GSs could be easily assembled into 3D porous graphene macroform (PGM), which shows excellent properties with high wettability, rich porous structures and large surface area, and provides more conductive channels for the access of the electrolyte ions onto the graphene surface. Ye et al. reported that PGM-based symmetric SCs exhibited good rate performance (Fig. 13(a,b)), confirmed by 60 per cent capacitance retention as the current densities increased 40 times (Fig. 13c). When the PGM was assembled with $Li_4Ti_5O_{12}$/carbon (LTO/C) anode, the resulting Li-HSCs could deliver high energy density of 72 Wh kg^{-1} at a power density of 650 W kg^{-1}, higher than that of commercial available activated carbon YP-17D based Li-HSCs (35 Wh kg^{-1}) (Fig. 13d) (Ye et al. 2015). Similar result was also demonstrated in high-energy and high-power Li-HSCs based on TiO_2 nanobelt array as anode and graphene hydrogel as cathode (Wang et al. 2015a,b).

Considering the fact that there are big differences of various graphene products in terms of chemical and physical properties, different graphene products could also be applied in one Li-HSC. As an example, Kang reported all-graphene based Li-HSCs, in which the functionalised graphene and pre-lithiated rGO were used as cathode and anode, respectively. As a result, Li-HSCs exhibited exceptionally high energy density of 130 Wh kg^{-1} at a power density of 2,150 W kg^{-1} in two-electrodes systems (Kim et al. 2014).

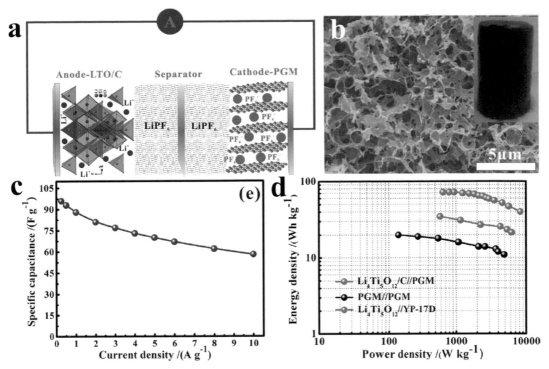

Fig. 13. (a) Illustration of Li-HSCs consisting of PGM cathode and LTO/C anode. (b) SEM image of PGM. (c) Specific capacitance at different current densities of the PGM-based supercapacitors. (d) Ragone plot of LTO/C//PGM Li-HSCs, PGM//PGM supercapacitors, and LTO//YP-17D Li-HSCs (Adapted with modification from Ye et al. 2015).

5.2.2 Graphene-based Hybrid Anodes

Generally, most anode materials with high capacity are not conductive, e.g., metal oxides, and their intercalation reactions are much slower than the surface reactions, and thus the imbalance in kinetics between cathode and anode hinders the fast development of Li-HSCs. Therefore, improving the intercalation reaction kinetics of anodes is an urgent task to achieve high-performance Li-HSCs. To this end, GSs have been used to effectively improve the conductivity and structure of anode materials, facilitating fast intercalation reactions and structural stability. For instance, high-capacitance molybdenum dioxide (MoO_2) nanoparticles were decorated on the surface of GSs, which served as an inhibitor to restrict the crystal growth and agglomeration of MoO_2 nanoparticles. The resulting hybrid Li-HSCs exhibited high specific capacitance up to 624.0 F g^{-1} at 50 mA g^{-1}, higher than that of pure MoO_2 electrode (269.2 F g^{-1}) (Han et al. 2013). LTO is another appealing anode material for Li-HSCs due to its stable operating voltage plateau and the negligible volume change during charge/discharge cycles. However, LTO suffers from poor electrical conductivity and low Li$^+$ diffusion seriously, hindering its application in Li-HSCs (Ye et al. 2015). When LTO was composited with GSs, the fabricated graphene-LTO (G-LTO) anode disclosed higher reversible capacity in comparison with the theoretical capacity (175 mAh g^{-1}) of pure LTO at low rates, and enhanced rate capability. Moreover, when combined with a graphene-sucrose (G-SU) cathode, the Li-HSC displayed ultrahigh energy density of 95 Wh kg^{-1} at a rate of 0.4 C (2.5 h) over a wide voltage range (0–3 V), and still retained an energy density of 32 Wh kg^{-1} at a high rate of 100 C, demonstrating the great role of graphene in boosting electrochemical performance of LTO (Leng et al. 2013).

Some metal oxides with capacitive-type Li^+ intercalation characteristics are promising anode materials in Li-HSCs. For example, the unique open channels of NbO_x sheets can effectively reduce the energy barrier and facilitate the local charge transfer between lithium and oxygen structures, endowing orthorhombic Nb_2O_5 (T-Nb_2O_5) capacitive-type Li^+ intercalation characteristics. However, the poor electric conductivity (3.4×10^{-6} S cm^{-1}) seriously limits its electrochemical utilisation and high-rate capability. When GSs were employed to hybridize with T-Nb_2O_5, and the as-prepared T-Nb_2O_5/graphene paper showed nanoporous layer-stacked structure with good ionic-electric conductive pathways. As a consequence, they presented high energy density of 47 Wh kg^{-1} and power density of 18 kW kg^{-1} (Kong et al. 2015).

5.3 Graphene-based Materials for Na-HSCs

Similar to Li-HSCs, one of the obstacles to develop Na-HSCs is the imbalance of kinetics from different charge storage mechanisms between the sluggish faradaic anode and the rapid non-faradaic capacitive cathode. To solve this, GSs with excellent properties were introduced to improve the electrochemical performance of sluggish faradaic anode. As an example, the aggregation of Nb_2O_5@C nanoparticles can be significantly inhibited during the insertion/extraction of Na-ions in the presence of rGO (Fig. 14(a–c)). Therefore, in comparison with Nb_2O_5@C nanoparticles, Nb_2O_5@C/rGO-50 showed not only considerably superior capacity but also excellent rate capability, as shown in Fig. 14d. Moreover, Na-HSCs using Nb_2O_5@C/rGO anode and activated carbon (MSP-20) cathode delivered high energy density of 6 Wh kg^{-1} and power densities of 2.1 kW kg^{-1}, outperforming those of similar Li-HSCs and Na-HSCs (Fig. 14f) (Lim et al. 2016). It is pointed out that solid electrolyte for Na-HSCs could effectively avoid the safety issues of liquid electrolytes, such as the flammability, possibility of leakage and internal short-circuits. For this purpose, the quasi-solid-state Na-HSCs constructed based on nanoporous carbon as negative electrode and macroporous graphene as positive electrode, using sodium ion conducting gel polymer as electrolyte could operate well at a cell voltage as high

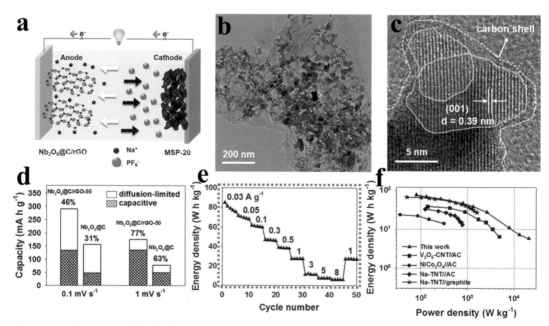

Fig. 14. (a) Illustration of Na-HSCs consisting of MSP-20 cathode and Nb_2O_5@C/rGO-50 anode. (b) TEM and (c) HR-TEM images of Nb_2O_5@C/rGO-50. (d) Comparison of the capacities of Nb_2O_5@C/rGO-50 and Nb_2O_5@C NPs at 0.1 and 1.0 mV s^{-1}. (e) Rate capability of Na-HSCs obtained at various current densities. (f) Ragone plot of the Na-HSCs (Adapted with modification from Lim et al. 2016).

as 4.2 V, and unprecedented energy density of 168 Wh kg^{-1}, which was well comparable with those of Li-HSCs (Wang et al. 2015a,b).

6. Graphene-based Fibers for Supercapacitors

With the rapid development of miniaturised, light-weight, flexible and wearable electronic devices, flexible energy storage devices are highly required in the integrated systems. Among different systems, fiber-based supercapacitors (FSCs) have been considered as one of the most promising candidates because of their significant advantages in power and energy densities, and unique properties of being flexible and wearable, lightweight, low-cost, and environmentally friendly (Yu et al. 2015). So far, great efforts have been devoted to FSCs to maximise their performance. In particular, developing new-type fiber-based electrode materials for flexible energy storage devices is a very important research direction. By virtue of the unique structure, excellent conductivity, favorable mechanical and outstanding electrochemical properties, graphene-based fibers (GFs) are expected to be a novel class of flexible fiber electrode materials for assembling high performance supercapacitors.

6.1 Preparation of Graphene Fibers

Recently, various strategies, such as wet-spinning (Xu et al. 2013), hydrothermal treatment (Yu et al. 2014), electrophoretic assembly (Jang et al. 2012), film conversion (Cruz-Silva et al. 2014), and substrate-assisted reduction and assembly (Fig. 15) (Hu et al. 2013), have been developed

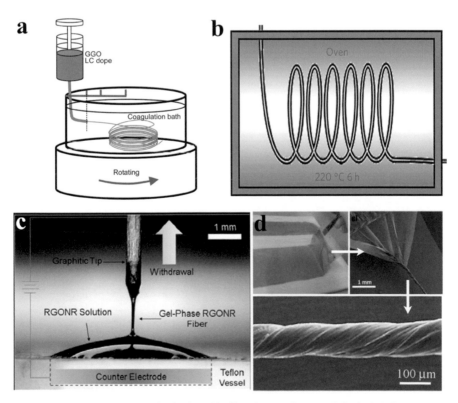

Fig. 15. (a) Wet-spinning set-up showing the fabrication of the fibers in a rotating coagulation bath (Adapted with modification from Xu et al. 2013). (b) The fabrication of GFs in dimensionally confined spaces using a fused silica capillary column (Adapted with modification from Yu et al. 2014). (c) A photograph showing the formation of gel-phase GFs on the graphitic tip drawn up with a certain electric field applied (Adapted with modification from Jang et al. 2012). (d) The fabrication of GO fibers twisted a long-strip GO film, which was cut from a GO film after evaporating the solvent from GO solution (Adapted with modification from Cruz-Silva et al. 2014).

to fabricate one-dimensional GFs. Among them, the wet-spinning technology usually allows the massive and continuous production of GO fibers from liquid crystalline GO dispersions. Meanwhile, GO liquid-crystalline solution can serve as a useful host for various foreign species, such as CNTs, metal nanowires, and conducting polymers, therefore, numerous graphene-based composite fibers with multifunctional properties have been fabricated. Furthermore, by modifying the structure of the spinneret and coagulant composition, hollow, core-shell, multiplied yarn, varying cross-section shapes (such as triangle, Y-shaped, pentagon, trefoil, four leaf, fan shaped), and porous GFs can be fabricated (Chen et al. 2016, Dong et al. 2012, Xu et al. 2015a,b). The concentration of GO spinning solution, the size of GO sheets, the composition of coagulation bath, and the techniques of spinning, play important roles in the formation and properties of GFs. Meanwhile, this method also has some disadvantages, such as extra post processing for transforming GO fibers into GFs, and the difficulty of completely removing residues of inorganic or organic ions from the coagulation bath on the surfaces of GFs.

Compared to the wet-spinning method, the dimensionally confined hydrothermal strategy can overcome these disadvantages, and the resulting fibers have relatively high electrical conductivity without further reduction treatment. In addition, foreign species can be also introduced into the GO solution for the fabrication of hybrid fibers. However, the limitation of this approach lies in the size and length of the fibers which is mainly depended on the pre-designed pipeline. This is not suitable for its practical application. In addition, it is difficult to synthesise GO fibers using hydrothermal strategy, and the process includes sealing tube terminals and heating for hours, making it inappropriate for continuous GF spinning. The methods of substrate-assisted reduction and assembly, film conversion, and electrophoretic assembly are also alternative approaches for the fabrication of high-performance GFs.

6.2 Graphene-based Fiber Supercapacitors

The configuration of the FSCs can be classified into three types of parallel, twisted and coaxial design (Fig. 16) (Yu et al. 2015). The parallel FSCs can be readily scaled up by integrating the fiber electrodes on the planar substrate to meet specific power and energy requirement for microscale electronics.

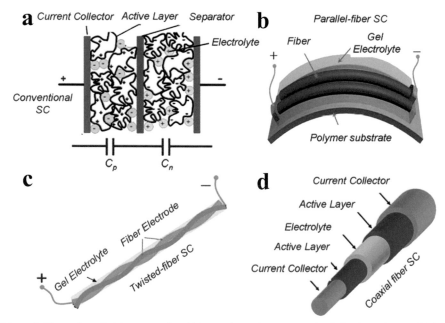

Fig. 16. Schematic illustration of (a) the conventional 2D two-electrode supercapacitor in its charged state with its equivalent circuit shown below, and (b–d) three typical device configurations of FSCs: (b) two parallel FSCs, (c) two twisted FSCs, and (d) single coaxial FSCs (Adapted with modification from Yu et al. 2015).

However, the use of a planar substrate may restrict their wearable applications of FSCs. In a sharp contrast, both the twisted and coaxial fiber configurations do not require a planar substrate. In general, the twisted FSCs can be produced by twisting two fiber electrodes together with a separator or solid state electrolyte between them, while the coaxial FSCs are fabricated by layer-by-layer assembling of a core fiber electrode, a separator or solid state electrolyte, and an outer electrode layer. So far, various GF electrodes for FSCs have been explored, including pure GFs, graphene-conducting polymer composite fibers, and graphene-metal oxide composite fibers.

Gao and his coworkers developed a coaxial wet-spinning assembly strategy to prepare polyelectrolyte-wrapped GFs using a double-inlet spinneret (Kou et al. 2014), in which GO was fed to the inner spinning dope and sodium carboxymethyl cellulose (CMC) as an ionic conductor of polyelectrolyte was fed to the sheath. The resulting core-sheath fibers efficiently prevented the twisting electrodes from short circuit and provided access for ion penetration. Further, the polyelectrolyte sheath enabled the supercapacitors exempt from the electrolyte leakage when the device was deformed. Thereby, the core-sheath fiber was flexible and robust enough to sustain the long-term and repeated deformation, and the capacitance of graphene FSCs was slightly fluctuated under continuous bending for 1,000 cycles.

Conducting polymers such as PANI, PPy, and poly (3,4-ethylenedioxythiophene) are widely used to deliver pseudocapacitance in supercapacitors. Compositing with graphene has been proved an effective way to enhance the mechanical strength and capacitance performance. Wang et al. fabricated high-performance fiber-shaped supercapacitors by doping PANI fibers with GO sheets (Wang et al. 2009). The resulting FSCs fabricated by twisting two composite fibers produced a specific capacitance of 531 F g^{-1}, which was attributed to the synergetic effect of the PANI and GO. Further, transition metal oxides such as MnO_2 and Co_3O_4 have been intensively studied as flexible and high energy density electrode materials. To overcome the disadvantages of their low electron conductivity, the conductive additives, such as graphene, are mixed with metal oxides to enhance electrochemical performance. Currently, graphene-metal oxide composite fibers have been explored by several groups for high electrochemical capacitive performance. For example, Yan et al. (Yan et al. 2010) found that MnO_2 remarkably enhances the porous GFs performance with respect to both specific capacitance (12.4 mF cm^{-2}) and cycling stability. Li et al. (Li et al. 2013) assembled a symmetric solid-state supercapacitor with two GFs immobilised by MnO_2 nanoparticles, which showed a maximum areal capacitance of 42.02 mF cm^{-2} at 10 mVs^{-1}, much higher than that of pure GFs of 1.4 mF cm^{-2}.

Graphene FSCs with its unique advantage have attracted widespread attention, and gained much progress. However, the R and D of these graphene FSCs is still facing the challenges. First, large scale production of graphene fibers with high conductivity and quality is still unsolved. Graphene FSCs exhibit high performance in small-scale tests. Considering its applications at an industrial level, there is still a lot of work to do to achieve high performance, chemical and electronic stable devices. Second, besides the conductivity and capacitance, the flexibility and stretchability of GFs are also the crucial factors for manufacturing flexible supercapacitors. Third, at present, most graphene FSCs, configured in symmetric type with limited voltage window, cannot meet the requirement of high energy density. Therefore, the development of asymmetric graphene FSCs in organic electrolytes is one of the important ways to improve performance. Fourth, considering the practical application of graphene FSCs, electrochemical stability is one key factor that must be considered. Fifth, in wearable applications, safety is the greatest concern for users. Now the electrolytes of graphene based FSCs in use mostly were corrosive. To consider the future development of device, developing green nontoxic electrolyte to improve the safety of the device will be one of the most key directions. Last but not the least; large scale production of FSCs is one of the biggest challenges in the future development.

7. Porous Graphene Film Electrodes for Micro-Supercapacitors

7.1 Micro-Supercapacitors

The current trend of multifunctional, portable and miniaturised integrated circuitry indeed lead to the ever-increasing demand for efficient, microscale and compact on-chip energy storage systems (Kyeremateng et al. 2017, Qi et al. 2017, Wu et al. 2014). However, the conventional energy storage systems, for example, electrolytic capacitors, supercapacitors and lithium ion batteries, are hard to miniaturise in size and shape (Wu et al. 2015, Xiong et al. 2014). The emerging on-chip micro-supercapacitors (MSCs) constructed with two electrodes, current collector, separator and electrolyte on one single substrate in an interdigitated planar form are well compatible with microscale integrated circuits. Moreover, the planar MSCs show numerous desirable properties, e.g., high-power density, excellent rate capability, high-frequency response and ultra-long lifetime. These superior characteristics, crucial for future application with miniaturised electronics, are attributed to the rapid accessibility of the electrolyte to electrodes in in-plane architecture and low ion transport resistance because of the absence of separator and short interspace between electrodes. Recently, graphene is demonstrated to be of great importance for MSCs owing to the superior match of graphene and in-plane geometry for maximising the performance. In particular, the channels in porous graphene, including micro-, meso-, and macro-porous structure in either graphene planes or between graphene sheets, can significantly enlarge the accessible surface area of the active electrodes, reduce ion transport resistance, and eventually improve the electrochemical performance of MSCs.

7.2 Porous Graphene Films for MSCs

To enhance the energy densities and rate capabilities of graphene-based MSCs, a rational way is to facilitate the electrolyte accessibility to the active electrodes and avoid the aggregation and restacking of graphene sheets. Kaner's group (El-Kady et al. 2012, El-Kady and Kaner 2013) reported the fabrication of porous graphene films with robust mechanical property, high electrical conductivity and large specific surface area of 1,520 m^2 g^{-1}. The produced porous films were fabricated by the direct reduction of the tight restacked graphite oxide films using a standard LightScribe DVD optical drive, which simplifies the fabrication process and leads to cost-effective MSCs (Fig. 17(a–c)). This technique can be readily scalable, since 112 MSCs were produced within 30 min or less (Fig. 17 (d, e)). The colour is changed from the brown yellow of GO film to black (Fig. 17f), revealing the efficient reduction of GO into graphene by direct irradiation of laser. The cross-section SEM image exhibited the formation of the unique porous network structure between graphene sheets. By virtue of open inner porous network and high conductivity, laser-scribed graphene MSCs (LSG-MSCs) with 16 electrodes per unit area displayed superior electrochemical performance in comparison of sandwich-type LSG supercapacitors, using poly(vinyl alcohol)/H_2SO_4 (PVA/H_2SO_4) as electrolyte. For example, LSG-MSCs delivered a volumetric capacitance of 3.05 F cm^{-3} at 16.8 mA cm^{-3}, which is much higher than that of sandwich-type LSG supercapacitors (~ 0.8 F cm^{-3}) (Fig. 17g). Furthermore, LSG-MSCs maintained 60 per cent of initial capacitance even operated at an ultrahigh current density of 1.84×10^4 mA cm^{-3}. This current density is at least two orders of magnitude higher than the conventional supercapacitors, indicative of excellent rate capability and high power density. Bode plots (Fig. 17h) further confirmed superior power performance of LSG-MSCs, displaying fast frequency response with a small resistance-capacitor (RC) time constant of 19 ms, lower than commercial active carbon supercapacitors, and well comparable to electrolytic capacitors. LSG-MSCs delivered an ultrahigh stacked power density of ~ 200 W cm^{-3} tested at ion liquid electrolyte with a cell voltage of 2.5 V (Fig. 17i). These outstanding results are well explained by (i) significant reduction of mean ionic diffusion pathway between two microelectrodes and (ii) miniaturised resistance of ion diffusion from electrolyte to electrode surface due to open porous network structure.

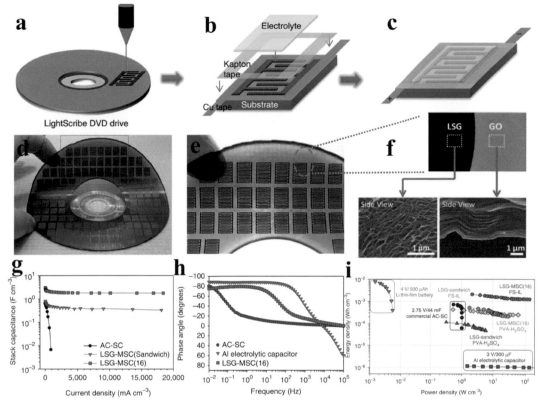

Fig. 17. Fabrication and electrochemical performance of LSG-MSCs. (a–c) Schematic illustration exhibiting the fabrication process for LSG-MSCs: (a) Interdigital graphene circuits were produced by laser, (b) Copper and Kapton tape were applied to define the interdigitated area, (c) Electrolyte overcoat, (d, e) Scalable fabrication of LSG-MSCs, (f) The colour of GO film changed from brown to black. Cross-section SEM images of the stacked GO film and exfoliated LSG film, (g) Specific capacitance of sandwich and interdigitated structure supercapacitors, (h) Impedance phase angle versus frequency, (i) Ragone plot of volumetric energy and power density of LSG-MSCs compared with commercial energy storage devices (Adapted with modification from El-Kady et al. 2013).

To further understand the importance of porosity for ion diffusion, Wu et al. (Wu et al. 2015) reported the formation of three-dimensional interconnected porous structure for alternating current line-filter that required less than 8.3 ms at 120 Hz. The porous network was immediately established through the reduction of GO in patterned metal interdigital electrodes at room temperature. Afterward, the MSCs composed of the patterned interdigitated graphene microelectrodes showed a phase angle of 75.4° at 120 Hz and a low RC time constant of 0.35 ms even based on ~ 100 μm-thick electrodes. In addition, the characteristic frequency at phase angle of –45° is 529 Hz. Correspondingly, the relaxation time constant (the minimum time required to release all the energy from the device with an efficiency of more than 50 per cent) was calculated to 1.9 ms, indicative of outstanding power density.

7.3 Porous Graphene Hybrid Film for MSCs

Porous graphene microelectrodes have demonstrated improved performance in terms of power density. However, further enhancements in energy density are required that do not sacrifice high power energy for actual application. Porous graphene-based hybrid films, as a composite film, have not only porous network for the accessibility of electrodes but also the synergistic effect of other active materials to offer high energy density. To improve energy density of MSCs based on porous graphene materials, Lee's group (Chang et al. 2015) reported a simple route of porous nanochanneled and poly (diallyldimethylammonium chloride) PDDA-mediated RGO (nc-PDDA-Gr) film using Cu(OH)$_2$ nanowires as a sacrificial template (Fig. 18a). The fabricated nc-PDDA-Gr with the help of vacuum filtration possessed high packing density and efficient ion transport pathways. The porous hybrid films were used to fabricate MSCs with photolithography technique, followed by O$_2$ plasma etching (Fig. 18(b,c)). As a result, the nc-PDDA-Gr film for MSCs delivered high volumetric capacitance of 348 F cm^{-3}, and high areal capacitance of 409 mF cm^{-2} for 11.8 µm-thick electrodes. Additionally, the fabricated devices exhibited high energy density of 6.7 mWh cm^{-3} and high power density of 20 W cm^{-3} as well as superior rate capability. These outstanding results are ascribed to the unique structural features of open porous nanochannels between graphene sheets, allowing for rapid ion diffusion and high packing density.

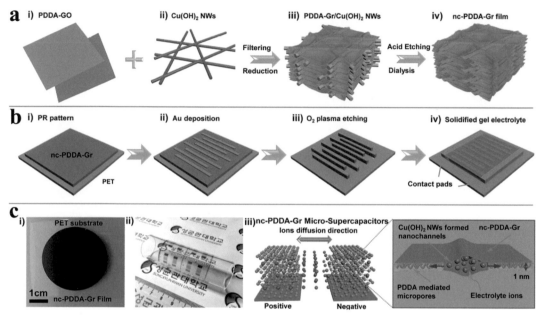

Fig. 18. Schematic of the fabrication of nc-PDDA-Gr MSCs. (a) The fabricated process of the nc-PDDA-Gr film, (b) MSCs fabrication process based on Au deposition and O$_2$ plasma etching, (c) The ion transport pathway in the as-fabricated MSCs (Adapted with modification from Chang et al. 2015).

Conclusion and Perspective

We overview the recent progress of different emerging graphene-based porous materials such as pure porous graphene, 2D graphene-based porous hybrid nanosheets, 3D graphene-based porous framework, graphene fibers and porous graphene films, as high-performance electrode candidates applied in various supercapacitors, including EDLCs, pseudocapacitors, ASCs, Li-HSCs, Na-HSCs, FSCs, and MSCs. In comparison of pure graphene, these porous graphene materials have demonstrated to present improved capacitance, rate capability, energy and power densities, and cyclability. Nevertheless, further improvement in packing density, electrical conductivity to reduce the contact resistance among graphene sheets and optimised hierarchical porosity to allow efficient and rapid ion transport and maximise the accessible surface area are highly needed to endow high power density and high energy density of these porous graphene-based supercapacitors.

Despite of huge advanced progress in this field, some key challenging issues are still not answered well as follows. (i) The first challenging issue is to prepare high-quality graphene products with controllable number of layers, surface properties, lateral sizes, defects and conductivity, which is the prerequisite for further processing. (ii) The second issue needed to be addressed is how to conduct large-scale fabrication of graphene-based materials with desired porous structures and high SSA at low cost for targeted supercapacitor applications. For example, highly porous and conductive graphene powder synthesised by chemical activation could be the best choice for high-energy bulk supercapacitors, while nanoporous yet continuous graphene film is possibly the right one for flexible supercapacitors and MSCs, and similarly, graphene-based fibers for 1D wearable FSCs. Therefore, different-type graphene products should be reasonably chosen taking into consideration of their corresponding supercapacitors. (iii) As a versatile building block, porous graphene and their derivatives (2D porous nanosheets or films, 3D porous frameworks) could construct various hybrid materials with specific functionalities, so a key issue considered is how to simultaneously achieve high volumetric and gravimetric capacitances of these supercapacitors by rational design and processing of these nanoporous and compact graphene-based electrodes. (iv) Systematical screening and integrity of different main device components (e.g., graphene electrodes, electrolyte, current collectors, separator, and package) for high-performance supercapacitors are greatly challenging. Specifically, enhancing the overall performance of supercapacitors is dependent not only on the intrinsic properties of graphene electrode materials, but also on the reasonable design, optimisation, and compatible combination of each individual device component into single cell.

Acknowledgments

This work was financially supported by the National Natural Science Foundation of China (Grant 51572259), National Key R&D Program of China (Grant 2016YFB0100100 and 2016YFA0200200), Natural Science Foundation of Liaoning Province (Grant 201602737), Recruitment Program of Global Expert (1000 Talent Plan), DICP (DICP ZZBS201708), China Postdoctoral Science Foundation (Grant 2016M601348, 2016M601349, 2017T100188), Exploratory Research Program of Shaanxi Yanchang Petroleum (Group) Co., LTD & DICP.

References

Aravindan, V., J. Gnanaraj, Y.S. Lee and S. Madhavi. 2014. Insertion-type electrodes for nonaqueous Li-ion capacitors. Chem. Rev. 114: 11619–11635.

Chang, J., S. Adhikari, T.H. Lee, B. Li, F. Yao, D.T. Pham et al. 2015. Leaf vein-inspired nanochanneled graphene film for highly efficient micro-supercapacitors. Adv. Energy Mater. 5: 1500003.

Chen, J., J. Xu, S. Zhou, N. Zhao and C.-P. Wong. 2015. Template-grown graphene/porous Fe_2O_3 nanocomposite: A high-performance anode material for pseudocapacitors. Nano Energy 15: 719–728.

Chen, L., Y. Liu, Y. Zhao, N. Chen and L. Qu. 2016. Graphene-based fibers for supercapacitor applications. Nanotechnology 27: 032001.

Chen, T. and L.M. Dai. 2013. Carbon nanomaterials for high-performance supercapacitors. Mater. Today 16: 272–280.

Cruz-Silva, R., A. Morelos-Gomez, H.-i. Kim, H.-k. Jang, F. Tristan and S. Vega-Diaz et al. 2014. Super-stretchable graphene oxide macroscopic fibers with outstanding knotability fabricated by dry film scrolling. ACS Nano 8: 5959–5967.

Dong, X.-C., H. Xu, X.-W. Wang, Y.-X. Huang, M.B. Chan-Park and H. Zhang et al. 2012. 3D graphene–cobalt oxide electrode for high-performance supercapacitor and enzymeless glucose detection. ACS Nano 6: 3206–3213.

Dong, Z., C. Jiang, H. Cheng, Y. Zhao, G. Shi and L. Jiang et al. 2012. Facile fabrication of light, flexible and multifunctional graphene fibers. Adv. Mater. 24: 1856–1861.

Dreyer, D.R., S. Park, C.W. Bielawski and R.S. Ruoff. 2010. The chemistry of graphene oxide. Chem. Soc. Rev. 39: 228–240.

El-Kady, M.F., V. Strong, S. Dubin and R.B. Kaner. 2012. Laser scribing of high-performance and flexible graphene-based electrochemical capacitors. Science 335: 1326–1330.

El-Kady, M.F. and R.B. Kaner. 2013. Scalable fabrication of high-power graphene micro-supercapacitors for flexible and on-chip energy storage. Nat. Commun. 4: 1475.

Geim, A.K. 2009. Graphene: Status and prospects. Science 324: 1530–1534.

Han, P., W. Ma, S. Pang, Q. Kong, J. Yao and C. Bi et al. 2013. Graphene decorated with molybdenum dioxide nanoparticles for use in high energy lithium ion capacitors with an organic electrolyte. J. Mater. Chem. A 1: 5949–5954.

Han, S., D.Q. Wu, S. Li, F. Zhang and X.L. Feng. 2014. Porous graphene materials for advanced electrochemical energy storage and conversion devices. Adv. Mater. 26: 849–864.

Hu, C.G., X.Q. Zhai, L.L. Liu, Y. Zhao, L. Jiang and L.T. Qu. 2013. Spontaneous reduction and assembly of graphene oxide into three-dimensional graphene network on arbitrary conductive substrates. Sci. Rep. 3: 2065.

Huang, X., X. Qi, F. Boey and H. Zhang. 2012. Graphene-based composites. Chem. Soc. Rev. 41: 666–686.

Jang, E.Y., J. Carretero-Gonzalez, A. Choi, W.J. Kim, M.E. Kozlov and T. Kim et al. 2012. Fibers of reduced graphene oxide nanoribbons. Nanotechnology 23: 8.

Jiang, L. and Z. Fan. 2014. Design of advanced porous graphene materials: From graphene nanomesh to 3D architectures. Nanoscale 6: 1922–1945.

Kim, H., K.-Y. Park, J. Hong and K. Kang. 2014. All-graphene-battery: Bridging the gap between supercapacitors and lithium ion batteries. Sci. Rep. 4: 5278.

Kong, L., C. Zhang, J. Wang, W. Qiao, L. Ling and D. Long. 2015. Free-standing T-Nb_2O_5/graphene composite papers with ultrahigh gravimetric/volumetric capacitance for li-ion intercalation pseudocapacitor. ACS Nano 9: 11200–11208.

Kotz, R. and M. Carlen. 2000. Principles and applications of electrochemical capacitors. Electrochim. Acta 45: 2483–2498.

Kou, L., T.Q. Huang, B.N. Zheng, Y. Han, X.L. Zhao and K. Gopalsamy et al. 2014. Coaxial wet-spun yarn supercapacitors for high-energy density and safe wearable electronics. Nat. Commun. 5: 3754.

Kyeremateng, N.A., T. Brousse and D. Pech. 2017. Microsupercapacitors as miniaturized energy-storage components for on-chip electronics. Nat. Nanotechnol. 12: 7–15.

Lee, J.H., W.H. Shin, M.-H. Ryou, J.K. Jin, J. Kim and J.W. Choi. 2012. Functionalized graphene for high performance lithium ion capacitors. ChemSusChem 5: 2328–2333.

Leng, K., F. Zhang, L. Zhang, T. Zhang, Y. Wu and Y. Lu et al. 2013. Graphene-based Li-ion hybrid supercapacitors with ultrahigh performance. Nano Research 6: 581–592.

Li, W., J. Liu and D. Zhao. 2016. Mesoporous materials for energy conversion and storage devices. Nat. Rev. Mater. 1: 16023.

Li, X., T. Zhao, Q. Chen, P. Li, K. Wang and M. Zhong et al. 2013. Flexible all solid-state supercapacitors based on chemical vapor deposition derived graphene fibers. Phys. Chem. Chem. Phys. 15: 17752–17757.

Lim, E., C. Jo, M.S. Kim, M.-H. Kim, J. Chun and H. Kim et al. 2016. High-performance sodium-ion hybrid supercapacitor based on Nb_2O_5@carbon core–shell nanoparticles and reduced graphene oxide nanocomposites. Adv. Funct. Mater. 26: 3711–3719.

Lin, Y., X.G. Han, C.J. Campbell, J.W. Kim, B. Zhao and W. Luo et al. 2015. Holey graphene nanomanufacturing: Structure, composition, and electrochemical properties. Adv. Funct. Mater. 25: 2920–2927.

Liu, J., L. Zhang, H.B. Wu, J. Lin, Z. Shen and X.W. Lou. 2014. High-performance flexible asymmetric supercapacitors based on a new graphene foam/carbon nanotube hybrid film. Energy Environ. Sci. 7: 3709–3719.

Liu, S., P. Gordiichuk, Z.-S. Wu, Z. Liu, W. Wei and M. Wagner et al. 2015. Patterning two-dimensional free-standing surfaces with mesoporous conducting polymers. Nat. Commun. 6: 8817.

Ma, Y., H. Chang, M. Zhang and Y. Chen. 2015. Graphene-based materials for lithium-ion hybrid supercapacitors. Adv. Mater. 27: 5296–5308.

Mei, Y., Y. Huang and X. Hu. 2016. Nanostructured Ti-based anode materials for na-ion batteries. J. Mater. Chem. A 4: 12001–12013.

Novoselov, K.S., A.K. Geim, S.V. Morozov, D. Jiang, Y. Zhang and S.V. Dubonos et al. 2004. Electric field effect in atomically thin carbon films. Science 306: 666–669.

Qi, D., Y. Liu, Z. Liu, L. Zhang and X. Chen. 2017. Design of architectures and materials in in-plane micro-supercapacitors: Current status and future challenges. Adv. Mater. 29: 1602802.

Ren, J.J., L.W. Su, X. Qin, M. Yang, J.P. Wei and Z. Zhou et al. 2014. Pre-lithiated graphene nanosheets as negative electrode materials for li-ion capacitors with high power and energy density. J. Power Sources 264: 108–113.

Shen, J., C. Yang, X. Li and G. Wang. 2013. High-performance asymmetric supercapacitor based on nanoarchitectured polyaniline/graphene/carbon nanotube and activated graphene electrodes. ACS Appl. Mater. Interfaces 5: 8467–8476.

Simon, P. and Y. Gogotsi. 2008. Materials for electrochemical capacitors. Nat. Mater. 7: 845–854.

Stoller, M.D., S. Murali, N. Quarles, Y. Zhu, J.R. Potts and X. Zhu et al. 2012. Activated graphene as a cathode material for Li-ion hybrid supercapacitors. Phys. Chem. Chem. Phys. 14: 3388–3391.

Sun, Y., Q. Wu and G. Shi. 2011. Graphene based new energy materials. Energy Environ. Sci. 4: 1113.

Tang, Q., M. Chen, L. Wang and G. Wang. 2015. A novel asymmetric supercapacitors based on binder-free carbon fiber paper@ nickel cobaltite nanowires and graphene foam electrodes. J. Power Sources 273: 654–662.

Wang, F., X. Wang, Z. Chang, X. Wu, X. Liu and L. Fu et al. 2015a. A quasi-solid-state sodium-ion capacitor with high energy density. Adv. Mater. 27: 6962–6968.

Wang, H., Q. Hao, X. Yang, L. Lu and X. Wang. 2009. Graphene oxide doped polyaniline for supercapacitors. Electrochem. Commun. 11: 1158–1161.

Wang, H., H. Yi, X. Chen and X. Wang. 2014. Asymmetric supercapacitors based on nano-architectured nickel oxide/graphene foam and hierarchical porous nitrogen-doped carbon nanotubes with ultrahigh-rate performance. J. Mater. Chem. A 2: 3223.

Wang, H., C. Guan, X. Wang and H.J. Fan. 2015b. A high energy and power li-ion capacitor based on a TiO_2 nanobelt array anode and a graphene hydrogel cathode. Small 11: 1470–1477.

Wang, X., L. Lv, Z. Cheng, J. Gao, L. Dong and C. Hu et al. 2016. High-density monolith of n-doped holey graphene for ultrahigh volumetric capacity of Li-ion batteries. Adv. Energy Mater. 6: 1502100.

Wu, Y., G. Gao and G. Wu. 2015. Self-assembled three-dimensional hierarchical porous V_2O_5/graphene hybrid aerogels for supercapacitors with high energy density and long cycle life. J. Mater. Chem. A 3: 1828–1832.

Wu, Z.-S., W. Ren, L. Xu, F. Li and H.-M. Cheng. 2011. Doped graphene sheets as anode materials with superhigh rate and large capacity for lithium ion batteries. ACS Nano 5: 5463–5471.

Wu, Z.-S., Y. Sun, Y.Z. Tan, S.B. Yang, X.L. Feng and K. Müllen. 2012a. Three-dimensional graphene-based macro- and mesoporous frameworks for high-performance electrochemical capacitive energy storage. J. Am. Chem. Soc. 134: 19532–19535.

Wu, Z.-S., Y. Sun, Y.Z. Tan, S.B. Yang, X.L. Feng and K. Müllen. 2012c. Three-dimensional graphene-based macro- and mesoporous frameworks for high-performance electrochemical capacitive energy storage. J. Am. Chem. Soc. 134: 19532–19535.

Wu, Z.-S., G. Zhou, L.-C. Yin, W. Ren, F. Li and H.-M. Cheng. 2012b. Graphene/metal oxide composite electrode materials for energy storage. Nano Energy 1: 107–131.

Wu, Z.-S., X. Feng and H.-M. Cheng. 2014. Recent advances in graphene-based planar micro-supercapacitors for on-chip energy storage. Natl Sci. Rev. 1: 277–292.

Wu, Z.-S., K. Parvez, S. Li, S. Yang, Z.Y. Liu and S.H. Liu et al. 2015. Alternating stacked graphene-conducting polymer compact films with ultrahigh areal and volumetric capacitances for high-energy micro-supercapacitors. Adv. Mater. 27: 4054–4061.

Wu, Z., L. Li, Z. Lin, B. Song, Z. Li and K.-S. Moon et al. 2015. Alternating current line-filter based on electrochemical capacitor utilizing template-patterned graphene. Sci. Rep. 5: 10983.

Xiong, G.P., C.Z. Meng, R.G. Reifenberger, P.P. Irazoqui and T.S. Fisher. 2014. A review of graphene-based electrochemical microsupercapacitors. Electroanalysis 26: 30–51.

Xu, Y., G. Shi and X. Duan. 2015a. Self-assembled three-dimensional graphene macrostructures: Synthesis and applications in supercapacitors. Acc. Chem. Res. 48: 1666–1675.

Xu, Y.X., Z.Y. Lin, X. Zhong, X.Q. Huang, N.O. Weiss and Y. Huang et al. 2014. Holey graphene frameworks for highly efficient capacitive energy storage. Nat. Commun. 5: 4554.

Xu, Z., H. Sun, X. Zhao and C. Gao. 2013. Ultrastrong fibers assembled from giant graphene oxide sheets. Adv. Mater. 25: 188–193.

Xu, Z. and C. Gao. 2015b. Graphene fiber: A new trend in carbon fibers. Mater. Today 18: 480–492.

Yabuuchi, N., K. Kubota, M. Dahbi and S. Komaba. 2014. Research development on sodium-ion batteries. Chem. Rev. 114: 11636–11682.

Yan, J., Z. Fan, T. Wei, W. Qian, M. Zhang and F. Wei. 2010. Fast and reversible surface redox reaction of graphene–MnO_2 composites as supercapacitor electrodes. Carbon 48: 3825–3833.

Yan, J., Z. Fan, W. Sun, G. Ning, T. Wei and Q. Zhang et al. 2012. Advanced asymmetric supercapacitors based on $Ni(OH)_2$/graphene and porous graphene electrodes with high energy density. Adv. Funct. Mater. 22: 2632–2641.

Yang, S.B., X.L. Feng and K. Müllen. 2011a. Sandwich-like, graphene-based titania nanosheets with high surface area for fast lithium storage. Adv. Mater. 23: 3575–3579.

Yang, X.R., H. Zhu, X.L. Wang and F. Yang. 2011b. Promising carbons for supercapacitors derived from fungi. Adv. Mater. 23: 2745–2748.

Ye, L., Q. Liang, Y. Lei, X. Yu, C. Han and W. Shen et al. 2015. A high performance li-ion capacitor constructed with $Li_4Ti_5O_{12}$/C hybrid and porous graphene macroform. J. Power Sources 282: 174–178.

Yoo, J.J., K. Balakrishnan, J. Huang, V. Meunier, B.G. Sumpter and A. Srivastava et al. 2011. Ultrathin planar graphene supercapacitors. Nano Lett. 11: 1423–1427.

Yu, D., K. Goh, H. Wang, L. Wei, W. Jiang and Q. Zhang et al. 2014. Scalable synthesis of hierarchically structured carbon nanotube-graphene fibres for capacitive energy storage. Nat. Nanotechnol. 9: 555–562.

Yu, D., Q. Qian, L. Wei, W. Jiang, K. Goh and J. Wei et al. 2015. Emergence of fiber supercapacitors. Chem. Soc. Rev. 44: 647–662.

Yun, S., S.O. Kang, S. Park and H.S. Park. 2014. CO_2-activated, hierarchical trimodal porous graphene frameworks for ultrahigh and ultrafast capacitive behavior. Nanoscale 6: 5296–5302.

Zhai, T., F. Wang, M. Yu, S. Xie, C. Liang and C. Li et al. 2013. 3D MnO_2-graphene composites with large areal capacitance for high-performance asymmetric supercapacitors. Nanoscale 5: 6790–6796.

Zhai, Y., Y. Dou, D. Zhao, P.F. Fulvio, R.T. Mayes and S. Dai. 2011. Carbon materials for chemical capacitive energy storage. Adv. Mater. 23: 4828–4850.

Zhang, H. 2015. Ultrathin two-dimensional nanomaterials. ACS Nano 9: 9451–9469.

Zhao, Y., J. Liu, Y. Hu, H. Cheng, C. Hu and C. Jiang et al. 2013. Highly compression-tolerant supercapacitor based on polypyrrole-mediated graphene foam electrodes. Adv. Mater. 25: 591–595.

Zheng, S., Z.-S. Wu, S. Wang, H. Xiao, F. Zhou and C. Sun et al. 2017. Graphene-based materials for high-voltage and high-energy asymmetric supercapacitors. Energy Storage Mater. 6: 70–97.

Zhu, Y., S. Murali, M.D. Stoller, K.J. Ganesh, W. Cai and P.J. Ferreira et al. 2011. Carbon-based supercapacitors produced by activation of graphene. Science 332: 1537–1541.

3

Building Porous Graphene Architectures for Electrochemical Energy Storage Devices

Yao Chen[1] and *George Zheng Chen*[1,2,3,*]

1. Introduction

Graphene is a flat monolayer of carbon atoms composing of two dimensional (2D) hexagonal lattices. It is a mother unit of zero dimensional fullerenes, one dimensional carbon nanotubes and three dimensional (3D) graphite (Geim and Novoselov 2007). Since graphene was obtained by mechanical exfoliation of highly oriented pyrolytic graphite in 2004 (Novoselov et al. 2004), a hypothesis has been overthrown that 2D crystals could not exist due to the belief that it would be thermodynamically unstable. It has been found that graphene possesses unique properties, such as large theoretical specific surface area (SSA) of 2,630 m^2/g (Peigney et al. 2001), high carrier mobility over 10,000 $cm^2/V/s$ (Zhang et al. 2005), and high Young's modulus of 1.0 TPa (Liu et al. 2007).

It is obvious that graphene monolayer is a 2D material. The question is how to define the boundary between graphene and graphite. The electronic structure is highly dependent on the number of layers. It is known that graphene monolayer and bilayer have no overlap between the conduction and valence bands, whilst the difference in band overlap between 11 or more graphene layers and bulk graphite is smaller than 10 per cent, suggesting that a structure with 10 graphitic layers or fewer can be accepted as graphene (Partoens and Peeters 2006).

Due to the excellent properties mentioned above, researchers all over the world have devoted efforts to make graphene via chemical approaches towards facile and scalable production, two of which are based on reduction of graphite oxide (Li et al. 2008, Stankovich et al. 2007) and chemical

[1] The State Key Laboratory of Refractories and Metallurgy, College of Materials and Metallurgy, Wuhan University of Science and Technology, Wuhan 430081, P. R. China.
E-mail: y.chen@wust.edu.cn
[2] Department of Chemical and Environmental Engineering, Faculty of Science and Engineering, University of Nottingham Ningbo China, Ningbo 315100, P. R. China.
[3] Department of Chemical and Environmental Engineering, Faculty of Engineering, University of Nottingham, Nottingham NG2 7RD, UK.
* Corresponding author: george.chen@nottingham.ac.uk

vapour deposition (CVD) (Li et al. 2009). Reduction of graphite oxide can produce graphene on a large scale, but the main obstacle to obtain individual graphene sheets is the restacking of graphene layers after removal of solvent due to the van der Waals force. This results in a smaller SSA, fewer active/anchoring sites, and lower electrical conductivity in the normal plane. Alternatively, better quality graphene films could be fabricated by CVD, but the yield is limited.

Since graphene was pioneered for use in supercapacitors (Stoller et al. 2008) and lithium ion batteries (Paek et al. 2009), enormous efforts have been devoted to the synthesis of graphene-based materials for application in electrochemical energy storage (EES) devices, mainly supercapacitors and rechargeable lithium batteries including lithium ion, lithium sulphur, and lithium air (oxygen) batteries. All EES devices require fast electrode process kinetics in actual applications, urging the conversion of 2D graphenes into 3D macroporous structures. Additionally, different EES devices have specific principles. For instance, supercapacitors severely rely on an electrode with high SSA; lithium sulphur batteries need a positive electrode with reservoirs to impede poly-sulfide ions shutting between the positive and negative electrodes; lithium air batteries impose the need for gas diffusion channels transporting the air to the positive electrode surface. Hence, micro-, meso- and macroporous graphene materials have been developed to meet all these requirements for efficient and effective EES devices.

Herein, principles of graphene electrode materials for supercapacitors, lithium ion batteries, lithium sulphur batteries and lithium oxygen batteries are firstly generalised. Four main synthetic methods of porous graphenes are discussed in terms of pore formation mechanisms. These are (1) microporous and mesoporous graphene powders by potassium hydroxide activation, (2) macroporous graphene aerogels by self-assembly, (3) variously porous graphenes by a porous template that is removed later, as well as (4) intercalated graphene films. Further, new trends of porous graphenes for EES devices are also be speculated.

2. Electrode Principles for Electrochemical Energy Storage Devices

There are various energy conversion and storage devices such as internal combustion engine (ICE), fly wheels, elastic springs, and various electrochemical energy storage (EES) devices whose applications are usually based on their performance characteristics. Of these, ESS devices, including rechargeable batteries, redox flow batteries, fuel cells, and supercapacitors, offer great versatility for various applications and occupy irreplaceable places in the market of, for example, personal electronic devices and communication technologies. Such commercial successes of EES devices result largely from their performance reliability, module-based scalability, low environmental impact, and convenient portability. A particular and promising trend of EES development is into the automobile market, which however depends on further improvement in terms of power capability and energy capacity. For comparison, Fig. 1 schematically presents the so called Ragone plots of various EES devices (Chen 2017), showing that ICEs still outperform all EES devices.

However, the room for technical improvement of EES devices is still large as shown in Fig. 1. For instance, fuel cells and redox flow batteries can offer very large energy capacity comparable with ICEs, whilst supercapacitors can match ICEs in power capability. It is thus desirable to combine the merits of different EES devices into one device, and a widely recognised key area for improvement is electrode materials. Thanks to their high electronic conductivity, great mechanical and chemical stability, and unique nanostructures, graphenes promise to be ideal building blocks for making high performance electrodes in EES devices.

In the current literatures, the reported EES devices with graphene based electrodes are mainly supercapacitors, lithium ion and lithium metal based rechargeable batteries as discussed in detail in the following sections.

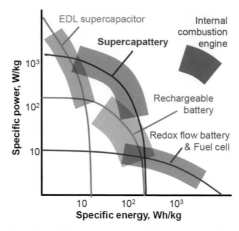

Fig. 1. Ragone plots for various electrochemical energy storage devices in comparison with internal combustion engines. Reproduced with permission from ref. (Chen 2017).

2.1 Supercapacitors

As a unique member of EES family, supercapacitors have the advantages of high power capability, long cycle life, a wide thermal operating range and low weight over rechargeable batteries, and of larger energy capacity over conventional solid state and electrolytic capacitors. In technical terms, supercapacitors play an important role in complementing or replacing some rechargeable batteries in the energy storage field. The first patent of a supercapacitor dates back to 1957 based on high surface area carbon (Becker 1957). Intrinsic supercapacitors store electric energy by potential-dependent accumulation of electrostatic charge from reversible adsorption of ions of the electrolyte into active electrode materials to form an electrical double layer (EDL) as illustrated by Fig. 2 (Pandolfo and Hollenkamp 2006). The EDL capacitance (C_{EDL}) is defined by Equation 1.

$$C_{EDL} = \frac{\varepsilon_0 \varepsilon_r A}{d} \tag{1}$$

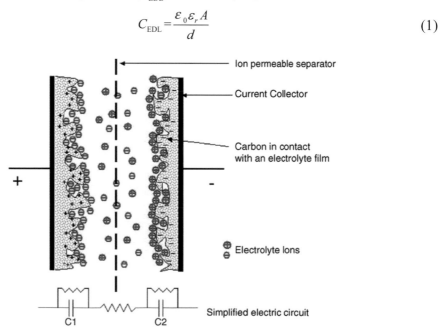

Fig. 2. Electrochemical double layer formed in the charged state of supercapacitors. Reproduced with permission from ref. (Pandolfo and Hollenkamp 2006), Copyright 2006 Elsevier.

where ε_r is the relative dielectric constant, ε_0 is the dielectric constant of the vacuum, A is the electrode surface area and d is the effective thickness of the EDL. Most EDL capacitors use electrodes made from carbon materials with high SSA to attain high capacitances.

Another type of supercapacitors, called pseudocapacitors by some researchers, is based on fast and reversible redox or Faradic reactions in the materials for charge storage. Both transition metal oxides (TMOs) and electronically conducting polymers (ECPs) can be used as active electrode materials in pseudocapacitors and can provide much higher energy densities than EDL capacitors due to the Faradic storage mechanism at molecular or atomic scales. However, Faradaic storage is always accompanied by ingression (or egression) of counter ions from (or into) the electrolyte into (or from) the electrode materials, like in rechargeable batteries, to maintain electric neutrality. As a result, repeated swelling and contraction occur within the electrodes of pseudocapacitors. Such ion movements in the solid pseudocapacitive material correspond to sluggish process kinetics, whilst the repeated stress variations upon charge-discharge cycling at molecular scales can eventually damage the material structure and result in very limited power capability and lifetimes for pseudocapacitors.

Commercially available activated carbons can possess a Brunauer-Emmett-Teller (BET) SSA in the range of 500 to 2,700 m^2/g (Inagaki et al. 2010) and have specific (or gravimetric) capacitances of 100 to 300 F/g in organic or alkaline aqueous electrolytes (Sevilla and Mokaya 2014). The area-normalised EDL capacitance of most activated carbons is not beyond 20 $\mu F/cm^2$ (Kotz and Carlen 2000, Pandolfo and Hollenkamp 2006). The relatively low area-normalised EDL capacitance was attributed to the existence of ultra-small micropores (< 1 nm) in activated carbon which were thought to be inaccessible to electrolyte ions with their solvation shells.

The specific energy of commercial supercapacitors based on porous activated carbon is about 5 Wh/kg at a specific power of 0.4 kW/kg (Burke 2007). The energy (E) is related to the total capacitance (C_T) of the supercapacitor cell and the cell voltage (U) which is dependent primarily on the electrolyte used in the device according to Equation 2.

$$E = \frac{C_T U^2}{2} \tag{2}$$

The maximum power (P_{max}) and actual power (P) of the device can be expressed by Equations 3 and 4, respectively.

$$P_{max} = \frac{U^2}{4R} \tag{3}$$

$$P = \frac{E}{t} \tag{4}$$

Where R is the total effective series resistance, including at least electrolyte separator, current collector, porous layer in contact with the current collector and other contact resistances, t is the discharge time.

Besides maximising the capacitance, also required is minimising the characteristic relaxation time constant (τ_0) for capacitors. τ_0 is the minimum time needed to discharge all the energy from the device with an efficiency of greater than 50 per cent, describing the frequency response and the frontier between the resistive and the capacitive behaviour (Taberna et al. 2003). It is associated with the diffusion length (L) and diffusion coefficient of the electrolyte (D) according to Equation 5.

$$\tau_0 = \frac{L^2}{D} \tag{5}$$

Practically made graphenes in laboratory had typical SSAs of only 400 m^2/g (Lv et al. 2009, Stankovich et al. 2007), but still exhibited capacitances of 135 ~ 279 F/g which are comparable with those of activated carbons in alkaline aqueous electrolyte at a small current densities or potential scan rates. However, the ion-accessible SSA of graphene may be increased by producing porosity on the basal plane of graphene via chemical activation. Although the conductivity of single-layer graphene along the plane direction was estimated to be 600,000 S/m, and even the restacked graphene sheets by reducing graphite oxide could reach 55,000 S/m (Mattevi et al. 2009), the electrode consisting of randomly arranged restacks of graphene sheets impedes electrolyte ions reaching the whole surface of graphene sheets at a high rate. Obviously, fast kinetics would require an electrical network that has high electronic conductivity and short ionic diffusion lengths, calling for a 3D graphene network with macroporous structure.

2.2 Rechargeable Lithium Batteries

In this chapter, the term 'rechargeable lithium batteries' is used to differentiate with the term 'lithium battery' which often refers to the primary lithium metal battery in literatures. Typical rechargeable lithium batteries include lithium ion, lithium sulphur and lithium oxygen batteries. They all have relatively higher energy density than other rechargeable batteries such as the Ni-MH, Ni-Cd and Pb-acid batteries. Rechargeable lithium batteries have been applied in portable electronic devices and promise a high potential for emerging electric vehicles and smart grid applications. TiS_2 was exploited as a positive electrode material, and used to assemble a rechargeable lithium battery with lithium as the negative electrode (Whittingham 1976). $LiCoO_2$, $LiMn_2O_4$ and $LiFePO_4$ were discovered to be stable positive electrode materials with relatively high positive potentials and conductivity (Mizushima et al. 1980, Padhi et al. 1997, Thackeray et al. 1984). Note that these materials are also often described as high voltage materials in some literatures, but it refers to the fact that when used to make the positive electrode, these materials enable high single cell voltages. After the intercalation of Li^+ ions into graphitic layers was realised, graphite had been exploited to be a new negative electrode in a commercial lithium ion battery (LIB) by Sony (Nagaura and Tozawa 1990), replacing metal lithium which has well known dendritic problem and severe safety issue. Since then, LIBs have been widely applied in portable electronic products. However, the energy density of current LIBs is limited by the intercalation chemistry occurring within the electrode materials. In order to significantly increase the energy storage capacity to meet the requirement of electric automobile, lithium sulphur (Li-S) batteries and lithium oxygen (Li-O_2) batteries are developing. The Li-S batteries use sulphur as the active material in the positive electrode and have a theoretical specific energy value of 2,567 Wh/kg, whilst the Li-O_2 batteries utilise the reduction reaction of the O_2 in the positive electrode, leading to 3,505 Wh/kg in theoretical specific energy (Choi, N.-S. et al. 2012). The configurations of all rechargeable lithium batteries are represented in Fig. 3 (Bruce et al. 2012).

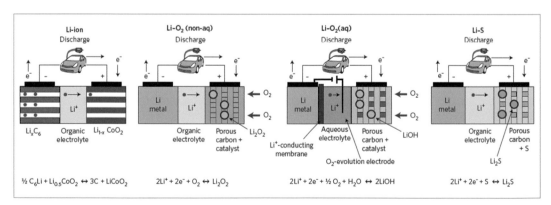

Fig. 3. Representative configurations of lithium-ion, non-aqueous and aqueous lithium oxygen, and lithium sulfur batteries. Reproduced with permission from ref. (Bruce et al. 2012), Copyright 2012 NPG.

2.2.1 Lithium Ion Batteries

A typical commercial lithium ion battery (LIB) consists of a positive electrode of $LiCoO_2$ and a negative electrode of graphite, together with an electrolyte-filled separator that allows lithium ion transfer but prevents electrodes from direct contact. The theoretical specific charge capacity of graphite is 372 mAh/g with a maximum amount of inserted lithium between graphitic layers to give rise to the so called stage I Li-graphite intercalation compound, LiC_6. Graphite materials with higher crystallinity absorb Li^+ ions at a potential closer to that of lithium metal, which in turn enables a higher energy capacity to meet the demand of LIBs. This is because lithium storage occurs between any two neighbouring layers in graphite, that effectively each graphitic layer corresponds to one layer of lithium. A graphene monolayer can absorb lithium on both sides of the graphitic basal plane, resulting in two layers of lithium per graphene sheet with the stoichiometry of Li_2C_6. This understanding is in line with the experimental observation that the actual reversible specific charge capacity of graphene at the first charge-discharge cycle could be much larger than the theoretical value of graphite (Murugan et al. 2009, Yoo et al. 2008).

Although LIBs are featured by their high charge storage capacity, they suffer from low charge-discharge rates because of slow electrode reaction kinetics. Fast kinetics requires a large accessible surface area for the transfer of both electrons and Li^+ ions, and high Li^+ ion diffusivity inside the electrodes. It in turn means that the kinetics of a properly structured graphene electrode can be much faster than that of a graphite electrode.

In a LIB, an irreversible capacity loss is observed commonly between the first intercalation and deintercalation of Li^+ ions between the graphitic layers. The instability of some electrolytes which cannot withstand the very negative electrode potential of metallic lithium or highly lithiated carbons may also be responsible for this initial irreversibility. Electrolytes are eventually reduced to form an electronically insulating solid electrolyte interphase (SEI) mostly on the surface of carbon (Fong et al. 1990). Consequently, a carbon electrode with a higher surface area can result in the formation of more SEI and hence more heat generation, at least due to the Joule heat produced by current passing through the resistive SEI. If not dissipated quickly, the generated heat accumulate to raise the temperature high enough to cause thermal decomposition of the SEI, which may lead to more dangerous exothermic processes within the cell, e.g., lithium reaction with components of electrolyte.

In comparison, a graphene electrode with a larger SSA should be more likely experiencing these problems than a graphite electrode. In other words, building a 3D graphene network with the macroporous structure enable high electronic conductivity, short ionic diffusion length and modest SSA, but it could also compromise the advantages in capacity and kinetics of ideally individual graphene sheets.

2.2.2 Lithium Sulphur and Lithium Oxygen Batteries

A typical lithium sulphur (Li-S) battery is composed of a positive electrode of a conducting matrix loaded with sulphur, a negative electrode of lithium metal, and an electrolyte-filled separator. With high conductivity, light weightiness and numerous cavities, porous carbon is the most popular matrix in positive electrodes of Li-S batteries. In a typical discharge course in the positive electrode, the reduced sulphur reacts with Li^+ ions to form polysulfide intermediates (Li_2S_x, $x = 2 \sim 8$) and to generate lithium sulfide (Li_2S) at the end of discharge. Actually, the polysulfide intermediates with long chains can dissolve into the organic electrolyte, migrate to the negative electrode and be reduced to lower order polysulfides ions and even the insoluble Li_2S_2 and Li_2S that then deposit as an insulating layer on the lithium metal. The lower order polysulfides ions at the negative electrode can also shuttle through the electrolyte to the positive electrode to be re-oxidised back into polysulfide intermediates, which markedly reduces the coulombic efficiency. The insulating layer of Li_2S_2 and Li_2S retained on the negative electrode enormously reduces not only the amount of accessible active materials on the negative electrode, but also that of the originally available active materials on the positive electrode.

The shuttle effect and the loss of accessible and available active materials become the main reasons for deteriorated performance of the Li-S cell upon repeated charge-discharge cycling.

It should be mentioned that the cavities of the porous carbon in the positive electrode provide reservoirs to refrain the dissolution of the polysulfide intermediates into the electrolyte and to accommodate the volume expansion resulting from the density difference between sulphur and Li_2S. Therefore, highly ordered mesoporous carbon can be used to achieve improved cycling performance and arouses the hope to develop Li-S batteries (Ji et al. 2009). Besides the effect of porous structure to refrain polysulfide intermediates, it was also found that the oxygen functional groups on the graphene oxides could play a significant role in binding polysulfide intermediates, providing a barrier to its dissolution (Ji et al. 2011). It is believed that porous graphene with heteroatoms can also be employed as an excellent matrix to load sulphur in the positive electrode of a Li-S battery.

Lithium oxygen (Li-O_2) batteries are categorised into aqueous and non-aqueous cells according to the type of electrolyte used. Non-aqueous Li-O_2 batteries are made of a positive electrode of a conducting porous matrix with catalysts, a negative electrode of lithium metal, a separator with a carbonate or ether electrolyte. The structure of aqueous lithium oxygen batteries is more complicated with an additional OH$^-$ conducting membrane covered on the positive electrode, a Li$^+$ conducting membrane covered on the negative electrode and an O_2 evolution electrode in the electrolyte.

In a non-aqueous electrolyte, the discharge reaction of the positive electrode is the reduction of O_2 to the superoxide ion, O_2^-, which is further reduced and reacts with Li$^+$ ions to form Li_2O_2 (Laoire et al. 2009). In practice, many side reactions have been discovered, including decomposition of the carbon electrode, the electrolyte and even the binder (Black et al. 2012, Ottakam Thotiyl et al. 2013a). As a result, porous non-carbon materials, including Au, TiC and Co_3O_4 were considered as the matrix in the positive electrodes (Ottakam Thotiyl et al. 2013b, Peng et al. 2012, Riaz et al. 2013). However, carbon and graphene that can be synthesised via controllable, facile and low cost processes are still popularly used as the positive electrode materials. The positive electrodes of Li-O_2 batteries require a porous structure to provide void volume to accommodate the discharge product of Li_2O_2, and to provide gas diffusion channels transporting the air (or O_2) to the electrode surface with a cooling effect. It has been suggested that a pore size distribution from 10 to 100 nm would be more desirable for Li-O_2 batteries (Choi, N.-S. et al. 2012) and a dual pore structure offers advantages to improve oxygen transport into the inner regions of the air (or O_2) electrode (Williford and Zhang 2009).

3. Synthesis, Formation Mechanisms and Application of Porous Graphene

The International Union of Pure and Applied Chemistry (IUPAC) categorises pores into micropores, mesopores and macropores according their diameters. A micropore is less than 2 nm in diameter, whilst that of a macropore larger than 50 nm, and that of a mesopore in between (Rouquerol et al. 1994). For all EES devices, macropores facilitate the penetration of electrolyte ions to the interior surface of the electrode, and hence can improve the rate capability. Mesopores provide a large electrolyte accessible surface to obtain a high performance for supercapacitors and LIBs. They can function as a reservoir to refrain the diffusion of polysulfide intermediates into electrolyte in Li-S batteries, or as the gas channels to transport O_2 to electrode surface in Li-O_2 batteries, and/or a void volume to accommodate the discharge products with volume expansion in both Li-S and Li-O_2 batteries. Micropores which are still substantially larger than the size of the electrolyte ions and their solvation shells greatly increase SSA to enable higher capacitances of both electrodes in supercapacitors.

Recently, four main methods have been developed to synthesise porous graphenes, which can be generalised as chemical activation, self-assembly, template and intercalation. The formation mechanisms and roles in performance of the pores in graphene electrodes are discussed below according to different synthesis methods.

3.1 Chemical Activation

Chemical activation is usually used to activate amorphous carbon derived from calcination or pyrolysis of carbon-rich organic precursors in presence of an activating agent at an appropriate temperature in an inert atmosphere, producing a porous network in the bulk of the carbon particles. KOH, H_3PO_4 and $ZnCl_2$ are representative of the conventional activating agents for chemical activation. The mechanism of KOH activation of carbon is globally based on Equation 6.

$$6KOH + 2C \triangleq 2K \uparrow +3H_2 \uparrow +2K_2CO_3 \qquad (6)$$

The practical effect of Reaction (*i*) is that KOH attacks the bulk carbon to generate pores in the carbon due to the oxidation or etching of the carbon into carbonate ions, starting at 400°C. K_2CO_3 can also be decomposed at above 700°C in the presence of carbon into K_2O with the evolution of CO_2 and CO. The formed K_2O can react with carbon to produce K vapour and CO at around 800°C, resulting in carbon lattice expansion. Thus activated carbon can possess a very high SSA. It is widely accepted that KOH is more efficient than NaOH to activate carbon, particularly that with a moderate-to-low structural order than graphitic materials (Raymundo-Piñero et al. 2005).

The Ruoff group prepared porous graphene powders by activation of microwave-exfoliated graphite oxide (MEGO) with the aid of KOH. The largest BET SSA of the porous graphene powders was observed to be 3,100 m^2/g (which is even higher than the theoretical value of graphene monolayer), apparently due to the high porosity resulting from the KOH activation. The electron microscopic images clearly indicated that MEGO was etched in the activation process to generate a network of micropores and mesopores ranging from 0.3 ~ 10 nm (Zhu et al. 2011). Actually, as the precursor for producing porous graphene, the MEGO powders had a relatively high SSA of 463 m^2/g (Zhu et al. 2010), which facilitated the access of KOH to the basal plane of graphene distorted by residual oxygen functional groups. The high SSA and decreased structural order from residual oxygen made the activation of MEGO successful. Additionally, the porous graphene powders had a high electrical conductivity of 500 S/m, low oxygen content with the C/O atomic ratio of 35, very low H content and the essential absence of dangling bonds, suggesting that carbon atoms were mostly sp^2-bonded in the porous graphene powders with very few edge atoms. As a result, a two-electrode symmetrical supercapacitor fabricated from the porous graphene powders with a SSA of 2,400 m^2/g showed a specific capacitance of 166 F/g at a mass normalised current of 5.7 A/g and a specific energy of 70 Wh/kg when an organic electrolyte of 1-butyl-3-methyl-imidazolium tetrafluoroborate in acetonitrile (AN) was used with a wide working voltage of 3.5 V. The area normalised EDL capacitance of the porous graphene powders was calculated to be 7 $\mu F/cm^2$, which is a typical value for carbon materials in organic electrolytes. The maximum specific power was calculated to be 250 kW/kg. In practical applications of supercapacitors, active electrode materials occupied 25 ~ 30 weight per cent of the whole mass of a supercapacitor, meaning that the specific energy and maximum specific power of the entire supercapacitor based on the porous graphene could have reached about 20 Wh/kg and above 60 kW/kg, respectively. These were about four and six times as those of commercial supercapacitors. Moreover, the porous graphene powders exhibited excellent cycling stability with retention of 97 per cent of the initial capacitance after 10,000 cycles.

Subsequently, the influences of activation temperature and KOH/MEGO ratio on SSA and specific capacitance of the porous graphene activated by KOH for 1 hr were investigated (Murali et al. 2012). The largest SSA of 3,100 m^2/g for the porous graphene powders were obtained at 800°C and a KOH/MEGO mass ratio of 6.5, giving a higher specific capacitance of 172 F/g at 1 A/g in a two-electrode cell as shown in Fig. 4a. The SSA and specific capacitance of the porous graphene powders increased with increasing the activation temperature till 800°C at a constant KOH/MEGO ratio of 6.5. When the activation temperature reached 1,000°C, both the SSA and specific capacitance diminished due to the collapse of the porous structure as a result of sintering at the high temperature.

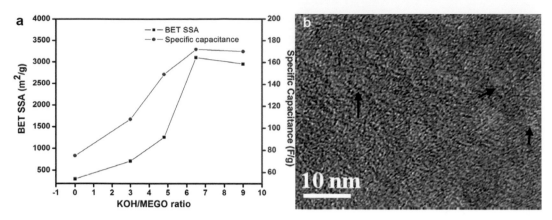

Fig. 4. (a) Effect of activation KOH/MEGO ratio on BET SSA and specific capacitance of the activated graphene powders at a temperature of 800°C. Reproduced with permission from ref. (Murali et al. 2012), Copyright 2012 Elsevier. (b) High-resolution TEM image of the activated graphene film, some uniform micropores smaller than 2 nm are clearly indicated by the black arrows. Reproduced with permission from ref. (Zhang et al. 2012), Copyright 2012 ACS.

Increasing the KOH/MEGO ratio to 6.5 can promote both the SSA and specific capacitance, whilst it is worth noting that non-conductive gray white powders with a low C/O atomic ratio of 0.3 were produced when the KOH/MEGO ratio was increased to beyond 9, indicative of a nearly complete etching of MEGO at a high KOH loading.

In another study, porous graphene films were fabricated and tested for supercapacitor applications. Firstly, the aqueous colloid of graphene oxide (GO) was partially reduced with KOH and form a paste by evaporating in a hot oil bath. The reduced mixture of GO-KOH was then deposited as a paste on a polytetrafluoroethylene membrane by vacuum filtration and desiccation to produce the reduced GO film. After activation at 800°C for 1 hr with the KOH/film ratio of 14, the resulting porous graphene film showed a continuous structure with dense micropores (Fig. 4b) and a remarkable SSA of 2,400 m²/g (Zhang et al. 2012). Note that this graphene film was formed without using any binder. The highly interconnected structure of the porous graphene film led to a maximum conductivity of 5,880 S/m among all tested films of different thicknesses and a very high maximum specific power of 500 kW/kg in an AN electrolyte of tetraethylammonium tetrafluoroborate (TEABF$_4$) with a 2.7 V voltage window. However, in spite of the high power (or rate) capability, this binder free and porous graphene film exhibited only a moderate capacitance of 120 F/g at 10 A/g with a specific energy of 26 Wh/kg.

Further the MEGO was activated with KOH at a mild temperature below 550°C for an extended time of three or 10 hr. A sample of the porous graphene powder produced at the activation temperature of 500°C for 10 hr exhibited a modest SSA of 1,532 m²/g, but a fairly high specific capacitance of 265 F/g at 1 A/g in the 6 M KOH aqueous electrolyte in a two-electrode supercapacitor (Wu et al. 2016). The flake-like morphology with a lamellar structure was well maintained under relatively mild activation conditions except for the sample prepared at 550°C for 10 hr which showed a chunk-like morphology in bulk, similar to that produced by high temperature activation. The high capacitance was attributed to the pores in the retained lamellar structure which should be more beneficial for electrolyte ions to access, suggesting that the low temperature activation built an efficient porosity and SSA for supercapacitors. Moreover, the activation under relatively mild conditions should benefit to increasing the density of the porous graphene powders. As a result, an electrode density of 0.70 g/cm³ was achieved, twice of the value of 0.35 g/cm³ for the sample prepared by high temperature activation, giving rise to a volumetric capacitance of 185 F/cm³ at 1 A/g in aqueous electrolyte. However, the data in organic electrolytes are missing for low temperature activation.

To summarise, the chemical activation method could produce porous graphene powders and films with very high SSA. The porous graphene films could offer very high rate capability due to their high conductivity, whilst the porous graphene powders were capable of outputting both high energy and power densities, promising a high potential for practical application in supercapacitors.

3.2 Self-Assembly

A graphene aerogel was developed with an interconnected 3D porous network via the hydrothermal route, followed by cryodesiccation (Xu et al. 2010). The pore sizes of the network were in the range of sub-micrometer to micrometer. The pore walls consisted of thin layers of stacked graphene sheets as shown in Fig. 5.

A self-assembly mechanism was proposed for the observed microstructure. The hydrothermal treatment could cause a deoxygenation effect on the GO (Nethravathi and Rajamathi 2008). The self-assembly involved partial overlapping and coalescing of graphene sheets via π–π stacking interactions. It was thought that the large amount of water was entrapped by the residual oxygenated functional groups on the graphene sheets in the produced sample. This effect induced the formation of the interconnected 3D porous network with macropores above a critical GO concentration, such as 1 g/L. The graphene aerogel showed excellent mechanical properties with compressive elastic modulus and yield stress being 290 and 24 kPa, respectively. The graphene aerogel could be used as the electrodes of supercapacitors without using any binding agent and conducting additive because of its excellent mechanical strength and high porosity. However, the average specific capacitance for the graphene aerogel was only 160 F/g at 1 A/g in 5 M KOH aqueous electrolyte, lower than that of the activated graphene powders, which was ascribed to the absence of micro- and mesopores.

Subsequently the graphene aerogel obtained by the hydrothermal treatment was further reduced by immersion in HI or $N_2H_4 \cdot H_2O$ in order to increase its conductivity (Zhang and Shi 2011). The BET SSA of the graphene aerogel before and after reduction with $N_2H_4 \cdot H_2O$ was 166 and 215 m²/g, respectively, suggesting that reduction did not lead to obvious change in SSA. The conductivities of the graphene aerogels after reduction were range between 3.2 and 2.7 S/m, which were 11 and nine times as that of the unreduced graphene aerogel. The higher conductivities after reduction by HI or $N_2H_4 \cdot H_2O$ resulted in a higher capacitance retention of 80 per cent and 76 per cent in the current range from 1 to 100 A/g, whilst it was 57 per cent for the unreduced graphene aerogel. However, the reduction did not give rise to a significantly higher capacitance at 1 A/g. The highest capacitance of 222 F/g was obtained by $N_2H_4 \cdot H_2O$ reduction for 8 hr, increasing by only 9 per cent over the unreduced graphene aerogel.

Chemical reduction of GOs at a mild temperature without stirring could also yield 3D graphene aerogels by the self-assembly method, demonstrating that hydrothermal treatment was not essential for the self-assembly course. Interestingly, it was found that the shapes of the as-prepared graphene

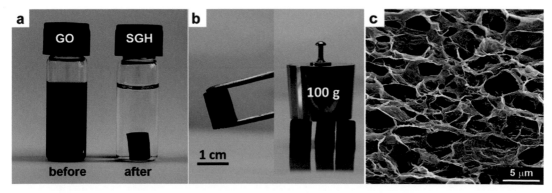

Fig. 5. (a) Photographs of a 2 mg/mL homogeneous GO aqueous dispersion before and after hydrothermal reduction at 180°C for 12 h. (b) Photographs of a graphene aerogel allowing easy handling and weight supporting. (c) SEM images of the interior microstructure of a graphene aerogel. Reproduced with permission from ref. (Xu et al. 2010), Copyright 2010 ACS.

hydrogels, the precursors of the aerogels, were analogous to those of the reactors, implying that the shrinkage of the as-formed graphene hydrogel by self-assembly was likely isotropic (Chen and Yan 2011).

When GO colloids at a high concentration of 4 g/L were reduced by L-ascorbic acid standing at 40°C, the graphene aerogel processed by supercritical CO_2 desiccation showed obvious differences from that by cryodesiccation (Zhang et al. 2011). The colour of the graphene aerogels by supercritical desiccation was darkish, whilst the graphene aerogels by cryodesiccation showed a metallic luster. The SSA of the graphene aerogel by supercritical desiccation reached 512 m^2/g with a combination of mesopores and macropores, much higher than the value of 11.8 m^2/g for the aerogel synthesised by cryodesiccation which exhibited only macropores. It was claimed that the formation of mesopores by supercritical desiccation arose from the inter-twisted pore walls. The absence of micropores of the aerogels by cryodesiccation was probably because the mesopores within the hydrogel precursors were fused together to form macropores impelled by gradual growth of ice crystals during cryodesiccation. However, the high SSA with mesopores merely corresponded to a low specific capacitance of 128 F/g at a very small current of 0.05 A/g. Similarly, the electrochemical reduction of graphene oxide dispersion in an aqueous $LiClO_4$ electrolyte finally deposited 3D graphene aerogel on a working electrode, which was also based on self-assembly method (Chen et al. 2012).

Hydroquinone was also used to reduce GO for the fabrication of the graphene aerogel. Alternatively, a functional graphene aerogel was produced by the absorption of hydroquinone on graphene, which was confirmed according to an extra UV-vis peak and increased height in atomic force microscopy (AFM) image (Xu et al. 2013). The freeze-dried functional graphene aerogel had a high SSA of 297 m^2/g with pore size distribution of 2 to 70 nm, suggesting that the functional graphene aerogels by chemical reduction and cryodesiccation had the combined structure of mesopores and macropores. The hydroquinone functionalised graphene hydrogel was used to assemble a flexible supercapacitor with a H_2SO_4-polyvinyl alcohol (PVA) electrolyte. The device exhibited a significantly increased storage capacity that can be attributed to the redox reaction of hydroquinone. When the mass normalised current increased from 1 to 20 A/g, 74 per cent of the storage capacity was retained. Besides hydrothermal treatment and standing chemical reduction, direct cryodesiccation of giant GO colloids can also produce ultra-light weight graphene aerogels with the lowest density of 0.16 mg/cm^3 by subsequent reduction (Sun et al. 2013).

The macropores present in the graphene aerogel could facilitate ion transport to increase the rate capability. However, the absence of mesopores would suppress the SSA and capacitance of the graphene aerogel. The 3D holey graphene aerogels synthesised by treating GO colloids with H_2O_2 under the hydrothermal condition were considered as a satisfactory candidate for supercapacitor electrodes (Xu et al. 2014). H_2O_2 can partially oxidise and etch the carbon atoms around the more active defective sites of GO to leave behind vacancies and form mesopores (Fig. 6).

The BET SSA of the holey graphene aerogel was 830 m^2/g, whilst the non-holey graphene aerogel had a SSA of only 260 m^2/g. The holey graphene film obtained by compressing the holey graphene hydrogel had a comparable SSA of 810 m^2/g. After the electrolyte, 1-ethyl-3-methylimidazolium tetrafluoroborate ($EMIBF_4$) in AN with 3.5 V of potential window, was integrated before compression of the holey graphene hydrogel, the compressed holey graphene film showed a very high specific capacitance of 298 F/g at 1 A/g, which was comparable to the value of 310 F/g obtained in the 6 M KOH aqueous electrolyte. At 100 A/g, a satisfactory capacitance of 202 F/g was maintained, indicative of a high rate capability. Therefore, an unprecedented specific energy of 127 Wh/kg was achieved based on two electrodes, which is comparable to the theoretical value (165 Wh/kg) of lead-acid battery electrodes. The volumetric capacitance and energy density were 212 F/cm^3 and 90 Wh/L, respectively. When the mass loading of electrodes increased from 1 to 10 mg/cm^2 for commercial consideration, the gravimetric capacitance of the holey graphene film electrode decreased to 262 F/g at 1 A/g. The practical gravimetric and volumetric energy of the supercapacitor devices sealed by parafilm could be up to 35 Wh/kg and 49 Wh/L, respectively. The high performance in energy and

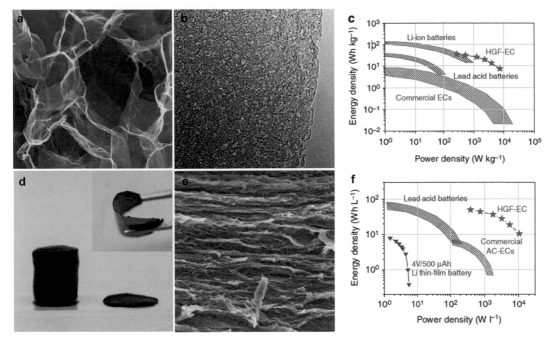

Fig. 6. (a) SEM image of interior microstructures of a holey graphene aerogel. Scale bar: 1 mm. (b) TEM image of graphene sheets in a holey graphene aerogel. Scale bar: 10 nm. (c) Ragone plots of specific energy versus specific power for a holey graphene aerogel supercapacitor in comparison with lead-acid batteries, lithium-ion battery, commercial ECs. (Note: In (c), the *y*-axis label should be Specific Energy and the *x*-axis label should be Specific Power.) (d) A photograph showing a holey graphene aerogel before and after mechanical compression with the flexibility of the compressed holey graphene aerogel film shown in the inset. (e) Cross-sectional SEM image of the compressed holey graphene aerogel film. Scale bar: 1 mm. (f) Ragone plots of energy density versus power density for a holey graphene aerogel supercapacitor in comparison with lead-acid batteries, lithium-ion battery, commercial ECs. The energy and power values are normalised by the actual weight or volume of the entire device stack including two electrodes, two current collectors, electrolyte, one separator and packaging. Reproduced with permission from ref. (Xu et al. 2014), Copyright 2014 NPG.

kinetics for supercapacitors were attributed to the high porosity (with mesopores and macropores) and high electrical conductivity of the network.

The incorporation of the Fe_3O_4@graphene core-shell sphere into the graphene aerogel was completed by the hydrothermal treatment of the Fe_3O_4@GO core-shell sphere, and then by *ex situ* electrostatic assembly with additional GO, followed by annealing in Ar (Wei et al. 2013). When tested in the LIB, the Fe_3O_4@graphene core-shell sphere exhibited a capacity of 744 mAh/g and could last for 50 charge-discharge cycles, whilst the Fe_3O_4-graphene aerogel showed a superior reversible capacity of 920 mAh/g, a stable life over 150 charge-discharge cycles, and better rate capability of 39 per cent from 0.15 to 4.8 A/g. The obvious improvement was thought to originate from the graphene frameworks which could suppress the pulverisation effect of Fe_3O_4 in the Fe_3O_4 loaded graphene aerogel, and facilitate the access of electrolyte ions to Fe_3O_4.

Sulphur was *in situ* loaded into the graphene aerogel by hydrothermal treatment of a mixed dispersion of GO in ethanol/water and sulphur powders in CS_2 for Li-S batteries (Zhou et al. 2013). The S-graphene aerogel electrode delivered a capacity of 1,160 mAh/g at 0.3 A/g based on the mass of sulphur and an overall capacity of 731 mAh/g based on the masses of both the sulphur and graphene. It could retain a capacity of 700 mAh/g at 0.3 A/g after 50 charge-discharge cycles. These data compare favourably with those of the S-graphene powders which exhibited only 350 mAh/g at 0.3 A/g after 50 cycles. Such a good charge-discharge cycling performance was associated with the second discharge plateau where the S–O bond was detected by X-ray photoelectron spectroscopy on the S 2p spectrum. It indicates that the oxygen functional groups on graphene could immobilise sulphur and the polysulfides produced during discharging. Similar performance was observed on the S-graphene aerogel prepared by *ex situ* loading sulphur into graphene aerogel (Xu et al. 2015).

To summarise, the reduction of GO colloids in a standing solution process followed by cryodesiccation could produce the graphene aerogels in the form of a macroporous network by self-assembly. Supercritical CO_2 desiccation could more perfectly maintain the previous morphology of hydrogel than cryodesiccation. The holey graphene aerogel by extra H_2O_2 etching had a combined structure of mesopores and macropores, giving rise to the state of the art supercapacitors with unprecedented specific energy of 127 Wh/kg for electrode and 35 Wh/kg for the whole device in practice. The graphene aerogel had a framework that could be loaded with metal oxide and sulphur, which promise applications of the graphene aerogel for LIB and Li-S batteries. The residual oxygen functional groups could absorb the sulphur and ploysulfide intermediates by forming the S–O bond.

3.3 Template

3.3.1 Ni Foams

Ni foams have 3D interconnected macropores and hence are often used as current collectors to enhance electrolyte access to the surface of activated materials loaded inside the macropores. An interesting way of loading graphene into the Ni foam was achieved by electrophoretic deposition. In this method, the stable functional graphene colloids, which were positively charged in organic solutions (Chen et al. 2009), were used as electrolyte. Graphene nanosheets were electrophoretically deposited onto the 3D Ni foams which were subsequently annealed in Ar (Chen et al. 2010).

A CVD method using mixed CH_4 in H_2 gases as the feed was applied to prepare graphene monolayer on Cu foil. Then the graphene monolayer could be transferred to any substrate via a surface catalysed process (Li et al. 2009). The CVD method could also be used to obtain free-standing graphene foams using Ni foams as templates and catalysts. The Ni foams were exposed in CH_4 and Ar/H_2 gas for 5 min at 1,000°C, followed by rapid cooling to form graphene on them. The free standing graphene foams could be obtained by dissolution of the Ni foam in aqueous HCl. In this procedure, a thin poly (methyl methacrylate) (PMMA) layer was drop-coated to cover the graphene on the nickel foams after CVD in order to prevent structural failure of the graphene foams when the nickel was etched away by HCl. The PMMA cover was etched away by acetone. The removal of PMMA led to the graphene foams became thinner than the original nickel foam due to the liquid capillary force caused during the acetone evaporation. The thickness of the graphene foams increases with increasing the number of graphene layers by introducing a higher concentration of CH_4 in Ar/H_2. In the Raman spectra of the graphene foams, ultrahigh intensity ratios of 2D to G bands without D band were observed. Combining the Raman results, the lattice fringe images of the edge of the graphene foams in Fig. 7 showed monolayers and few layers of graphene with high quality (Chen et al. 2011). The graphene foam with three layers on average obtained with a 0.7 volume per cent of CH_4 had a very high SSA of 850 m²/g and ultralow density of 5 mg/cm³, corresponding to a porosity as high as 99.7 per cent. The maximum electrical conductivity of 1,000 S/m for the graphene foams was attained at an optimal average number of graphene layers of five. Actually, ethanol vapour can replace CH_4 for fabrication of the graphene foams by CVD (Maiyalagan et al. 2012).

When the exposure time was prolonged to 60 min without rapid cooling, ultrathin graphite foams with wall thickness of tens of nanometers were formed on nickel foams, instead of graphene foams (Ji et al. 2012). After removal from the template, the graphite foam could be used as the current collector to load LiFePO$_4$ for testing as the positive electrode of LIB. A specific capacity of 70 mAh/g was attained at a current of 1.28 A/g, whereas the LiFePO$_4$ loaded on an Al electrode failed at such a high rate. The obvious advantage in rate capability of the graphite foam based electrode was attributed to the 3D highly conductive network that could benefit faster kinetics. Considering the whole electrode including the graphite foam current collector, the superiority in capacity of

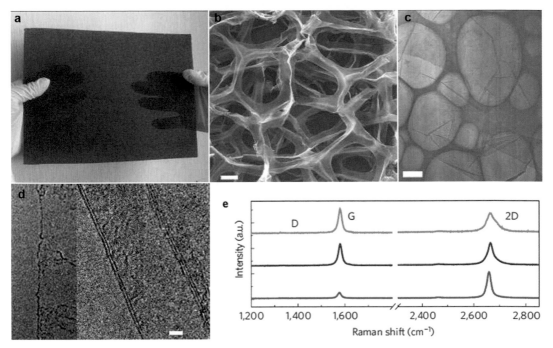

Fig. 7. (a) Photograph of a free-standing graphene foam. (b) SEM image of a graphene foam. (c) Low-magnification TEM image of a graphene foam. (d) High-resolution TEM images of graphene sheets with different numbers of layers in a graphene foam. (e) Typical Raman spectra of a graphene foam. The bottom spectrum and the other two correspond to monolayer and few-layer graphenes, respectively. Reproduced with permission from ref. (Chen et al. 2011), Copyright 2011 NPG.

102 mAh/g for LiFePO$_4$ on the graphite foam was also apparent, which was 23 per cent higher than that of LiFePO$_4$ on Al and 170 per cent higher than that of LiFePO$_4$ on Ni foam.

N-doped graphite foams could also be deposited on macroporous Ni sheets by CVD using pyridine as the carbon and nitrogen sources in the atmosphere of mixed H$_2$ and Ar gases. RuO$_2$ nanoparticles chemically plated on the surface of the N-doped graphite foams were used as catalysts for applications in Li-O$_2$ batteries (Guo et al. 2015). Repeated charging-discharging of RuO$_2$ on N-doped graphite foams as the positive electrode only lasted for 62 cycles at a curtailing capacity of 2,000 mAh/g due to coarsening of the catalysts. A repeated CVD course could deposit 2 ~ 3 layers of graphene to encapsulate the RuO$_2$ nanoparticles with N-doped graphite foams after removal of the macroporous Ni sheets. The macroporous graphite and graphene with encapsulated RuO$_2$ had a SSA of 189 m^2/g and the macropores of 258 nm. After initial full discharge, the macropores in the electrode were filled with toroid Li$_2$O$_2$ with sizes ranging from 200 to 500 nm, suggesting that the macropores of graphite foams could provide oxygen diffusion channels and voids to accommodate the discharged product. The covered graphene layers effectively restrained the coarsening of RuO$_2$, resulting in better cyclic performance of more than 110 cycles.

3.3.2 Metal Oxides

Although Cu and Ni had been demonstrated to be suitable for deposition of graphene by CVD via surface catalysed processes, mesoporous graphene produced by CVD at a large scale is absent. Calcination of Mg(OH)$_2$ layers formed by boiling MgO powders could yield laminated MgO templates. It was discovered that one to two graphene layers with mesopores could be deposited on the surface of the porous MgO templates which also functioned as catalysts and could be removed later by CH$_4$ cracking (Ning et al. 2011). The graphene powders with pores of 10 nm were found to form along the MgO (200) plane, having a similar polygonal morphology to the porous MgO layers. Its SSA was about 1,654 m^2/g, arising from the mesopores and the graphene basal plane, indicating that the average number of layers was 2, according to the theoretical SSA of graphene monolayer. The EDL capacitance was measured to be 14 μF/cm^2 in a three electrode cell.

It is worth noting that the mesoporous graphene powders were not deposited on the pore walls but only on the upper and lower surfaces of the porous MgO template. Carbonisation of coal tar pitch in tetrahydrofuran impregnated into the porous MgO could form carbon not only on the upper and lower surfaces but also on the pore walls of the templates (Fan et al. 2012).

Inspired by these designs for deposition of graphene on the surfaces and carbon on the pore walls of the MgO templates, unstacked double layer graphenes were produced using a new template of MgAl-layered double oxide (LDO) flakes by CVD (Zhao et al. 2014). The precursors of the LDO templates were MgAl-layered double hydroxides which are anionic hydrotalcite-type materials consisting of positively charged layers and charge-balancing anionic interlayer, present in a hexagonal morphology with a thickness of 10 nm. Calcination of MgAl-layered double hydroxides at 950°C gave the porous LDO templates with a perfectly preserved laminated structure. The main components were MgO and MgAl$_2$O$_4$. The LDO templates were uniformly single crystalline structure, which is quite different from the MgO templates made up of aggregate of MgO nanocrystals by calcination of Mg(OH)$_2$ sheets. The unique mesoporous and laminated structure of the LDO templates catalysed the formation of uninterrupted and unstacked double layers of graphene powders with protuberances in the interlayer when introducing CH$_4$ as shown in Fig. 8. Mesopores within graphene basal plane and monolayers of graphene on the edge could be seen in Fig. 8d.

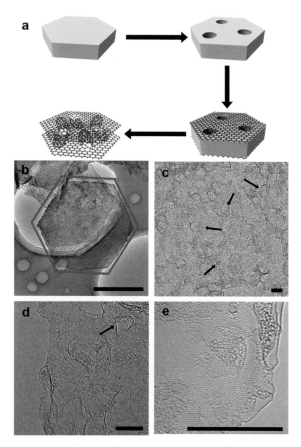

Fig. 8. (a) Scheme for the synthesis of unstacked double layer graphene. (b, c) TEM images of the cast unstacked double layer graphene flakes. (d, e) High-resolution TEM images of unstacked double layer graphene with protuberances extending from the graphene sheets and the graphene layer for unstacked double layer graphene, the scale bars in (b) and (c–e) are 1 mm and 10 nm, respectively. Reproduced with permission from ref. (Zhao et al. 2014), Copyright 2014 NPG.

A height of 10 nm was exhibited in the image of the as-prepared graphene from atomic force microscopy (AFM), whilst the appearance of broadly increased intensity in the small angle X-ray diffraction region instead of the obvious peak corresponding to the graphite (002) plane is an indication of an enlarged d-space between graphene layers. Combining with the morphologic observation, it could be concluded that the mesopores in graphene arose from the graphene protuberance between interlayers of the unstacked double layer graphene, rather than cavities in stacked bilayer graphene. The unstacked double layer graphene showed an SSA of 1,628 m^2/g with a large quantity of mesopores ranging from 2 to 7 nm and a conductivity of 43,800 S/m. The unstacked double layer graphene was used as the matrix of the positive electrode in Li-S batteries to load sulphur as high as 64 per cent by hydrothermal diffusion. Sulphur was incorporated not only in the mesoscale protuberances but also in the interlayer space, instead of the external surface of the unstacked double layer graphene. An interesting and relevant observation was that the volatilisation temperature was delayed for the sulphur in the unstacked double layer graphene matrix comparing with that of pure sulphur, suggesting that the accommodation of sulphur in tens of nanometer scale could reduce the dissolution of the polysulfides into the electrolyte during the discharge course. A reversible capacity of 1,200 mAh/g at 0.5 C and an excellent rate capability of only 16 per cent decrease from 1 to 5 C were achieved. A severe capacity fade of 35 per cent and a relative low coulombic efficiency of 90 per cent occurred at 1 C after 200 cycles. When increasing the rate to 5 C, the cyclic performances showed only 20 per cent decay after the same number of cycles and higher coulombic efficiency of 96 per cent. The variation in lifetime and coulombic efficiency with different rate proved that the shuttling effect was effectively prevented with increasing current rate in faster charge/discharge process. Even after 1,000 cycles, high reversible capacities of 530 and 380 mAh/g were still retained at 5.0 C and 10.0 C, respectively. The long lifetime of the Li-S battery were attributed to the unique unstacked structure of double layer graphene to impede the dissolution of the polysulfides into the electrolyte.

It is well known that burning Mg metal in a CO_2 atmosphere can produce carbon materials and the process can be expressed as Reaction 7.

$$Mg + CO_2 = MgO + C \qquad (7)$$

In a unique study of the magnesiothermic reaction with CO_2, by igniting Mg ribbons inside a dry ice bowl covered by another dry ice slab, a black product was collected after the combustion, and it was confirmed to be composed of few layer graphenes (Chakrabarti et al. 2011). In another study, similar few layer graphenes with mesopores was produced by burning Mg ribbons in CO_2 gas with subsequent quenching in NH_4HCO_3 solution. The small angle X-ray diffraction revealed that mesoporous structure arose from Mg and produced MgO (Zhang et al. 2013b). The quench decomposed NH_4HCO_3 into the NH_3 gas which played a role on the exfoliation of graphene. Afterwards, the conventional furnace annealing method was employed to enable the reaction of Mg (Zhang et al. 2013a) or mixture of Mg and Zn powders (Xing et al. 2015) in a CO_2 atmosphere for a large productivity of porous graphene. The stacked mesoporous graphene-like nanosheets could be produced below 700°C, otherwise tubular or caged carbon emerged at higher temperature. When using mixed Zn and Mg, the SSA of the produced stacked porous graphene-like nanosheets increased from ca. 800 to 1,900 m^2/g, corresponding to a structure with mesopores and micropores arising from the produced MgO and ZnO respectively. The stacked porous graphene-like nanosheets had a good capacitance of 190 F/g at a large current of 10 A/g due to their high conductivity.

Recently, a method called 'self-propagating high-temperature synthesis (SHS)' was employed to produce scalable mesoporous few layer graphene by igniting Mg in the mixture of Mg and MgO via an initially stimulated combustion wave in a sealed CO_2 atmosphere (Li et al. 2017). A current of 3 A was exerted through the tungsten coil embedded into the Mg/MgO for only 5 s to produce a combustion wave which provided a sufficiently large driving force to trigger the SHS. The SHS-made graphene had a SSA of 709 m^2/g and unimodal pore distribution at 4 nm which is consistent with uniform diameters of 4 ~ 5 nm of MgO grains. More importantly, besides providing a template to yield homogeneous mesopores, MgO acted as spacers to suppress the restack of graphene as shown in Fig. 9.

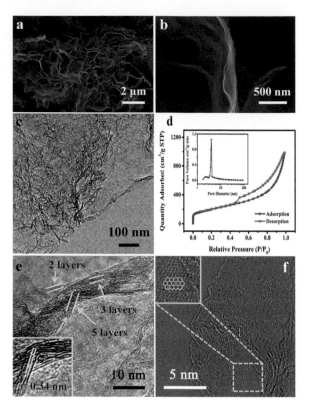

Fig. 9. Characterisation of the SHS made graphene. (a) SEM image of the microstructures. (b) High magnification SEM image showing the separated graphene layers and the interlayer space. (c) TEM image. (d) N_2 adsorption/desorption isotherm with the inset presenting the pore size distribution according to the Barret–Joyner–Halenda method. (e) HR-TEM with the inset demonstrating the few-layer feature. (f) Spherical aberration-corrected HR-TEM image and magnified area on graphene plane in the inset. Reproduced with permission from ref. (Li et al. 2017), Copyright 2017 Wiley.

The amount of sp^2 hybridisation in the SHS-made graphene was up to 98 per cent, resulting in high electrical conductivity of 13,000 S/m. At 2 A/g, the maximum capacitance was measured to be 172 and 190 F/g in $EMIBF_4$ and ethyl-methylimidazolium bis(trifluormethylsulfonyl)imide (EMIMTFSI) as electrolyte with potential windows being 3.5 and 4.0 V, respectively. High specific energy of 136 Wh/kg in EMIMTFSI was achieved at an actual specific power of 10 kW/kg in the two electrode cell containing 70 per cent graphene. More significantly, unprecedentcd actual specific power of 1,000 kW/kg and specific energy of 60 Wh/kg were fulfilled, which corresponded to a current drain time of 0.2 s. The relaxation time constant τ_0 was calculated to be 0.012 s in $EMIMBF_4$. As the mass loading increased from 0.5 to 3 mg/cm^2, the highest specific energy could still be 114 Wh/kg in EMIMTFSI, corresponding to a decrease of 16 per cent.

3.3.3 NH_4Cl

A new approach to build the 3D graphene network was discovered by calcination of glucose and NH_4Cl together, inspired by an ancient food art of blown sugar (Wang et al. 2013). Released gases from NH_4Cl blew glucose-derived polymers into numerous large bubbles and made the bubble walls gradually thinner. A high temperature graphitised the very thin bubble walls into graphene membranes which were inlayed in the borders of the bubbles which were called as struts. The struts constituted

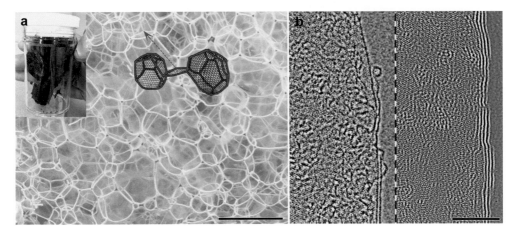

Fig. 10. (a) SEM image and optical photograph of a strutted graphene piece obtained at a heating rate of 4°C/min for decomposing NH_4Cl. The inset is a reconstructed topology corresponding to the region marked in red for the two connected decahedron–dodecahedron bubbles faced by eight pentagons and two quadrangles, and by eight pentagons, three quadrangles and one heptagon, respectively. (b) High-resolution TEM images of a one- to two-layered graphene (left), and a three- to four-layered graphene (right). Scale bar: 5 nm. Reproduced with permission from ref. (Wang et al. 2013), Copyright 2013 NPG.

the skeletons of the bubbles and had an average width of 3.5 μm. However, the graphene membranes were 2.2 nm in thickness, corresponding to six layers of graphene as shown in Fig. 10. The sp^2 fraction of the graphene membranes was 99 per cent, rendering high conductivity of 20,000 S/m. The material as a whole was called strutted graphene, giving an SSA of 1,005 m^2/g with a multimodal pore distribution. The heat rate applied to decompose NH_4Cl was 4°C/min, which played a paramount role in increasing the fraction of graphene membranes in the entire material. The capacitance was measured as high as 250 F/g at 1 A/g and decreased to 130 F/g at a higher current of 100 A/g with an impressive maximum specific power of 893 kW/kg owing to the high SSA and high conductivity.

3.3.4 SiO_2 and Polystyrene

Mesoporous SiO_2 is a classic template for synthesis of ordered mesoporous carbon, and the template can be removed by etching with HF (Lee et al. 2006). Similarly, the methyl group grafted hollow or solid SiO_2 spheres with different sizes as templates could be used to accurately fabricate mesoporous and macroporous graphene by mixing graphene oxide and functional SiO_2 spheres in solution and annealing in Ar. The obtained mesoporous graphene was composed of three layers of graphene, which agreed very well with the SSA of 851 m^2/g. Although a large reversible capacity of 750 mAh/g was delivered at the first cycle at 0.3 A/g, the poor cyclic performance and the absence of an obvious discharge plateau hindered the application of such obtained pure graphene for use as the negative electrode material of an LIB. A similar porous graphene prepared by removing methyl group grafted SiO_2 templates with loading of Ru was used as the positive electrode of $Li-O_2$ batteries (Sun et al. 2014). The mesoporous and macroporous graphenes had SSAs of 242 and 374 m^2/g with pore sizes of 62 and 250 nm, respectively. The macroporous graphene showed a larger discharge capacity. The Ru catalyst loaded on the macroporous graphene reduced the charge/discharge over potential from 1.579 to 0.355 V, giving energy efficiency of 77.8 per cent and 200 cycles at a curtailing capacity of 500 mAh/g. In analogy, polystyrene (PS) spheres, like SiO_2, could also be used as macroporous templates to fabricate macroporous graphene films for the LIB with a reversible capacity of 1,108 mAh/g, a coulombic efficiency of 86.2 per cent and 90.6 per cent capacity retention after 50 cycles (Choi, B.G. et al. 2012).

3.3.5 Water

In a different study, the so called breath-figure method was employed to prepare macroporous graphene films. PS chains, not PS spheres, grafted GO platelets were readily dispersed in benzene. The dispersion was cast onto a suitable substrate and then exposed to a stream of humidified air. The evaporation of benzene offered an endothermic condition to assist the condensation of water vapour to form the droplets which were closely packed on the surface of PS chains grafted GO in benzene. The macroporous graphene structure was formed after removing water by drying and PS chains by pyrolysis (Lee et al. 2010). The macropores were tunable by adjusting the chain length of PS and concentration of graphene oxide but the capacitance of the obtained graphene films was as low as 89 F/g as measured in a three-electrode cell, suggesting that pore walls comprised of thick stacks of graphene.

To summarise, the CVD method using 3D Ni foams as templates could produce 3D macroporous graphene foams with high quality with fewer layers and defects. Porous and layered metal oxides could also be used as templates for the synthesis of unstacked double layer graphene by CVD with an interspace of 10 nm to load sulphur and reduce dissolution of polysulfides for application in Li-S batteries. Igniting Mg in the CO_2 atmosphere with MgO as templates could yield mesoporous graphene powders at a large scale with single peak distribution, showing ultrahigh actual specific power of 1,000 kW/kg and τ_0 of 0.012 s for supercapacitors. Blowing glucose by the gases produced from decomposition of NH_4Cl could give the so called strutted graphene in large quantity with an unparalleled conductivity of 20,000 S/m and a maximum specific power of 893 kW/kg for supercapacitors.

3.4 Intercalation

Initially heterogeneous entities, such as Pt nanoparticles, were used as separators to exfoliate graphene during the reduction of graphene oxide (Si and Samulski 2008). Numerous metal oxides as separators were also deposited on graphene sheets to form metal oxide/graphene powders, where metal oxides not only acted as separators to prevent the restacking of graphene but also contributed to the extra pseudocapacitance. In contrast to graphene powders, graphene films or papers could be better components of batteries and supercapacitors without using a binder and even conductive additives (Dikin et al. 2007). An ammonia-dispersion strategy was firstly contrived on the basis of ζ potential to gain stable graphene colloids and the corresponding graphene films (Li et al. 2008). However, oriented graphene films in the dry state after removing water between interlayers inclined to stack, which was not recoverable due to the dominant π–π interactions instead of repulsive hydration force. Water was utilised as a soft matter existing in as-prepared graphene hydrogel films just after vacuum filtration to break through the bottleneck (Yang et al. 2011). The graphene hydrogel films were exchanged with the electrolyte solution before assembly. The obtained graphene hydrogel film in an acid aqueous electrolyte possessed a specific capacitance of 215 F/g, and even outperformed the freeze-dried graphene films, suggesting that the pores were naturally more continuous and larger to facilitate the diffusion of electrolyte ions. The cyclic charge-discharge performance was also verified to be 97 per cent over 10,000 cycles even under a high mass normalised current of 100 A/g. When the water in the graphene hydrogel films was exchanged with $EMIBF_4$, a gravimetric capacitance of 273 F/g, very high specific energy of 151 Wh/kg and maximum specific power of 777 kW/kg were obtained.

However, the graphene hydrogel films within which large interspace exists had a low packing density of 0.069 g/cm^3 and volumetric capacitance of 18 F/cm^3. The electrolyte-exchanged graphene hydrogel films were further developed to electrolyte-incorporated graphene films by evaporating the volatile solvent and retaining nonvolatile liquids within electrolyte-exchanged graphene hydrogel films as shown in Fig. 11 (Yang et al. 2013). By changing the ratio of volatile solvent and nonvolatile solute, the packing density of electrolyte-incorporated graphene films could be controlled from 0.13 to 1.33 g/cm^3, much larger than that of graphene hydrogel films and slightly smaller than 1.49 g/cm^3

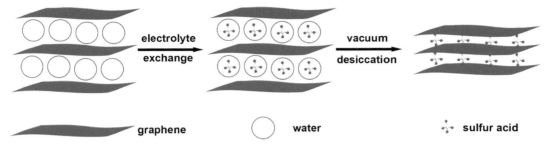

Fig. 11. A schematic illustration of fabrication process of electrolyte-incorporated graphene films containing H_2SO_4.

of fully dried graphene films. At 0.1 A/g, the gravimetric capacitances of the electrolyte-incorporated graphene films were measured to be 192 and 209 F/g at the packing density of 1.33 and 1.25 g/cm^3, giving rise to volumetric capacitance of 256 and 261 F/cm^3 in H_2SO_4 and $EMIBF_4$ as electrolytes, respectively, whilst the dried graphene film showed a capacitance of 155 F/g. The capacitive difference between the electrolyte-incorporated graphene film and the dried graphene film was obvious where the specific capacitance of the electrolyte-incorporated graphene films was 100 F/g whereas that of dried graphene film was 10 F/g at 100 A/g. Increasing the loading to 10 mg/cm^2, the electrolyte-incorporated graphene films showed an energy density of 60 Wh/L with an power density of 8.6 kW/L, in contrast to 18 Wh/L for dried graphene films estimated for the whole device.

4. Perspective

Graphene with a large theoretical specific surface area, high conductivity and excellent mechanical property is a promising new material in various applications. It is desirable to produce porous graphene materials with high performance to meet the requirements of EES devices. Four main approaches to produce porous graphene have been elucidated here, which are chemical activation, self-assembly, template and intercalation. Generally, chemical activation yields microporous and mesoporous graphene powders, self-assembly produces 3D macroporous graphene aerogel, whilst template brings out various porous graphene and intercalation makes large interlayer for graphene films. Different types of EES devices require different porous structures. Macropores are paramount to obtain fast kinetics for all EES devices. Appropriate micropores can satisfy the needs of supercapacitors and lithium sulphur batteries. Mesoporous materials are popular in supercapacitors, Li-S batteries and Li-O_2 batteries.

Among those synthesis strategies in the assembly method, the holey graphene hydrogel prepared by hydrothermal treatment in aqueous H_2O_2 solution possessed the structure of combined mesoporosity and macroporosity. The mesopores in the holey graphene hydrogel increased the SSA and hence enhanced the capacitance of the graphene hydrogel only by hydrothermal treatment. The holey graphene hydrogel drew lessons from pre-incorporation of the electrolyte into electrode and were compressed into the holey graphene hydrogel film, showing an ultrahigh volumetric capacitance. This good example illuminates that combining different methods to breed optimal and desirable porosity is one of the most effective strategies to achieve high performance for these EES devices. For another example, chemical activation of the self-assembled graphene hydrogel can also produce micropores and mesopores in a macroporous network of graphene.

Besides the optimal porosity, bulk production and high quality have to be considered in the synthesis strategies of porous graphene for future applications. High quality of graphene means fewer layers and higher electrical conductivity, which can facilitate power and rate in EES devices. Template methods combining with furnaces are the most possible method to realise scalability of production of porous graphene, whilst further effort is needed in balancing between scalability and quality with either a new strategy or a substantial improvement towards to fewer layers and higher electrical conductivity of graphene.

It is worth noting that some strategies to increase the energy capacity of EES devices can also be developed from the synthesis of porous graphene, mainly including hetero-atom doping and composite. N, S and O atoms on graphene have been demonstrated to provide absorptivity to polysulfide for Li-S batteries. Appropriate doping hetero-atoms as well as forming composites can also increase the charge storage capacity of graphene. These approaches may be also incorporated into the porous graphene to exert the synergistic effects. The pores on graphene favour the refinement and immobilisation of nanoparticles with small sizes which have more active sites. The advantages can even be applied in the catalysis filed.

A totally different structure of porous graphene from those described in previous sections is always sought after for testing in the EES research community. One illusion is the possible graphene arrays and porous graphene arrays using templates such as TiO_2. With the unremitting efforts from multidiscipline, it is believed that the porous graphene materials can open up significant vista for the next generation EES devices in the near future.

References

Becker, H.I. 1957 Low voltage electrolytic capacitor. U.S. Patent # 2,800,616.

Black, R., S.H. Oh, J.-H. Lee, T. Yim, B. Adams and L.F. Nazar. 2012. Screening for superoxide reactivity in Li-O_2 batteries: effect on Li_2O_2/LiOH crystallization. J. Am. Chem. Soc. 134: 2902–2905.

Bruce, P.G., S.A. Freunberger, L.J. Hardwick and J.-M. Tarascon. 2012. Li-O_2 and Li-S batteries with high energy storage. Nat. Mater.. 11: 19–29.

Burke, A. 2007. R&D considerations for the performance and application of electrochemical capacitors. Electrochim. Acta 53: 1083–1091.

Chakrabarti, A., J. Lu, J.C. Skrabutenas, T. Xu, Z. Xiao, J.A. Maguire et al. 2011. Conversion of carbon dioxide to few-layer graphene. J. Mater. Chem. 21: 9491–9493.

Chen, G.Z. 2017. Supercapacitor and supercapattery as emerging electrochemical energy stores. Int. Mater. Rev. 62: 173–202.

Chen, K., L. Chen, Y. Chen, H. Bai and L. Li. 2012. Three-dimensional porous graphene-based composite materials: electrochemical synthesis and application. J. Mater. Chem. 22: 20968–20976.

Chen, W. and L. Yan. 2011. *In situ* self-assembly of mild chemical reduction graphene for three-dimensional architectures. Nanoscale 3: 3132–3137.

Chen, Y., X. Zhang, P. Yu and Y. Ma. 2009. Stable dispersions of graphene and highly conducting graphene films: a new approach to creating colloids of graphene monolayers. Chem. Commun. 4527–4529.

Chen, Y., X. Zhang, P. Yu and Y. Ma. 2010. Electrophoretic deposition of graphene nanosheets on nickel foams for electrochemical capacitors. J. Power Sources 195: 3031–3035.

Chen, Z., W. Ren, L. Gao, B. Liu, S. Pei and H.-M. Cheng. 2011. Three-dimensional flexible and conductive interconnected graphene networks grown by chemical vapour deposition. Nat. Mater. 10: 424–428.

Choi, B.G., S.-J. Chang, Y.B. Lee, J.S. Bae, H.J. Kim and Y.S. Huh. 2012. 3D heterostructured architectures of Co_3O_4 nanoparticles deposited on porous graphene surfaces for high performance of lithium ion batteries. Nanoscale 4: 5924–5930.

Choi, N.-S., Z. Chen, S.A. Freunberger, X. Ji, Y.-K. Sun, K. Amine et al. 2012. Challenges facing lithium batteries and electrical double-layer capacitors. Angew. Chem. Int. Ed. 51: 9994–10024.

Dikin, D.A., S. Stankovich, E.J. Zimney, R.D. Piner, G.H.B. Dommett, G. Evmenenko et al. 2007. Preparation and characterization of graphene oxide paper. Nature 448: 457–460.

Fan, Z., Y. Liu, J. Yan, G. Ning, Q. Wang, T. Wei et al. 2012. Template-directed synthesis of pillared-porous carbon nanosheet architectures: high-performance electrode materials for supercapacitors. Adv. Energy Mater. 2: 419–424.

Fong, R., U. von Sacken and J.R. Dahn. 1990. Studies of lithium intercalation into carbons using nonaqueous electrochemical cells. J. Electrochem. Soc. 137: 2009–2013.

Geim, A.K. and K.S. Novoselov. 2007. The rise of graphene. Nat. Mater. 6: 183–191.

Guo, X., P. Liu, J. Han, Y. Ito, A. Hirata, T. Fujita et al. 2015. 3D nanoporous nitrogen-doped graphene with encapsulated RuO_2 nanoparticles for Li-O_2 batteries. Adv. Mater. 27: 6137–6143.

Inagaki, M., H. Konno and O. Tanaike. 2010. Carbon materials for electrochemical capacitors. J. Power Sources 195: 7880–7903.

Ji, H., L. Zhang, M.T. Pettes, H. Li, S. Chen, L. Shi et al. 2012. Ultrathin graphite foam: a three-dimensional conductive network for battery electrodes. Nano Lett. 12: 2446–2451.

Ji, L., M. Rao, H. Zheng, L. Zhang, Y. Li, W. Duan et al. 2011. Graphene oxide as a sulfur immobilizer in high performance lithium/sulfur cells. J. Am. Chem. Soc. 133: 18522–18525.

Ji, X., K.T. Lee and L.F. Nazar. 2009. A highly ordered nanostructured carbon-sulphur cathode for lithium-sulphur batteries. Nat. Mater. 8: 500–506.

Kotz, R. and M. Carlen. 2000. Principles and applications of electrochemical capacitors. Electrochim. Acta 45: 2483–2498.

Laoire, C.O., S. Mukerjee, K.M. Abraham, E.J. Plichta and M.A. Hendrickson. 2009. Elucidating the mechanism of oxygen reduction for lithium-air battery applications. J. Phys. Chem. C 113: 20127–20134.

Lee, J., J. Kim and T. Hyeon. 2006. Recent progress in the synthesis of porous carbon materials. Adv. Mater. 18: 2073–2094.

Lee, S.H., H.W. Kim, J.O. Hwang, W.J. Lee, J. Kwon, C.W. Bielawski et al. 2010. Three-dimensional self-assembly of graphene oxide platelets into mechanically flexible macroporous carbon films. Angew. Chem. Int. Ed. 49: 10084–10088.

Li, C., X. Zhang, K. Wang, X. Sun, G. Liu, J. Li et al. 2017. Scalable self-propagating high-temperature synthesis of graphene for supercapacitors with superior power density and cyclic stability. Adv. Mater. 29: 1604690.

Li, D., M.B. Muller, S. Gilje, R.B. Kaner and G.G. Wallace. 2008. Processable aqueous dispersions of graphene nanosheets. Nat. Nanotechnol. 3: 101–105.

Li, X., W. Cai, J. An, S. Kim, J. Nah, D. Yang et al. 2009. Large-area synthesis of high-quality and uniform graphene films on copper foils. Science 324: 1312–1314.

Liu, F., P. Ming and J. Li. 2007. Ab initio calculation of ideal strength and phonon instability of graphene under tension. Phys. Rev. B 76: 064120.

Lv, W., D.-M. Tang, Y.-B. He, C.-H. You, Z.-Q. Shi, X.-C. Chen et al. 2009. Low-temperature exfoliated graphenes: vacuum-promoted exfoliation and electrochemical energy storage. ACS Nano 3: 3730–3736.

Maiyalagan, T., X. Dong, P. Chen and X. Wang. 2012. Electrodeposited Pt on three-dimensional interconnected graphene as a free-standing electrode for fuel cell application. J. Mater. Chem. 22: 5286–5290.

Mattevi, C., G. Eda, S. Agnoli, S. Miller, K.A. Mkhoyan, O. Celik et al. 2009. Evolution of electrical, chemical, and structural properties of transparent and conducting chemically derived graphene thin films. Adv. Funct. Mater. 19: 2577–2583.

Mizushima, K., P.C. Jones, P.J. Wiseman and J.B. Goodenough. 1980. Li_xCoO_2 $(0<x<-1)$: a new cathode material for batteries of high energy density. Mater. Res. Bull. 15: 783–789.

Murali, S., J.R. Potts, S. Stoller, J. Park, M.D. Stoller, L.L. Zhang et al. 2012. Preparation of activated graphene and effect of activation parameters on electrochemical capacitance. Carbon 50: 3482–3485.

Murugan, A.V., T. Muraliganth and A. Manthiram. 2009. Rapid, facile microwave-solvothermal synthesis of graphene nanosheets and their polyaniline nanocomposites for energy strorage. Chem. Mater. 21: 5004–5006.

Nagaura, T. and K. Tozawa. 1990. Lithium-ion rechargeable battery. Prog. Batt. Solar Cells 9: 209–217.

Nethravathi, C. and M. Rajamathi. 2008. Chemically modified graphene sheets produced by the solvothermal reduction of colloidal dispersions of graphite oxide. Carbon 46: 1994–1998.

Ning, G., Z. Fan, G. Wang, J. Gao, W. Qian and F. Wei. 2011. Gram-scale synthesis of nanomesh graphene with high surface area and its application in supercapacitor electrodes. Chem. Commun. 47: 5976–5978.

Novoselov, K.S., A.K. Geim, S.V. Morozov, D. Jiang, Y. Zhang, S.V. Dubonos et al. 2004. Electric field effect in atomically thin carbon films. Science 306: 666–669.

Ottakam Thotiyl, M.M., S.A. Freunberger, Z. Peng and P.G. Bruce. 2013a. The carbon electrode in nonaqueous $Li-O_2$ Cells. J. Am. Chem. Soc. 135: 494–500.

Ottakam Thotiyl, M.M., S.A. Freunberger, Z. Peng, Y. Chen, Z. Liu and P.G. Bruce. 2013b. A stable cathode for the aprotic $Li-O_2$ battery. Nat. Mater. 12: 1050–1056.

Padhi, A.K., K.S. Nanjundaswamy and J.B. Goodenough. 1997. Phospho-olivines as positive-electrode materials for rechargeable lithium batteries. J. Electrochem. Soc. 144: 1188–1194.

Paek, S.-M., E. Yoo and I. Honma. 2009. Enhanced cyclic performance and lithium storage capacity of SnO_2/graphene nanoporous electrodes with three-dimensionally delaminated flexible structure. Nano Lett. 9: 72–75.

Pandolfo, A.G. and A.F. Hollenkamp. 2006. Carbon properties and their role in supercapacitors. J. Power Sources 157: 11–27.

Partoens, B. and F.M. Peeters. 2006. From graphene to graphite: electronic structure around the K point. Phys. Rev. B 74: 075404.

Peigney, A., C. Laurent, E. Flahaut, R.R. Bacsa and A. Rousset. 2001. Specific surface area of carbon nanotubes and bundles of carbon nanotubes. Carbon 39: 507–514.

Peng, Z., S.A. Freunberger, Y. Chen and P.G. Bruce. 2012. A reversible and higher-rate $Li-O_2$ battery. Science 337: 563–566.

Raymundo-Piñero, E., P. Azaïs, T. Cacciaguerra, D. Cazorla-Amorós, A. Linares-Solano and F. Béguin. 2005. KOH and NaOH activation mechanisms of multiwalled carbon nanotubes with different structural organisation. Carbon 43: 786–795.

Riaz, A., K.-N. Jung, W. Chang, S.-B. Lee, T.-H. Lim, S.-J. Park et al. 2013. Carbon-free cobalt oxide cathodes with tunable nanoarchitectures for rechargeable lithium-oxygen batteries. Chem. Commun. 49: 5984–5986.

Rouquerol, J., D. Avnir, C.W. Fairbridge, D.H. Everett, J.H. Haynes, N. Pernicone et al. 1994. Recommendations for the characterization of porous solids. Pure Appl. Chem. 66: 1739–1758.

Sevilla, M. and R. Mokaya. 2014. Energy storage applications of activated carbons: supercapacitors and hydrogen storage. Energy Environ. Science 7: 1250–1280.

Si, Y. and E.T. Samulski. 2008. Exfoliated graphene separated by platinum nanoparticles. Chem. Mater. 20: 6792–6797.

Stankovich, S., D.A. Dikin, R.D. Piner, K.A. Kohlhaas, A. Kleinhammes, Y. Jia et al. 2007. Synthesis of graphene-based nanosheets via chemical reduction of exfoliated graphite oxide. Carbon 45: 1558–1565.

Stoller, M.D., S. Park, Y. Zhu, J. An and R.S. Ruoff. 2008. Graphene-based ultracapacitors. Nano Lett. 8: 3498–3502.

Sun, B., X. Huang, S. Chen, P. Munroe and G. Wang. 2014. Porous graphene nanoarchitectures: an efficient catalyst for low charge-overpotential, long life, and high capacity lithium-oxygen batteries. Nano Lett. 14: 3145–3152.

Sun, H., Z. Xu and C. Gao. 2013. Multifunctional, ultra-flyweight, synergistically assembled carbon aerogels. Adv. Mater. 25: 2554–2560.

Taberna, P.L., P. Simon and J.F. Fauvarque 2003. Electrochemical characteristics and impedance spectroscopy studies of carbon-carbon supercapacitors. J. Electrochem. Soc. 150: A292–A300.

Thackeray, M.M., P.J. Johnson, L.A. de Picciotto, P.G. Bruce and J.B. Goodenough. 1984. Electrochemical extraction of lithium from $LiMn_2O_4$. Mater. Res. Bull. 19: 179–187.

Wang, X., Y. Zhang, C. Zhi, X. Wang, D. Tang, Y. Xu et al. 2013. Three-dimensional strutted graphene grown by substrate-free sugar blowing for high-power-density supercapacitors. Nat. Commun. 4: 2905.

Wei, W., S. Yang, H. Zhou, I. Lieberwirth, X. Feng and K. Müllen. 2013. 3D graphene foams cross-linked with pre-encapsulated Fe_3O_4 nanospheres for enhanced lithium storage. Adv. Mater. 25: 2909–2914.

Whittingham, M.S. 1976. Electrical energy storage and intercalation chemistry. Science 192: 1126–1127.

Williford, R.E. and J.-G. Zhang. 2009. Air electrode design for sustained high power operation of Li/air batteries. J. Power Sources 194: 1164–1170.

Wu, S., G. Chen, N.Y. Kim, K. Ni, W. Zeng, Y. Zhao et al. 2016. Creating pores on graphene platelets by low-temperature KOH activation for enhanced electrochemical performance. Small 12: 2376–2384.

Xing, Z., B. Wang, W. Gao, C. Pan, J.K. Halsted, E.S. Chong et al. 2015. Reducing CO_2 to dense nanoporous graphene by Mg/Zn for high power electrochemical capacitors. Nano Energy 11: 600–610.

Xu, C., Y. Wu, X. Zhao, X. Wang, G. Du, J. Zhang et al. 2015. Sulfur/three-dimensional graphene composite for high performance lithium-sulfur batteries. J. Power Sources 275: 22–25.

Xu, Y., K. Sheng, C. Li and G. Shi. 2010. Self-assembled graphene hydrogel via a one-step hydrothermal process. ACS Nano 4: 4324–4330.

Xu, Y., Z. Lin, X. Huang, Y. Wang, Y. Huang and X. Duan. 2013. Graphene hydrogels: functionalized graphene hydrogel-based high-performance supercapacitors. Adv. Mater. 25: 5828–5828.

Xu, Y., Z. Lin, X. Zhong, X. Huang, N.O. Weiss, Y. Huang et al. 2014. Holey graphene frameworks for highly efficient capacitive energy storage. Nat. Commun. 5: 4554.

Yang, X., J. Zhu, L. Qiu and D. Li. 2011. Bioinspired effective prevention of restacking in multilayered graphene films: Towards the next generation of high-performance supercapacitors. Adv. Mater. 23: 2833–2838.

Yang, X., C. Cheng, Y. Wang, L. Qiu and D. Li. 2013. Liquid-mediated dense integration of graphene materials for compact capacitive energy storage. Science 341: 534–537.

Yoo, E., J. Kim, E. Hosono, H.-S. Zhou, T. Kudo and I. Honma. 2008. Large reversible Li storage of graphene nanosheet families for use in rechargeable lithium ion batteries. Nano Lett. 8: 2277–2282.

Zhang, H., X. Zhang, X. Sun and Y. Ma. 2013a. Shape-controlled synthesis of nanocarbons through direct conversion of carbon dioxide. Sci. Rep. 3: 3534.

Zhang, H., X. Zhang, X. Sun, D. Zhang, H. Lin, C. Wang et al. 2013b. Large-scale production of nanographene sheets with a controlled mesoporous architecture as high-performance electrochemical electrode materials. ChemSusChem. 6: 1084–1090.

Zhang, L. and G. Shi. 2011. Preparation of highly conductive graphene hydrogels for fabricating supercapacitors with high rate capability. J. Phys. Chem. C 115: 17206–17212.

Zhang, L.L., X. Zhao, M.D. Stoller, Y. Zhu, H. Ji, S. Murali et al. 2012. Highly conductive and porous activated reduced graphene oxide films for high-power supercapacitors. Nano Lett. 12: 1806–1812.

Zhang, X., Z. Sui, B. Xu, S. Yue, Y. Luo, W. Zhan et al. 2011. Mechanically strong and highly conductive graphene aerogel and its use as electrodes for electrochemical power sources. J. Mater. Chem. 21: 6494–6497.

Zhang, Y., Y.-W. Tan, H.L. Stormer and P. Kim. 2005. Experimental observation of the quantum Hall effect and Berry's phase in graphene. Nature 438: 201–204.

Zhao, M.-Q., Q. Zhang, J.-Q. Huang, G.-L. Tian, J.-Q. Nie, H.-J. Peng et al. 2014. Unstacked double-layer templated graphene for high-rate lithium-sulphur batteries. Nat. Commun. 5: 3410.

Zhou, G., L.-C. Yin, D.-W. Wang, L. Li, S. Pei, I.R. Gentle et al. 2013. Fibrous hybrid of graphene and sulfur nanocrystals for high-performance lithium-sulfur batteries. ACS Nano 7: 5367–5375.

Zhu, Y., S. Murali, M.D. Stoller, A. Velamakanni, R.D. Piner and R.S. Ruoff. 2010. Microwave assisted exfoliation and reduction of graphite oxide for ultracapacitors. Carbon 48: 2118–2122.

Zhu, Y., S. Murali, M.D. Stoller, K.J. Ganesh, W. Cai, P.J. Ferreira et al. 2011. Carbon-based supercapacitors produced by activation of graphene. Science 332: 1537–1541.

4

Role of Heteroatoms on the Performance of Porous Carbons as Electrode in Electrochemical Capacitors

Ramiro Ruiz-Rosas,[1,a] *Edwin Bohórquez-Guarín,*[2] *Diego Cazorla-Amorós*[1,b] and *Emilia Morallón*[2,*]

1. Introduction

Electrochemical capacitors or supercapacitors are electrochemical energy storage systems where energy is stored in an electrode/electrolyte interface by means of the electrostatic attraction between the polarised surface of the electrodes and the ions in the electrolyte forming the electrical double layer (Inagaki et al. 2010, Yu et al. 2013). They have recently received special attention due to their highly interesting properties such as: (i) they are able to store medium energy densities, in the range of several Watts-hour per kilogram, (ii) they can be charged in few seconds, (iii) they are able to provide high power densities, in the range of few kiloWatts per kilogram, (iv) they can stand power peaks and be charged and discharged intermittently without being damaged, and (v) they have long cycle life, from ten thousand to more than one million of charge-discharge cycles.

Figure 1 presents a scheme of a supercapacitor. The main components are: (i) the electrodes, which are usually composed of porous carbon particles and a small amount of additives, such as conductivity promoters (usually a carbon black, novel nanostructured carbon materials, such as carbon nanotubes (Portet et al. 2005), have also been used) and a binder, mostly polytetrafluoroethylene, PTFE, and polyvinylidene difluoride, PVDF (Zhong et al. 2015), although other polymeric matrixes can be employed for this purpose (Böckenfeld et al. 2013, Aslan et al. 2014), and even binderless electrodes of improved conductivity can be used (Berenguer et al. 2016); (ii) the electrolyte, a solution of high conductivity that provides the ions that will form the electrical double layer; (iii) the separator, a porous non-conductive membrane that isolates both electrodes. Other necessary components as collectors or caskets are not considered in this chapter.

[1] Materials Institute and Inorganic Chemistry Department, University of Alicante, Ap. 99, 03080, Alicante, Spain.
[a] E-mail: ramiro@ua.es
[b] E-mail: cazorla@ua.es
[2] Materials Institute and Physical Chemistry Department, University of Alicante, Ap. 99, 03080, Alicante, Spain.
  E-mail: edwbog@hotmail.com
* Corresponding author: morallon@ua.es

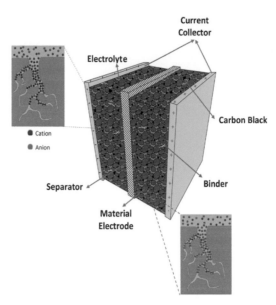

Fig. 1. Scheme of a supercapacitor cell.

When the supercapacitor is loaded, the ions of the electrolyte migrate within the porosity of the electrode to compensate the charge in the electrical double layer formed between the electrolyte and electrode surface. The capacitance of the electrical double layer, considering a plane-parallel electrical capacitor, is proportional to the electrode surface, and inversely proportional to the separation distance between the ions and the surface of the electrode (Conway 1991, Inagaki et al. 2010, Béguin et al. 2014). Consequently, porous carbon materials, with their right conjugation of properties, such as large available surface area, adequate electrical conductivity and high electrochemical stability, constitute the most employed electrode materials in the field of supercapacitors.

1.1 Porous Carbon Materials

Porous carbon materials are solids with a high content of carbon atoms that exhibit large apparent surface areas and pore volumes (Bansal et al. 1988, Marsh and Reinoso 2006, Lozano-Castelló et al. 2010). Most of these solids have a wide pore size distribution (PSD), and can be obtained by different techniques and with different morphologies, such as powder, pellets, granules, fibers, monoliths, etc.

In this chapter, we will pay attention to the properties of the porous carbons that determine their performance when used as electrode in supercapacitors. Most of these properties are influenced by the surface chemistry of the carbon materials that can be modified through different strategies.

Porosity in carbon materials is responsible for their adsorption properties. The pore width is the most relevant porosity parameter and it determines the adsorption mechanism. The most accepted classification of pores according to their sizes is that recommended by the International Union of Pure and Applied Chemistry (Sing 1985). In it, micropores are those of width less than 2 nm. Mesopores are those of width between 2 and 50 nm and macropores are those of width higher than 50 nm.

The presence of macropores is mostly undesirable, since they do not provide as much capacitance as smaller pores while they deliver a lower packing density of the material. The narrow mesopores can contribute to the capacitance. They can also improve ion mobility by acting as electrolyte reservoirs, bringing the ions closer to the surface where the electrical double layer is formed. Finally, micropores are the main contributors to the double layer capacitance since they provide the higher contribution to surface area. The micropore size can affect the capacitance of the double layer. For instance, ion sieving mechanism, i.e., ions being unable to enter into pores smaller than their size, can appear when micropores are too narrow (Eliad et al. 2001). However, micropores of sizes similar to the size of unsolvated ions are known to enhance the capacitance (Chmiola et al. 2006).

The connectivity, tortuosity and arrangement of pores also play an important role on the performance of a porous carbon as an electrode. In order to provide a high-power output, the ion mobility within the porosity of the electrode needs to be fast. Otherwise, ions will not be able to migrate to the electrode-electrolyte interface on time to compensate the charge of the electrode surface when the electrode operates at high rate (i.e., at high power or current). Several features can enhance the mass transfer and ion mobility in the porosity (Bleda-Martínez et al. 2010, Li et al. 2012), such as: (i) a well-interconnected porous structure, (ii) the presence of narrow mesopores connected to the micropores, (iii) a low tortuosity of the micropores and (iv) a proper arrangement of porosity like in the case of hierarchical porous carbon. These are necessary characteristics required in porous carbon materials employed as electrodes of supercapacitor for high power applications.

Activated carbons (ACs) are the most commonly used electrodes in supercapacitors. The synthesis of these ACs can be done by the so-called physical activation process in which a carbon-rich precursor is submitted to a controlled gasification at high temperatures (700–1000°C) mainly with carbon dioxide and steam, although other gases that can produce controlled carbon gasification can be used (Ahmadpour and Do 1996, Marsh and Reinoso 2006). This method can be employed for the production of microporous ACs with surface areas over 1500 $m^2\ g^{-1}$.

On the other hand, chemical activation can be directly conducted over lignocellulosic precursors (Gonzalez-Serrano et al. 1997, Jagtoyen and Derbyshire 1998, Linares-Solano et al. 2008, Rosas et al. 2009) and coal (Ahmadpour and Do 1996, Lozano-Castelló et al. 2001, Linares-Solano et al. 2008). In this procedure, the carbon precursor is mixed with the chemical agent, carbonised at temperatures and times usually lower than those employed in physical activation, and thoroughly washed in order to recover the chemical agent (phosphoric acid, zinc chloride, alkaline hydroxides, etc.) and release the porosity of the resulting AC. Thus, ACs with surfaces over 3000 $m^2\ g^{-1}$ and tunable pore size distributions that can contain micropores, mesopores or both of them, can be prepared (Ahmadpour and Do 1996, Jagtoyen and Derbyshire 1998, Lozano-Castelló et al. 2001, Rodríguez-Reinoso and Molina-Sabio 1992, Lillo-Ródenas et al. 2003, 2004, Linares-Solano et al. 2008).

Chemical activation presents several advantages over physical activation (Linares-Solano et al. 2008): (i) it is usually conducted in one stage, (ii) yields are higher, (iii) lower temperatures and holding times are required and (iv) porosity development and size distribution is much more tunable. However, it also presents several disadvantages, the most important being the need of a washing step after the heat treatment for releasing the porosity and recovering the chemical agent. Moreover, the use of strong acids and alkalis makes it necessary to employ corrosion-resistance equipment.

Porous carbon materials can be also obtained by template methods. In them, a carbon precursor is mixed with a templating agent, in order to fill either the porosity of the template or the space within the template particles. The mixture is later carbonised, obtaining a carbon-template composite. After the removal of the templating agent from the composite, a negative carbon replica of the template structure is obtained.

These methods can be divided in soft templating and hard templating, depending on the use of solids or liquids, respectively, as the porogen agent (Kyotani 2003, Nishihara and Kyotani 2012). Hard templating procedures usually employ porous crystalline solids, such as ordered mesoporous silicas (Ryoo et al. 1999), zeolites (Ma et al. 2000) and anodic aluminum oxides (Kyotani et al. 1995). These templates are filled by a carbon rich precursor via infiltration, polymerisation or by chemical vapour deposition. After the porosity is filled by carbon, the template is removed (usually by means of a strong acid washing), releasing a carbon material that has replicated the inverse of the structure of the template. This strategy can be also implemented with nanosized amorphous solids, such as colloidal silica (Zakhidov et al. 1998). These solids are imbedded within a carbon matrix and serve as template in order to cast large mesopores in them. As for soft templating procedures, the use of block-copolymers or surfactants (Chuenchom et al. 2012), which can form micelles of controllable diameter that drives the pore size of the resulting solid, are some of the most commonly employed strategies.

Templated carbons have unique properties, such as large structural order, high surface areas, ordered porosity arrangement and connectivity. These properties are extremely valuable for electrochemical applications. However, their high production cost makes their commercialisation difficult.

Carbon aerogels are porous carbon solids that are produced from the carbonisation of organic aerogels. They are prepared by the polymerisation of resorcinol with formaldehyde in a sol-gel process (Pekala 1989, Pekala et al. 1998). The resulting gel goes through a drying step that is critical for maintaining the porosity, which is usually conducted with supercritical carbon dioxide. The aerogel is subsequently pyrolysed at high temperatures (higher than 900°C). Carbon aerogels have surface areas ranging from 500 to 1000 m² g⁻¹, controllable and uniform pores of sizes between 2 and 50 nm, have a high electrical conductivity and can be shaped in very different morphologies, even in the form of monolith or seamless electrodes (Pandolfo and Hollenkamp 2006, Calvo et al. 2010, 2013). However, activation of the aerogel in order to develop microporosity is usually required for increasing the surface area and their performance as electrodes of supercapacitors (Pröbstle et al. 2003).

1.2 Parameters Governing the Performance of Porous Carbon based Supercapacitors

The most important characteristics of electrochemical capacitors or supercapacitors are their capacitance, energy density, power density and durability. Other parameters, such as energy efficiency or self-discharge, should also be considered.

It has been previously mentioned that supercapacitors are constructed using two electrodes submerged in an electrolyte, being isolated by a wetted ion permeable separator, Fig. 1. Thus, the supercapacitor consists of two capacitors arranged in series. Hence, the capacitance of the full device, C, can be related to the capacitances of the negative (C^-) and positive (C^+) electrodes through the following equation:

$$\frac{1}{C} = \frac{1}{C^-} + \frac{1}{C^+}$$

Capacitances of each electrode and the supercapacitor are commonly expressed as gravimetric capacitance, C_g. Since in most of the cases the space occupied by the device can limit its application, especially in the case of mobile systems, volumetric capacitance is one of the most important parameters from an application point of view. Volumetric capacitance, C_v, can be determined from the gravimetric capacitance and the density of the electrode, ρ. For instance, excessive porosity or a large contribution of wide pores in the carbon materials employed as electrodes could lead to a low packing density and would render wasted spaces (Ruiz-Rosas et al. 2014). Capacitance is usually evaluated as the total charge retrieved from the device, Q, divided by the operating voltage, U, during a galvanostatic discharge cycle or by cyclic voltammetry (CV) (Gryglewicz et al. 2005, Pandolfo and Hollenkamp 2006).

A relevant part of the research involving porous carbons for supercapacitors is developed correlating the capacitance with porous texture and surface area (Lozano-Castelló et al. 2003); however, it has been demonstrated that the expected linear relationship between capacitance and total surface area does not always occur. Important deviations from this trend confirm that there are other factors that contribute significantly to the capacitance; for example, not all the surface area determined by gas adsorption experiments is electrochemically accessible or homogeneous in terms of the energy involved during the formation of the electrical double layer. Essentially, the capacitance of porous carbons is strongly dependent on porosity, surface chemistry, electrical conductivity and preparation method of the electrode and electrolyte characteristics (Bleda-Martínez et al. 2005, Gryglewicz et al. 2005).

The total energy density (E) of an ideal supercapacitor (i.e., all charge is stored through double layer formation and capacitance is independent of the potential) can be derived from the next equation:

$$E = \int_0^U Q/w \cdot dU = \int_0^U C_g \, U \cdot dU = \frac{1}{2} \cdot C_g \cdot U^2$$

In it, w and U stands for the weight of the electrodes and the working voltage, respectively. A supercapacitor can be theoretically loaded up to the thermodynamic stability limit of the electrolyte, being 1.23 V for aqueous electrolytes, 2.5–2.7 V for propylene carbonate and acetonitrile, and higher than 3 V for ionic liquids (Pandolfo and Hollenkamp 2006, Peng et al. 2010, Gao et al. 2012, Béguin et al. 2014, Piñeiro-Prado et al. 2016). Energy density is usually measured by galvanostatic charge-discharge experiment.

The power density of ideal supercapacitors, P_g, is usually estimated as follows:

$$P_g = \frac{U^2}{4 \cdot w \cdot R_{cell}}$$

In it, R_{cell} represents the cell internal resistance. This resistance is usually described as the electrical serial resistances of the system. It results from the combination of several factors, namely the collector-electrode contact resistance, the electrode resistance and the electrolyte resistance (Kötz and Carlen 2000, Taberna et al. 2003, Fletcher et al. 2013). Some of these resistances relies on a good construction of the cell and of the electrode, whereas others are inherent to the active phase, as the electrode resistance, which is related to the electrical conductivity of the porous carbon, and the electrolyte resistance within the pore system, which depends on the mobility of ions within the porosity of the porous carbon. A high internal resistance would limit the power of the supercapacitor and, consequently, its application. Resistance can be evaluated by either galvanostatic charge-discharge experiments or by electrochemical impedance spectroscopy.

Finally, durability is the ability of a supercapacitor for maintaining the capacitance, energy and power characteristics after thousands of charge-discharge cycles. Durability is usually assessed by following these parameters during a series of galvanostatic charge-discharge experiments at different current densities or power densities. These tests are considered to be successfully passed when the cells are able to retain most of their performance parameters (i.e., more than 80 per cent of the capacitance or a resistance increase lower than 100 per cent) (Ratajczak et al. 2014). Accelerated tests where the capacitors are submitted to dramatic conditions during several hours or even days, such as high temperatures or high voltages (i.e., floating tests), are also proposed (Niu et al. 2006, Cazorla-Amorós et al. 2010, Weingarth et al. 2013, Salinas-Torres et al. 2015).

1.3 Properties of Porous Carbons Related to their Performance as Electrodes in Supercapacitor

The capacitor performance parameters mentioned in Section 1.2 are affected by the properties of porous carbons like wettability with the electrolyte, electrical conductivity, redox activity and electrochemical stability.

1.3.1 Electrical Conductivity

The electrical conductivity of porous carbons depends on the raw material and the method employed during their synthesis. Those carbons prepared from precursors which are poor in aromatic carbon content, such as biomass wastes, will render low electrical conductivity carbons when they are obtained at low carbonisation temperatures (300–600°C). At these temperatures, the predominant carbonisation reactions are aromatic condensation and cyclisation (Inagaki 2010), and solids with low aromatic carbon content will need the occurrence of polycondensation, demethanation and dehydrogenation reactions in order to develop graphene-like polyaromatic structures with high electrical conductivity. Enhanced electrical conductivity can be achieved by using higher carbonisation/ activation temperatures (Sánchez-González et al. 2011, Ruiz-Rosas et al. 2014), chemical activation methods based on alkaline hydroxides (Lozano-Castelló et al. 2003) or graphitisation agents such as transition metal compounds (Chang et al. 2015). When high-rank coals, mesophase pitch and other highly ordered carbon precursors with pre-graphitic structures, such as highly conductive carbon fibers, are employed (Bleda-Martínez et al. 2005), electrical conductivity of the obtained porous carbons is higher.

The presence of heteroatoms in edge or in-plane sites of the graphene layers can also modify the electrical conductivity of porous carbon materials. For instance, the withdrawing character of certain functional groups, like carboxylic groups, reduces the electron mobility on the π-cloud; while nitrogen doping can enhance conductivity via the injection of additional π-electrons. The effect of surface oxygen functional groups and nitrogen functionalities will be detailed in the next sections.

1.3.2 Wettability by the Electrolyte

It has been previously mentioned that the capacitance of a porous carbon material is expected to be proportional to their surface area, so it is easy to conclude that the surface area of the electrodes must be maximised in order to enhance the energy density of supercapacitors. However, a lack of direct relationship between capacitance and BET surface area is frequently reported in the literature (Barbieri et al. 2005, Centeno and Stoeckli 2006). In a first approximation, one could expect that this phenomenon can be explained by ion sieving effects on pores with sizes lower than those of solvated ions (Eliad et al. 2001). Again, the literature is rich in experimental evidences about the opposite, i.e., capacitance can be even higher in narrow micropores with sizes similar to that of unsolvated ions (Chmiola et al. 2006, 2008).

This lack of relationship should be explained considering also the concept of wettability, that is, the ability of the electrolyte to maintain contact with the surface of the electrode. When wettability of the porosity by the solvent is high, most of the specific surface area of activated carbons will be available for the formation of the electrical double layer, leading to a positive relationship between surface area and capacitance. Otherwise, only a fraction of the porosity will be wetted by the electrolyte, and a clear trend of capacitance improvement with surface area development maybe will not be observed or this tendency cannot reach the maximum value according to the porosity of the material.

Since the surface of carbon materials is known to be highly hydrophobic, the wettability of pure carbon materials with electrolytes of non-polar character would be high, ensuring that the electrolyte reaches all the porosity that is accessible considering the size of the molecules and ions of the electrolyte. However, the polar character of most electrolytes may render inaccessible a considerable fraction of the internal surface of pure carbon materials. Wettability on such scenario needs to be improved by the introduction of surface functionalities with polar character on activated carbons (Lozano-Castelló et al. 2003, Bleda-Martínez et al. 2005, Centeno and Stoeckli 2006).

For instance, Szubzda et al. analysed the influence of pore structure and wettability in the capacitance of carbon aerogels (Szubzda et al. 2012). They evaluated water and propylene carbonate adsorption in aerogels prepared with and without the addition of a surfactant that increased the hydrophobicity of their carbons. They found that, no matter the microporous to mesoporous ratio of these aerogels, gravimetric capacitance in 6 M KOH was determined by water wettability, whereas the addition of the surfactant leaded to higher capacitances in 1 M $TEABF_4$ propylene carbonate.

The addition of heteroatoms is a straightforward method for enhancing the wettability of carbon surface. Certain oxygen and nitrogen functional groups can increase the polarity of the surface, while nitrogen and boron doping is known to facilitate the wettability of the electrolyte/electrode interface (González-Gaitán et al. 2015). The effect of heteroatoms on wettability will be detailed in the foregoing sections.

1.3.3 Pseudocapacitance

The predominant charge storage in supercapacitors is via electrostatic interaction; however, it is also possible to store energy by means of fast, reversible redox processes. This charge storage can imply faradaic reactions similar in kinetics to those employed in the storage mechanism of batteries, but it can also include much faster mechanisms, as electroactive adsorption processes, faradaic reactions without phase change or ion intercalation in conducting polymers (Conway 1991). When the charge storage by this mechanism renders an electrochemical behaviour similar to that of a capacitive material (because the implied redox reactions have a strong dependence on potential), the phenomenon is named as pseudocapacitance (Conway et al. 1997). Pseudocapacitance is one of the most promising approaches for achieving energy densities in aqueous electrolyte that can be comparable to those of organic ones. Capacitance can be increased by several hundreds of farads per gram (Conway 1991). However, huge challenges to permit their future applications, like extending the lifetime of pseudocapacitive materials and reducing their cost, still remain unaddressed.

Pseudocapacitance is usually related to metal oxides that show several overlapping redox processes, as in the case of ruthenium or manganese oxides, among others (Conway and Pell 2003, Wang et al. 2013), that involves proton and electron injection in a highly reversible process where no phase changes arise. The modification of the electrode surface with electroactive polymers such as polyaniline or polypyrrole is also known to develop pseudocapacitive behavior due to the redox processes that take place on these polymers at certain potentials (Arbizzani et al. 1996, Salinas-Torres et al. 2013).

Reversible redox reactions of surface functional groups are also known to increase the capacitance of carbon electrodes in aqueous electrolyte, and especially in acid or basic media (Conway et al. 1997), and they involve heteroatoms such as oxygen (Bleda-Martínez et al. 2006) and nitrogen (Hulicova et al. 2006). Then, the insertion of heteroatoms in the composition of porous carbons in form of different surface functional groups will modify their surface chemistry (González-Gaitán et al. 2015). The large variety of these functional groups makes necessary to characterise the surface chemistry in detail in order to identify and quantify the generated functionalities and discriminate how they can lead to different modifications of the properties that determine the electrochemical behaviour. Thus, and also as in any other of their potential applications, the performance of porous carbons is not only defined by their porosity, but also by their surface chemistry.

The surface functional groups with pseudocapacitive behaviour, the reaction mechanism and their applicability will be addressed in the following sections.

1.3.4 Electrochemical Stability

The lifetime of supercapacitors is also related to the electrochemical stability of porous carbon materials. When they are polarised, they can suffer from carbon electrochemical gasification (Kinoshita and Bett 1973a, Kinoshita 1988). They can also promote the electrochemical decomposition of the solvent, eventually producing a cell failure due to the formation of gases or solid residues that block the porosity of the electrodes (Zhu et al. 2008, Cazorla-Amorós et al. 2010, Kurzweil et al. 2015, Salinas-Torres et al. 2015).

The electrochemical stability of carbon materials is known to be highly related to the presence of electroactive carbon atoms in the edge sites of the graphene layers (Radovic 2010). Unsaturated edge sites in the form of carbene-like zigzag and carbine-like armchair sites has been proven to exist (Radovic and Bockrath 2005, Ishii et al. 2014) and to show notable differences on the electrochemical behaviour with the basal plane. They have one order of magnitude higher capacitance (Randin and Yeager 1971), an improved electron transfer rate (Banks et al. 2004), and they are more easily electrochemically oxidised (Berenguer et al. 2013). This last feature can explain the reasons behind the most intense ageing of the positive electrodes of supercapacitors in aqueous media; quinone-like surface oxygen groups on free edge sites are generated upon positive polarisation, which are subsequently oxidised to carboxylic moieties that later evolve as CO_2. The formation of electron-withdrawing groups, gases and electrolyte decomposition leads to capacitor failure (He et al. 2016).

In this sense, the presence of heteroatoms plays a dominant role in modulating the stability of carbon materials. Certain oxygen functionalities can facilitate the decomposition and polymerisation of organic solvent at excessively positive potentials (Cazorla-Amorós et al. 2010), while nitrogen and phosphorus functionalities can enhance the electrochemical stability in both aqueous and organic electrolytcs (Berenguer et al. 2015, Salinas-Torres et al. 2015, Mostazo-López et al. 2016). Given the preponderant role of heteroatoms in this field, this effect on electrochemical stability will be discussed in the following sections.

2. Oxygen-containing Functional Groups and their Role on the Electrochemical Behavior of Porous Carbons

Oxygen is the most frequently found heteroatom in porous carbon materials. Their presence is difficult to be avoided, since the dangling bonds at edge sites of the surface of freshly prepared porous carbons are highly reactive and get oxidised when the material is exposed to air.

Oxidation treatments are the preferred methods for a controlled introduction of surface oxygen groups. Two groups of methods can be distinguished, consisting in, essentially, chemical reactions in gas (most common treatment consisting in air oxidation) or liquid (wet oxidation) phases. These treatments employ an appropriate oxidising or reducing agent and, in some cases, further heat treatments in an inert atmosphere in order to selectively remove certain functionalities (Bandosz 2008). Plasma treatment is also employed for this purpose, though their use is much less frequent (García 1998, Paredes et al. 2000). In general, the influence of oxidative treatments on the textural and structural properties greatly depends on the reactivity of the carbon material. The selectivity of these methods is usually low, while the amount of surface oxygen groups depends on the oxidising power and concentration of the reagent, the temperature and the time of treatment. Wet oxidation treatments usually achieve a higher generation of surface oxygen groups of acid character (Moreno-Castilla et al. 2000), whereas air treatment produces a higher amount of basic or neutral surface groups (Polovina et al. 1997, Paredes et al. 2000). These treatments show some disadvantages, such as the aforementioned low selectivity, low control of the introduced amount of functionalities, important blockage of porosity if an excessive amount of surface groups is generated in microporous solids, or carbon gasification if too aggressive conditions are selected on air treatments.

Recently, studies on the electrochemical modification of porous carbon materials have revealed that electrochemical techniques can be employed to attain a more selective and controlled modification of the surface chemistry of carbon materials (González-Gaitán et al. 2015). Berenguer et al. reported the modification of a granular activated carbon by means of galvanostatic treatment in a filter press cell using acid, neutral and alkaline electrolytes (Berenguer et al. 2009). They demonstrated that, upon anodic polarisation, the textural properties of the activated carbons are preserved. However, different amounts and kinds of surface oxygen groups were introduced. Thus, the selection of the current and the electrolyte allows tailoring the surface chemistry achieved by the treatment. This treatment can be extended to other carbon morphologies, such as activated carbon cloths (Tabti et al. 2013, 2014), and even to different porous carbon materials, such as zeolite templated carbons (Berenguer et al. 2013) or hierarchical porous carbons (Ruiz-Rosas et al. 2014).

Oxygen functionalities have been studied for a long time. Figure 2 illustrates the rich variety of surface oxygen groups that can be found in carbon materials. There exist well established techniques

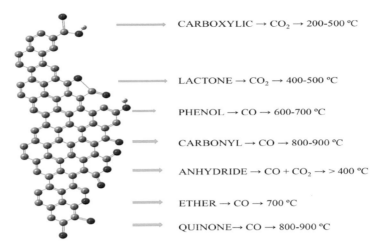

CARBOXYLIC → CO_2 → 200-500 °C

LACTONE → CO_2 → 400-500 °C

PHENOL → CO → 600-700 °C

CARBONYL → CO → 800-900 °C

ANHYDRIDE → CO + CO_2 → > 400 °C

ETHER → CO → 700 °C

QUINONE → CO → 800-900 °C

Fig. 2. Surface oxygen groups of carbon (grey spheres: carbon atoms; red spheres: oxygen atoms; white spheres: hydrogen atoms) and their decomposition temperatures.

that can be employed for their classification and quantification, such as Fourier Transform Infrared Spectroscopy (FTIR), X-Ray photoelectron spectroscopy (XPS), potentiometric titrations and Boehm titration and Temperature Programmed Desorption (TPD) (Boehm 1994, Salame and Bandosz 2001, Boehm 2002, Figueiredo et al. 1999).

TPD is one of the most powerful techniques available for this purpose (Otake and Jenkins 1993, Román-Martínez et al. 1993). In this technique, the carbon material is submitted to a thermal treatment in inert atmosphere at a constant heating rate. As the temperature increases, the functional groups on the surface of carbon materials are thermally decomposed, releasing certain gases at different temperatures, which depend on the kind of functional group. In the case of surface oxygen groups, they decompose mainly as CO and CO_2. In general, CO-evolving groups are those of neutral or slightly basic character, whereas CO_2-evolving ones are those of strong acid character. Furthermore, surface oxygen groups are distinguished by their decomposition temperature (Boehm 2002, Figueiredo et al. 1999, Otake and Jenkins 1993, Román-Martínez et al. 1993). Thus, carboxylic acids decompose as CO_2 at low temperatures (200–400°C). Anhydrides and lactones decompose as CO_2 at medium temperatures (400–500°C). Decomposition of anhydrides also releases an additional CO molecule. Phenols evolve as CO at higher temperatures (600–700°C), while carbonyl, ether and quinones are reported to be more stable, decomposing at temperatures over 700°C. Groups inserted in the graphene layer, such as pyrones, are expected to be even more thermally stable.

The modification of electrochemical behaviour of carbon materials by the introduction of oxygen heteroatoms is known from a long time. In a pioneer work, Kinoshita and Bett reported the electrochemical behaviour of carbon black with different degrees of oxidation by performing cyclic voltammetry (CV) in 1 M H_2SO_4 (Kinoshita and Bett 1973b). They observed the appearance of redox processes after the oxidation treatment, where the current of the redox peaks increased with the concentration of surface oxides. This feature has been reported in a large number of occasions so far, and is common for all kind of carbon materials acting as electrodes in acid media, such as carbon blacks, carbon nanotubes (Chen et al. 2002), graphene (Kuila et al. 2012), activated carbons (Tabti et al. 2013) or templated carbons (Berenguer et al. 2013). This behaviour, as well as other changes induced on the electrochemical properties in aqueous electrolyte by surface oxygen functionalities, has motivated several seminal works (Hsieh and Teng 2002a, Bleda-Martínez et al. 2005, 2006, Seredych et al. 2008).

Next, we will discuss their effect on the most relevant properties from the point of view of supercapacitors.

2.1 Effect on Electrical Conductivity

The conductivity of carbon materials strongly depends on the content of heteroatoms. On the graphitic carbon surfaces, edge sites are more reactive than basal sites as they are often associated with unpaired electrons, and the oxygen functionalities formed at these edges sites may increase the barrier for the electron mobility and consequently increase the resistivity of carbon material (Pandolfo and Hollenkamp 2006). The electron-withdrawing or donating character of functional groups can also tune the electrical conductivity of carbon materials. Thus, the presence of a large amount of electron-withdrawing functional groups, such as carboxylic acids, anhydrides and mostly CO_2-evolving groups during TPD, can drain delocalised л-electrons, resulting in a loss of electrical conductivity. Therefore, the electron delocalisation is favored by the removal of strong electron-withdrawing groups, such as those previously mentioned. In that sense, the removal of CO_2-type groups should improve capacitance retention and power characteristics of carbon electrodes used for electrochemical energy storage (Bleda-Martínez et al. 2006).

Tabti et al. (Tabti et al. 2013) conducted a series of electrochemical treatments over a granular activated carbon and an activated carbon cloth using a filter-press cell and analysed the generation of surface oxygen groups and their effect in the electrochemical behaviour of the obtained samples. Both carbon materials were electrochemically modified in a filter-press electrochemical cell, using 0.5 M NaCl as electrolyte. The two pristine materials have a similar pore size distribution, showing a similar content of narrow microporosity, which accounts for about 80 per cent of the total micropore volume. After the most aggressive electrochemical treatment, under anodic conditions, the surface area and the micropore volume of both materials only decreased ca. 20 per cent, while the cathodic treatment did not significantly modify the textural properties for both porous carbons. Therefore, not important changes were observed in the porosity after the electrochemical oxidation treatments.

The surface chemistry of these samples was evaluated by means of TPD. The amount of CO and CO_2 evolved after each treatment is plotted versus the specific charge (charge divided by weight of sample) employed in the treatment, Fig. 3. The anodically modified samples presented an important increase on the oxygen content. The increments in the surface oxygen groups grow logarithmically with the specific charge, demonstrating that for higher charges, a saturation of the carbon surface with surface oxygen groups is probably attained. It can also be observed that CO-type functionalities are transformed into CO_2-type ones when a larger specific current is employed as deduced from the

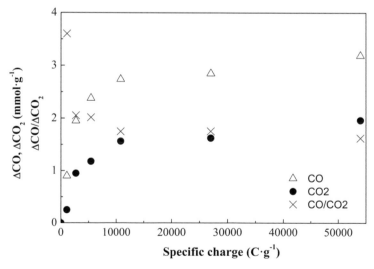

Fig. 3. Relationship between the increment of CO and CO_2 evolution as well as $\Delta CO/\Delta CO_2$ ratio and the specific charge applied during anodic treatments of porous carbons.

decrease on $\Delta CO/\Delta CO_2$ ratio (Fig. 3), which points out the possibility of tailoring the predominant functional groups by adjusting the time or the current of the treatment.

The electrochemical behaviour of these samples was studied before and after the electrochemical treatments. Figure 4 includes as an example the CVs obtained for a pristine activated carbon and after the highest electrooxidation treatment. The figure clearly shows that an extensive electrooxidation treatment produces an important distortion of the voltammograms, from a near rectangular to a tilted one. Tilted voltammograms are usually associated with ohmic losses owing to poor wettability or decreased electrical conductivity. The surface wettability is expected to be promoted by acidic, ionisable surface groups attached to the surface by the electrooxidation treatments, thus the distortion can be associated to the high amount of electron withdrawing surface groups.

The above results confirm that a severe oxidation would increase the number of CO_2-type groups, which are electron withdrawing, producing a loss in the conductivity by strong charge localisation on those surface oxygen complexes.

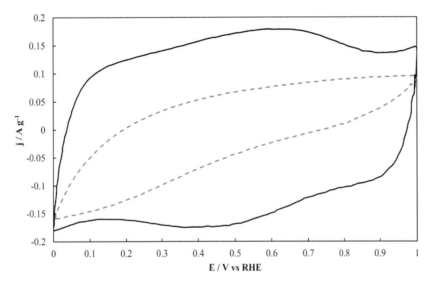

Fig. 4. Steady state voltammograms of the parent (solid line) and anodically treated porous carbon cloths (dashed line). Electrolyte: 0.5 M H_2SO_4. Sweep rate: 1 mVs^{-1}.

2.2 Effect on Wettability by the Electrolyte

Depending on the nature of the electrolyte used in the supercapacitor (organic or aqueous), different requirements arise for the surface chemistry because of the predominant character of the solvent and the different polarities of functional groups (Frackowiak and Béguin 2001).

The non-polar nature of carbon materials makes wettability with aqueous electrolytes low; however, the presence of oxygen functional groups facilitates the interaction with the water molecules. Thus, the wettability of carbon materials immersed in aqueous electrolyte is determined by the interaction of the surface charge and the dipolar moment of water (Hsieh and Teng 2002b). Treatments that generate oxygen groups on the surface of carbon materials could improve their wettability and therefore their interaction with aqueous-based electrolytes resulting in a better penetration of ions within the porous structure (Gryglewicz et al. 2005, Pandolfo and Hollenkamp 2006). In practice, the measurement of capacitance of porous carbon materials should be done after the complete soaking of the porosity of the electrodes by the electrolyte (Frackowiak and Béguin 2001).

Lozano-Castelló et al. (Lozano-Castelló et al. 2003) prepared activated carbon samples from three precursors: (i) seven activated carbons prepared from an anthracite (Al-A7), (ii) two activated carbons prepared from a bituminous coal (Bl, B2), (iii) an activated carbon prepared from a lignite coke (L) and, additionally, a commercial activated carbon with the highest surface area reported so far was also used (Maxsorb-A). The chemical activation of the samples was done with KOH, using different preparation conditions. All the samples presented type I isotherms, characteristic of microporous solids, and additionally the samples A7 and L presented isotherms characteristic of samples containing mesoporosity. Most of the functional groups at the surface of the activated carbons were of oxygen nature. All these samples were analysed as electrodes in organic media (1.0 M $LiClO_4$ in propylene carbonate).

Table 1 depicts the relationship between the BET surface area and the capacitance determined in the potential range 2–4 V vs. Li/Li^+. A general trend between capacitance and BET surface area for all the samples was observed, and a capacitance to surface area ratio between 0.06 and 0.08 F m^{-2} can be calculated for most of them. However, some deviations of this tendency were attributed to other characteristics of porous carbon materials such as pore size distribution (samples A1, A2) and surface chemistry (sample A4). Of interest is the sample A4, which, when compared to the sample A5 that has a similar pore size distribution, provides a higher capacitance than expected. The authors found that the amount of oxygen evolved in the TPD experiments (Table 1) is the highest for sample A4 and considered as the most plausible explanation for such capacitance increase, that the oxygen groups could improve the wettability of the carbon surface, which is of paramount importance to maximise the access of the electrolyte to the whole surface of carbon.

A similar finding to that of Lozano-Castello et al. was reported by Bleda-Martínez et al. for activated carbons electrochemically characterised in aqueous media (Bleda-Martínez et al. 2006), where the authors also attributed the increased wettability generated by surface oxygen groups as one of the reasons for the increased capacitance in activated carbons with larger amount of such functionalities.

This work proposed the characterisation by chemical and electrochemical techniques of a large collection of activated carbons with very similar porous texture and widely different surface chemistry. Among the samples one activated carbon (CA) was thermally treated in H_2 flow to remove most of the oxygen functionalities (CAH). This sample CAH was further subjected to a thermal treatment in air flow to recover part of the surface oxygen groups (CAHOx). Changes in porosity caused by oxidation and thermal treatments are slight and hence, they will not affect the electrochemical behaviour in an important way. Table 2 shows the amount of oxygen groups desorbed in the TPD for these three

Table 1. Porous texture (S_{BET} and micropore volume (V_{DR})), total amount of oxygen groups (O_{TPD}) obtained from TPD and gravimetric capacitance (Cg) for KOH-activated carbons.

Sample	S_{BET} m^2g^{-1}	V_{DR} cm^3g^{-1}	C_g Fg^{-1}	Cg/S_{BET} Fm^{-2}	O_{TPD}* $mmolg^{-1}$
A1	730	0.33	18	0.025	–
A2	950	0.43	43	0.045	2.2
A3	1450	0.62	114	0.079	5.9
A4	2650	1.19	220	0.083	7.6
A5	3010	1.33	204	0.068	3.8
A6	3500	1.51	215	0.061	3.6
A7	2180	1.01	126	0.058	2.9
B1	2100	0.94	167	0.080	4.1
B2	2120	0.93	161	0.076	4.3
L	880	0.43	65	0.074	1.3
MAXSORB	3100	1.38	217	0.070	6.1

*: determined from amounts of CO and CO_2 evolved during TPD experiments.

Table 2. Micropore volume (V_{DR}), amount of CO and CO_2 evolving surface oxygen groups obtained from TPD and gravimetric capacitance (C_g) in 1 M H_2SO_4 of KOH-activated carbon after thermal treatments under different atmospheres.

Sample	V_{DR} cm^{3}g^{-1}	CO^{TPD} mmolg^{-1}	CO$_2$^{TPD} mmolg^{-1}	C_g Fg^{-1}
CA	1.39	2.82	1.04	250
CAH	1.33	0.13	0.16	Negligible
CAHOx	1.22	3.70	0.89	191

activated carbons. The main difference between sample CA and CAH is that the treatment in H_2 removes most of the surface oxygen groups and saturates carbon bonds with hydrogen atoms. Thus, this sample is quite useful to test the role of surface chemistry in the wettability of the carbon material. Sample CAHOx has a higher amount of CO and CO_2 desorbing groups because the treatment in air at 450°C has generated an important number of surface oxygen groups.

The electrochemical behaviour of the samples was measured by CVs in 1M H_2SO_4. The values of capacitance in Table 2, shows that sample CAH, the most hydrophobic one, presents a negligible value of capacitance using the same conditions for impregnating the electrolyte as with the other samples. Interestingly, when this sample is reoxidised in air flow (CAHOx), it recovers the capacitance to a value close to the original sample. These results evidence that surface oxygen groups (or the dangling bonds created after decomposition in inert atmosphere) have a key role in improving carbon wettability in aqueous electrolyte; thus, their presence on the carbon surface is essential to take profit of the large electrical double layer contribution to the capacitance of the porous carbons which is associated to their high porosity.

2.3 Effect on Pseudocapacitance

Pseudocapacitance from surface oxygen groups has been known for a long time. In the decade of 1970, Kinoshita (Kinoshita and Bett 1973b) and Yeager (Randin and Yeager 1971) reported the electrochemical activity of surface oxygen complexes found on the surface of carbon materials, and they attributed such activity to the presence of certain quinone-like functional groups, that in presence of protons could undergo the well-known oxidation/reduction of hydroquinone/quinone groups.

$$>C=O + H^+ + e^- \rightarrow >C\text{-}OH$$

Where >C=O represents quinone-type groups and >C-OH hydroquinone-type group, with the redox reaction occurring approximately at 0.7 V vs. NHE. Two decades after, Conway et al. proposed that fast redox reactions are one of the mechanisms that can be employed for charge storage through the so-called pseudocapacitance (Conway et al. 1997). In that case, the redox reactions take place in species that are chemically anchored onto the surface of the electrode, thus avoiding any diffusional control and therefore bringing a much large electron transfer rate, similar to that of capacitive processes (Conway et al. 1997). Later, Teng et al. (Hsieh and Teng 2002a, Cheng and Teng 2003), Okajima et al. (Okajima et al. 2005) and Bleda-Martinez et al. (Bleda-Martínez et al. 2005, 2006) pointed out in different works that the presence of quinone-like species, and more generally all surface oxygen groups that evolve as CO during TPD, in the surface of activated carbons rendered an increase of capacitance, which could be ascribed to Conway's concept of pseudocapacitance. Since then, many works have been interested in increasing the pseudocapacitance of electrodes based on carbon materials in aqueous electrolyte by generating electroactive surface oxygen groups, including different materials as activated carbons (Raymundo-Piñero et al. 2006), zeolite templated carbons (Ania et al. 2007, Itoi et al. 2013), carbon nanotubes (Frackowiak and Béguin 2002) or graphene (Oh et al. 2014), among others.

The redox reactions involving surface oxygen groups in aqueous media seem to be suppressed in neutral electrolyte, and are especially relevant in acid media, what points out that protons are involved in them. Andreas et al. carried out a deep analysis of the effect of pH in the capacitance of activated carbon cloths (Andreas and Conway 2006). As the pH increased from 0 to 11, a capacitance decrease of 30 per cent is observed, and they attributed it partially to the disappearance of the pseudocapacitance contribution from the redox reactions of the surface quinone groups. Interestingly, as pH increased above 11, capacitance increased again, although only 20 per cent. The authors related this increase of capacitance to the electrochemical activity of certain oxygen groups as pyrones and derivatives, although no clear assignment was done.

Pseudocapacitance is also claimed in organic media. In this sense, Nueangnoraj et al. studied the electrochemical behaviour of a zeolite templated carbon (ZTC) in 1 M tetraethyl fluoroborate in propylene carbonate and in 1 M lithium hexafluorophosphate in a mixture of ethylene carbonate and diethyl carbonate (Nueangnoraj et al. 2015). This porous carbon is well known for its extremely high pseudocapacitance in acid electrolyte thanks to the large amount of redox-active surface oxygen groups that are easily generated by the electrochemical oxidation of edge sites upon anodic polarisation (Itoi et al. 2013). The electrochemical study demonstrated the appearance of clear redox processes at –0.6 and 0.4 V vs. Ag/AgClO$_4$, which have been related to the electrochemical activity of anion radical of quinone and cation radial of furan-type ether, respectively. However, most porous carbons are not as rich in functional groups as ZTCs and do not show any relevant pseudocapacitance contribution in organic electrolytes, and the occurrence of oxygen groups in this electrolyte is not desired in most cases because they compromise the durability of supercapacitors, as will be detailed in Section 2.4.

Bleda-Martinez et al. prepared a large collection of porous carbon materials using different precursors such as anthracite, general-purpose carbon fibres and high performance carbon fibres, which were activated by KOH, NaOH, CO$_2$ and steam (Bleda-Martínez et al. 2005). These materials were electrochemically characterised in H$_2$SO$_4$ aqueous medium, and the composition and amount of surface oxygen groups were characterised by TPD experiments. The specific capacitance divided by BET surface area had a good correlation with the amount of surface oxygen groups desorbing as CO divided by BET surface area (see Fig. 5). This suggests that the contribution of the CO-type groups to the capacitance of porous carbon is positive.

Moreover, Bleda-Martínez et al. studied the effect of thermal treatments at different temperatures on the electrochemical behaviour of activated carbons in order to analyse the effect of oxygen surface

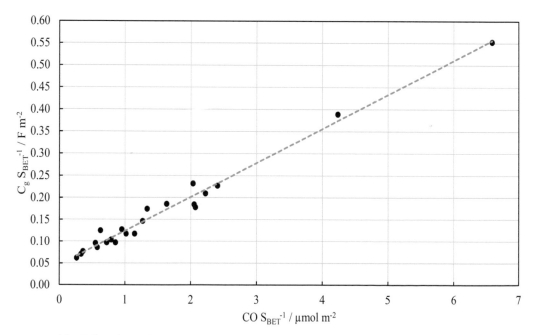

Fig. 5. Capacitance divided by BET surface area versus CO content divided by BET surface area.

groups (Bleda-Martínez et al. 2006). Table 3 shows the values of capacitance and the amount of oxygen surface groups for activated carbons prepared by KOH activation (sample AKN) and NaOH activation (sample ANaN). It is important to remark that the pristine activated carbon samples were chemically oxidised with HNO_3 (sample AKN) and then submitted to thermal treatments at different temperatures (i.e., from 250 to 900°C) in N_2 to get carbon materials containing different amounts of surface oxygen complexes (the nomenclature includes the treatment temperature). All the activated carbons have similar porosities, and almost no changes in pore volume and surface area occur, caused by oxidation and thermal treatments as deduced from the similar values of micropore volumes (V_{DR}).

The voltammograms for these materials show an oxidation peak at 0.5 V and a second one at 0.63 V that are related to pseudocapacitive behaviour (Bleda-Martínez et al. 2006). This contribution is especially intense for samples AKN, ANaN, ANaN-250, AKN-450 and ANaN-450, that is, for materials in which CO-type complexes have not been removed and are present with the highest amounts. Then, the higher the amount of surface oxygen groups of CO-type, the higher the capacitance. The peak at 0.63 V/RHE is more intense for the most oxidised samples (AKN and ANaN), and after thermal treatments at 750°C or 900°C, most or even all of these redox processes disappear. These results indicate that CO-type groups such as quinones and carbonyls, play a special role in the redox contribution of the surface oxygen complexes in the capacitance.

Itoi et al. deepened in the use of pseudocapacitance as a way to enhance the energy storage of carbon materials in aqueous electrolyte by the controlled electrochemical oxidation of a zeolite templated carbon (ZTC) in order to generate electroactive CO-evolving groups (Itoi et al. 2013). The ZTC obtained by the method proposed by Kyotani et al. (Ma et al. 2000) is a nano-sized carbon material with tailored and ordered pore network, which is synthesised using a zeolite as hard template. This porous carbon has a uniform pore size of 1.2 nm, a high surface area (close to 4000 m^2g^{-1}) and a large amount of carbon edge sites highly available to the electrolyte owing to the ordered structure of the solid. Thanks to this structure, ZTC constitutes a suitable platform for fundamental studies. Moreover, the presence of such a high amount of edge sites makes possible to prepare a super porous carbon-oxygen alloy (theoretical calculations shown that oxidised ZTC could reach oxygen contents higher than 25 per cent wt., Berenguer et al. 2013) by a controlled electrochemical oxidation of these sites. This ZTC shows a large anodic current in the first positive scan above 0.4 V/Ag/AgCl that corresponds to a strong electrochemical oxidation in the carbon framework; however, this anodic current disappears and the voltammogram is almost stable by the fourth cycle. This electrochemical behaviour is related to the direct electrochemical oxidation of ZTC upon anodic polarisation (Berenguer et al. 2013). A redox process develops into well-defined voltammetric peaks and the capacitance measured from the CV reaches 500 F g^{-1}. TPD measurements demonstrated that

Table 3. Micropore volume (V_{DR}), amount of CO and CO_2 evolving surface oxygen groups obtained from TPD and gravimetric capacitance (C_g) in 1 M H_2SO_4 of KOH- and NaOH-activated carbons after thermal treatments under different atmospheres.

Sample	V_{DR} cm^3g^{-1}	CO^{TPD} $mmolg^{-1}$	CO_2^{TPD} $mmolg^{-1}$	C_g Fg^{-1}
AKN	1.05	3.55	2.06	254
AKN-450	1.11	3.33	0.57	232
AKN-750	1.09	0.33	0.10	189
AKN-900	0.98	0.14	0.11	167
ANaN	1.19	2.38	1.73	238
ANaN-250	1.09	2.19	1.31	242
ANaN-450	1.09	2.12	0.65	202
ANaN-750	1.09	0.32	0.15	171
ANaN-900	1.00	0.07	0.24	148

the oxygen content at this point is 15 per cent wt., and the ratio between CO and CO_2 evolution after the electrochemical treatment was 7.9, demonstrating the high selectivity of the treatment towards the formation of CO-evolving groups.

Taking advantage of this high capacitance, a hybrid-asymmetric supercapacitor using electrochemically oxidised ZTC as positive electrode and highly stable superporous carbon as negative electrode was constructed, reaching energy density higher than 25 Wh kg^{-1} in 1 M H$_2$SO$_4$ (Nueangnoraj et al. 2014), a value in the range of organic-based supercapacitors. This example clearly illustrates the huge importance of surface oxygen groups to raise the energy density of supercapacitors based on porous carbons in aqueous electrolyte.

2.4 Effect on Durability

The relationship between the durability of carbon materials and the presence of surface oxygen groups in aqueous electrolyte is straightforward. Several studies about the kinetics and mechanism of carbon electrooxidation propose that the oxidation rate is related to the concentration of carboxylic acid-like species (Choo et al. 2007, Maass et al. 2008), which act as the initiators of the electrochemical evolution of CO_2. Since anodic polarisation is known to produce electrochemical oxidation of the carbon surface (Kinoshita and Bett 1973b, Berenguer et al. 2009), rendering the formation of CO-evolving species that can be subsequently oxidised towards CO_2-evolving ones, an extensive electrochemical oxidation would lead to carbon degradation, what undeniably shortens the lifetime of a supercapacitor. Even before a massive carbon loss is achieved, the formation of a large amount of carboxylic groups via electrochemical oxidation of the anode can cause a drastic decrease of electrical conductivity. A large increase of the ohmic drop is also considered a proof for the end of service of a supercapacitor.

In organic media, surface oxygen functionalities are also considered the mediators of the degradation of the solvent. The decomposition of propylene carbonate and other carbonate-based electrolytes over carbon electrodes has been analysed, and it renders the evolution of CO_2 on the positive electrode and propene and H$_2$ in the negative electrode as the main gases (Hahn et al. 2005), while the formation of solids consisting in organic acids and polymer products renders a porosity blockage that decreases the capacitance, a phenomena that is more intense in the positive electrode (Cazorla-Amorós et al. 2010). Similar behaviour is observed in the case of acetonitrile (Azaïs et al. 2007, Kurzweil and Chwistek 2008). It has been found that a low amount of surface oxygen groups is a requirement for a better life cycle of supercapacitors (Azaïs et al. 2007).

Morallón et al. identified a loss of mass on a commercial activated carbon electrode when it was submitted to CV in a wide potential window in aqueous electrolyte (Morallón et al. 2009). The electrochemical experiments were performed on a sodium chloride electrolyte at pH 12 (0.02 M NaCl + 0.01 M KOH), using an electrochemical quartz crystal microbalance (EQCM) to monitor changes of the electrode mass with ng sensitivity. This loss was due to a reduction process that was related to the carbon-oxygen surface groups previously formed during the anodic scan. CVs for the carbon electrode in the Fig. 6a reflects an essentially capacitive process. However, the mass profile of the carbon electrode versus the applied potential showed interesting changes, Fig. 6b.

As the potential applied to the electrode during the voltammograms is lowered during the cathodic scan, the weight recorded by the ECQM remained constant and, after a certain potential, it decreased. The weight increased again on the next anodic scan, but a net weight loss was registered at the end of the cathodic scan. This continuous decrease of the mass was observed for five consecutive cycles, with a loss per cycle of approximately 14–18 ng/cycle. The experiments were also conducted continuously under similar conditions for up to 30 cycles, while exhibiting very similar steady mass losses, possibly due to a net mass loss of carbon from redox processes, the largest mass losses occurring from –0.2 V/Ag/AgCl to lower potentials during the cathodic sweep.

In the anodic sweep, a mass gain was attributed to oxidation processes producing carbon-oxygen surface complexes. Under anodic conditions in alkaline solution, carbon oxidation can occur via adsorption of hydroxide and electrochemical oxidation of carbon following the next reaction scheme

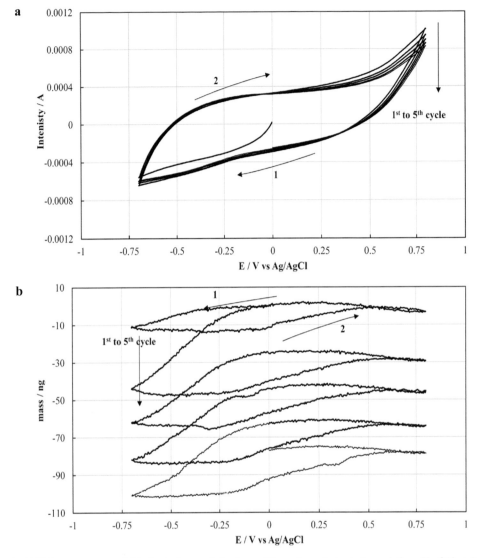

Fig. 6. (a) Voltammograms and (b) electrode mass behaviour of the carbon electrode in a solution pH of 12 at 50 mV/s in the NaCl electrolyte for five CVs.

(in it, the carbon–oxygen surface complexes can be semiquinones, phenols, other CO-desorbing groups in the first reaction, or carboxyls, anhydrides or lactones, i.e., CO_2 desorbing groups, in the second reaction):

$$C() + 2OH^- \rightarrow C(O) + H_2O + 2e$$

and

$$2C() + 3OH^- \rightarrow C(O_2) + C(H) + H_2O + 3e$$

Where $C()$ represents a reactive site and $C(O)$ a carbon-oxygen surface complex.

Under cathodic conditions in alkaline solution, carbon–oxygen surface complexes can also be reduced:

$$C(O) + 2H_2O + 2e \rightarrow CO + 2C(H) + 2OH^-$$
$$or\ CO + 2C() + H_2 + 2OH^-$$

and

$$C(O_2) + 3H_2O + 3e \rightarrow CO_2 + 3C(H) + 3OH^-$$
$$or\ CO_2 + C() + H_2 + 2OH^-$$

Thus, the products of the electrochemical reduction are proposed to be carbon monoxide or carbon dioxide and possibly H_2. These electrogasification reactions do not only cause a loss of carbon from the electrode in form of CO and CO_2, but also create additional active sites, explaining why the weight loss is still detected on successive voltammograms.

In a subsequent work, Leyva et al. monitored the mass changes suffered by a zeolite templated carbon (ZTC) in acid electrolyte (1 M H_2SO_4) using the same procedure (Leyva-García et al. 2015). This material is known to be highly reactive and very sensitive towards oxidation (Berenguer et al. 2013), so it was selected for this fundamental study in order to observe in more detail the electrochemical oxidation and gasification processes. ZTC is significantly electrochemically oxidised in the first voltammetric cycles under the potential intervals from –0.10 V up to 0.80 V and 1.20 V (vs. Ag/AgCl) as shown by the oxidation current and the large mass increase recorded at high potential values. Gravimetric capacitance changed from 175 to 353 F g^{-1} after such treatments, confirming that pseudocapacitance was generated as a result of the introduction of electroactive CO-evolving groups. Since the mass increase is lower than that expected from the increase in the specific capacitance value and from the TPD results, it seems that the electrochemical oxidation and gasification of the ZTC take place simultaneously. Under more severe electrochemical conditions (i.e., by performing large number of CVs between –0.10 V and 1.40 V (vs. Ag/AgCl) and potentiostatic experiments at 1.40 V and –0.10 V (vs. Ag/AgCl)), a large decrease in the current density demonstrates that the ZTC has suffered a remarkable degradation that involves a noticeable loss of conductivity and of the porous structure. Moreover, a net mass decrease is recorded, showing that electrochemical gasification of the sample occurs to a considerable extent.

Cazorla-Amorós et al. analysed the durability of two KOH-activated carbons in 1 M $TEABF_4$ in propylene carbonate as electrolyte (Cazorla-Amorós et al. 2010). The ACUA-1 sample corresponds to the activated carbon with a total oxygen content of 6140 μmol g^{-1}. On the other hand, the ACUA-2 sample corresponds to the same activated carbon after a 'purification' treatment in which most of the oxygen groups and inorganic impurities of the surface of the carbon material have been removed. After the purification, the total O content was 1960 μmol g^{-1}. The porous texture of the parent activated carbon is unmodified after the purification treatment (micropore volume of 1.24 and 1.18 cm^3 g^{-1} before and after the purification treatment, respectively).

Figure 7 shows the capacitance of the two activated carbon samples tested in two electrode cells for 600 galvanostatic charge-discharge cycles. From cycle 200 to cycle 500 drastic conditions of voltage (3 V) and temperature (70ºC) were imposed. The last 100 cycles were performed at same conditions than the initial 200 cycles in order to compare the performance before and after the durability test. The starting capacitance of the supercapacitor built with ACUA-1 electrodes was higher than that of ACUA-2. However, the final capacitance for both samples after the durability tests is similar. The comparison of the porous texture and surface chemistry before and after the durability test pointed out that the microporosity decreased about 33 per cent for the ACUA-1 sample and 20 per cent for the ACUA-2 sample (determined as an average of the positive and negative electrode). This pore blockage is explained considering the undesirable electrolyte degradation reactions, where the decomposition of propylene carbonate and even $TEABF_4$ may happen. This effect is also more pronounced on the positive electrodes. As for the surface chemistry, the number of oxygen groups evolving as CO_2 during TPD increases and those decomposing as CO decrease after the durability test for the positive electrode in sample ACUA-1. Interestingly, the positive electrode of the ACUA-2 sample showed no significant changes in its surface chemistry. In the case of the negative electrode there was a significant decrease in the amount of surface oxygen groups for both samples. Since the evolution of propene and hydrogen on the negative electrode has been demonstrated in the literature,

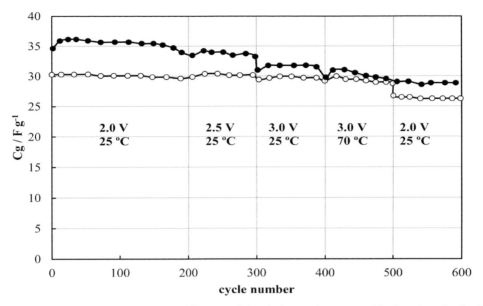

Fig. 7. Capacitance by two-electrode cell under different conditions (voltage and temperature) for the activated carbon before the 'purification' treatment (sample ACUA-1, solid dots) and after the 'purification' treatment (sample ACUA-2, empty dots).

it is possible that these reducing agents, or even a direct electrochemical reduction, could produce an important decrease in surface oxygen groups at the negative electrodes.

The changes on surface chemistry of the electrodes detected by TPD cannot explain by themselves the porosity blockage suffered by the electrodes. Attending to the larger porosity blockage suffered by ACUA-1, the authors propose that the electrochemical decomposition of propylene carbonate could be promoted by the presence of surface oxygen functionalities. A possible degradation reaction of propylene carbonate, forming hydroxyalkyl derivatives and CO_2 evolution, is schematised in Fig. 8. Thus, the surface functional groups such as phenol (first reaction) or carboxylic acids (second reaction), which are found in large amounts on ACUA-1 activated carbon, can promote these reactions and further propylene carbonate polymerisation, that would render a blockage of the porosity. Moreover, the evolution of CO_2 combined with the porosity blockage may also have a strong influence in the capacitance drop caused by the durability test. In summary, these results point out that the removal of oxygen groups from the surface of activated carbons electrodes is positive to enhance the durability of supercapacitors in organic electrolyte.

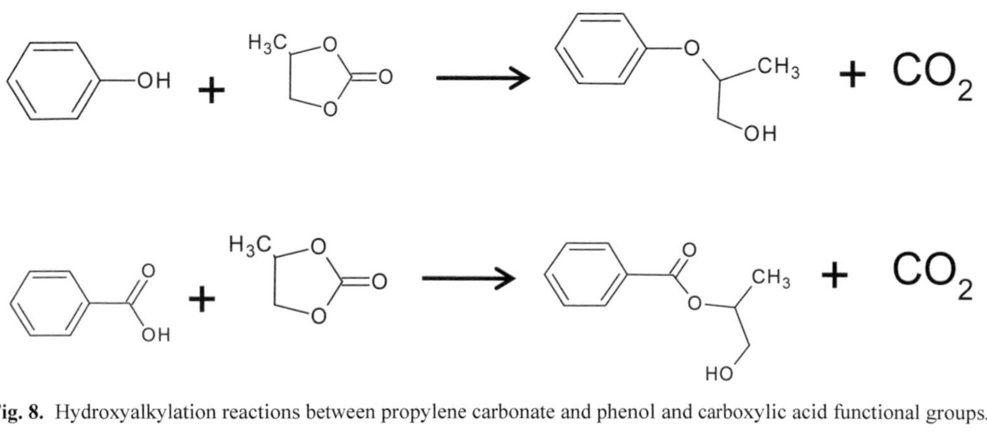

Fig. 8. Hydroxyalkylation reactions between propylene carbonate and phenol and carboxylic acid functional groups.

3. Nitrogen-containing Functional Groups and their Role on the Electrochemical Behaviour of Porous Carbons

Nitrogen functionalities can be produced onto the surface of carbon materials by several methods (Boehm 1994, Raymundo-Piñero et al. 2002, 2003, Matter et al. 2006, White et al. 2009, Shen and Fan 2012, Figueiredo 2013, Deng et al. 2016), being the most frequent:

- Reaction between a carbon material and nitrogen-containing reagents, either in gas phase or liquid phase, i.e., ammonia, urea or NO, between others.
- Converting carboxyl functionalities into amides by acyl chloride activation of the carboxyl group after $SOCl_2$ treatment.
- Thermal decomposition of a nitrogen-containing precursor or polymer (melamine, polyacrylonitrile, polypyrrole, polyaniline, etc.) in presence of the carbon material.
- Using a nitrogen-containing material as carbon precursor or as additive during the preparation of the carbon material.
- Hydrothermal carbonisation of nitrogen-containing biomass precursors.

Nitrogen heteroatoms form different functionalities on the surface of carbon materials, what mostly depends on the method employed for their generation. These functionalities can be classified according to the number of nitrogen-carbon bonds and their position in the graphene layer (Raymundo-Piñero et al. 2002, 2003).

Figure 9 shows the most frequently found nitrogen functionalities. Positively charged quaternary nitrogen (N-Q in Fig. 9) can be found within the graphene layer. Pyridine (N-6) and pyrrole (N-5) like functionalities, where nitrogen is located either in six-member or in five-member rings, respectively, are usually emplaced in edge sites, substituting one carbon atom. They can be also found within the graphene layer, where their presence generates a defect in form of a carbon vacancy. Pyridones (N-C-OH) are obtained when a hydroxyl group is attached to a carbon atom next to the pyiridinic function, and are known to possess a chemical environment similar to that of pyrrolic nitrogen (Pels et al. 1995). Oxidised forms of nitrogen, such as oxidised piridine (N-O), can be obtained when N-containing porous carbon are exposed to air at medium temperatures. Finally, nitro (NO_2), amines (NH_2), amides ($CONH_2$), secondary amines (NH) and cyano (CN) groups can be also found covalently attached in edge sites or forming bridges between aromatic units (such as the amine/imine bridges in conducting polymers).

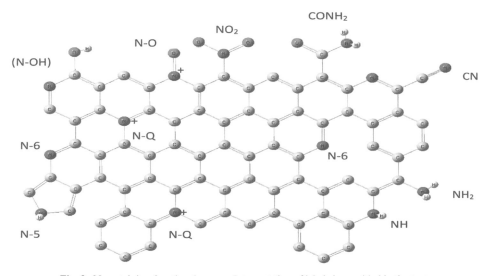

Fig. 9. N-containing functional groups. Interpretation of labels is provided in the text.

Nitrogen doped porous carbon materials have been extensively employed as supercapacitor electrodes during the last years. In one of the pioneer works about this topic, Jurewicz et al. synthesised 3 per cent wt. nitrogen doped activated carbon by ammoxidation of brown coals (Jurewicz et al. 2002). They found a huge capacitance improvement in both acid and alkaline electrolyte, which is connected to the formation of different types of nitrogen functionalities.

In a different work, Hulicova et al. reported the electrochemical performance of a nitrogen-doped carbon prepared from a melamine residue. They observed an astonishing increase of two orders of magnitude in the aerial capacitance (i.e., Farads per square meter of apparent area, a value which is considered to be ca. 100 mF m^{-2} in most activated carbons in aqueous electrolyte) for their nitrogen-containing carbons (Hulicova et al. 2005).

Thus, the presence of nitrogen functionalities on the surface of porous carbon material has an important effect on their electrochemical properties. Next, the effect of these functionalities in the properties that affect the performance of porous carbon materials as electrodes of supercapacitor will be discussed.

3.1 Effect on Electrical Conductivity

The conductivity of carbon materials has been expected to increase when nitrogen is inserted within the graphene layer due to the incorporation of an electron from the nitrogen atom to the π-electronic system of graphene layers. The general consensus relies on electrical conductivity being improved by the presence of N-Q groups. Quaternary nitrogen decreases the energy gap between lowest unoccupied molecular orbitals (LUMO) and highest occupied molecular orbitals (HOMO), and their donor properties are responsible for filling the conduction band with electrons (Lota et al. 2007). However, there is some controversy about these assessments.

For instance, Terrones et al. claimed that both divalently bonded and substituted nitrogen can form impurity levels in the vicinity of the Fermi level (Terrones et al. 2002). Nishihara et al. prepared carbon-coated anodised aluminum oxide seamless electrodes where they were able to incorporate substituted nitrogen and boron atoms. They reported a resistivity decrease of more than 50 per cent in nitrogen-substituted carbon film, what they related to the presence of quaternary nitrogen and pyrrolic species (Kwon et al. 2009). On the other hand, Kang et al. reported that conductivity is only improved when substituted nitrogen is inserted in the graphene layer (Kang and Jeong 2004). Another positive presumption is provided by the theoretical studies of Strelko et al., where they showed that pyrrolic nitrogen improves the charge mobility in a carbon matrix by introducing electron-donor characteristics (Strelko et al. 2000).

Differently, Ozaki et al. found a conductivity decrease after nitrogen doping, which they explained to be a result of the induction of disorder and defects on the graphene layers of their materials (Ozaki et al. 2007). Similarly, Zhu et al. claimed that conductivity is decreased because nitrogen substitution inside the graphene layer reduces the π-electron conjugation, whereas capacitance could be improved due to an increase on the HOMO energy of a graphene layer, a result that they verified experimentally with nitrogen-doped carbon aerogels (Zhu et al. 2005).

To summarise, it seems that nitrogen substitution as quaternary nitrogen and pyrrolic functionalities renders an improvement on electrical conductivity. However, the generation of structural disorder or electron-withdrawing functionalities could be detrimental for this property.

In a recent work, Mostazo-López et al. reported a nitrogen functionalisation route based on amidation reaction and Hofmann rearrangement that was applied to a highly microporous activated carbon (>3000 m^2 g^{-1}) (Mostazo-López et al. 2015).

The functionalisation scheme includes: (i) a chemical oxidation with concentrated nitric acid, in order to generate carboxylic functions, (ii) amidation treatment, where the previously generated carboxylic acids are converted to acyl groups that are subsequently modified to amide groups, (iii) amination by Hofmann rearrangement, where the amide functions are transformed into amines by a bromine and sodium methoxide treatment and (iv) a final hydrolysis step.

The porosity and the surface chemistry of the activated carbon were determined on the pristine material and after each treatment (oxidation, amidation, amination). Table 4 includes the amount and kind of incorporated nitrogen, the BET surface area (S_{BET}) and micropore volume (V_{DR}). It is important to note that after the amidation treatment, more than 3 per cent wt. of nitrogen was incorporated to the activated carbon, while a large fraction of the porosity was kept intact (pore volume decrease of 25 per cent from the oxidation to the amidation step). Interestingly, XPS revealed that the treatment delivered a surface chemistry more complex than expected, revealing the formation of aromatic nitrogen functions in edge position, such as pyridine, pyridone and pyrrole groups. The generation of aromatic functionalities is the outcome of the highly heterogeneous surface of activated carbons, and has been explained in terms of condensation and cyclation reactions of amide groups with CO-evolving oxygen functionalities in their surroundings, which are largely available in the oxidised carbon (Mostazo-López et al. 2015, 2016). The formation of aromatic nitrogen usually requires the use of high temperature treatment over porous carbon filled with a nitrogen-rich precursor, causing a porosity drop that is avoided using the route proposed by Mostazo-López et al., since it is done at low temperatures. After amination process, most of the porosity of the oxidised activated carbon is recovered thanks to the similar sizes of carboxylic and amine groups, while the aromatic functions remain unmodified, and the hydrolysable nitrogen functions, such as imines, are removed.

The electrochemical behaviour of these electrodes was evaluated in acid aqueous electrolyte and the capacitance values are included in Table 4. The capacitance increased in the amidated and aminated carbons compared to the oxidised activated carbon, what is related to an increase on wettability and pseudocapacitance contribution on those materials. Since all these materials share the same origin, differences in capacitance retention are related to the effect of the surface chemistry on electrical conductivity. Initially, the oxidation treatment leads to a reduction on capacitance retention.

As has been described in detail in Section 2 of this chapter, this is an expectable outcome due to the presence of electron-withdrawing carboxylic functionalities (Hsieh and Teng 2002a, Bleda-Martínez et al. 2005). Surprisingly, the amidation treatment results in a huge improvement on capacitance retention. This effect can be related to the formation of pyrrole and pyridone functionalities in this sample. However, the aminated sample does not show the same capacitance retention while having the same amount of these groups. Hence, the improvement on electrical conductivity seems to be connected to the presence of cyclic amides (such as lactams and imides), that are lost during the amination treatment.

Table 4. Porous texture and nitrogen species of a microporous activated carbon after nitrogen functionalisation and capacitance measured in acidic electrolyte at 0.5 A g^{-1} (Cg^{05}) and 20 A g^{-1} (Cg^{20}).

Sample	S_{BET} m^{2}g^{-1}	V_{DR} cm^{3}g^{-1}	N_{XPS} at.%	N-5, N-C-OH	NH$_2$, CONH$_2$	N-6, imines	Cg^{05} F g^{-1}	Cg^{20} F g^{-1}
Raw AC	3140	1.10	0.3	–	–	0.3	268	82
Oxidation	2680	0.96	–	–	–	–	138	6
Amidation	2080	0.74	3.8	0.7	1.9	1.2	206	125
Amination	2450	0.91	3.0	0.7	1.6	0.7	209	17

3.2 Effect on Wettability by the Electrolyte

The incorporation of nitrogen functionalities enhances the wettability of porous carbons. The introduction of nitrogen functionalities increases the number of hydrophilic sites. This effect is considered to be especially significant in the case of nitrogen atoms located at the edges of graphene layers (Lota et al. 2005). Wetting of the carbon pores by organic solvents can be also improved by nitrogen incorporation. Candelaria et al. added nitrogen functionalities to a porous carbon by filling the pores with hexamine and pyrolysing it at high temperatures. They found a noticeable increment of capacitance in 1 M tetraethylammonium tetrafluoroborate in a 1:1 mixture of propylene carbonate–dimethylcarbonate on the nitrogen-modified carbon (Candelaria et al. 2012). They demonstrated that this capacitance enhancement is related to an improved wettability, as was observed by the measurement of the contact angle after the nitrogen modification. Similarly, Kawaguchi et al. reported the preparation of nitrogen containing carbon rich in pyridinic and quaternary nitrogen by pyrolysis of diaminomaleonitrile at 750°C. This material showed a similar capacitance in 1 M H_2SO_4 as an activated carbon with much larger porosity development, what they related to a wettability improvement due to the presence of nitrogen functionalities (Kawaguchi et al. 2010). Han et al. reported the preparation of highly porous nitrogen-doped aerogels from graphene oxide and ammonia as carbon and nitrogen source, respectively (Sui et al. 2015). They measured the wettability by means of a simple test, 5 microliters of water were dropped on top of the surface of carbon pellets. They found that the pellet consisting in nitrogen-doped carbon aerogel needed a much shorter time (i.e., only 25 per cent of the time) to be fully infiltrated by the water than the undoped carbon aerogel.

3.3 Effect on Pseudocapacitance

Pseudocapacitance of nitrogen functionalities is the motivation behind most of the research devoted to nitrogen functionalisation of porous carbons. In general, nitrogen functionalities at edge position of graphene layers are proposed to have electroactivity, whereas nitrogen in valley position or within the graphene layer is expected to be electrochemically inactive (Ornelas et al. 2014). The type of redox reactions that could face these groups is poorly described so far. Lota et al. mentioned that the origin of nitrogen induced pseudocapacitance could be the reactions of oxidation/reduction of pyridinic ring, related to pyridinic, pyridonic and pyridine N-oxide groups (Lota et al. 2005). Kwon et al. prepared N-doped templated carbons from anodic aluminum oxides, and after discounting the effects of the wettability and the electrical conductivity, they concluded that pseudocapacitance is the most important factor that enhances the capacitance by N-doping, although they did not assess to a certain functionality or redox reaction the origin of this pseudocapacitance (Kwon et al. 2009). Seredych et al. studied the relationship between the gravimetric capacitance and the surface properties of biomass-derived activated carbons subjected to oxidation and treated with melamine and urea (Seredych et al. 2008). When they normalised the capacitance by the micropore volume, a clear relationship was stablished between the population of quaternary nitrogen (N-Q) and nitrogen oxides (N-O), however they did not propose any redox reaction for the contribution of N-Q to capacitance, and instead of that, they assign this effect to an improved electron transfer mobility (i.e., electrical conductivity).

There are only few studies regarding pseudocapacitance of nitrogen groups in organic electrolyte. Recently, Itoi et al. reported the electrochemical behaviour in organic electrolyte of ordered microporous carbons doped with oxygen, nitrogen and boron groups (Itoi et al. 2016). They found a large capacitance increase in the –0.5/0 V (vs. Ag/AgClO$_4$) potential range for the nitrogen-doped carbon, what they have assigned to the pseudocapacitive effect of nitrogen groups.

As for aqueous media, pseudocapacitance of nitrogen groups is proposed to be higher in alkaline electrolyte, especially at cathodic potentials (Pietrzak et al. 2007). Ornelas et al. synthesised CNTs and nitrogen-doped CNTs as model materials for differentiating the effect on the electrochemical behaviour of nitrogen groups in acidic and alkaline electrolytes (Ornelas et al. 2014). They proposed that the additional electron of pyridine groups could be involved in electron transfer process in alkaline media, while pyridine could be protonated in acid electrolyte, rendering them inactive in such media. Redox reactions explaining the larger capacitance of N-doped carbons in alkaline media are usually not mentioned (Andreas and Conway 2006, Hulicova-Jurcakova et al. 2009a,b).

In acid electrolyte, the enhancement of the capacitance values has been interpreted by pseudofaradaic reactions of nitrogen functionalities. Lee et al. compiled some of the proposed reactions so far (Lee et al. 2013), consisting in the pyridine and pyrrolic protonation, with the participation of one electron, and in the reduction of the hydroxyl function of pyridone in a 2-electron process, although not clear evidences supporting these reactions have been provided yet.

Apart from the nitrogen functionalities located in the graphene layer, certain nitrogen functional groups such as amines and imines could also be of interest for increasing capacitance. In most cases, these groups are located within a conducting polymer, such as polyaniline (Salinas-Torres et al. 2013). The formation of electroactive polymers constitutes one of the most straightforward methods for generating pseudocapacitance in porous carbon electrodes. Since conducting polymers usually have a limited potential window where they are electrochemically active in aqueous medium, they are typically used in asymmetric capacitors (Frackowiak et al. 2006, Salinas-Torres et al. 2013). Moreover, it is possible to anchor different species containing electrochemically active nitrogen functionalities different to conducting polymers. In the next example, one of such species is introduced in a zeolite templated carbon in order to generate pseudocapacitance.

González-Gaitán et al. reported the electrochemical functionalisation with nitrogen functionalities of a zeolite templated carbon (ZTC) (González-Gaitán et al. 2016). In this work, the functionalisation of ZTC with 2- and 4-aminobenzoic acids was explored using potentiodynamic techniques. The experimental conditions were selected in order to perform a successful functionalisation without the loss of the unique structure of the ZTC, what has been corroborated by XPS, XRD and FTIR.

XPS analyses demonstrated the addition of 1.4 per cent at of N after the functionalisation treatment, being primarily neutral and positively charged amines. Thus, the redox reactions observed in the CV are attributed to the self-doping effect of the carboxylic group close to the amine group (Benyoucef et al. 2005). The capacitance value increased from 399 to 427 F g^{-1} after 2-aminobenzoic acid functionalisation. Interestingly, this increase in capacitance is maintained when the scan rate is increased from 1 to 100 mV s^{-1}, what points out the fast electron transfer in the generated functional groups. These results point out that these nitrogen functionalities can be easily generated by means of electrochemical treatments, and they can be employed for enhancing capacitance by fast redox reactions.

3.4 Effect on Durability

Durability of carbon electrodes is mediated by the reactivity of the carbon surface. Since most of nitrogen heteroatoms introduced in porous carbon material are located in edge sites, they could have a noticeable effect on the reactivity of such sites. Although there are experimental results showing that nitrogen functional groups improve the stability of supercapacitors (Salinas-Torres et al. 2015), little is known about the role of nitrogen functional groups of porous carbon electrodes on the stabilisation of organic electrolytes, or in the electrochemical oxidation of carbon materials. These factors, as was mentioned before, play a central role in the degradation of carbon electrodes in aqueous electrolyte. Next, two examples are provided regarding the durability of N-doped carbon materials in aqueous and organic electrolyte.

Mostazo-López et al. modified the surface chemistry by the introduction of nitrogen functionalities on a highly microporous KOH-activated carbon (Mostazo-López et al. 2016). They built supercapacitor cells in acid electrolyte and studied the changes on the performance after nitrogen doping. In the previous work, an oxidation step was performed before the amidation treatment in order to favour the generation of carboxylic acids that are later transformed into amides. However, a detailed analysis of the surface chemistry of the activated carbon before and after amidation revealed that CO-evolving groups were also active and formed cyclic amides (Mostazo-López et al. 2015). Since the surface of the parent carbon is rich in this kind of functionalities, the amidation protocol was directly implemented on the parent activated carbon. XPS data revealed that 4.1 per cent at. of nitrogen is introduced after this treatment, a similar amount to that obtained when the porous carbon was previously oxidised. However, by avoiding the oxidation treatment, it was possible to fully preserve the porosity of the parent carbon (S_{BET} of 3450 m^2g^{-1} and V_{DR} of 1.19 cm^3 g^{-1} for both activated carbons).

Cyclic voltammetry in 1 M H_2SO_4 of these electrodes showed that the pseudocapacitance contribution attributed to quinone/hydroquinone redox processes at ca. 0.35 V/RHE on the parent activated carbon is lost after the amidation step. It is also important to remark that the oxidation peak at the upper potential limit has a lower intensity on the amidated carbon. These two factors point out that the concentration of electroactive surface oxygen groups is lower, and that the electrochemical oxidation reaction of the carbon electrodes at high potential, which is the starting point of the degradation of the electrode, is inhibited.

Asymmetric supercapacitors were constructed for these carbons and submitted to a durability test consisting in 5,000 cycles at 1 A g^{-1}. The asymmetric design was done aiming to achieve a voltage of 1.4 V in acid electrolyte. In the durability test, gravimetric capacitance of the cell was estimated from the discharge curve. The relationship between energy and power (known as Ragone plot) of both capacitors was studied before and after the durability test, and it was found that the maximum deliverable power was reduced by 62 per cent for the parent carbon, while only 14 per cent was lost for the amidated carbon (Mostazo-López et al. 2016). These results confirm the improved durability in aqueous electrolyte of nitrogen-containing activated carbons.

Salinas-Torres et al. synthesised N-doped activated carbon fibers by using polymerisation of aniline as nitrogen feedstock (Salinas-Torres et al. 2015). This procedure rendered the deposition of a thin polyaniline (PANI) film of around 0.5 nm inside the porosity of A20 (Salinas-Torres et al. 2012). The sample was submitted to a carbonisation in N$_2$ atmosphere at two different temperatures, 600 and 800°C.

The generation of PANI inside the porosity of A20 produces a 50 per cent decrease on the surface area and porosity, Table 5. The carbonisation of A20/PANI releases only a part of the occupied porosity, demonstrating that an important residue is produced by the carbonisation of PANI. The presence of nitrogen was confirmed by XPS, even after the carbonisation at 600°C and 800°C, Table 5, confirming that N-doped carbon materials were obtained.

XPS of carbonised samples revealed that the breakage of PANI bonds and the occurrence of cross-linking reactions during carbonisation render the formation of nitrogen groups which are different to those of PANI. After carbonisation, amine groups from PANI are transformed into positively charged N species (like pyridone/pyrrole and quaternary N) and pyridine groups.

These materials were employed for the construction of supercapacitor cells in organic electrolyte (1 M triethylmethylammonium tetrafluoroborate, TEMA-BF$_4$, in propylene carbonate, PC) for analysing the effect of the different functionalities on the durability of the capacitor. For this purpose, a highly demanding stability test was performed over the cells using two steps. On the first one, the capacitor was kept at 40°C in galvanostatic charge-discharge experiments for five cycles up to 2.5 V. Then, five cycles were done at 40°C, another five cycles at 70°C, during the next five cycles temperature was kept at 70°C while the operating voltage was raised up to 3.2V. At this point, the cell voltage was held at 3.2 V for 100 hours. Afterwards, five galvanostatic charge-discharge cycles were performed at the same temperature at 2.5 V, and the last five cycles were done at 40°C in order to compare the performance with that measured at the beginning of the test. This constant-voltage hold test is a useful tool to do a faster durability test of the capacitor.

At the initial conditions, all samples behave similarly in terms of efficiency. When the temperature or the operating voltage increases, the efficiency decays for the A20/PANI sample that has not been carbonised, while the carbonised ones are able to retain most of their capacitance and efficiency. The floating test at 3.2 V produces huge changes in the capacitance. In this sense, Table 5 compiles the

Table 5. Porous texture, nitrogen content, capacitance and retention of capacitance for N-doped A20 activated carbon fibers.

Sample	S_{BET} m^2 g^{-1}	V^{DR}_{N2} cm^3 g^{-1}	N_{XPS} % at.	C_g^0 F g^{-1}	C_g/C_g^0 %
A20	2300	1.06	–	26	71
A20/PANI	1155	0.42	9.0	27	56
A20/PANI@600°C	1310	0.54	6.7	22	82
A20/PANI@800°C	1355	0.52	3.8	22	80

initial capacitance (C_g^0) values as well as the capacitance retention after the holding test. Initially, supercapacitor made with A20-PANI has a similar capacitance as A20, while the samples carbonised at 600 and 800°C show a 20 per cent decrease in capacitance with respect to that of A20.

After the durability test, the A20 sample loaded with PANI suffers a severe capacitance decrease, Table 5, which is presumably due to the reaction between the amine groups in PANI and propylene carbonate, what results in propylene carbonate-ring opening reactions. Interestingly, the N-doped samples obtained after carbonisation of the A20/PANI material, are able to retain more capacitance than the parent activated carbon fibers. The presence of positively charged nitrogen groups reduces the possibility of nucleophilic attack to propylene carbonate, increasing the stability of the carbon electrode. Additionally, the strength of C-N bonds on aromatic nitrogen groups is higher than in the C-N bond of the amine groups. Hence, these results confirm that certain nitrogen functional groups can enhance the durability of supercapacitors in organic electrolyte.

4. Effect of Other Heteroatom Doping

Phosphorus, boron and sulphur are the most frequent doping agents apart from nitrogen and oxygen, although they have been much less used in this field.

Boron is a highly interesting doping element that has been employed in the past for modulating the semiconducting properties of carbon materials. As nitrogen, boron can be inserted within the graphene layer. As a substituting atom in the carbon framework it acts as an electron acceptor. This causes a shift in the Fermi level to the conduction band of the graphene layer, improving the charge transfer, a highly desirable feature in electrochemical applications. Boron doping is claimed to be beneficial for similar reasons as nitrogen doping (Paraknowitsch and Thomas 2013).

Most boron doping strategies of porous carbon materials consists in co-doping with nitrogen (Guo and Gao 2009, You et al. 2016) and, when boron is introduced alone, it is doped over graphene or carbon nanotubes in most of the studies (Han et al. 2013, Fujisawa et al. 2014). Few examples about B-doping in porous carbon materials exist in the literature. Wang et al. prepared B-doped mesoporous carbon using a hard template method where boric acid is added to sucrose and the mixture is infiltrated within the mesoporosity of SBA-15 (Wang et al. 2008). The authors claim that the boron doping modified the electronic structure of space charge layer, increasing the capacitance in both alkaline and acid aqueous electrolytes. The effect of B-doping is clearly seen in alkaline electrolyte, where the voltammograms revealed a larger capacitance in the whole potential range. Kwon et al. prepared a B-doped carbon layer over an anodised aluminum oxide as template (Kwon et al. 2009). They compared the electrochemical behaviour in acid and organic electrolyte of this carbon layer and the one obtained without doping, which allows them to ascribe the observed changes to the presence of boron atoms. It was found that B-doping enhanced the electrical conductivity and generated pseudocapacitance in both media, delivering a larger performance improvement than N-doping of a carbon layer prepared by the same procedure. Enterría et al. prepared hierarchical porous carbons with different boron contents by the combination of hydrothermal and soft template treatments (Enterría et al. 2015). The addition of boric acid during the polymerisation of the resorcinol-formaldehyde mixture allowed to obtain B-doped resins that were carbonised at 800°C. The electrochemical characterisation in acid electrolyte using a 2-electrode configuration revealed that boron doping enhanced the performance of the capacitor due to a superior conductivity and wettability of the carbon electrodes. Similarly, Gao et al. synthesised boron-doped ordered mesoporous carbon using silica KIT-6 as template, furfuryl alcohol as carbon source and boric acid as the doping agent (Gao et al. 2016). The characterisation of B-doped and B-free ordered mesoporous carbon in supercapacitors demonstrated that the capacitance increases from 52 to 70 F g^{-1} on the B-doped carbon electrodes in 6 M KOH solution.

Sulphur doping in carbon materials is less common. Sulphur doping in aromatic rings is able to induce a positive charge on neighbouring carbon atoms (Liang et al. 2012). Oxidised sulphur species as sulfones and sulfoxides are proposed to be electrochemically active, producing pseudocapacitance.

Sulphur doping in the field of supercapacitors is a recent subject. One of the first studies seems to be the work by Hasegawa et al., who prepared S-doped activated carbon monoliths by physical activation with CO_2 of a sulfonated polydivinylbenzene network (Hasegawa et al. 2011). The electrochemical characterisation in alkaline aqueous electrolyte showed promising results regarding the gravimetric capacitance and the high electrical conductivity. However, the amount of sulphur was low (only 0.83 per cent at.), making difficult to assess clearly what is the role played by sulphur on the electrochemical behavior of the S-doped carbon monolith. In this sense, Zhao et al. prepared a series of ordered mesoporous carbon (OMC) with different amounts of sulphur, from 0 to 1.5 per cent wt., and similar porosities (Zhao et al. 2012). They also employed a hydrogen peroxide treatment in order to oxidise the surface of the S-doped carbon. The combined used of TPD and XPS revealed the presence of several sulphur species, such as aromatic sulfide in the parent S-doped samples and sulfoxide and sulfones in the oxidised ones. The electrochemical characterisation in 6 M KOH revealed an increase in the capacitance for the S-doped OMCs when compared to the undoped sample. This capacitance enhancement was larger for the H_2O_2-oxidised S-doped samples. The authors attributed the improvement on S-doped samples to the n-doping character of aromatic sulfide, whereas reversible redox reactions of sulfones and sulfoxides were proposed to explain the larger capacitance of oxidised S-doped samples. Seredych and Bandosz reported the electrochemical behaviour of porous S-doped carbon-graphene composite using graphite oxide and sulfonated polystyrene as raw materials (Seredych and Bandosz 2013). They claimed that sulphur species are unevenly distributed on the porous carbon, with sulfones and sulfoxides being placed on mesoporosity, where they contribute to pseudocapacitance, and sulphur in thiophene-like configuration being located in narrow microporosity. The latter species are hydrophobic and also bring positive charge to the surface of the micropores, increasing the selective electroadsorption of ions and increasing capacitance. Huang et al. also analysed the electrochemical behaviour of S-doped activated carbons in organic electrolyte (Huang et al. 2014). These materials were synthesised by the pyrolytic decomposition at 400°C of sulphur flakes mixed with a commercial activated carbon. The electrochemical characterisation revealed a larger capacitance in the sulphur-containing sample. The authors related this feature to the presence of thiophene-like functional groups. They proposed that a lone electron pair of sulphur atom is able to overlap with π-cloud of the graphene layer to form an extended system with a filled valence band, resulting in a conductivity improvement. They also suggest that pseudocapacitance may arise from redox reactions induced by the lone electron pair from the sulphur groups interacting with the cations in the electrolyte. Finally, co-doping strategies where sulphur is added to carbon materials along with nitrogen (Li et al. 2016) and even with nitrogen and phosphorus (Hasegawa et al. 2015) are also reported in the literature.

Phosphorus is an element that has been traditionally introduced in carbon materials in order to improve their oxidation resistance. In the case of electrical energy storage, computational studies have predicted that phosphorus doping could modify the band gap of graphene, being more effective than sulphur for that purpose. In addition, its incorporation should need a lower energy amount (Paraknowitsch and Thomas 2013). Calculations have shown that phosphorus doping should improve the electron-donor properties of carbon materials. In the case of supercapacitors, the first studies reporting the possible influence of phosphorus groups on the electrochemical behaviour of activated carbon were published by Hulicova-Jurcacova et al. (Hulicova-Jurcakova et al. 2009, Huang et al. 2013, Huang et al. 2014, Wen et al. 2015).

The most straightforward method for the introduction of phosphorus into the carbon matrix is the polymerisation and carbonisation of a carbon precursor in presence of phosphorus compounds, such as phosphoric acid. Thus, Hulicova-Jurcacova et al. prepared P-containing microporous carbons using phosphoric acid activation of three carbon precursors (Hulicova-Jurcakova et al. 2009). The electrochemical characterisation of these activated carbons in 1 M H_2SO_4 demonstrated that they were highly stable, being able to work at potentials larger than the thermodynamic decomposition potential of water. This ability was profited for increasing the cell voltage up to 1.3 V, what enhanced the energy density more than 200 per cent with respect to that of a commercial activated carbon operating at 1 V. In a subsequent work, they activated waste coffee grounds with phosphoric acid (Huang et al. 2013). The presence of phosphorus was again crucial to enhance the electrochemical stability of carbon materials, as demonstrated by the high capacitance retention of 82 per cent achieved by a supercapacitor cell operated at 1.5 V for 10,000 cycles (although a high specific current of 5 A g^{-1} was employed in the test).

Berenguer et al. analysed the enhanced electrooxidation resistance of P-containing activated carbons in acid aqueous electrolyte (Berenguer et al. 2015). They employed chemical activation of olive stone by H_3PO_4 for the preparation of the activated carbons, whereas physical activation of the same precursor was employed for preparing a P-free activated carbon of similar surface area. The P-containing activated carbon showed a lower oxidation current when exposed to high potentials. The surface chemistry of the activated carbons exposed to severe electrochemical oxidation treatments was studied by XPS and TPD. The results demonstrated that surface phosphorus groups mediate in the electrochemical oxidation of the carbon surface. These groups are preferentially electrooxidised instead of the carbon surface, and later they transfer oxygen to the carbon surface, but at a lower rate than that observed on the direct electrochemical oxidation of P-free activated carbon. Moreover, the formation of CO_2-evolving oxygen groups, which are the initiators of the electrochemical carbon degradation, is hindered in the activated carbons that have phosphorus groups on their surface.

Like in the case of boron and sulphur, phosphorus groups are also co-doped with nitrogen or sulphur in order to improve the electrochemical behaviour of carbon materials (Wen et al. 2016, Yu et al. 2016).

Conclusions

Porous carbon materials are the main component of the supercapacitors electrodes, thanks to their high porosity development, high electrochemical stability, reasonable cost and tunable surface chemistry. These materials have pores of different sizes and morphologies. The presence of micropores (large capacitance) with interconnected narrow mesopores (ion reservoirs) and hierarchical porous structures (improved ion mobility), is highly desirable for maximising the performance of supercapacitors.

The performance of supercapacitors is usually reported in terms of capacitance, energy, power and durability. All these parameters are related to different properties of porous carbons, such as wettability, electrical conductivity, redox ability and electrochemical stability. These properties are greatly affected by the surface chemistry of carbon materials, which is in turn controlled by the presence of heteroatoms in their surface.

Surface oxygen groups are the most frequent functionalities on carbon materials. In terms of their electrochemical behaviour, they can be distinguished between those that evolve as CO during a thermal treatment and those that thermally decompose forming CO_2. The presence of all of them improves the wettability of carbon materials, especially in aqueous electrolytes. The electrical conductivity of carbon materials is hampered by the presence of CO_2-evolving groups, which have an electron-withdrawing character. Pseudocapacitance due to fast redox reactions is observed in carbon materials containing CO-evolving functionalities, especially in acid electrolyte. As for the durability, the presence of surface oxygen groups is detrimental. In aqueous media, they are an intermediate for the carbon gasification reaction, whereas in organic electrolyte they also favour the decomposition and polymerisation of the solvent.

The second most common heteroatom studied in supercapacitor research is nitrogen. Nitrogen functionalities are distinguished in terms of their position in the graphene layer, with those located in the graphene layer, either inside or in edge positions, having the most important effect on the electrochemical properties of carbon materials. The polar character of most of these groups increases the wettability of the surface and can increase the number of ions within the electrical double layer of porous carbons. The presence of quaternary nitrogen, pyrroles and certain amide functionalities enhances the electrical conductivity of carbon materials thanks to the additional electrons that they bring into the π-cloud of the graphene layer. Pseudocapacitance effects due to the presence of nitrogen groups are observed in acid and in alkaline aqueous media although no clear assignments of redox reactions have been made. The durability of carbon materials is greatly improved when nitrogen functionalities are included, either in aqueous or organic electrolyte. In aqueous electrolyte, they seem to stabilise the carbon surface against electrochemical oxidation. In organic electrolyte, they decrease the reactivity of the carbon surface towards the electrolyte.

Boron, phosphorus and sulphur are other heteroatoms that can have an interesting influence on the performance of carbon materials in supercapacitors. However, they have not been studied in such a detail as with N and O. Boron seems to have similar effects to those found for nitrogen. Sulphur can improve electrical conductivity and contribute to pseudocapacitance, whereas phosphorus seems to enhance the electrochemical stability of activated carbons.

Acknowledgements

This work was supported by the Ministry of Economy and competitiveness of Spain (MINECO) and FEDER (CTQ2015-66080-R (MINECO/FEDER) and MAT2016-76595-R), Generalitat Valenciana (PROMETEOII/2014/010)). EBG thanks Universidad de Alicante for a Banco Santander grant for Master studies.

References

Ahmadpour, A. and D.D. Do. 1996. The preparation of active carbons from coal by chemical and physical activation. Carbon 34: 471–479.

Andreas, H.A. and B.E. Conway. 2006. Examination of the double-layer capacitance of an high specific-area C-cloth electrode as titrated from acidic to alkaline pHs. Electrochim Acta 51: 6510–6520.

Ania, C.O., V. Khomenko, E. Raymundo-Piñero, J.B. Parra and F. Béguin. 2007. The large electrochemical capacitance of microporous doped carbon obtained by using a zeolite template. Adv. Funct. Mater 17: 1828–1836.

Arbizzani, C., M. Mastragostino and L. Meneghello. 1996. Polymer-based redox supercapacitors: A comparative study. Electrochim Acta 41: 21–26.

Aslan, M., D. Weingarth, N. Jäckel, J.S. Atchison, I. Grobelsek and V. Presser. 2014. Polyvinylpyrrolidone as binder for castable supercapacitor electrodes with high electrochemical performance in organic electrolytes. J. Power Sources 266: 374–383.

Azaïs, P., Duclaux, L., Florian, P., Massiot, D., Lillo-Rodenas, M.-A., Linares-Solano, A. et al. 2007. Causes of supercapacitors ageing in organic electrolyte. J Power Sources 171: 1046–1053.

Bandosz, T.J. 2008. Surface chemistry of carbon materials. pp. 45–92. *In*: Serp, P. and J.L. Figueiredo (eds.). Carbon Materials for Catalysis. John Wiley & Sons, Inc., Hoboken, NJ, USA.

Banks, C.E., R.R. Moore, T.J. Davies and R.G. Compton. 2004. Investigation of modified basal plane pyrolytic graphite electrodes: definitive evidence for the electrocatalytic properties of the ends of carbon nanotubes. Chem. Commun. 1804–1805.

Bansal, R.C., J.B. Donnet and F. Stoeckli. 1988. Active Carbon. Marcel Dekker, New York, USA.

Barbieri, O., M. Hahn, A. Herzog and R. Kötz. 2005. Capacitance limits of high surface area activated carbons for double layer capacitors. Carbon 43: 1303–1310.

Béguin, F., V. Presser, A. Balducci and E. Frackowiak. 2014. Carbons and electrolytes for advanced supercapacitors. Adv. Mater 26: 2219–2251.

Benyoucef, A., F. Huerta, J.L. Vázquez and E. Morallon. 2005. Synthesis and *in situ* FTIRS characterization of conducting polymers obtained from aminobenzoic acid isomers at platinum electrodes. Eur. Polym. J. 41: 843–852.

Berenguer, R., F.J. García-Mateos, R. Ruiz-Rosas, D. Cazorla-Amorós, E. Morallón, J. Rodríguez-Mirasol et al. 2016. Biomass-derived binderless fibrous carbon electrodes for ultrafast energy storage. Green Chem. 18: 1506–1515.

Berenguer, R., J.P. Marco-Lozar, C. Quijada, D. Cazorla-Amorós and E. Morallón. 2009. Effect of electrochemical treatments on the surface chemistry of activated carbon. Carbon 47: 1018–1027.

Berenguer, R., H. Nishihara, H. Itoi, T. Ishii, E. Morallón, D. Cazorla-Amorós et al. 2013. Electrochemical generation of oxygen-containing groups in an ordered microporous zeolite-templated carbon. Carbon 54: 94–104.

Berenguer, R., R. Ruiz-Rosas, A. Gallardo, D. Cazorla-Amorós, E. Morallón, H. Nishihara et al. 2015. Enhanced electro-oxidation resistance of carbon electrodes induced by phosphorus surface groups. Carbon 95: 681–689.

Bleda-Martínez, M.J., D. Lozano-Castelló, D. Cazorla-Amorós and E. Morallón. 2010. Kinetics of double-layer formation: Influence of porous structure and pore size distribution. Energy Fuels 24: 3378–3384.

Bleda-Martínez, M.J., D. Lozano-Castelló, E. Morallón, D. Cazorla-Amorós and A. Linares-Solano. 2006. Chemical and electrochemical characterization of porous carbon materials. Carbon 44: 2642–2651.

Bleda-Martínez, M.J., J.A. Maciá-Agulló, D. Lozano-Castelló, E. Morallón, D. Cazorla-Amorós and A. Linares-Solano. 2005. Role of surface chemistry on electric double layer capacitance of carbon materials. Carbon 43: 2677–2684.

Böckenfeld, N., S.S. Jeong, M. Winter, S. Passerini and A. Balducci. 2013. Natural, cheap and environmentally friendly binder for supercapacitors. J. Power Sources 221: 14–20.

Boehm, H.P. 1994. Some aspects of the surface chemistry of carbon blacks and other carbons. Carbon 32: 759–769.

Boehm, H.P. 2002. Surface oxides on carbon and their analysis: a critical assessment. Carbon 40: 145–149.

Calvo, E.G., C.O Ania, L. Zubizarreta, J.A. Menéndez and A. Arenillas. 2010. Exploring new routes in the synthesis of carbon xerogels for their application in electric double-layer capacitors. Energy Fuels 24: 3334–3339.

Calvo, E.G., F. Lufrano, P. Staiti, A. Brigandì, A. Arenillas and J.A. Menéndez. 2013. Optimizing the electrochemical performance of aqueous symmetric supercapacitors based on an activated carbon xerogel. J. Power Sources 241: 776–782.

Candelaria, S.L., B.B. Garcia, D. Liu and G. Cao. 2012. Nitrogen modification of highly porous carbon for improved supercapacitor performance. J. Mater Chem. 22: 9884–9889.

Cazorla-Amorós, D., D. Lozano-Castelló, E. Morallón, M.J. Bleda-Martínez, A. Linares-Solano and S. Shiraishi. 2010. Measuring cycle efficiency and capacitance of chemically activated carbons in propylene carbonate. Carbon 48: 1451–1456.

Centeno, T.A. and F. Stoeckli. 2006. The role of textural characteristics and oxygen-containing surface groups in the supercapacitor performances of activated carbons. Electrochim Acta 52: 560–566.

Chang, B., Y. Guo, Y. Li, H. Yin, S. Zhang, B. Yang et al. 2015. Graphitized hierarchical porous carbon nanospheres: simultaneous activation/graphitization and superior supercapacitance performance. J. Mater Chem. A 3: 9565–9577.

Chen, J.., W. Li, D. Wang, S. Yang, J. Wen and Z. Ren. 2002. Electrochemical characterization of carbon nanotubes as electrode in electrochemical double-layer capacitors. Carbon 40: 1193–1197.

Cheng, P.-Z. and H. Teng. 2003. Electrochemical responses from surface oxides present on HNO_3-treated carbons. Carbon 41: 2057–2063.

Chmiola, J., C. Largeot, P.-L. Taberna, P. Simon and Y. Gogotsi. 2008. Desolvation of ions in subnanometer pores and its effect on capacitance and double-layer theory. Angew. Chem. Int. Ed. 47: 3392–3395.

Chmiola, J., G. Yushin, Y. Gogotsi, C. Portet, P. Simon and P.L. Taberna. 2006. Anomalous Increase in Carbon Capacitance at Pore Sizes Less Than 1 Nanometer Science 313: 1760–1763.

Choo, H.-S., T. Kinumoto, S.-K. Jeong, Y. Iriyama, T. Abe and Z. Ogumi. 2007. Mechanism for electrochemical oxidation of highly oriented pyrolytic graphite in sulfuric acid solution. J. Electrochem. Soc. 154: B1017–B1023.

Chuenchom, L., R. Kraehnert and B.M. Smarsly. 2012. Recent progress in soft-templating of porous carbon materials. Soft Matter 8: 10801–10812.

Conway, B.E. 1991. Transition from 'Supercapacitor' to 'Battery' behavior in electrochemical energy storage. J. Electrochem. Soc. 138: 1539–1548.

Conway, B.E., V. Birss and J. Wojtowicz. 1997. The role and utilization of pseudocapacitance for energy storage by supercapacitors. J. Power Sources 66: 1–14.

Conway, B.E. and W.G. Pell. 2003. Double-layer and pseudocapacitance types of electrochemical capacitors and their applications to the development of hybrid devices. J. Solid State Electrochem. 7: 637–644.

Deng, Y., Y. Xie, K. Zou and X. Ji. 2016. Review on recent advances in nitrogen-doped carbons: preparations and applications in supercapacitors. J. Mater Chem. A 4: 1144–1173.

Eliad, L., G. Salitra, A. Soffer and D. Aurbach. 2001. Ion sieving effects in the electrical double layer of porous carbon electrodes: Estimating effective ion size in electrolytic solutions. J. Phys. Chem. B 105: 6880–6887.

Enterría, M., M.F.R. Pereira, J.I. Martins and J.L. Figueiredo. 2015. Hydrothermal functionalization of ordered mesoporous carbons: The effect of boron on supercapacitor performance. Carbon 95: 72–83.

Figueiredo, J.L., M.F.R. Pereira, M.M.A. Freitas and J.J.M. Órfão. 1999. Modification of the surface chemistry of activated carbons. Carbon 37: 1379–1389.

Figueiredo, J.L. 2013. Functionalization of porous carbons for catalytic applications. J. Mater Chem. A 1: 9351–9364.

Fletcher, S., V.J. Black and I. Kirkpatrick. 2013. A universal equivalent circuit for carbon-based supercapacitors. J. Solid State Electrochem. 18: 1377–1387.

Frackowiak, E. and F. Béguin. 2001. Carbon materials for the electrochemical storage of energy in capacitors. Carbon 39: 937–950.

Frackowiak, E. and F. Béguin. 2002. Electrochemical storage of energy in carbon nanotubes and nanostructured carbons. Carbon 40: 1775–1787.

Frackowiak, E., V. Khomenko, K. Jurewicz, K. Lota and F. Béguin. 2006. Supercapacitors based on conducting polymers/nanotubes composites. J. Power Sources 153: 413–418.

Fujisawa, K., R. Cruz-Silva, K.-S. Yang, Y.A. Kim, T. Hayashi, M. Endo et al. 2014. Importance of open, heteroatom-decorated edges in chemically doped-graphene for supercapacitor applications. J. Mater Chem. A 2: 9532–9540.

Gao, J., X. Wang, Y. Zhang, J. Liu, Q. Lu and M. Liu. 2016. Boron-doped ordered mesoporous carbons for the application of supercapacitors. Electrochim. Acta 207: 266–274.

Gao, Q., L. Demarconnay, E. Raymundo-Piñero and F. Béguin. 2012. Exploring the large voltage range of carbon/carbon supercapacitors in aqueous lithium sulfate electrolyte. Energy Environ. Sci. 5: 9611–9617.

García, A. 1998. Modification of the surface properties of an activated carbon by oxygen plasma treatment. Fuel. 77: 613–624.

González-Gaitán, C., R. Ruiz-Rosas, E. Morallón and D. Cazorla-Amorós. 2015. Electrochemical methods to functionalize carbon materials. pp. 231–262. *In*: Thakur, V.K. and Thakur, M.K. (eds.). Chemical Functionalization of Carbon Nanomaterials: Chemistry and Applications. CRC Press, Taylor & Francis Group, Boca Ratón, FL, USA.

González-Gaitán, C., R. Ruiz-Rosas, H. Nishihara, T. Kyotani, E. Morallón and D. Cazorla-Amorós. 2016. Successful functionalization of superporous zeolite templated carbon using aminobenzene acids and electrochemical methods. Carbon 99: 157–166.

Gonzalez-Serrano, E., T. Cordero, J. Rodriguez-Mirasol and J.J. Rodriguez. 1997. Development of porosity upon chemical activation of kraft lignin with ZnCl2. Ind. Eng. Chem. Res. 36: 4832–4838.

Gryglewicz, G., J. Machnikowski, E. Lorenc-Grabowska, G. Lota and E. Frackowiak. 2005. Effect of pore size distribution of coal-based activated carbons on double layer capacitance. Electrochim. Acta 50: 1197–1206.

Guo, H. and Q. Gao. 2009. Boron and nitrogen co-doped porous carbon and its enhanced properties as supercapacitor. J. Power Sources 186: 551–556.

Hahn, M., A. Würsig, R. Gallay, P. Novák and R. Kötz. 2005. Gas evolution in activated carbon/propylene carbonate based double-layer capacitors. Electrochem. Commun. 7: 925–930.

Han, J., L.L. Zhang, S. Lee, J. Oh, K.-S. Lee, J.R. Potts et al. 2013. Generation of B-doped graphene nanoplatelets using a solution process and their supercapacitor applications. ACS Nano 7: 19–26.

Hasegawa, G., M. Aoki, K. Kanamori, K. Nakanishi, T. Hanada and K. Tadanaga. 2011. Monolithic electrode for electric double-layer capacitors based on macro/meso/microporous S-containing activated carbon with high surface area. J. Mater Chem. 21: 2060–2063.

Hasegawa, G., T. Deguchi, K. Kanamori, Y. Kobayashi, H. Kageyama, T. Abe et al. 2015. High-level doping of nitrogen, phosphorus, and sulfur into activated carbon monoliths and their electrochemical capacitances. Chem. Mater 27: 4703–4712.

He, M., K. Fic, E. Frckowiak, P. Novák and E.J. Berg. 2016. Ageing phenomena in high-voltage aqueous supercapacitors investigated by *in situ* gas analysis. Energy Environ. Sci. 9: 623–633.

Hsieh, C.-T. and H. Teng. 2002a. Influence of oxygen treatment on electric double-layer capacitance of activated carbon fabrics. Carbon 40: 667–674.

Huang, C., A.M. Puziy, T. Sun, O.I. Poddubnaya, F. Suárez-García, J.M.D. Tascón et al. 2014. Capacitive behaviours of phosphorus-rich carbons derived from lignocelluloses. Electrochim. Acta 137: 219–227.

Huang, C., T. Sun and D. Hulicova-Jurcakova. 2013. Wide electrochemical window of supercapacitors from coffee bean-derived phosphorus-rich carbons. ChemSusChem. 6: 2330–2339.

Huang, Y., S.L. Candelaria, Y. Li, Z. Li, J. Tian, L. Zhang et al. 2014. Sulfurized activated carbon for high energy density supercapacitors. J. Power Sources 252: 90–97.

Hulicova, D., M. Kodama and H. Hatori. 2006. Electrochemical performance of nitrogen-enriched carbons in aqueous and non-aqueous supercapacitors. Chem. Mater 18: 2318–2326.

Hulicova, D., J. Yamashita, Y. Soneda, H. Hatori and M. Kodama. 2005. Supercapacitors prepared from melamine-based carbon. Chem. Mater 17: 1241–1247.

Hulicova-Jurcakova, D., M. Kodama, S. Shiraishi, H. Hatori, Z.H. Zhu and G.Q. Lu. 2009a. Nitrogen-enriched nonporous carbon electrodes with extraordinary supercapacitance. Adv. Funct. Mater 19: 1800–1809.

Hulicova-Jurcakova, D., A.M. Puziy, O.I. Poddubnaya, F. Suárez-García, J.M.D. Tascón and G.Q. Lu. 2009b. Highly stable performance of supercapacitors from phosphorus-enriched carbons. J. Am. Chem. Soc. 131: 5026–5027.

Inagaki, M. 2010. Structure and texture of carbon materials. pp. 37–76. *In*: Béguin, F. and E. Frackowiak (eds.). Carbons for Electrochemical Energy Storage and Conversion Systems. CRC Press, Taylor & Francis group, Boca Raton, FL, USA.

Inagaki, M., H. Konno and O. Tanaike. 2010. Carbon materials for electrochemical capacitors. J. Power Sources 195: 7880–7903.

Ishii, T., S. Kashihara, Y. Hoshikawa, J. Ozaki, N. Kannari, K. Takai et al. 2014. A quantitative analysis of carbon edge sites and an estimation of graphene sheet size in high-temperature treated, non-porous carbons. Carbon 80: 135–145.

Itoi, H., H. Nishihara, T. Ishii, K. Nueangnoraj, R. Berenguer and T. Kyotani. 2013. Large pseudocapacitance in quinone-functionalized zeolite-templated carbon. Bull. Chem. Soc. Jpn. 87: 250–257.

Itoi, H., H. Nishihara and T. Kyotani. 2016. Effect of heteroatoms in ordered microporous carbons on their electrochemical capacitance. Langmuir 32: 11997–12004.

Jagtoyen, M. and F. Derbyshire. 1998. Activated carbons from yellow poplar and white oak by H3PO4 activation. Carbon 36: 1085–1097.

Jurewicz, K., K. Babeł, A. Ziółkowski, H. Wachowska and M. Kozłowski. 2002. Ammoxidation of brown coals for supercapacitors. Fuel. Process Technol. 77–78: 191–198.

Kang, H.S. and S. Jeong. 2004. Nitrogen doping and chirality of carbon nanotubes. Phys. Rev. B 70: 233411.

Kawaguchi, M., T. Yamanaka, Y. Hayashi and H. Oda. 2010. Preparation and capacitive properties of a carbonaceous material containing nitrogen. J. Electrochem. Soc. 157: A35–A40.

Kinoshita, K. 1988. Carbon: electrochemical and physicochemical properties. Wiley, New York, USA.

Kinoshita, K. and J. Bett. 1973a. Electrochemical oxidation of carbon black in concentrated phosphoric acid at 135°C. Carbon 11: 237–247.

Kinoshita, K. and J.A.S. Bett. 1973b. Potentiodynamic analysis of surface oxides on carbon blacks. Carbon 11: 403–411.

Kötz, R. and M. Carlen. 2000. Principles and applications of electrochemical capacitors. Electrochim. Acta 45: 2483–2498.

Kuila, T., S. Bose, A.K. Mishra, P. Khanra, N.H. Kim and J.H. Lee. 2012. Chemical functionalization of graphene and its applications. Prog. Mater Sci. 57: 1061–1105.

Kurzweil, P. and M. Chwistek. 2008. Electrochemical stability of organic electrolytes in supercapacitors: Spectroscopy and gas analysis of decomposition products. J. Power Sources 176: 555–567.

Kurzweil, P., B. Frenzel and A. Hildebrand. 2015. Voltage-dependent capacitance, aging effects, and failure indicators of double-layer capacitors during lifetime testing. ChemElectroChem. 2: 160–170.

Kwon, T., H. Nishihara, H. Itoi, Q.-H. Yang and T. Kyotani. 2009. Enhancement mechanism of electrochemical capacitance in nitrogen-/boron-doped carbons with uniform straight nanochannels. Langmuir 25: 11961–11968.

Kyotani, T. 2003. Chapter 7—Porous carbon. pp. 109–127. *In*: Yasuda, E., M. Inagaki, K. Kaneko, M. Endo, A. Oya and Y. Tanabe (eds.). Carbon Alloys. Elsevier Science, Oxford, UK.

Kyotani, T., L. Tsai and A. Tomita. 1995. Formation of ultrafine carbon tubes by using an anodic aluminum oxide film as a template. Chem. Mater 7: 1427–1428.

Lee, Y.-H., K.-H. Chang and C.-C. Hu. 2013. Differentiate the pseudocapacitance and double-layer capacitance contributions for nitrogen-doped reduced graphene oxide in acidic and alkaline electrolytes. J. Power Sources 227: 300–308.

Leyva-García, S., K. Nueangnoraj, D. Lozano-Castelló, H. Nishihara, T. Kyotani, E. Morallón et al. 2015. Characterization of a zeolite-templated carbon by electrochemical quartz crystal microbalance and *in situ* Raman spectroscopy. Carbon 89: 63–73.

Li, Y., Z.-Y. Fu and B.-L. Su. 2012. Hierarchically structured porous materials for energy conversion and storage. Adv. Funct. Mater 22: 4634–4667.

Li, Y., G. Wang, T. Wei, Z. Fan and P. Yan. 2016. Nitrogen and sulfur co-doped porous carbon nanosheets derived from willow catkin for supercapacitors. Nano Energy 19: 165–175.

Liang, J., Y. Jiao, M. Jaroniec and S.Z. Qiao. 2012. Sulfur and nitrogen dual-doped mesoporous graphene electrocatalyst for oxygen reduction with synergistically enhanced performance. Angew. Chem. Int. Ed. 51: 11496–11500.

Lillo-Ródenas, M.A., D. Cazorla-Amorós and A. Linares-Solano. 2003. Understanding chemical reactions between carbons and NaOH and KOH: An insight into the chemical activation mechanism. Carbon 41: 267–275.

Lillo-Ródenas, M.A., J. Juan-Juan, D. Cazorla-Amorós and A. Linares-Solano. 2004. About reactions occurring during chemical activation with hydroxides. Carbon 42: 1371–1375.

Linares-Solano, A., D. Lozano-Castelló, M.A. Lillo-Ródenas and D. Cazorla-Amorós. 2008. Carbon activation by alkaline hydroxides preparation and reactions, porosity and performance. pp. 1–62. *In*: Radovic, L.R. (ed.). Chemistry and Physics of Carbon: Volume 30. CRC Press, Taylor & Francis group, New York, USA.

Lota, G., B. Grzyb, H. Machnikowska, J. Machnikowski and E. Frackowiak. 2005. Effect of nitrogen in carbon electrode on the supercapacitor performance. Chem. Phys. Lett. 404: 53–58.

Lota, G., K. Lota and E. Frackowiak. 2007. Nanotubes based composites rich in nitrogen for supercapacitor application. Electrochem. Commun. 9: 1828–1832.

Lozano-Castelló, D., M.A. Lillo-Ródenas, D. Cazorla-Amorós and A. Linares-Solano. 2001. Preparation of activated carbons from Spanish anthracite: I. Activation by KOH. Carbon 39: 741–749.

Lozano-Castelló, D., D. Cazorla-Amorós, A. Linares-Solano, S. Shiraishi, H. Kurihara and A. Oya. 2003. Influence of pore structure and surface chemistry on electric double layer capacitance in non-aqueous electrolyte. Carbon 41: 1765–1775.

Lozano-Castelló, D., F. Suárez-García, D. Cazorla-Amorós and A. Linares-Solano. 2010. Porous texture of carbons. pp. 115–162. *In*: Béguin, F. and E. Frackowiak (eds.). Carbons for Electrochemical Energy Storage and Conversion Systems. CRC Press, Taylor & Francis group, New York, USA.

Ma, Z., T. Kyotani and A. Tomita. 2000. Preparation of a high surface area microporous carbon having the structural regularity of Y zeolite. Chem. Commun. 2365–2366.

Maass, S., F. Finsterwalder, G. Frank, R. Hartmann and C. Merten. 2008. Carbon support oxidation in PEM fuel cell cathodes. J. Power Sources 176: 444–451.

Marsh, H. and F.R. Reinoso. 2006. Activated Carbon. Elsevier Science, Oxford, UK.

Matter, P.H., E. Wang, M. Arias, E.J. Biddinger and U.S. Ozkan. 2006. Oxygen reduction reaction catalysts prepared from acetonitrile pyrolysis over alumina-supported metal particles. J. Phys. Chem. B 110: 18374–18384.

Morallón, E., J. Arias-Pardilla, J.M. Calo and D. Cazorla-Amorós. 2009. Arsenic species interactions with a porous carbon electrode as determined with an electrochemical quartz crystal microbalance. Electrochim. Acta 54: 3996–4004.

Moreno-Castilla, C., M.V. López-Ramón and F. Carrasco-Marín. 2000. Changes in surface chemistry of activated carbons by wet oxidation. Carbon 38: 1995–2001.

Mostazo-López, M.J., R. Ruiz-Rosas, E. Morallón and D. Cazorla-Amorós. 2015. Generation of nitrogen functionalities on activated carbons by amidation reactions and Hofmann rearrangement: Chemical and electrochemical characterization. Carbon 91: 252–265.

Mostazo-López, M.J., R. Ruiz-Rosas, E. Morallón and D. Cazorla-Amorós. 2016. Nitrogen doped superporous carbon prepared by a mild method. Enhancement of supercapacitor performance. Int. J. Hydrog. Energy 41: 19691–19701.

Nishihara, H. and T. Kyotani. 2012. Templated nanocarbons for energy storage. Adv. Mater 24: 4473–4498.

Niu, J., W.G. Pell and B.E. Conway. 2006. Requirements for performance characterization of C double-layer supercapacitors: Applications to a high specific-area C-cloth material. J. Power Sources 156: 725–740.

Nueangnoraj, K., H. Nishihara, T. Ishii, N. Yamamoto, H. Itoi, R. Berenguer et al. 2015. Pseudocapacitance of zeolite-templated carbon in organic electrolytes. Energy Storage Mater 1: 35–41.

Nueangnoraj, K., R. Ruiz-Rosas, H. Nishihara, S. Shiraishi, E. Morallón, D. Cazorla-Amorós et al. 2014. Carbon–carbon asymmetric aqueous capacitor by pseudocapacitive positive and stable negative electrodes. Carbon 67: 792–794.

Oh, Y.J., J.J. Yoo, Y.I. Kim, J.K. Yoon, H.N. Yoon, J.-H. Kim et al. 2014. Oxygen functional groups and electrochemical capacitive behavior of incompletely reduced graphene oxides as a thin-film electrode of supercapacitor. Electrochim. Acta 116: 118–128.

Okajima, K., K. Ohta and M. Sudoh. 2005. Capacitance behavior of activated carbon fibers with oxygen-plasma treatment. Electrochim. Acta 50: 2227–2231.

Ornelas, O., J.M. Sieben, R. Ruiz-Rosas, E. Morallón, D. Cazorla-Amorós, J. Geng et al. 2014. On the origin of the high capacitance of nitrogen-containing carbon nanotubes in acidic and alkaline electrolytes. Chem. Commun. 50: 11343–11346.

Otake, Y. and R.G. Jenkins. 1993. Characterization of oxygen-containing surface complexes created on a microporous carbon by air and nitric acid treatment. Carbon 31: 109–121.

Ozaki, J., N. Kimura, T. Anahara and A. Oya. 2007. Preparation and oxygen reduction activity of BN-doped carbons. Carbon 45: 1847–1853.

Pandolfo, A.G. and A.F. Hollenkamp. 2006. Carbon properties and their role in supercapacitors. J. Power Sources 157: 11–27.

Paraknowitsch, J.P. and A. Thomas. 2013. Doping carbons beyond nitrogen: an overview of advanced heteroatom doped carbons with boron, sulphur and phosphorus for energy applications. Energy Environ. Sci. 6: 2839–2855.

Paredes, J.I., A. Martínez-Alonso and J.M.D. Tascón. 2000. Comparative study of the air and oxygen plasma oxidation of highly oriented pyrolytic graphite: a scanning tunneling and atomic force microscopy investigation. Carbon 38: 1183–1197.

Pekala, R.W. 1989. Organic aerogels from the polycondensation of resorcinol with formaldehyde. J. Mater Sci. 24: 3221–3227.

Pekala, R.W., J.C. Farmer, C.T. Alviso, T.D. Tran, S.T. Mayer, J.M. Miller et al. 1998. Carbon aerogels for electrochemical applications. J. Non-Cryst. Solids 225: 74–80.

Pels, J.R., F. Kapteijn, J.A. Moulijn, Q. Zhu and K.M. Thomas. 1995. Evolution of nitrogen functionalities in carbonaceous materials during pyrolysis. Carbon 33: 1641–1653.

Peng, C., S. Zhang, X. Zhou and G.Z. Chen. 2010. Unequalisation of electrode capacitances for enhanced energy capacity in asymmetrical supercapacitors. Energy Environ. Sci. 3: 1499–1502.

Pietrzak, R., K. Jurewicz, P. Nowicki, K. Babeł and H. Wachowska. 2007. Microporous activated carbons from ammoxidised anthracite and their capacitance behaviours. Fuel. 86: 1086–1092.

Piñeiro-Prado, I., D. Salinas-Torres, R. Ruiz-Rosas, E. Morallón and D. Cazorla-Amorós. 2016. Design of activated carbon/ activated carbon asymmetric capacitors. Front Mater 3: 16.

Polovina, M., B. Babić, B. Kaluderović and A. Dekanski. 1997. Surface characterization of oxidized activated carbon cloth. Carbon 35: 1047–1052.

Portet, C., P.L. Taberna, P. Simon and E. Flahaut. 2005. Influence of carbon nanotubes addition on carbon–carbon supercapacitor performances in organic electrolyte. J. Power Sources 139: 371–378.

Pröbstle, H., M. Wiener and J. Fricke. 2003. Carbon aerogels for electrochemical double layer capacitors. J. Porous Mater 10: 213–222.

Radovic, L.R. 2010. Surface chemical and electrochemical properties of carbons. pp. 163–219. *In*: Béguin, F. and E. Frackowiak (eds.). Carbons for Electrochemical Energy Storage and Conversion Systems. CRC Press, Taylor & Francis group, Boca Raton, FL, USA.

Radovic, L.R. and B. Bockrath. 2005. On the chemical nature of graphene edges: Origin of stability and potential for magnetism in carbon materials. J. Am. Chem. Soc. 127: 5917 5927.

Randin, J.-P. and E. Yeager. 1971. Differential capacitance study of stress-annealed pyrolytic graphite electrodes. J. Electrochem. Soc. 118: 711–714.

Ratajczak, P., K. Jurewicz, P. Skowron, Q. Abbas and F. Béguin. 2014. Effect of accelerated ageing on the performance of high voltage carbon/carbon electrochemical capacitors in salt aqueous electrolyte. Electrochim. Acta 130: 344–350.

Raymundo-Piñero, E., D. Cazorla-Amorós and A. Linares-Solano. 2003. The role of different nitrogen functional groups on the removal of SO2 from flue gases by N-doped activated carbon powders and fibres. Carbon 41: 1925–1932.

Raymundo-Piñero, E., D. Cazorla-Amorós, A. Linares-Solano, J. Find, U. Wild and R. Schlögl. 2002. Structural characterization of N-containing activated carbon fibers prepared from a low softening point petroleum pitch and a melamine resin. Carbon 40: 597–608.

Raymundo-Piñero, E., F. Leroux and F. Béguin. 2006. A high-performance carbon for supercapacitors obtained by carbonization of a seaweed biopolymer. Adv. Mater 18: 1877–1882.

Rodríguez-Reinoso, F. and M. Molina-Sabio. 1992. Activated carbons from lignocellulosic materials by chemical and/or physical activation: an overview. Carbon 30: 1111–1118.

Román-Martínez, M.C., D. Cazorla-Amorós, A. Linares-Solano and C.S.-M. de Lecea. 1993. Tpd and TPR characterization of carbonaceous supports and Pt/C catalysts. Carbon 31: 895–902.

Rosas, J.M., J. Bedia, J. Rodríguez-Mirasol and T. Cordero. 2009. HEMP-derived activated carbon fibers by chemical activation with phosphoric acid. Fuel 88: 19–26.

Ruiz-Rosas, R., M.J. Valero-Romero, D. Salinas-Torres, J. Rodríguez-Mirasol, T. Cordero, E. Morallón et al. 2014. Electrochemical performance of hierarchical porous carbon materials obtained from the infiltration of lignin into zeolite Templates. ChemSusChem. 7: 1458–1467.

Ryoo, R., S.H. Joo and S. Jun. 1999. Synthesis of highly ordered carbon molecular sieves via template-mediated structural transformation. J. Phys. Chem. B 103: 7743–7746.

Salame, I.I. and T.J. Bandosz. 2001. Surface chemistry of activated carbons: Combining the results of temperature-programmed desorption, boehm, and potentiometric titrations. J. Colloid Interface Sci. 240: 252–258.

Salinas-Torres, D., S. Shiraishi, E. Morallón and D. Cazorla-Amorós. 2015. Improvement of carbon materials performance by nitrogen functional groups in electrochemical capacitors in organic electrolyte at severe conditions. Carbon 82: 205–213.

Salinas-Torres, D., J.M. Sieben, D. Lozano-Castelló, D. Cazorla-Amorós and E. Morallón. 2013. Asymmetric hybrid capacitors based on activated carbon and activated carbon fibre-PANI electrodes. Electrochim. Acta 89: 326–333.

Salinas-Torres, D., J.M. Sieben, D. Lozano-Castello, E. Morallón, M. Burghammer, C. Riekel et al. 2012. Characterization of activated carbon fiber/polyaniline materials by position-resolved microbeam small-angle X-ray scattering. Carbon 50: 1051–1056.

Sánchez-González, J., F. Stoeckli and T.A. Centeno. 2011. The role of the electric conductivity of carbons in the electrochemical capacitor performance. J. Electroanal. Chem. 657: 176–180.

Seredych, M. and T.J. Bandosz. 2013. S-doped micro/mesoporous carbon–graphene composites as efficient supercapacitors in alkaline media. J. Mater Chem. A 1: 11717–11727.

Seredych, M., D. Hulicova-Jurcakova, G.Q. Lu and T.J. Bandosz. 2008. Surface functional groups of carbons and the effects of their chemical character, density and accessibility to ions on electrochemical performance. Carbon 46: 1475–1488.

Shen, W. and W. Fan. 2012. Nitrogen-containing porous carbons: Synthesis and application. J. Mater Chem. A 1: 999–1013.

Sing, K.S.W. 1985. Reporting physisorption data for gas/solid systems with special reference to the determination of surface area and porosity (Recommendations 1984). Pure Appl. Chem. 57: 603–619.

Strelko, V.V., V.S. Kuts and P.A. Thrower. 2000. On the mechanism of possible influence of heteroatoms of nitrogen, boron and phosphorus in a carbon matrix on the catalytic activity of carbons in electron transfer reactions. Carbon 38: 1499–1503.

Sui, Z.-Y., Y.-N. Meng, P.-W. Xiao, Z.-Q. Zhao, Z.-X. Wei and B.-H. Han. 2015. Nitrogen-doped graphene aerogels as efficient supercapacitor electrodes and gas adsorbents. ACS Appl. Mater Interfaces 7: 1431–1438.

Szubzda, B., A. Szmaja and A. Halama. 2012. Influence of structure and wettability of supercapacitor electrodes carbon materials on their electrochemical properties in water and organic solutions. Electrochim. Acta 86: 255–259.

Taberna, P.L., P. Simon and J.F. Fauvarque. 2003. Electrochemical characteristics and impedance spectroscopy studies of carbon-carbon supercapacitors. J. Electrochem. Soc. 150: A292–A300.

Tabti, Z., R. Berenguer, R. Ruiz-Rosas, C. Quijada, E. Morallón and D. Cazorla-Amorós. 2013. Electrooxidation methods to produce pseudocapacitance-containing porous carbons. Electrochemistry 81: 833–839.

Tabti, Z., R. Ruiz-Rosas, C. Quijada, D. Cazorla-Amorós and E. Morallón. 2014. Tailoring the surface chemistry of activated carbon cloth by electrochemical methods. ACS Appl. Mater Interfaces 6: 11682–11691.

Terrones, M., P.M. Ajayan, F. Banhart, X. Blase, D.L. Carroll, J.C. Charlier et al. 2002. N-doping and coalescence of carbon nanotubes: synthesis and electronic properties. Appl. Phys. A 74: 355–361.

Wang, D.-W., F. Li, Z.-G. Chen, G.Q. Lu and H.-M. Cheng. 2008. Synthesis and electrochemical property of boron-doped mesoporous carbon in supercapacitor. Chem. Mater 20: 7195–7200.

Wang, F., S. Xiao, Y. Hou, C. Hu, L. Liu and Y. Wu. 2013. Electrode materials for aqueous asymmetric supercapacitors. RSC Adv. 3: 13059–13084.

Weingarth, D., A. Foelske-Schmitz and R. Kötz. 2013. Cycle versus voltage hold—Which is the better stability test for electrochemical double layer capacitors? J. Power Sources 225: 84–88.

Wen, Y., T.E. Rufford, D. Hulicova-Jurcakova and L. Wang. 2016. Nitrogen and phosphorous co-doped graphene monolith for supercapacitors. ChemSusChem. 9: 513–520.

Wen, Y., B. Wang, C. Huang, L. Wang and D. Hulicova-Jurcakova. 2015. Synthesis of phosphorus-doped graphene and its wide potential window in aqueous supercapacitors. Chem. – Eur. J. 21: 80–85.

White, R.J., M. Antonietti and M.M. Titirici. 2009. Naturally inspired nitrogen doped porous carbon. J. Mater Chem. 19: 8645–8645.

You, B., F. Kang, P. Yin and Q. Zhang. 2016. Hydrogel-derived heteroatom-doped porous carbon networks for supercapacitor and electrocatalytic oxygen reduction. Carbon 103: 9–15.

Yu, A., V. Chabot and J. Zhang. 2013. Electrochemical supercapacitors for energy storage and delivery: Fundamentals and applications. CRC Press, Taylor & Francis Group, Boca Ratón, FL, USA.

Yu, X., Y. Kang and H.S. Park. 2016. Sulfur and phosphorus co-doping of hierarchically porous graphene aerogels for enhancing supercapacitor performance. Carbon 101: 49–56.

Zakhidov, A.A., R.H. Baughman, Z. Iqbal, C. Cui, I. Khayrullin, S.O. Dantas et al. 1998. Carbon structures with three-dimensional periodicity at optical wavelengths. Science 282: 897–901.

Zhao, X., Q. Zhang, C.-M. Chen, B. Zhang, S. Reiche, A. Wang et al. 2012. Aromatic sulfide, sulfoxide, and sulfone mediated mesoporous carbon monolith for use in supercapacitor. Nano Energy 1: 624–630.

Zhong, C., Y. Deng, W. Hu, J. Qiao, L. Zhang and J. Zhang. 2015. A review of electrolyte materials and compositions for electrochemical supercapacitors. Chem. Soc. Rev. 44: 7484–7539.

Zhu, M., C.J. Weber, Y. Yang, M. Konuma, U. Starke, K. Kern et al. 2008. Chemical and electrochemical ageing of carbon materials used in supercapacitor electrodes. Carbon 46: 1829–1840.

Zhu, Z.H., H. Hatori, S.B. Wang and G.Q. Lu. 2005. Insights into hydrogen atom adsorption on and the electrochemical properties of nitrogen-substituted carbon materials. J. Phys. Chem. B 109: 16744–16749.

5

Three-Dimensional Nanostructured Electrode Architectures for Next Generation Electrochemical Energy Storage Devices

Terence K.S. Wong

1. Introduction

The energy transition—an electricity supply grid-based on burning the fossil fuels in centralised power stations to a smart grid with multiple distributed renewable energy sources and embedded information network is one of the greatest challenges facing the world today (Smalley 2005). The electricity grid in its present legacy network structure has the advantage of supplying power on demand and is generally highly reliable (Amin and Stringer 2008). However, the large-scale combustion of fossil fuels (coal, natural gas) in conventional power stations can result in particulate matter air pollution, associated health issues and most importantly, the release of carbon dioxide into the atmosphere. The trapping of heat by this greenhouse gas has already resulted in a consistent pattern of warming of the Earth in the 21st century. The adverse effect of carbon dioxide on the natural environment is a more immediate problem than the long-term sustainability of fossil fuels (Strom 2007).

The basic concept of the smart grid is to decentralise the power generation nodes of the electricity network to include a portfolio of renewable energy sources such as photovoltaics, solar thermal, wind and hydropower (Amin and Stringer 2008). In addition, a bidirectional data network will be integrated with the electricity supply network to enhance grid reliability and efficiency by providing real time data about electricity pricing and demand (Harris and Meyers 2010). Although the amount of solar power received by the Earth (162,000 TW) greatly exceeds the daily power demand by the world population (Ginley et al. 2008) and renewable sources do not emit carbon dioxide, the extensive deployment of renewable sources can result in uncertainty in the electricity supply in the long term. This is because both solar and wind energy are inherently intermittent sources of energy with low capacity factors. The capacity factor refers to the percentage of the rated power that is equal to the average annual generated power of an intermittent renewable energy source (Twidell and Weir 2006). When an electrical grid consists of many low capacity factor renewable sources, the electricity supply uncertainty will increase greatly from the legacy grid. Supply disruption has already occurred in

School of Electrical and Electronic Engineering, Nanyang Technological University, Singapore 639798.
E-mail: ekswong@ntu.edu.sg

South Australia in 2016 (Reese 2017). In this incident, a significant amount of wind power sources is connected to the electricity grid and a conventional power plant was decommissioned when storms caused the wind turbines to be offline. The intermittency of renewable energy sources such as solar and wind is the main obstacle in the transition to a smart grid. Although a large scale photovoltaic power plant with stabilised output has been demonstrated in a research project (Ueda et al. 2009), but still a lot of research and development work remains to be carried out.

A critical and indispensable components of the future smart grid are energy storage devices, which are key to overcoming the intermittency of renewable sources (Harris and Meyers 2010). As shown in Table 1 and discussed in (Francois et al. 2015), there are no fewer than nine energy storage technologies. Some of these technologies are not yet commercialised. At present, grid scale energy storage is implemented mainly by: (i) hydraulic storage, (ii) compressed air energy storage (CAES) and (iii) sodium sulphur batteries. Hydraulic storage is the oldest energy storage technology and is still by far the most important in terms of energy storage capacity. The principle of hydraulic storage is straightforward. Surplus electricity is used to pump water to a reservoir sited above a power plant and energy is stored in the form of gravitational potential energy. When the demand for electricity increases, water from the reservoir flows back downhill through turbines to generate electricity. This type of energy storage technology, however, is not widely applicable because it requires sites with a plateau or mountain and the construction of a reservoir.

In CAES, surplus generated electricity is used to drive a compressor to pump air into large underground caverns. Since air heats up when compressed, the air needs to pass through a heat exchanger before storage. Energy is stored temporarily in underground in the form of air under high pressure. During periods of higher electricity demand, the compressed air is reheated by burning natural gas and the heated air is used to drive a gas turbine to generate electricity (Francois et al. 2015). As with pumped hydroelectric storage, compressed air storage requires sites that fulfil specific geological requirement such as non-permeable rock formation that allows the air pressure to be maintained during storage. As a result, this form of grid energy storage is also not common and is limited to certain locations.

A more widely applicable grid energy storage technology is the sodium sulphur (NaS) battery (Dunn et al. 2011). This electrochemical energy storage (EES) device is a secondary battery that makes use of the reversible redox reaction between the alkali metal sodium and sulphur. In a typical NaS battery, the anode is in the form of molten liquid sodium contained within a solid electrolyte made of the ceramic β-alumina. The molten sulphur electrode occupies the space between the electrolyte and a hermetically sealed metal container, which also serves as a current collector. Although the NaS battery has been installed in numerous locations around the world especially in Japan, its energy capacity is limited to the range of 1.4–14.6 MWh (Doughty et al. 2010). In particular, the NaS battery must be maintained at a temperature of about 300°C during normal operation. This requirement makes it costlier to operate the NaS battery and there can be concerns about leakage of Na from the battery.

Table 1. Energy Storage Technologies.

Energy Storage Technology	Storage Mechanism
Hydraulic storage	Height difference between reservoir and generator
Compressed air storage	Internal energy of air under pressure in underground cavities
Sensitive-heat storage	Temperature increase of molten salt
Latent-heat storage	Solid to liquid phase transition
Electrochemical storage	Chemical bonds of electrode material
Electrostatic storage	Electric field of electrical double layer
Kinetic storage	Rotation of a mass (flywheel)
Hydrogen storage	Covalent bond of molecular hydrogen
Electromagnetic storage	Magnetic field of electric current in a superconducting coil

The above status summary of current grid energy storage technologies show that EES devices are the only long term viable solution to the storage requirement of the future smart grid. It is also important to point out that EES devices are vitally important to the development of electric transportation and the reduction of road side carbon emission from diesel and internal combustion engine vehicles. At present, both the plug-in hybrid electric vehicle (PHEV) and electric vehicle (EV) are small cars with very limited driving range. They are used only as private vehicles and are incapable of carrying heavy loads. This is because the PHEV and EV make use of the lithium ion battery (LIB) as their main or only power source respectively (Cairns and Albertus 2010). The limited driving range between charge is a major impediment that deters consumers from purchasing such vehicles. It is especially significant at a time when the charging infrastructure is not yet fully developed in most countries. While the LIB based on the Li ion intercalation principle is adequate for mobile devices and small electrical appliances, both the energy density and power density of LIBs are inadequate for the more demanding electric transportation application. As a result, there is a critical need to develop new high performance EES devices (beyond the NaS and LIB batteries) that are scalable, reliable, and simple to operate. The key performance characteristics for EES devices are: energy density, power density, cycle lifetime, self-discharge rate, roundtrip efficiency and safety.

In this chapter, it will be shown that the fulfilment of this challenging goal in the field of EES will require radically new EES design and new electrode structure to be developed. In particular, the integration of porous nanomaterials into an electrode structure with a three dimensional (3D) architecture can bring about enhancements in both energy density and power density (Arthur et al. 2011). These characteristics are difficult to be realised simultaneously in conventional EES devices with two dimensional (2D) electrodes. This chapter is different from previous reviews in that it is not focused exclusively on small foot print 3D batteries for powering microelectromechanical systems (Long et al. 2004) and on multidimensional materials and their device architecture (Lukatskaya et al. 2016). In addition, it considers the electrochemical capacitors (EC) as well as secondary (rechargeable) batteries.

In the next section, the difference in basic operation between the EC and the secondary battery is briefly discussed with emphasis on the fabrication and limitation of conventional 2D EES electrodes. The concept of the 3D electrode in EES devices is then defined followed by a comprehensive survey of the existing fabrication techniques. The fabrication techniques are divided into those yielding apcriodic and periodic porous electrodes and infiltration methods. Examples of EES devices with these 3D electrodes from the literature are included for illustration.

2. EES Devices with 2D Electrodes

In this section, we first briefly describe the principle operation of conventional secondary batteries and ECs so that the reader can better appreciate the limitation of conventional 2D electrodes and the rationale behind recent research efforts on 3D EES electrodes. Several excellent references are available on EES devices with 2D electrodes (Yu et al. 2013, Conway 1999).

Figure 1 shows the Ragone plot for the EES devices which are the subject matter of this chapter. The Ragone plot is a double common logarithmic plot of the specific power versus the specific energy of different EES devices. The specific power (energy) is defined as the power (energy) per unit mass of the electrode material and has the unit of W/kg (Wh/kg). Although the dielectric capacitor is not an EES device, it is included in the Ragone plot to show that the ECs bridge the specific energy gap between the rechargeable batteries and the dielectric capacitors. There are two important features about the conventional Ragone plot that should be clarified. First, as plotted in this conventional way, it implies that the EC and the rechargeable battery are separate categories of EES devices. ECs are devices with higher specific power (up to 10^6 W/kg) but their specific energy are considerably lower than rechargeable batteries. For batteries, their specific energy is higher (up to ~ 500 Whr/kg)

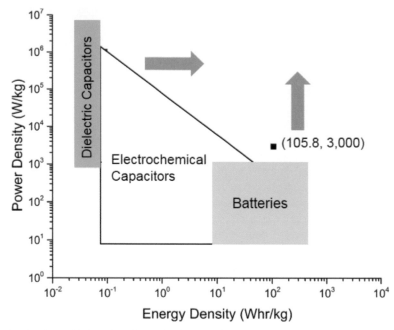

Fig. 1. Ragone plot of rechargeable batteries, electrochemical capacitors and dielectric capacitors. The arrows indicate possible directions of future research. The data point with coordinates indicated is based on (Zhou et al. 2016).

but their specific power is inferior to the ECs. It will be shown that this distinction is not absolute and that in future, this status quo may change. The second important feature is an empty quarter in the upper right corner of the Ragone plot. This means that there is no EES device at present which possesses both high specific energy (battery like) and high specific power (EC like) characteristics. In order to reach the empty quarter of the Ragone plot, one can either start from the EC region and increase the specific energy or alternatively, one can start from the battery region and increase the specific power. For both these approaches, the fabrication of highly porous and open 3D electrodes is essential to the further development of high performance EES devices.

2.1 Rechargeable Batteries

A rechargeable or secondary battery consists of a number of electrochemical cells connected together. Each electrochemical cell has two dissimilar electrodes called cathode and anode that are inserted into a liquid or solid electrolyte. In addition, there is usually a thin membrane separator between the two electrodes to prevent an internal electronic conduction (short circuit) of the battery. Conductors called current collectors are also attached to the cathode and anode to enable the battery to be connected to an electrical load.

Both primary and secondary batteries make use of Faradaic or oxidation-reduction (redox) reactions to store electronic charges. The term redox reactions refer to a class of chemical reactions in which electron transfer takes place between participating molecules. The atom or molecule that gives up one or more electrons in a redox reaction is the reducing agent and is said to be oxidised after the reaction. For each reducing agent in such a reaction, there has to be a complementary oxidising agent that is reduced by electron transfer. In some redox reactions such as combustion of hydrocarbons, the electron transfer takes place during the atomic rearrangement of the reacting molecules and no electrical current is involved. On the other hand, in an electrochemical cell, the reducing and oxidising agents in the form of the two electrodes are physically separated from one another and electron transfer manifests as an electrical current between the two electrodes. During discharge of a battery, the anode loses electrons and is oxidised. The cathode gains electrons and is chemically reduced. The electrons are transferred from the anode to cathode via an external circuit and deliver energy to a load such as a resistor or light emitting diode. The energy of the electrons is related to the free energy of the redox reaction involved in the battery and can be calculated using the Nernst equation (Conway 1999). In the rechargeable batteries which are of interest here, the redox reactions are all reversible. During charging of the battery by an external power supply, the anode gains electrons (reduction) while the cathode loses electrons (oxidation). As a result, the anode of a secondary battery is always the negative terminal while the cathode is always the positive terminal regardless of the mode of use. It is also important to point out that the battery is an asymmetric EES device with different cathode and anode materials. Electric charge is stored within the bulk volume of the cathode and anode.

Current conduction within a battery involves ionic transport and is the reason why an electrolyte is needed between the cathode and anode. An electrolyte can be easily prepared by either dissolving an ionic salt such as KCl in an aqueous solvent or by dissolving an organic salt in an organic solvent such as acetonitrile (CH_3CN). The organic solvents in general exhibit greater electrochemical stability than the aqueous solvents at higher potentials. However at lower potentials, both types of solvents are electrochemically stable.

In recent years, ionic liquids have also been used as electrolytes for electrochemical applications. In some batteries, the ions within the electrolyte directly participate in the electrochemical reactions at the cathode and anode. In such battery chemistries, the diffusion rate of the ions in the electrodes can become the limiting factor in the power performance of the battery. A good example of this is the lead acid battery where the H_2SO_4 of the electrolyte is directly involved in the reactions at the Pb and PbO_2 electrodes. On the other hand, in batteries like the LIB, the electrolyte merely provides a conducting medium for the Li ions diffusing back and forth between the electrodes. Nevertheless, the diffusion rate of the Li ions in the electrodes and the electrolyte is also crucial to the power performance of the battery.

The NaS battery mentioned in the introduction has an especially straightforward electrochemistry and is given here to illustrate the basic battery concepts (Soloveichik 2011). The anode of the NaS battery is molten Na and the reversible reactions at the anode are given by:

$$Na_{(l)} \leftrightarrows Na_{(l)}^+ + e \tag{i}$$

During discharge, the monovalent Na atoms oxidise to Na^+ ions and diffuse through the solid β-alumina electrolyte to reach the cathode. The subscript *l* refers to the liquid state. At the cathode, which is molten sulphur, the following reversible reactions take place:

$$S_{x(l)} + 2e \leftrightarrows S_{x(l)}^{2-} \tag{ii}$$

Consistent with the anode equations, the left to right reaction takes place during discharge. The subscript *x* (~ 3 to 5) in the above equation represents the typical length of S oligomer chains in the cathode. The S oligomer chains are reduced to S_x^{2-} ions and these combine with Na^+ to form the ionic compound Na_2S_x. The overall reaction for the NaS battery is thus:

$$2Na_{(l)} + S_{x(l)} \leftrightarrows Na_2S_{x(l)} \tag{iii}$$

The unique high temperature operation requirement of the NaS battery is because the diffusion rate of Na^+ ions in the solid β-alumina electrolyte is quite low at room temperature. In order for the charging and discharging rates of the NaS battery to be acceptable, temperature activation of the Na^+ ion transport is essential. Since the β-alumina electrolyte is an electronic insulator, it can also fulfil the function of the separator.

Amongst the four constituent parts of the battery, the most critical part is the electrodes. This is because the electrodes are where electrical energy is stored. The structure of the electrochemically active electrode materials is determined by the fabrication method used during manufacturing and will determine the specific energy and specific power of the battery. Since the operating principle of the battery is based on redox reactions at the electrodes, both electronic transport and ionic transport are involved. In order to facilitate electron transport, the electrode material must have sufficiently low resistivity. This is important to reduce the internal resistance of the battery when a load is connected to the battery.

For ionic transport, the electrode must present a large surface area at the electrode/electrolyte interface so that ions in the electrolyte can readily access the electrochemically active material both inside and at the surface of the electrode. This is especially important because charges are stored throughout the volume of the battery electrodes. If the ions cannot readily access the interior of the electrodes, both the energy capacity and the rate of charge and discharge will be adversely affected. This requirement is fulfilled in practice by introducing porosity into the electrode material.

Until very recently, the electrodes of most batteries are fabricated from composite powder slurry pasted onto the current collectors as thick films. The composite thick film typically consists of a mixture of the electrochemically active material, a conductive filler, a polymer binder and possibly other proprietary additives. The electrochemically active material is usually in the form of a metal oxide powder or graphitic carbon. For example in the LIB, the active cathode material can be $LiMO_x$, where M is a transition metal or $LiFePO_4$. The particle size of these metal oxide powders is in the micrometre range and these powders are blended with the other components of the composite. The conductive filler is usually graphitic carbon and its role is to reduce the resistivity of the composite. This is because many metal oxides have fairly high resistivity and alternate conductive pathways within the electrode need to be provided. Since more than one material is present in the electrode, a polymer binder is needed to yield a continuous thick film. As mentioned above, during fabrication of the electrode layer, it is important to choose conditions that will result in an open porous film structure to facilitate ion transport.

In this chapter, the term two dimensional (2D) electrodes refers to this type of conventional battery electrode where there is no significant change in the structure of the electrode material in the direction perpendicular to the plane of the electrode. The 2D electrode as defined here has a finite thickness (and volume) and has no connotation with 2D materials such as graphene and MoS_2. The essence of the 2D electrode is that there is no intentional structuring of the electrode material in the out of plane direction. Thus, if one were to observe the surface and sub-surface region of the 2D electrode by cross sectional microscopy, there will be little difference between the two.

In the following section, we will proceed to show that by structuring the 2D electrode in the out of plane direction, a 3D electrode with superior battery performance can be realised.

2.2 Electrical Double Layer Capacitors

The electrical double layer capacitor (EDLC) is a type of EC used for pulsed power applications and for heavy electric vehicles (Yu et al. 2013). Like the battery, the EDLC consists of two electrodes with associated current collectors, an electrolyte and a separator. However, unlike the battery, the two EDLC electrodes are made of identical materials and the EDLC is therefore a symmetrical EES device. The principle of operation of the EDLC is the electrical double layer (EDL) or Helmholtz layer that is formed whenever an electrode is immersed into a liquid electrolyte (Hamann et al. 2007).

Due to the different affinity of the electrode for the cations and anions in the electrolyte, an isolated electrode will be surrounded by one type of ion. This sheath of ions attracts another layer of ions of the opposite charge and forms an EDL. The separation between the two layers of ions is of order of 1 nm. When a voltage source is connected across two identical isolated electrodes and energy is used to transfer electrons from one electrode to the other, two EDLs which are electrically connected in series are formed. The EDL in this case consists of the net charge in the electrode and the ionic charge of the opposite sign from the electrolyte. The total capacitance of the EDLC is the series capacitance of the two EDLs. Since the separation between the two layers of charge in the EDL is again of order 1 nm, the capacitance of the EDLC is much greater than the dielectric capacitor where an applied electric field is used to induce polarisation charge in a dielectric layer sandwiched between two parallel metal plates. However, both the EDLC and the dielectric capacitor store energy in the form of an electric field between two sheets of charge.

The capacitance of the EDL in an EDLC is given by the same parallel plate capacitor formula as for dielectric capacitors:

$$C = \varepsilon_0 \, \epsilon_r \, \frac{A}{d} \qquad \text{(iv)}$$

In this formula, C is the capacitance; ε_0 is the permittivity of free space; ε_r is the dielectric constant of the electrolyte; A is the area and d is the thickness of the EDL. Thus, having an electrode with a very large EDL area is critical to achieving a high C for the EDLC. When C is divided by the mass of the electrode, the resulting quantity is called the specific capacitance and has the unit of F/g. The energy stored in an EDLC is given by:

$$E = \frac{1}{2} C V^2 \qquad \text{(v)}$$

where E is the stored energy and V is the voltage across the two electrodes. This equation is also applicable to the dielectric capacitor and shows that the maximum amount of energy stored depends on the capacitance of the EDLC and the maximum applied voltage. The latter depends on the electrochemical stability of the electrolyte and is about 1 V for aqueous electrolytes and ~ 2.3V for organic electrolytes. The quadratic dependence of the stored energy on applied voltage shows that increasing the maximum voltage is an effective way to increase the amount of stored energy in an EDLC.

At present, the most common electrode material for EDLCs is still the 2D activated carbon (AC) electrode. This is a synthetic material processed from biomaterials such as coconut shells and was first developed by the Sohio company in the 1950s (Conway 1999). The AC synthesis process typically involves heating biomass in an inert atmosphere until all the hydrogen and oxygen atoms are volatised and a porous low density graphitic carbon skeleton is left behind. The AC electrode has a planar configuration and is characterised by a high surface area per unit mass (specific surface area). Since the AC electrode is in direct contact with the electrolyte, the latter must not react chemically with the AC material during charge and discharge of the EDLC. A typical organic electrolyte will consist of an organic salt dissolved in acetonitrile.

The charge-discharge characteristic of an EDLC can be studied by using an electroanalytical technique called cyclic voltammetry (CV). In CV measurements, a potentiostat applies a time varying voltage sweep to an electrochemical device and the current is measured. For a constant voltage ramp rate and an EDLC, the CV curve is box like (i.e., without peaks). This is because capacitance is given by: C = Q/V, where Q is the charge and V the voltage. The current is constant during the forward sweep and backward sweep when the capacitance of the EDLC is constant.

2.3 Pseudocapacitors

The second class of EC was discovered by Conway of the University of Ottawa in 1975 (Conway 1999). In this more recent type of EC, the capacitance is enhanced by incorporating an electrode with a surface layer that can undergo the Faradaic (redox) reactions. Since the redox reactions are processes occurring within a battery, this type of capacitor has battery like characteristics and is therefore referred to as having pseudocapacitance. The surface redox reactions increase the specific capacitance of the pseudocapacitor and the amount of energy stored relative to the EDLC. A diverse range of energy materials have been investigated for application as pseudocapacitors. These include oxides of various transition metals (Ru, V, Ti, Mn, Nb), organic conjugated polymers and nanostructured forms of these materials. Both the EDLC and the pseudocapacitor have been commercialised.

An early demonstration of pseudocapacitance involved the transition metal oxide ruthenium oxide (RuO_2) (Zheng et al. 1995a,b). Ruthenium is in the same group as iron (Fe) and osmium (Os) in the periodic table and is an expensive metal with dye applications in the dye sensitised solar cell as well. Like the EDLC, the RuO_2 pseudocapacitor is a symmetrical EES device with two RuO_2 electrodes. The RuO_2 can be synthesised by solution reactions and is applied to the current collectors as a planar electrode. A separator soaked with an aqueous electrolyte is placed between the RuO_2 electrodes. As discussed in (Long et al. 2011), the following reversible electrochemical reaction occurs during the charge-discharge of the RuO_2 pseudocapacitor:

$$RuO_2 + xH^+ \leftrightarrows Ru_{1-x}Ru_xO_2H_x \qquad \text{(vi)}$$

During charging, when electrons are stored in the surface region of RuO_2, the oxidation state of some Ru ions are reduced from +4 to +3. In order to maintain charge neutrality, protons from the aqueous electrolyte has to be adsorbed onto the RuO_2 surface and forms a mixed valence ruthenium oxy-hydride. The electrons that are involved in this reaction comes from the complementary oxidation reaction of the ruthenium oxy-hydride to RuO_2 at the other electrode. The amount of charge that can be stored in a RuO_2 pseudocapacitor depends on the structure of the electrodes and the specific capacitance can range from 10 to more than 100 F/g. The highest specific capacitance for a RuO_2 pseudocapacitor was reported by Zheng et al. 1995a,b. Using solution synthesised hydrated RuO_2, a specific capacitance of 720 F/g was demonstrated (Zheng et al. 1995a,b).

Although the RuO_2 pseudocapacitor shows high specific capacitance, its practical application is constrained by the high cost and supply of Ru metal. This led to the development of the manganese oxide (MnO_2) pseudocapacitor (Lee and Goodenough 1999). In the periodic table, Mn is a transition metal in the group adjacent to Fe, Ru and Os. Like Ru, Mn has multiple oxidation states that can be utilised in pseudocapacitor devices to store charge but in comparison, Mn is more abundant and cheaper. The structure of the MnO_2 pseudocapacitor is basically the same as the RuO_2 pseudocapacitor. From detailed compositional and electrochemical characterisation experiments, the following reversible reaction has been proposed as the pseudocapacitance mechanism for MnO_2 devices (Kuo and Wu 2006):

$$MnO_2 + X^+ + e \leftrightarrows MnOOX \qquad \text{(vii)}$$

In this equation, X^+ represents one of the following ions: H^+, Na^+, K^+ or Li^+. During charging, one of the MnO_2 electrodes is reduced and the oxidation state of manganese is lowered from +4 to +3. In order to compensate charge, cations in the form of protons, Na^+ or K^+ have to be inserted into the electrode and this results in the mixed oxides of MnOOH, MnOONa and MnOOK respectively. The reverse of this reaction occurs at the other electrode and the mixed oxide is oxidised back to MnO_2.

Note that the MnO_2 redox reaction involves the electrode and the electrolyte. By preparing MnO_2 electrodes in the form of thin films and composite powders, and determining the specific capacitance resulting from these electrodes, it has been found that the thin film MnO_2 electrode has a higher specific capacitance (Long et al. 2011). This suggests that the redox reaction of the MnO_2 electrode occurs primarily near the surface of the electrode.

An excellent and thorough review on the MnO_2 pseudocapacitor has been published recently (Wei et al. 2011).

3. EES Devices with 3D Electrodes

EES devices with 2D electrodes have a number of major limitations. One especially serious drawback is the power performance of rechargeable batteries. Since ions have to diffuse into and out of the electrode in a battery and ionic diffusivity in many solids are generally low, the specific power of rechargeable batteries with 2D electrodes are limited to those shown in Fig. 1. In addition, for the LIB, which depends on the intercalation of Li^+ ions, the volume change accompanied by the intercalation can cause high stress within the planar electrode (Harris et al. 2010). This imposes limits on both the specific energy and the cycle lifetime of the battery. The latter means that after repeated charge and discharge, defects will develop in the electrode and electrode failure eventually occurs. A more flexible and compliant electrode structure will be beneficial to the cycle lifetime of LIBs (Polat and Keles 2015). However, this is difficult to be realised using a 2D electrode structure. For ECs and the pseudocapacitor in particular, a planar electrode severely limits the specific capacitance and therefore the energy stored because the charge storage reactions occur only at or near the electrode surface. A non-planar electrode can greatly enhance the electrochemically active surface area used for storage of charge.

Initial efforts to improve EES device performance involved using nanostructured electrode materials instead of conventional powder composite slurry (Wang et al. 2011). The rationale is that by synthesising the EES electrode material as nanoparticles in the 1–100 nm size range, one can greatly increase the surface area to volume ratio of the particles. This can be easily demonstrated by considering the example of a spherical particle. Such a particle with a 1 μm radius has a surface area to volume ratio of 3×10^6. This ratio increases to 3×10^8 for a 10 nm particle radius because the surface area to volume ratio varies inversely with the particle radius for a spherical particle. For nanoparticles with other morphologies such as nanorods, nanowires, nanoplates, the surface area to volume ratio will similarly increase with decreasing particle size. Since both the battery and the EC depend on a large surface area for their energy storage function, the use of nanomaterials in EES devices has been studied extensively in recent years. A comprehensive review of the application of nanomaterials in EES devices can be found in (Arico et al. 2005). However, in most of these studies, the nanostructured materials are fabricated as 2D electrodes and thus the benefit of a large surface area to volume ratio is not fully utilised.

3.1 Definition of 3D Electrodes

The concept of 3D EES devices was first put forward by Long and Rolison of the United States Naval Research Laboratory and collaborators (Long et al. 2004). In this initial period, the research was primarily motivated by the need to develop small foot print power sources for microelectromechanical (MEMS) systems. Although MEMS fabrication techniques have by then advanced to the point where sensors and actuator devices can be integrated with silicon transistors in compact packages, it is difficult to integrate a power source with the MEMS system so that it can be deployed to operate autonomously. This power source integration is especially critical for wireless sensor network applications. The difficulty arises because of the limited energy capacity of batteries with the conventional device structure and the slow pace of innovation in battery research. One solution to this power supply dilemma for microsystems was to adopt a non-planar geometry for the battery electrode. By leveraging on the out of plane dimension, a battery with a 3D structure can store more energy per unit foot print area than a conventional battery.

In the 2004 review article by Long, the 3D battery electrode was described as an electrode with a non-planar geometry characterised by periodic high aspect ratio structures or an aperiodic porous architecture. Despite the morphological complexity of the electrode, the ion and charge transport at the microscopic level remain essentially one dimensional (1D) (Long et al. 2011). This important characteristic enables the 3D battery to potentially outperform the 2D battery in both specific energy and specific power. For low power MEMS systems, the enhanced energy density results in a small battery foot print area and integration of the power source with the MEMS system may become feasible.

In this chapter, we adopted a modified definition of the 3D EES electrode, which takes into account the most recent developments in this field. A 3D electrode in an EES device is an electrode consisting of an open, interconnected scaffold structure made from an electrochemically active nanoscale material (Fig. 2). This scaffold structure is intermingled with a complementary interconnected network of nanoscale voids (porosity) which can be filled conformally with electrolyte or another electrode

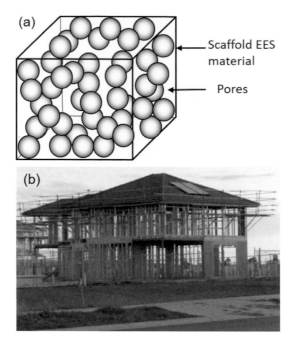

Fig. 2. (a) Schematic diagram of a 3D EES electrode comprising a scaffold electrochemical material and pores; (b) architectural analogy of 3D electrode.

material. In other words, a 3D EES electrode is a nanocomposite of electrochemical materials and voids. Under this general definition, there is no restriction on the periodicity of the scaffold structure. The 3D electrode geometry can be either periodic or aperiodic. As with the earlier definition, both ionic transport and charge transport are 1D at the microscopic length scale. This is why it is important to ensure that any deposition into the scaffold is conformal and does not interrupt the initial continuity of the pore network.

In the literature, the 3D EES electrode has been referred to as 'nanoarchitecture'. This terminology has its origins in an architectural analogy discussed at length in the article (Long and Rolison 2007). As shown in Fig. 2, the scaffold structure of the 3D electrode resembles the timber frame of the building under construction. The conduits and plumbing of this building correspond to the voids in the 3D EES electrode because fluids can flow freely through the interconnected voids. The roof and building envelope material correspond to the electrochemical material deposited subsequently into the 3D electrode to complete the EES device.

In order to maximise both the specific energy and specific power of a 3D EES device, it is essential for the voids to be in the nanometre length scale so that there is a large surface area to volume ratio. In addition, the voids should be interconnected to form a continuous void network because any voids that are closed are unavailable for ionic transport. Nanoscale voids or porosity is described by three standardised terminology: micropores, mesopores and macropores. Micropores refers to pores with a diameter below 2 nm. Mesopores have a diameter in the range of 2 nm–50 nm while macropores refer to pores with diameter above 50 nm.

It should be noted that while the use of electrodes with engineered porosity is relatively recent in the field of EES, the use of such materials is extremely widespread in other fields of technology. Examples include: catalysis, water purification and semiconductor manufacturing. For the latter, porous low permittivity dielectrics are currently being used as the intermetal dielectric in the interconnects of a complementary metal oxide semiconductor (CMOS) integrated circuit (Yu et al. 2002). This is because porosity reduces the dielectric constant of the insulator and reduces the parasitic resistance capacitance delay of the digital signals in the interconnects.

4. Fabrication Techniques of 3D EES Electrodes

In the remainder of this chapter, we focus on the techniques for the fabrication of 3D EES electrodes. Based on the definition of 3D electrodes stated above, the fabrication process can be divided into formation of the scaffold structure and the conformal deposition or infiltration of the other electrochemical phases. The earlier technique of using porous alumina membranes as a template for fabricating periodic porous structures will not be covered because it has previously been reviewed (Long et al. 2004, Arthur et al. 2011). As the degree of porosity in an electrode is increased, the percentage of atoms that are located near the surface will increase for both the battery and EC. As a result, it is anticipated that there will be a blurring of the distinction between these two EES devices (Simon et al. 2014). Since 3D architectures can result in improvement in both specific energy and specific power, we will not distinguish between batteries and ECs in the examples given for the surveyed techniques.

4.1 Fabrication of Porous Scaffold Structure

The porous scaffold structure of a 3D EES electrode can be fabricated by several bottom-up techniques in which the final structure is assembled from small molecular units in solution. These include: sol-gel solution synthesis, mesoporous template assembly, three-dimensional ordered macroporous (3-DOM) solid and electrodeposition. After these more established methods are discussed, we will describe several interesting emerging techniques.

4.2 Sol-gel Process

A versatile and facile technique for synthesising the open, porous scaffold network of 3D EES electrodes is the sol-gel process (Chow and Gonsalves 1998). The sol-gel process is a solution synthesis method that can be carried out at room temperature and pressure using simple laboratory glassware and mild reaction conditions if supercritical drying is not used (see below). Both the thin films and bulk samples can be prepared by the sol-gel method. The term sol-gel implies that there are two stages in the synthesis (Fig. 3). The first stage is the preparation of the sol which refers to a colloidal mixture of an aqueous solvent and the particles that will react in the solvent to form the gel

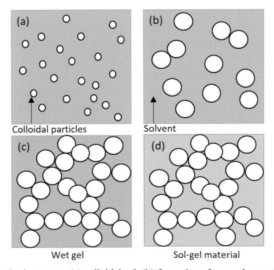

Fig. 3. Stages of the sol-gel synthesis process: (a) colloidal sol, (b) formation of network material by solution phase reactions during aging, (c) crosslinking and gelation and (d) solvent removal and drying of wet gel.

(Chow and Gonsalves 1998). The reaction that takes place during sol formation is called hydrolysis. During hydrolysis, water molecules are used to react with precursor molecules of the sol to generate the sol particles.

The sol-gel synthesis of battery relevant transition metal oxide materials has been reviewed in (Livage et al. 1998). Here, we use a more straightforward example which is the sol-gel synthesis of silicon oxide (SiO_2), an intermetal dielectric used in integrated circuits. For SiO_2, the precursor molecule is tetraethyl orthosilicate (TEOS). The TEOS precursor is added to water under acidic conditions and with the aid of stirring, the precursor is hydrolysed to the sol and ethanol. The hydrolysis basically involves cleavage of the ethoxy group in the TEOS molecule ($Si(OC_2H_5)_4$) and the formation of a silanol group in the molecule (Yu et al. 2002):

$$\equiv Si(OC_2H_5) + H_2O \rightarrow \, \equiv SiOH + C_2H_5OH \tag{viii}$$

When sufficient sol molecules have been generated from the precursor, the second stage of the sol-gel reaction will occur. In this stage, the sol molecules will react with one another by condensation reactions to form an open low density skeletal network of the sol molecules. The condensation reaction can be considered as the reverse of hydrolysis in which two identical molecules react to form a larger molecule and a small molecule such as H_2O. For SiO_2, the condensation reaction involving water can be written as (Yu et al. 2002):

$$\equiv Si - OH + HO - Si \equiv \rightarrow \equiv Si - O - Si \equiv + H_2O \tag{ix}$$

The condensation reaction involves the reaction of silanol groups to form siloxane and water. Condensation reactions leading to ethanol generation can also occur (Yu et al. 2002). With an adequate supply of sol molecules, the condensation reactions can be repeated recursively and this will result in an increasing number of condensation products in the aqueous solvent. The siloxane consisting of alternating covalently bonded Si and O atoms is the skeletal structure of the sol-gel derived silicon oxide. The period in which the condensation takes place is called the aging stage. During aging, multiple condensation products will be growing within the solvent. Eventually these condensation products will cross link sufficiently to form a wet gel. The gel is wet because during the condensation and crosslinking reactions, aqueous solvent will be trapped within the space between the skeletal structure. The onset of gelation is marked by a sharp change in the rheological properties of the sol. Prior to the gelation point, the sol is a colloid and for an aqueous solvent, the viscosity is relatively low. After the gelation point, however, the wet gel becomes more viscous and resembles jelly.

The final stage of the sol-gel synthesis process is the drying out of the wet gel. This step is usually the most challenging because the solvent trapped in the spaces between the skeleton of the condensation network is the template for the pores in the desired porous material. The solvent needs to be removed without affecting the pore structure in the dried gel. If the solvent is removed by evaporation in the ambient, the resulting sol-gel material is referred to as a xerogel (Pierre 2011a). The term xerogel was coined by Freundlich to describe those gels that shrink upon drying. Xerogels are often formed when a high surface tension polar solvent such as water and an alcohol is used for the sol-gel synthesis. Surface tension arises from the intermolecular forces exerted on the liquid molecules near the surface and is responsible for the shape of water droplets (Ho 2012). For trapped liquids inside pores, surface tension results in capillary forces at the liquid-gas interface. In inkjet printing, it is an important property of the ink that must be carefully tailored for successful printing. For sol-gel synthesis, capillary forces due to surface tension at the liquid-vapour interface can be considerable during drying because as mentioned earlier, the surface area to volume ratio increases with decreasing pore size. The capillary forces can cause the collapse of the pores and the ensuing shrinkage results in xerogels having a pore size distribution comprising mainly of micropores and very few mesopores or macropores. A transmission electron microscope (TEM) image of a silica xerogel can be found in Fig. 1 of the reference (Yu et al. 2002). For most applications including 3D

EES devices, xerogels are not useful because of their higher density and lack of mesopores (Long et al. 2004).

In order to produce a low density porous skeletal structure by the sol-gel method, a low surface tension solvent will be needed. When an organic solvent such as the alkane, hexane (C_6H_{18}) is used, the resulting sol-gel material is called an ambient pressure dried gel or ambigel (Long et al. 2004). Although there is also shrinkage during the evaporative drying of the wet gel, the extent of the densification is lesser due to the weaker capillary forces. As a result, there are more mesopores in an ambigel than in a xerogel.

An example of an ambigel is the vanadium pentoxide (V_2O_5) cathode reported in (Coustier et al. 1998). These authors refer to their cathode material as aerogel-like because the synthesis process does not involve supercritical drying. The sol-gel V_2O_5 was prepared by the protonation of an aqueous sodium metavanadate solution. After the V_2O_5 gel has been formed, the water in the wet gel was replaced by acetone which has a lower surface tension. This solvent exchange was repeated a second time by the replacement of acetone by hexane which has an even lower surface tension. The reason for having two consecutive solvent exchange is because water and hexane are immiscible liquids. At the end of the second ion exchange, the hexane in the V_2O_5 gel was evaporated first in dry air and then in vacuum. The resulting V_2O_5 ambigel did not show significant pore collapse and was used to form a composite LIB cathode. The general morphology of a V_2O_5 ambigel is shown in the TEM image Fig. 7b of the reference (Rolison and Dunn 2001).

A more complicated drying approach involves the use of a supercritical fluid (SCF) as the solvent for the wet gel. A SCF is formed when the pressure and temperature of a substance is increased beyond the critical point of the phase diagram. The critical point can be defined as the endpoint of a phase transition boundary in an equilibrium phase diagram. Below this point, the liquid phase and the gas phase have markedly different densities. However, beyond the critical point, there is no abrupt change in density between the liquid and gas phase and the two becomes one phase called the SCF. As a result, drying in a SCF does not involve capillary forces that result in pore collapse. The latent pores present in the wet gel are preserved and there is no shrinkage in the sol-gel material. The SCF phase can be reached by applying pressure to the wet gel in an autoclave and raising the temperature above the critical point of the fluid (Vega 2015). The most commonly used supercritical fluid is carbon dioxide (CO_2). When such a solvent is used for sol-gel synthesis, the material obtained after drying is called an aerogel. The first report of aerogel fabrication was made by Kistler using silica as the skeletal material (Kistler 1931). Today, the aerogels constitute an important class of ultra-low density material with extremely low thermal conductivity (Pierre 2011b). There are both mesopores and macropores in the aerogel structure (Long et al. 2004). The scanning electron microscope (SEM) image in Fig. 1 of the reference (Rolison and Dunn 2001) illustrates the surface morphology of a silica aerogel.

Aerogels made from conductive skeletal materials have found applications as 3D electrodes for EES devices because of their extremely large surface area and a high degree of connectivity between the pores. One important example are carbon aerogels which can be derived from resorcinol-formaldehyde (RF) aquagels. Aquagels are also sometimes called hydrogels. The sol-gel synthesis of RF aquagels was studied in detail by Tamon et al. based on earlier research by Pekala et al. (Tamon et al. 1997). We will discuss this synthesis in more detail to better illustrate the aerogel drying process.

The RF aquagel is an organic polymer skeletal material synthesised by polycondensation from two precursors: resorcinol ($C_6H_4(OH)_2$) and formaldehyde (HCHO) in aqueous conditions with sodium carbonate as a catalyst. Since the RF aquagel synthesis is carried out using water as a solvent, the water must be replaced by CO_2 prior to supercritical drying. Since water and CO_2 are not miscible, the water has to be first removed by immersing the wet aquagel in acetone. The aquagel and acetone are loaded together into an autoclave which is essentially a stainless steel pressure vessel with inlet and outlet piping and a pressure controller. Pressurised liquid CO_2 is preheated before entering the autoclave. Initially, the liquid CO_2 is used to purge the acetone from the RF aquagel. When most of the acetone has been removed and the aquagel is filled with liquid CO_2, the temperature and pressure within the autoclave is raised to above 304 K and 7.4 MPa which are the critical temperature and pressure of CO_2 respectively (Vega 2015). This converts the CO_2 into a SCF and purging continues until all acetone has been removed. Finally, the autoclave is depressurised to atmospheric pressure and the supercritical CO_2 is vented through the outlet valve to leave behind the RF aerogel. The final step is the pyrolysis of the RF aerogel in an inert atmosphere to remove the hydrogen and oxygen from the skeletal material as volatile species. This can be carried out in a conventional tube furnace and the material that remains is the carbon aerogel. In a subsequent section, we will describe how carbon aerogels are used in the fabrication of 3D ECs.

In addition to carbon aerogels, V_2O_5 aerogels have been fabricated by a sticky-carbon electrode technique (Dong et al. 2000). A specific capacitance of 1300F/g was obtained which was higher than the corresponding ambigels. Electrically conducting transition metal oxide aerogels has been thoroughly reviewed (Rolison and Dunn 2001). As discussed in this article, V_2O_5 aerogels prepared from vanadium containing alkoxy precursors and supercritical CO_2 can have specific surface area of 280 m²/g, average pore diameter of 8 nm and pore volume of 0.5 cm³/g. The V_2O_5 network of the aerogel is nanocrystalline and a TEM image of this is shown in Fig. 7a of the reference (Rolison and Dunn 2001). MoO_3 aerogels prepared by a similar process can have specific surface area of 180 m²/g, specific pore diameter of 30 nm and pore volume of 0.6 cm³/g. Both V_2O_5 and MoO_3 aerogels exhibit reversible Li ion intercalation properties which make them suitable for LIB applications.

Although the aerogel process can result in high porosity skeletal materials, it requires the use of high cost specialised supercritical drying equipment and liquid CO_2. An alternative to supercritical drying which nevertheless preserves the majority of the mesopores and macropores is freeze drying. Sol-gel materials prepared by this drying method are called cryogels. Yamamoto and co-workers used freeze dyring to prepare mesoporous carbon cryogels (Yamamoto et al. 2001). First, a RF hydrogel was synthesised from resorcinol and formaldehyde as discussed earlier for ambigels. The water trapped within the hydrogel was then replaced by t-butanol by repeated immersion. After the t-butanol was frozen by lowering the temperature of the gel to 243 K, the t-butanol was removed by sublimation, which is a direct solid to vapour phase transition. Since the drying of the t-butanol does not involve a liquid-vapour interface, there is a little shrinkage of the cryogel due to capillary stress. However, when freeze drying is used, it is important to ensure that all the frozen solvent is completely sublimed. Otherwise, when the temperature is eventually increased back to room temperature, the remaining liquid organic solvent will cause some pores to collapse.

The porosity properties of sol-gel materials discussed in this section are summarised in Table 2.

Table 2. Summary of porosity properties of sol-gel derived materials.

Sol-gel Material	Type	Specific Surface Area (m²/g)	Specific Pore Volume (cm³/g)
SiO_2 (Yu et al. 2003)	Xerogel	156–430	0.0238–0.112
V_2O_5 (Coustier et al. 1998)	Ambigel	200	–
RF (Tamon et al. 1997)	Aerogel	355–834	0.397–3.07
Carbon (Tamon et al. 1997)	Aerogel	104–192	0.744–2.41
V_2O_5 (Rolison and Dunn 2001)	Aerogel	280	0.50
MoO_3 (Rolison and Dunn 2001)	Aerogel	180	0.6
Carbon (Yamamoto et al. 2001)	Cryogel	498–569	0.99–1.15

4.3 Mesoporous 3D Electrodes by Self-Assembled Soft Template

Despite the versatility of the sol-gel method, the supercritical drying of aerogels or the solvent exchange in ambigel preparation are quite complicated. In addition, the controlling or tailoring of the pore size distribution of sol-gel material is not straightforward. These issues can be overcome by using the mesoporous template synthesis approach (Brezesinski et al. 2010). In this bottom up synthesis method, the template for the pores of the mesoporous film is assembled from amphiphilic molecular building blocks such as ionic surfactants, diblock and triblock copolymers. These soft molecular units readily form micelle structures when dissolved in a solvent (Chow and Gonsalves 1998). When the solvent of the solution is evaporated, these molecular units will spontaneously organise themselves into ordered hexagonal or cubic structures in a process known as evaporation induced self-assembly (EISA) (Rauda et al. 2013). The supramolecular assemblies produced by the EISA process can then serve as a template to guide the growth of the inorganic phase such as a transition metal oxide or carbon (Fig. 4a). If the EISA template is synthesised first and is followed by the synthesis of the inorganic phase, the synthesis is referred to as transcriptive synthesis (Long et al. 2004).

An example of this is the templated synthesis of mesoporous carbon (Ryoo et al. 1990). Here, the pore template, which is assembled first is filled with organic precursor molecules to form an organic inorganic (hybrid) composite material. After reaction, the organic polymeric material is pyrolysed in an inert atmosphere to leave behind a carbon matrix. Finally, the pore template is also removed to form a mesoporous carbon material with well-defined pore size. The TEM and SEM images of an ordered mesoporous carbon molecular sieve synthesised by this process can be found in Fig. 1 of the reference (Ryoo et al. 1990).

The assembly of the pore template and the sol-gel synthesis of the inorganic phase can also be carried out concurrently and this complementary approach is known as synergistic synthesis (Long et al. 2004). In a typical synergistic synthesis, the amphiphilic building blocks and the precursors of the mesoporous material are dissolved in a common polar solvent. When this sol is spin coated or casted onto a substrate, both the EISA process and the hydrolysis and condensation reactions of the precursors are initiated because of solvent evaporation. The assembled template restricts the growth of the sol-gel material and results again in a hybrid composite. Finally, the organic template is pyrolysed to form a mesoporous film.

The synergistic template synthesis method is illustrated in Fig. 4b. It has been implemented using porogen templates (Yu et al. 2004). The porogen template is an organic dendrimer molecule which has a monodisperse molecular diameter. Its role in the synthesis is simply to define the final pore size. The adjective sacrificial is often used because the porogen is not present in the final mesoporous material. Like its linear copolymer counterpart, the dendrimer is a high molecular weight organic molecule.

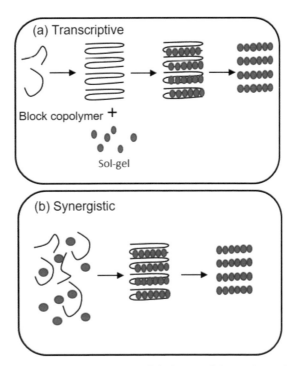

Fig. 4. (a) Transcriptive template synthesis of mesoporous materials; (b) synergistic template synthesis of mesoporous materials.

However, its molecular structure consists of repeated branching from the centre. An example of a porogen is polyamidoamine (PAMAM).

During a synergistic synthesis, the sacrificial porogen is added to the solution of the precursor of the sol-gel material. The weight ratio of the sol-gel precursor and the porogen determines the degree of templated porosity in the final material. Since the sol-gel precursor and the porogen have different hydrophilicity, a surfactant is often needed as well. After gelation and drying, the porogen which is now embedded within the sol-gel skeletal material will be pyrolysed by heating in an oxidising ambient. The heating temperature must be above the thermal decomposition temperature of the porogen and the volatile decomposition products of the porogen will escape via the pores of the material into the ambient. The decomposition temperature of the porogen can be found by the thermogravimetric technique. This temperature should not be so high that the pyrolysis step adversely affects the properties of the remaining mesoporous skeletal material. The TEM image of a low dielectric constant organosilicate prepared from hydrogen silisesquioxane and PAMAM template is shown in Fig. 9 of the reference (Yu et al. 2004).

4.4 3-DOM Solid

This is a template method for fabricating periodic macroporous 3-D electrodes (Arthur et al. 2011). Monodisperse silica or polymeric spheres with submicron size are first assembled into an ordered colloidal crystal (Fig. 5). The spheres are initially dispersed in a solvent as a colloid and the colloidal spheres are subsequently assembled into a close packed 3D crystalline template by casting or other

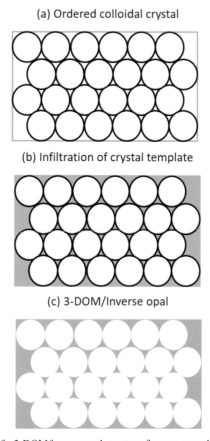

(a) Ordered colloidal crystal

(b) Infiltration of crystal template

(c) 3-DOM/Inverse opal

Fig. 5. Fabrication process of a 3-DOM/inverse opal structure from an assembled colloidal crystal template.

methods (O'Hanlon et al. 2017). From the atomic packing fraction of the body centred cubic (bcc) and face centred cubic (fcc) crystalline structures, the voids which are interconnected account for 32 per cent and 26.34 per cent of the total unit cell volume respectively. This template is then infiltrated and filled up with the electrode material desired by techniques such as sol-gel reaction, electrodeposition, atomic layer deposition or chemical vapour deposition. The final step in the material preparation is the removal of the templating spheres by etching or pyrolysis in an inert ambient. The macroporous structure that results from this process is sometimes called an inverse opal structure. The inverse opal consists of a scaffold of an electrochemical material and an interconnected ordered array of submicron pores. The SEM images of an inverse opal structure fabricated on fluorine doped tin oxide (FTO) and stainless steel can be found in Fig. 1b of the reference (O'Hanlon et al. 2017). One important advantage of the 3-DOM method is that it permits a hierarchy of pore size to be readily built into the electrode structure (Long et al. 2004). This is because the scaffold material can be synthesised to contain mesopores. If this is the case, the 3-DOM solid will consist of both mesopores and ordered macropores.

The 3-DOM fabrication process is illustrated by a macroporous carbon anode used for LIB applications (Ergang et al. 2007). The anode of this 3D electrochemical cell was fabricated by using monodisperse poly(methyl methacrylate) (PMMA) building blocks in a sol comprising RF, deionised water and sodium carbonate. The sol-gel reaction results in the RF polymer gel as discussed earlier. However, in this case, the RF filled the voids of the PMMA colloidal crystal. The carbon 3-DOM anode was formed subsequently by pyrolysis of the RF gel and PMMA at 900°C in an inert atmosphere. A 3D interpenetrating LIB was fabricated by infiltrating the 3-DOM anode with polyphenylene oxide (PPO) and sol-gel V_2O_5. The SEM image of V_2O_5 infiltration of a carbon 3-DOM anode is shown in Fig. 2 of the reference (Ergang et al. 2007). A reversible charge/discharge capacity of 350 μAh/g was reported for this LIB (Ergang et al. 2007).

4.5 Electrodeposition

Electrochemical deposition or electrodeposition was used to fabricate the 3D Cu_2Sb anode of a 3D micro-battery (Perre et al. 2010). In 2014, another group at Kyoto University published a simple one-step electrodeposition method that leads to the formation of 3D copper (3D Cu) for energy device applications (Arai and Kitamura 2014). The deposition can be carried out in a commercial electrolytic cell with electrodes made of Cu and Cu/P plates under galvanostatic (constant current) conditions. The electrolyte inside electrolytic cell consisted of H_2SO_4, $CuSO_4$ and polyacrylic acid. The polyacrylic acid in the electrolyte acts as a forming agent during the electrodeposition and results in a 3D Cu morphology consisting of stacks of Cu sheets. The electrodeposited 3D Cu is however mechanically rather fragile and can deform for thicker deposits. In order to improve properties, the same group of investigators added carbon nanotubes (CNT) to the electrolyte and electrodeposited 3D Cu/CNT composites. The SEM images of 3D Cu and 3D Cu/CNT composites showing the difference in morphology can be found in Fig. 3 of the reference (Arai et al. 2016). Figure 6 of this reference shows the cross sectional SEM image of 3D Cu/CNT composite. These composites were found to have higher porosity for the same normalised mass than 3D Cu and better mechanical properties (Arai et al. 2016).

The electrodeposition technique has also been used for forming the 3-DOM/inverse opal structure. In the study (Armstrong et al. 2015), a colloidal crystal of submicron polystyrene (PS) spheres was first deposited by electrophoresis onto various conducting substrates. After annealing to improve adhesion, the colloidal PS crystal template was used as the working electrode of a 3-electrode cell and V_2O_5 was electrodeposited from an aqueous electrolyte into the PS template. Finally, the PS was decomposed to gaseous products by heating to form the V_2O_5 3-DOM/inverse opal structure. The SEM images of a PS macroporous structure partially filled with V_2O_5 and the V_2O_5 3-DOM structure after PS removal are shown in Fig. 4 of the reference (Armstrong et al. 2015).

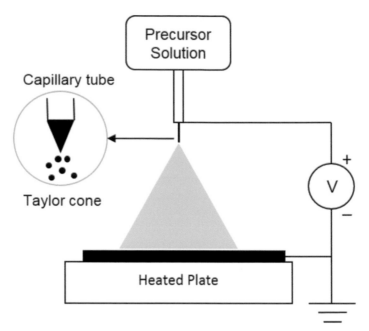

Fig. 6. Schematic diagram of the ESD process for fabricating 3D EES electrodes.

4.6 Electrostatic Spray Deposition

The electrostatic spray deposition (ESD) method was first reported by a research group at Delft University (Chen et al. 1996). In this method, the metallic precursors of the EES electrode material to be deposited is dissolved in a polar solvent such as ethanol. The solution is flowed through a metallic capillary tube which is maintained at a high positive potential relative to a heated and grounded metallic substrate (Fig. 6). The electric field due to the positive bias on the capillary tube acts on the cations of the precursors and forces these ions to the surface of the solution at the nozzle. The electrostatic force which is directed outwards is opposite to the surface tension force of the liquid. At sufficiently high fields, this leads to the Taylor-Rayleigh instability. The onset of the Taylor-Rayleigh instability is marked by the appearance of a Taylor cone. Since the surface electric field of the Taylor cone is inversely proportional to the square root of the cone radius (Chen et al. 1996), the electric field increases rapidly towards the tip apex and droplets containing the positive ions are ejected from the cone. Note that this process is also the mechanism of droplet generation in the continuous inkjet printer.

The aerosol droplets from the capillary tube traverses the space between the nozzle and the substrate mainly under the action of the electrostatic force. When they arrive at the substrate, these charged droplets will transfer their charge to the substrate and any solvent that has not evaporated during the transit will spread on the substrate. The extent of this wetting, which has a bearing on the morphology of the ESD film depends on the spreading coefficient and the viscosity of the solvent. The final step of the ESD process is the drying of the residual solvent and the chemical reaction of the precursors to form the electrode material. In the original study (Chen et al. 1996), $LiCoO_2$ was deposited by ESD using the organometallic precursor $Li(CH_3COO).2H_2O$ and $Co(NO_3)_2.6H_2O$ dissolved in ethanol. By varying the solution composition and the electrostatic conditions, various surface morphologies ranging from dense films to fractal like porous films were observed. More recently, the ESD method has been applied successfully to fabricate porous SnO_2 films on Ni foams (Yu et al. 2009). The SEM image of porous SnO_2 deposited by ESD can be found in Fig. 2 of this reference. When used as the anode of a LIB, a reversible charge capacity of 689 mAh/g was measured for a rate of 0.5C and the capacity retention was 94.8 per cent (Yu et al. 2009). SnO_2/CNT composite films with improved properties have also been used as LIB anodes (Dhanabalan et al. 2010).

4.7 Glancing Angle Deposition

Glancing angle deposition (GLAD) is a physical vapour deposition technique that was applied very recently to the deposition of 3D nano-helices as LIB anodes. In this emerging technique, the electrode materials are thermally evaporated inside a vacuum chamber. Unlike a typical thermal evaporation run, the sample holder and the substrate are tilted such that the angle of incidence of the vapour flux from the crucibles is at more than 70° from the substrate surface normal (Fig. 7). The angle of incidence is varied by using a stepper motor. A second stepper motor is used to control the azimuthal rotational speed of the sample holder so that 3D helical structures can be formed. In the GLAD study (Polat and Keles 2015), Si and Cu were evaporated in separate graphite crucibles. Prior to thermal evaporation, an Ar^+ ion beam was first used to clean the substrate by ion beam sputtering. After substrate cleaning, the ion beam energy was reduced and the ion beam assisted thermal evaporation was commenced. At an angle of incidence of 80° to the substrate normal and an azimuthal rotation speed of 0.2 rpm, well defined helical structures with a diameter of 100–300 nm were obtained with nanoscale pores between the Si/Cu nano-helices.

The top view and cross sectional SEM images of Si/Cu nano-helices can be found in Fig. 2 of the reference (Polat and Keles 2015). The authors argued that these electrodes are more mechanically flexible than solid Si anodes during Li ion intercalation and the conductivity is also improved by the co-deposition of Cu in the helices. Experimental LIB was fabricated using Li metal cathode, SiCu nano-helix anode and porous polypropylene separator soaked with organic $LiPF_6$ electrolyte. For a copper content of 30 per cent, a discharge capacity of 1697 mAh/g was measured after 100 charging cycles corresponding to a charge capacity retention of 49 per cent (Polat and Keles 2015). Both the discharge capacity and capacity retention depend on the Cu content of the SiCu nano-helices.

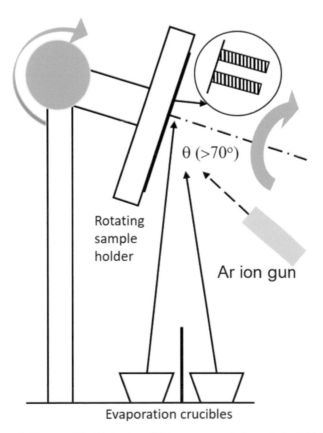

Fig. 7. Schematic diagram of the fabrication of 3D nanostructured electrodes by GLAD technique.

The properties of 3D electrode materials prepared by fabrication techniques other than sol-gel synthesis is presented in Table 3.

Table 3. Summary of porosity properties of 3D electrode materials prepared by various fabrication techniques.

3D Electrode Material	Fabrication Method	Pore Diameter (nm)	Porosity (%)	Specific Surface Area (m²/g)
Carbon (Ryoo et al. 1990)	EISA	3; 0.5–0.8	–	1380
Nb$_2$O$_5$ (Rauda et al. 2013)	EISA	13–15	–	–
TiO$_2$ (Rauda et al. 2013)	EISA	20–25; 1–4	–	–
V$_2$O$_5$ (O'Hanlon et al. 2017)	3-DOM/inverse opal	460–480	–	–
Cu (Arai et al. 2016)	Electrodeposition	–	64	–
Cu/CNT (Arai et al. 2016)	Electrodeposition	–	66–74	–
LiCoO$_2$ (Chen et al. 1996)	ESD	8000	–	–

4.8 Infiltration Techniques

After the 3D EES electrode scaffold has been fabricated by one of the above techniques, the void space within this porous structure will need to be back-filled with electrolyte and another electrode material. A critical requirement for this infiltration process is that the deposition must be precisely controllable and conformal. In this context, conformal means the topology of the pores must be followed. Any deviation from a conformal deposition will result in incomplete infiltration, preferential deposition on the surface of the scaffold and disruption of the connected pore space that is originally present. As a result of these requirements, standard semiconductor processing techniques such as evaporation, magnetron sputtering and chemical vapour deposition are not applicable. In this section, we discuss infiltration processes that have been used successfully for 3D EES devices.

4.9 Electroless Deposition

In this method, the pre-fabricated 3D electrode is simply immersed into an aqueous solution containing the reactants that form the desired electroactive material. During soaking, solution reactions occur and lead to precipitation onto the porous scaffold. The soaking time determines the thickness of the deposited material. Although the process is simple in principle, it is important to control the electroless deposition conditions such as the pH of the solution. In an EC study by (Fischer et al. 2007), amorphous MnO$_2$ was deposited by the electroless method onto carbon nanofoam substrates. Sulphuric acid (H$_2$SO$_4$) and sodium permanganate (Na$_2$MnO$_4$) were first vacuum infiltrated into the carbon nanofoam. This was followed by soaking in an aqueous mixture of either H$_2$SO$_4$/Na$_2$MnO$_4$ or Na$_2$SO$_4$/Na$_2$MnO$_4$.

It was found that the conformality of the deposition depends crucially on the solution pH. For acidic conditions (H$_2$SO$_4$/Na$_2$MnO$_4$), the deposition rate was higher and there was preferential deposition onto the top surface of the nanofoam causing a crust to be formed. Full permeation by the reactants (Na$_2$SO$_4$/Na$_2$MnO$_4$) could only be obtained in neutral pH condition. This was confirmed by SEM and energy dispersive spectroscopy (EDS) as shown in Fig. 1 and Fig. 2 of the reference (Fischer et al. 2007). Nitrogen porosimetry data showed that the specific pore volume decreased from 0.62 cm³/g for the bare carbon nanofoam to 0.31 cm³/g for MnO$_2$ deposited at neutral conditions. This decrease is due to the infiltration of MnO$_2$. Furthermore, the peak of the pore size distribution (45 nm) for neutral deposited MnO$_2$ is smaller than that of the bare carbon nanofoam (64 nm). ECs with neutral electroless deposited MnO$_2$-carbon electrodes have a specific capacitance of 110 F/g. (Mass refers to MnO$_2$ and carbon.) The specific capacitance is higher than both the bare carbon nanofoam electrode and MnO$_2$ deposited under acidic conditions.

4.10 Atomic Layer Deposition

The atomic layer deposition (ALD) technique is a method developed originally in the 1970s for the deposition of thin film electroluminescent materials such as ZnS. During the 1990s, it was adopted by the semiconductor industry to deposit ultrathin metallic barrier layers into high aspect ratio structures such as the dual Damascene copper interconnects. The ALD process is based on the sequential, self-limiting deposition of gas phase organometallic precursors onto a temperature controlled substrate (Elam et al. 2011). It is important to note that unlike chemical vapour deposition or plasma enhanced chemical vapour deposition, only one gas phase reactant is present in the ALD deposition chamber at any one time. The reactant or precursor reacts chemically with the available atomic sites on the substrate by chemisorption. When all available atomic sites on the substrate have undergone chemisorption, the reaction will automatically terminate and any unused reactants have to be purged. The next reactant is introduced and the process is repeated. Ideally, the ALD cycle can be used to control the deposited layer thickness with monolayer precision. Another key advantage of ALD is its conformality. Since the ALD reaction is surface controlled and self-limited, the precursors can permeate a scaffold as well as high aspect ratio structures and should result in a highly conformal deposition.

In the field of EES, ALD has been used to fabricate nanotubular arrays of metal insulator metal (MIM) capacitors (Banerjee et al. 2009). The substrate is a template with high aspect ratio holes to increase the area of the MIM capacitor array. ALD has also been used to deposit ultrathin alumina (Al_2O_3) onto LIB electrodes to mitigate the problem of capacity fading (CF) in the LIB (Elam et al. 2011). CF refers to the progressive deterioration in charge storage capacity of a LIB upon repeated charge and discharge cycles. It has been found that when 0.3–0.4 nm of Al_2O_3 is deposited by ALD onto a $LiCoO_2$ electrode, the CF can be reduced by half when compared with the same electrode without Al_2O_3. Finally, the ALD deposition of lanthanum titanate and lithium lanthanum titanante thin films used in Li ion thin film batteries has recently been reported (Aaltonen et al. 2010). However, due to the chemical reactivity of Li, the deposition of Li containing materials by ALD has thus far been difficult.

5. Hybrid EES Devices with 3D Electrodes

The most recent and exciting development is the realisation of hybrid EES devices. As alluded to earlier, this type of device represents a merging of the traditional EC and the pseudocapacitor (Long et al. 2011). It is an asymmetric energy storage device in which one electrode is supercapacitor (EDLC) like and the other electrode is either a battery or pseudocapacitor electrode. This novel design leverages on the higher power density of the supercapacitor and the higher energy density of the battery. Due to the asymmetric structure, the specific charge stored in each electrode may not be equal. In addition, the higher energy density of hybrid EES allows the use of safer and environmentally more benign aqueous electrolytes. Three dimensional structuring of the electrodes of hybrid EES devices has recently been reported (Zhou et al. 2016). These investigators describe their hybrid device as a supercabattery. The positive electrode of the supercabattery consisted of MnO_x deposited on aligned carbon nanotubes (ACNT) and operates by Li^+ insertion mechanism. The negative electrode comprised of carbon deposited on MnO_y/ACNT and is an EDLC like electrode. The SEM images of these two electrodes can be seen in Fig. 1 of the reference (Zhou et al. 2016). Both electrodes had to be pre-lithiated before assembly into the supercabattery with a lithium containing organic electrolyte. After 1,000 cycles of charging and discharging, the energy density remained at 105.8 Whr/kg and the power density was 3 kW/kg (Zhou et al. 2016). When plotted in the Ragone plot of Fig. 1, this data point is situated outside the regions for the batteries and ECs. This shows that by using the 3D electrode architecture, encouraging progress has been made in the pursuit of the ultimate goal of high energy and high power density EES devices. The electrochemical performance of the supercabattery and several other EES devices incorporating 3D electrodes is summarised in Table 4.

Table 4. Electrochemical properties of selected EES devices with 3D electrodes.

EES Device	Electrode Fabrication Method	Electrode Materials	Electrochemical Properties
EC (Fischer et al. 2007)	Electroless deposition	MnO_2 + C	Specific capacitance: 110 F/g (normalised to MnO_2+C) Area normalised capacitance: 1.5 F/cm^2
Battery (Ergang et al. 2007)	3-DOM/inverse opal	C, V_2O_5	Reversible gravimetric charge/discharge capacity: 0.35 mAh/g
Battery (Zheng and Wang 2012)	Sol-gel	$LiCoO_2$	Discharge capacity: 115 mAh/g at C/20 and 21 mAh/g at 1 C
Battery (Dhanabalan et al. 2010)	ESD	SnO_2/CNT	Discharge capacity: 450 mAh/g for 10 wt% CNT and 250°C
Battery (Polat and Keles 2015)	GLAD	Cu/Si	First discharge capacity: 3389 mAh/g for 30 at.% Cu. Capacity retention after 100 cycles: 49%
Supercabattery (Zhou et al. 2016)	Deposition on aligned CNT	MnO_x	Maximum discharge capacity: 217 mAh/g Energy density after 1000 cycles: 105.8 Wh/kg Power density after 1000 cycles: 3 kW/kg

References

Aaltonen, T., M. Alnes, O. Nilsen, L. Costelle and H. Fjelvag. 2010. Lanthanum titanate and lithium lanthanum titanate thin films grown by atomic layer deposition. J. Mater. Chem. 20: 2877–2881.

Amin, M. and J. Stringer. 2008. The electricity power grid: today and tomorrow. MRS Bull. 33: 399–407.

Arai, S. and T. Kitamura. 2014. Simple method for fabrication of three-dimensional (3D) copper nanostructured architecture by electrodeposition. ECS Electrochem. Lett. 3: D7–D9.

Arai, S., M. Ozawa and M. Shimizu. 2016. Fabrication of three-dimensional (3D) copper/carbon nanotube composite film by one-step electrodeposition. J. Electrochem. Soc. 163: D774–D779.

Arico, A.S., P. Bruce, B. Scrosati, J.-M. Tarascon and W.V. Schalkwijk. 2005. Nanostuctured materials for advanced energy conversion and storage devices. Nat. Mater. 4: 366–377.

Armstrong, E., M. O'Sullivan, J. O'Connell, J.D. Holmes and C. O'Dwyer. 2015. 3D vanadium oxide inverse opal growth by electrodeposition. J. Electrochem. Soc. 162: D605–D612.

Arthur, T.S., D.J. Bates, N. Cirigliano, D.C. Johnson, P. Malati, J.M. Mosby et al. 2011. Three-dimensional electrodes and battery architectures. MRS Bull. 36: 523–531.

Banerjee, P., I. Perez, L. Henn-Lecordier, S.B. Lee and G.W. Rubloff. 2009. Nanotubular metal-insulator-metal capacitor arrays for energy storage. Nat. Nanotechnol. 4: 292–296.

Brezesinski, K., J. Wang, J. Haetge, C. Reitz, S.O. Steinmueller, S.H. Tolbert et al. 2010. Pesudocapacitive contributions to charge storage in highly ordered mesoporous group V transition metal oxides with iso-oriented nanocrystalline domains. J. Am. Chem. Soc. 132: 6982–6990.

Cairns, E.J. and P. Albertus. 2010. Batteries for electric and hybrid-electric vehicles. Annu. Rev. Chem. Biomol. Eng. 1: 299–320.

Chen, C., E.M. Kelder, P.J.J.M. van der Put and J. Schoonman. 1996. Morphology control of thin $LiCoO_2$ films fabricated using the electrostatic spray deposition (ESD) technique. J. Mater. Chem. 6: 765–771.

Chow, G.M. and K.E. Gonsalves. 1998. Particle synthesis by chemical routes. pp. 55–72. In: Edelstein, A.S. and R.C. Cammarata [eds.]. Nanomaterials Synthesis, Properties and Applications. Institute of Physics Publishing, Bristol, UK.

Conway, B.E. 1999. Electrochemical supercapacitors scientific fundamentals and technological applications. Kluwer Academic/Plenum Publishers, New York.

Coustier, F., S. Passerini and W.H. Smyrl. 1998. A 400 mAh/g aerogel-like V_2O_5 cathode for rechargeable lithium batteries. J. Electrochem. Soc. 145: L73–L74.

Dhanabalan, A., Y. Yu, X. Li, W. Chen, K. Bechtold, L. Gu et al. 2010. Porous SnO_2/CNT composite anodes: influence of composition and deposition temperature on the electrochemical performance. J. Mater. Res. 25: 1554–1560.

Dong, W., D.B. Rolison and B. Dunn. 2000. Electrochemical properties of high surface area vanadium oxide aerogels. Electrochem. Sol. State Lett. 3: 457–459.

Doughty, D.H., P.C. Butler, A.A. Akhil, N.H. Clark and J.D. Boyes. 2010. Batteries for large-scale stationary electrical energy storage. Electrochem. Soc. Inter. 19: 49–53.

Dunn, B., H. Kamath and J.-M. Tarascon. 2011. Electrical energy storage for the grid: a battery of choices. Science 334: 928–935.

Elam, J.W., N.P. Dasgupta and F.B. Prinz. 2011. ALD for clean energy conversion, utilization, and storage. MRS Bull. 36: 899–906.

Ergang, N.S., M.A. Fierke, Z. Wang, W.H. Smyrl and A. Stein. 2007. Fabrication of a fully infiltrated three-dimensional solid-state interpenetrating electrochemical cell. J. Electrochem. Soc. 154: A1135–A1139.

Fischer, A.E., K.A. Pettigrew, D.R. Rolison, R.M. Stroud and J.W. Long. 2007. Incorporation of homogeneous nanoscale MnO_2 within ultraporous carbon structures via self-limiting electroless deposition: Implications for electrochemical capacitors. Nano. Lett. 7: 281–286.

Francois, B., G. Delille and C. Saudemont. 2015. Energy storage in electric power grids. Wiley, New York.

Ginley, D., M.A. Green and R. Collins. 2008. Solar energy conversion towards 1 Terawatt. MRS Bull. 33: 355–364.

Hamann, C.H., A. Hamnett, W. Vielstich. 2007. Electrochemistry. Wiley-VCH, Weinheim.

Hanlon, O., D. McNulty and C. O'Dwyer. 2017. The influence of colloidal opal template and substrate type on 3D microporous single and binary vanadium oxide inverse opal electrodeposition. J. Electrochem. Soc. 164: D111–D119.

Harris, C. and J.P. Meyers. 2010. Working smarter, not harder: an introduction to the "smart grid". Electrochem. Soc. Inter. 19: 45–48.

Harris, S.J., R.D. Deshpande, Y. Qi, I. Dutta and Y.-T. Cheng. 2010. Mesopores inside electrode particles can change the Li-ion transport mechanism and diffusion-induced stress. J. Mater. Res. 25: 1433–1440.

Ho, M.-W. 2012. Living rainbow H_2O. World Scientific, Singapore.

Kistler, S.S. 1931. Coherent expanded aerogels and jellies. Nature 127: 741.

Kuo, S.-L. and N.-L. Wu. 2006. Investigation of pseudocapacitive charge-storage reaction of $MnO_2.nH_2O$ supercapacitors in aqueous electrolytes. J. Electrochem. Soc. 153: A1317–A1324.

Lee, H.Y. and J.B. Goodenough. 1999. Supercapacitor behaviour with KCl electrolyte. J. Solid State Chem. 144: 220–223.

Livage, J., C. Sanchez and F. Babonneau. 1998. Molecular precursor routes to inorganic solids. pp. 389–448. *In:* Interrrante, L.V. and M.J. Hampden-Smith [eds.]. Chemistry of Advanced Materials. Wiley-VCH, New York, NY, USA.

Long, J.W., B. Dunn, D.R. Rolison and H.S. White. 2004. Three-dimensional battery architectures. Chem. Rev. 104: 4463–4492.

Long, J.W. and D.R. Rolison. 2007. Architecture design, interior decoration and three-dimensional plumbing en route to multifunctional nanoarchitectures. Acc. Chem. Res. 40: 854–862.

Long, J.W. D. Belanger, T. Brousse, W. Sugimoto, M.B. Sassin and O. Crosnier. 2011. Asymmetric electrochemical capacitors—stretching the limits of aqueous electrolytes. MRS Bull. 36: 513–522.

Lukatskaya, M.R., B. Dunn and Y. Gogotsi. 2016. Multidimensional materials and device architectures for future hybrid energy storage. Nature Comm. 7: 12647.

Perre, E., P.L. Taberna, D. Mazouzi, P. Poizot, T. Gustafsson, K. Edstrom et al. 2010. Electrodeposited Cu_2Sb as anode material for 3-dimensional Li-ion microbatteries. J. Mater. Res. 25: 1485–1491.

Pierre, A.C. 2011a. History of aerogels. pp. 3–18. *In:* Aegerter, M.A., N. Leventis and M.M. Koebel [eds.]. Aerogels Handbook. SpringerLink, Heidelberg, Germany.

Pierre, A.C. and A. Rigacci. 2011b. SiO_2 aerogels. pp. 21–45. *In:* Aegerter, M.A., N. Leventis and M.M. Koebel [eds.]. Aerogels Handbook. SpringerLink, Heidelberg, Germany.

Polat, B.D. and P. Keles. 2015. Evaluation of composite helices used as anodes for rechargeable lithium ion batteries. ECS Transactions USA 66(9): 3–15.

Rauda, I.E., V. Augustyn, B. Dunn and S.H. Tolbert. 2013. Enhancing pseudocapacitance charge storage in polymer template mesoporous materials. Acc. Chem. Res. 46: 1113–1124.

Reese, A. 2017. Blackouts cast Australia's green energy in dim light. Science 355: 1001–1002.

Rolison, D.R. and B. Dunn. 2001. Electrically conductive oxide aerogels: new materials in electrochemistry. J. Mater. Chem. 11: 963–980.

Ryoo, R., S.H. Joo and S. Jun. 1990. Synthesis of highly ordered carbon molecular sieves via template mediated structural transformation. J. Phys. Chem. B 103: 7743–7746.

Simon, P., Y. Gogotsi and B. Dunn. 2014. Where do batteries end and supercapacitors begin? Science 343: 1210–1211.

Smalley, R.E. 2005. Future global energy prosperity: the terawatt challenge. MRS Bull. 30: 412–417.

Soloveichik, G.L. 2011. Battery technologies for large-scale stationary energy storage. Annu. Rev. Chem. Biomol. Eng. 2: 503–527.

Strom, R. 2007. Hot house global climate change and the human condition. Prasis Publishing, New York.

Tamon, H., H. Ishizaka, M. Mikami and M. Okazaki. 1997. Porous structure of organic and carbon aerogels synthesized by sol-gel polycondensation of resorcinol with formaldehyde. Carbon 35: 791–796.

Twidell, J. and T. Weir. 2006. Renewable energy resources second edition. Taylor & Francis, Oxford.

Ueda, Y., S. Suzuki and T. Ito. 2009. Grid stabilization by use of an energy storage system for a large-scale PV generation plant. ECS Transactions 16(34): 17–25.

Vega, L.F. 2015. Fundamentals of supercritical fluids and the role of modelling. pp. 19–42. *In:* Domingo, C. and P. Subre-Paternault [eds.]. Supercritical Fluid Nanotechnology: Advances and Applications in Composites and Hybrid nanomaterials. Pan Stanford, Singapore.

Wang, Y., C. Cai and D. Guan. 2011. New developments in nanostructured electrode materials for advanced Li-ion batteries. *In:* Iniewski, K. [ed.]. Nanoelectronics Nanowires, Molecular Electronics, and Nanodevices. McGraw Hill, New York, USA.

Wei, W., X. Cui, W. Chen and D.G. Ivey. 2011. Manganese oxide-based materials as electrochemical supercapacitor electrodes. Chem. Soc. Rev. 40: 1697–1721.

Yamamoto, T., T. Nishimura, T. Suzuki and H. Tamon. 2001. Effect of drying conditions on mesoporosity of carbon precursors prepared by sol-gel polycondensation and freeze drying. Carbon 39: 2374–2376.

Yu, A., V. Chabot and J. Zhang. 2013. Electrochemical supercapacitors for energy storage and delivery fundamentals and applications. CRC Press, Roca Baton.

Yu, S., T.K.S. Wong, K. Pita and X. Hu. 2002. Synthesis of organically modified mesoporous silica as a low dielectric constant intermetal dielectric. J. Vac. Sci. Technol. B20: 2036–2042.

Yu, S., T.K.S. Wong, X. Hu and K. Pita. 2003. The effect of TEOS/MTES ratio on the structural and dielectric properties of porous silica films. J. Electrochem. Soc. 150: F16–F121.

Yu, S., T.K.S. Wong, X. Hu, K. Pita and V. Ligatchev. 2004. Synthesis and characterization of templating low dielectric constant organosilicate films. J. Electrochem. Soc. 151: F123–F127.

Yu, Y., A. Dhanabalan, C.-H. Chen and C. Wang. 2009. Three-dimensional porous amorphous SnO_2 thin films as anodes for Li-ion batteries. Electrochem. Acta 54: 7227–7230.

Zheng, J.P. and T.R. Jow. 1995a. A new charge storage mechanism for electrochemical capacitors. J. Electrochem. Soc. 142: L6–L7.

Zheng, J.P., P.J. Cygan and T.R. Jow. 1995b. Hydrous ruthenium oxide as an electrode material for electrochemical capacitors. J. Electrochem. Soc. 142: 2699–2703.

Zheng, Z. and Y. Wang. 2012. 3D structure of electrode with inorganic solid electrolyte. J. Electrochem. Soc. 159: A1278–A1282.

Zhou, H., X. Wang, E. Sheridan, H. Gan, J. Du, J. Yang et al. 2016. Boosting the energy density of 3D dual-manganese oxides-based Li-ion supercabattery by controlled mass ratio and charge injection. J. Electrochem. Soc. 163: A2618–A2622.

6

Three Dimensional Porous Binary Metal Oxide Networks for High Performance Supercapacitor Electrodes

Balasubramaniam Saravanakumar,[1] Tae-Hoon Ko,[2] Jayaseelan Santhana Sivabalan,[1] Jiyoung Park,[2,4] Min-Kang Seo[3] and Byoung-Suhk Kim[2,1,]*

1. Introduction

In this decade, energy crisis is one of the most challenging problems, which is influencing the global economy. Due to rapid growth in human population and industrialisation demand for energy sources have increased significantly. This crisis has been aggravated with the depletion of fossil fuels. Thus, higher priority has given to the development of high performance energy conversion as well as energy storage devices. The availability of renewable energy sources like solar, wind, hydro and tidal are limitless. However, these energy sources are time and place dependent, which requires systems to have advanced intermediate energy storage devices. Further, rapid growth and miniaturisation of electronic devices opens up the development of new flexible and wearable electronic gadgets which equally requires equally high performance energy storage devices (Miller and Simon 2008, Arico et al. 2005, Hall et al. 2010). The performance of the electrochemical energy storage devices are evaluated and compared with their energy and power densities and expressed together as Ragone plot (Simon and Gogotsi 2008). Figure 1 shows the Ragone plot of various energy storage devices like batteries, supercapacitors and fuel cells. From this plot, it is clearly visible that the batteries have higher energy density compared to supercapacitor but lacks in power density. Even though batteries have higher energy density, it still requires higher power density for number of practical applications (Dunn et al. 2011).

Ultracapacitors or supercapacitors attempts to bridge the gap between batteries and dielectric capacitor through their higher power density. But it still lacks in required energy density, which hinders its practical application. Thus efforts have been made to develop a high energy density

---

[1] Department of BIN Convergence Technology, Chonbuk National University, 567 Baekje-daero, Deokjin-gu, Jeonju-si, Jeollabuk-do 54896, Republic of Korea.

[2] Department of Organic Materials and Fiber Engineering, Chonbuk National University, 567 Baekje-daero, Deokjin-gu, Jeonju-si, Jeollabuk-do 54896, Republic of Korea.

[3] Korea Institute of Carbon Convergence Technology, Jeonju 54852, Republic of Korea.

[4] Convergence Technology Division, Industry Convergence Technology Center, Korea Conformity Laboratories, Seoul 08503, Republic of Korea.

* Corresponding author: kbsuhk@jbnu.ac.kr

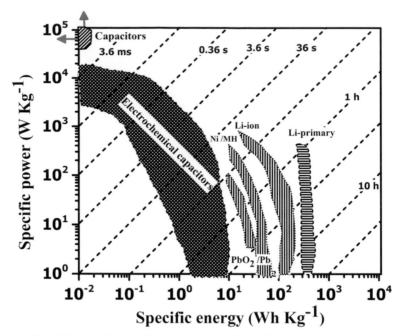

Fig. 1. Ragone plot of various energy storage devices (Simon and Gogotsi 2008).

supercapacitors without sacrificing their power density. However, supercapacitors are advantageous than batteries because of their, fast charge/discharge capability, long life cycle (> 100,000 cycles) and low maintenance cost (Zhang and Zhao 2009, Conway 1999).

The energy density of the supercapacitor effectively increased with increasing either or both of the capacitance and operating potential as per the following Eq. 1.

$$E = \tfrac{1}{2}CV^2 \tag{1}$$

where E is the energy density and is directly related to the capacitance (C) and operating potential (V) of the supercapacitor device.

For this purpose, two important things have to be followed: first to develop a high performance electrochemical materials which includes electrode, electroactive material and electrolyte (Fig. 2); second, have a controlled assembly of supercapacitor device based on symmetric and asymmetric configurations, i.e., electric double layer capacitor (EDLC), pseudocapacitor and hybrid capacitor (Fig. 3) (Zhong et al. 2015, Wang et al. 2016a,b).

In general, the electrochemical supercapacitors consist of current collector, electroactive material and electrolytes in addition to the separator used to separate anode and cathode. Greater attention has been paid to the development of higher capacitive electroactive materials with various dimensions from 0D to 3D nanostructures (Lukatskaya et al. 2016). A lot of electroactive materials have been tested for energy storage application from carbon families (activated carbon, porous carbon, graphene, CNT, carbides) (Lei et al. 2015), various forms of metal hydroxides/oxides, sulfites, carbides and conducting polymers (Radhakrishnan et al. 2016, Ko et al. 2017, Wang et al. 2016a,b, Ko et al. 2016, Devarayan et al. 2015).

In recent time, design and development of new types of current collectors have also gained increasing interest to enhance not only the electrical conductivity but also improve the electrolyte accessible surface area through porous hierarchal structures. These hierarchal pores can reduce the diffusion path length and improve the rate kinetics. The electrolyte is another important component which controls the rate of electrochemical reaction and widens the operating potential of the supercapacitor device (Zhong et al. 2015, Wang et al. 2016a,b, Wang et al. 2012). Several types of electrolytes are used for

Advanced Electrode materials

- High surface area material & porosity
- High conductivity
- Electrochemically stable
- 0D, 1D, 2D, 3D & hierarchical Materials
- EDLC, faradic and hybrid materials

Material Design

- Flexible
- Electrochemically stable
- Low interfacial resistance
- Higher contact area
- Hierarchical and 3D structures

- Aqueous electrolyte (Acid, Alkaline, Neutral)
- Non-aqueous electrolyte (Organic, Ionic)
- Gel electrolyte (Dry, Gel, inorganic)
- Redox-active electrolyte

Current Collector   **Electrolyte**

Fig. 2. Materials design for the fabrication of supercapacitor.

Electric double layer

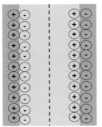

- Electrostatic double layer
- Long cycle life
- Higher power density
- High surface area & porosity
- Symmetric assembly
- Carbon materials

Pseudo capacitor

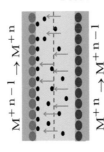

$M^{+n-1} \rightarrow M^{+n}$ $M^{+n} \rightarrow M^{+n-1}$

- Reversible Faradic reaction
- High energy density
- Symmetric & asymmetric assembly
- Metal oxides, hydroxides, conducting polymer, composite materials

Hybrid capacitor

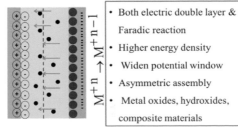

$M^{+n} \rightarrow M^{+n-1}$

- Both electric double layer & Faradic reaction
- Higher energy density
- Widen potential window
- Asymmetric assembly
- Metal oxides, hydroxides, composite materials

Fig. 3. Classification and assembly of supercapacitor device.

the fabrication of supercapacitor devices which include aqueous (acid, alkaline, neutral), organic, ionic liquid electrolytes, gel electrolyte and redox additive electrolytes. The operating potential of the water based aqueous electrolytes is limited to 1.3 V due to splitting of water resulting in H_2/O_2 generation, but using organic and ionic liquid electrolytes can extend the operating voltage to 2.5–3 V and ~ 4 V, respectively (Zhong et al. 2015, Wang et al. 2016a,b, Wang et al. 2012).

Based on the energy storage mechanism, supercapacitors can be classified as electrochemical double layer capacitor (EDLC), pseudocapacitor and hybrid capacitor. In EDLC device, the charges are stored by the formation of electrostatic double layer at the interface between electroactive material and electrolyte by adsorption/desorption process. Mostly, carbon materials with higher surface area are utilised for EDLC devices, which are symmetric devices. The performances of the EDLC highly rely on the active surface area and nature of pore structure in the electrode materials. The higher active surface area and hierarchal pore nature of the electrode allows ions to move freely to the inner parts the electrode and facilitates the formation of electric double layer on the pore walls. This results in higher charge/discharge capability due to faster ion movement and shorter diffusion path length. The pseudocapacitor is another class of supercapacitor device, which stores the charge through fast, reversible faradic reactions.

Three kinds of faradic reactions are involved in the charge storage mechanism in pseudocapacitors. In first type, the faradic reaction takes places at the surface of the electroactive material by fast reversible adsorption/desorption process. The surface redox reaction does not change any structural and crystallographic nature of the electrode material, which facilitates higher cyclic stability. The surface faradic reactions are usually observed in pseudo capacitive materials like RuO_2, MnO_2, NiO, etc. The following gives a sample representative reaction for RuO_2 electrode.

$$RuO_2 + xH^+ + xe^- \leftrightarrow RuO_{2-x}(OH)_x \qquad \text{(i)}$$

In the second type of faradic reaction, intercalation of electrolytic ion within electroactive materials occurs resulting in structural changes. Mostly metal oxides, layered materials as well as sulfides follow this mechanism to store charges. The following gives a sample representative reaction for Li^+ intercalation in Nb_2O_5 electrode.

$$Nb_2O_5 + nLi^+ + ne^- \leftrightarrow Li_xNb_2O_5 \qquad \text{(ii)}$$

Finally, reversible faradic reaction can also occur due to electrochemical doping and de-doping of ions in the electroactive materials such as conductive polymer electrode. The following mechanism that stores the electrical charges in conducting polymer is shown below. The whole reaction is expressed as,

$$CP \leftrightarrow CP^{n+}(A^-)_n + ne^- \qquad \text{(iii)}$$

$$CP + ne^- \leftrightarrow CP_{n-}(B^+)_n \qquad \text{(iv)}$$

where CP is conducting polymer, A$^-$, B$^+$ are electrolyte ions. The pseudocapacitive electrodes are assembled as symmetric and asymmetric device to improve the operating potential and energy density.

Recently, hybrid supercapacitors have been developed by combining electric double layer and faradic reactions in a single device. It is similar to an asymmetric supercapacitor, in which one electrode is made by redox materials (battery electrode) and another electrode by EDLC materials. The hybrid supercapacitors have gained much interest in energy storage applications due to its large operating potential and higher specific capacitance (El-Kady et al. 2016, Zhao and Zheng 2015, Wang 2016, Sun et al. 2016, Dubal 2015). However, developing a high energy density supercapacitor still remains a challenging task.

The solution to this problem lies in the detailed understanding of the electrochemical kinetics of the novel electrodes which can exhibit enhanced ion and electron transport. Various strategies are implemented to enhance the electrochemical kinetics which includes integration of high capacitive (pseudocapacitive) materials with highly porous carbon material (carbon aerogel) and direct growth of electroactive material in highly conductive, open porous current collectors.

Based on these strategies, the following sections looks into (i) the fabrication and electrochemical characterisation of $ZnCo_2O_4$ nanosheets decorated three dimensional, porous metal network electrode, (ii) fabrication and electrochemical characterisation of $NiCo_2O_4$ nano-needle decorated Ni@Ni foam electrode, (iii) fabrication and electrochemical characterisation of multiwall carbon nanotube-$NiCo_2O_4$ composite aerogel electrode and (iv) finally, the challenges and prospective of the porous materials as well as current collector for future energy storage device applications.

2. Fabrication and Electrochemical Properties of Three Dimensional (3D), Porous Metal Network with $ZnCo_2O_4$

Porous, three dimensional (3D) materials such as metallic foams and honeycomb like structures have created more interest due to their large active surface area and open pore structures which enables a higher access to ions and electron transport over electroactive materials (Franceschini et al. 2011). Accordingly, the energy density of the supercapacitor can be enhanced by tuning the physico-chemical properties of electroactive materials as well as electrolyte and designing a hierarchical current collector with low interfacial resistance. The direct growth of electroactive materials on current collectors reduces the ohmic resistance of the capacitor through strong inter-phase adhesion (Yan et al. 2014, Jiang et al. 2012, Xu et al. 2014). Mostly, nickel foam (Kundu and Liu 2013, Gao et al. 2010), 3D graphene (Dong et al. 2012, Wang et al. 2014) are used as 3D substrates for various electrochemical applications, which effectively improves the performance of the device by enhancing the electrochemical reaction rate, and effective charge transport (Zhang et al. 2011). Further, the hierarchal nanostructures (Gao et al. 2015, Qiu et al. 2014, Deng et al. 2013) as well as core-shell structures (Huang et al. 2014, Chen et al. 2013, Lu et al. 2013) are used as a current collector to improve the performance of the supercapacitor. There are various methods adopted to fabricate porous electrodes such as lyotropic liquid crystalline phases method (Attard et al. 1997, Bartlett et al. 2002), electrochemical dealloying (Pickering and Wagner 1967, Erlebacher et al. 2001, Deng et al. 2008), and electrodeposition within the interstitial spaces between colloidal spheres (Velev and Kaler 2000, Bartlett et al. 2002) and gas self-supported electrodeposition (Shin et al. 2003, Shin and Liu 2004, Zach and Penner 2000).

Compared to the other methods, gas bubble template electrodeposition is an easy and economical method. By employing this method, it is possible to fabricate foam structure that are mechanically robust which can be tuned for the pore size and pore wall thickness by varying the deposition parameters (Shin et al. 2003). The walls of metal foams are constructed by numerous interconnected metal particles of different sizes ranging from nano to micro, which facilitates to form additional micro- and mesoporous structures between interconnected metal particles. The formation of micro- and mesopores improves surface area, which are higher than commercially available 3D metal foams. Different types of metal foams such as Cu (Shin and Liu 2005), Ni (Zach and Penner 2000, Yang et al. 2013, Ashassi-Sorkhabi et al. 2016), Au (Cherevko and Chung 2011), Ag (Cherevko et al. 2010), Pd (Yang et al. 2011) and Pt (Ott et al. 2011) have been prepared via hydrogen bubble template electrodeposition methods. However there are only few studies showing 3D structured electroactive materials such as $NiCo_2S_4$, $Co(OH)_2$, $Ni(OH)_2$, MnOx, and polypyrrole being used for supercapacitor electrodes (Sun et al. 2013, Xia et al. 2011).

Last decade, transition metal oxides and sulfides have been widely explored for this application. This was mainly due to the fact that these materials showed higher electrochemical redox properties, higher theoretical capacitance and low-cost. But low electrical conductivity of these electrode materials reduces its rate capability. Recently, binary transition metal oxides created more interest because of higher conductivity, rich electrochemical redox reaction sites and multiple oxidation states compare to the mono metal oxides. Binary transition metal oxides such as $ZnCo_2O_4$ has shown huge potential as an electrode material for supercapacitor applications. $ZnCo_2O_4$ materials have been studied for different nanostructures like nanorods (Liu et al. 2013), nanowires (Wang et al. 2014), nanotubes (Guan et al.

2014), porous microspheres (Wang et al. 2015a/b), nanosheets (Cheng et al. 2015), and core–shell structures (Niu et al. 2015, Ma et al. 2015, Qiu et al. 2015). The following sections discusses the performance of $ZnCo_2O_4$ nanoflakes coated over 3D-Ni current collector by electrodeposition method.

2.1 Mechanism of Porous, Three Dimensional (3D) Ni Depositions

The porous 3D-Ni network was successfully fabricated by simple dynamic hydrogen bubble template assisted cathodic deposition. The deposition was carried out at a higher current density of 2.5 A/cm² in 2 M NH_4Cl and 0.1 M $NiCl_2$ electrolyte solutions for 60s. The stainless steel (SS, 1 cm × 1 cm) was used as a cathode and platinum plate as an anode with the separation of ~ 1 cm. The H_2 gas bubble was used as a dynamic template to deposit porous Ni film. The H_2 bubble size increases from cathode surface to outer wards (Plowman et al. 2015), which is schematically explained in Fig. 4. Generally, it has three different stages such as (i) formation of H_2 bubble, (ii) growth and coalescence and (iii) H_2 bubble release. At applied higher current density, H_2 gas bubble gets generated at the cathode surface in electrolyte solution due to over potential. The rate of H_2 evolution at cathode varies with various metals which is estimated from exchange current densities vs. hydrogen chemisorption energy plot. The estimated values ($\log i_o / A.cm^{-2}$) of various metal substrates are Pt (–3.1), Pd (–3.0), Ni (–5.21), Cu (–5.37), Au (–5.4), Ag (–7.85) (Norskov et al. 2005, Trasatti 1972). Compared to the other metals, Pt and Pd have low values which indicates that it can generate a higher volume of H_2 gas at over potential. The generated H_2 bubbles at cathode start to nucleate over surface irregularities (Vogt et al. 2004), defects and edges (Nikolić et al. 2007). In addition to water hydrolysis, the inclusion of acids like H_2SO_4, HNO_3, and NH_4Cl solution generates additional H_2 gas over the cathode surface (reaction (v)).

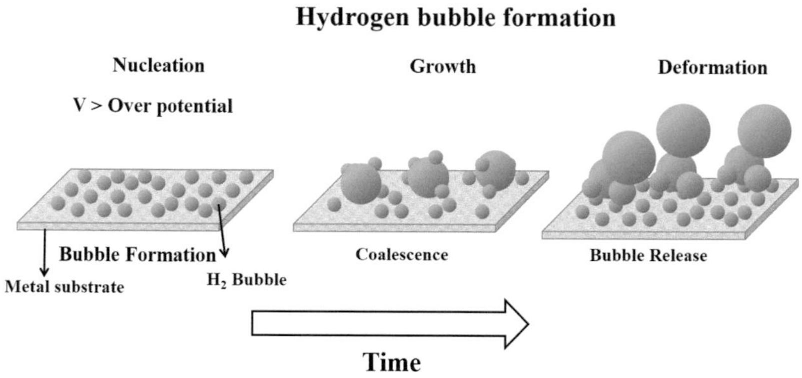

Hydrogen bubble formation

Fig. 4. Schematic representation of H_2 evaluation stages.

The H_2 gas generation over the cathodic surface is expressed as,

$$H^+(aq) + e^- \rightarrow H(ads) \tag{v}$$

$$H^+(aq) + H(ads) + e^- \rightarrow H_2 \tag{vi}$$

$$2H^+(aq) + 2e^- \rightarrow H_2 \tag{vii}$$

$$2H(ads) \rightarrow H_2 \tag{viii}$$

$$2NH_4^+ + 2e^- \rightarrow H_2 + 2NH_3 \tag{ix}$$

The generated H_2 bubbles escapes from the cathodic surface through electrolyte solution by diffusion and hold at electrode/electrolyte under super-saturation condition. At super-saturation condition, H_2 bubble starts to grow over the cathode surface by combining nearby bubbles which reduces its surface energy. The size of bubble keeps on increasing until the surface energy minimises. The accumulation of H_2 bubbles reduces the effective surface area for metal ion reduction (reaction (x)) by surface blocking effect (Ferna´ndez et al. 2014, Van Damme et al. 2010). Further, the size of the pores increases from cathodic surface to outward direction, which is schematically represented in Fig. 5. The metal deposition at bubbles interface creates the ramified structure with various sizes.

$$Ni^{2+}(aq) + 2e^- \rightarrow Ni(s) \qquad \text{(x)}$$

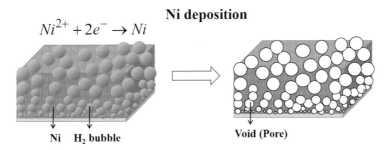

Fig. 5. Schematic diagram of Ni deposition on H_2 template metal surface.

The surface morphology of the fabricated porous film was analysed through field-effect scanning electron microscopy (FE-SEM). The results indicated the formation of uniform sized pores. The higher magnified images also indicated the formation of dendritic structured pore walls, as shown in Fig. 6. These dendritic structures comprised of numerous interconnected and interspaced Ni particles with various sizes of 200–800 nm. The pore size and thickness of the film was measured to be in the range of ~ 5–25 µm and ~ 12–14 µm, respectively.

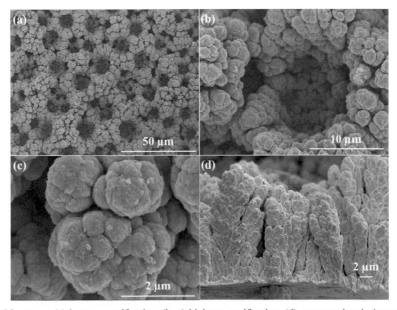

Fig. 6. FE-SEM image at, (a) lower magnification, (b, c) higher magnification, (d) cross sectional view of porous 3D-Ni network showing the formation of dendritic like structures.

2.2 Fabrication of Binder-Free, High Performance ZnCo₂O₄ Nanoflakes Decorated Porous, 3D Ni Electrode

The $ZnCo_2O_4$ nanoflakes were coated over 3D-Ni current collector by electrodeposition method. The electrodeposition was carried out in three electrode configurations with a potential of –1.2 to 0.2 V at a scan rate of 50 mV s⁻¹ for 2 cycles in an electrolyte solution containing 20 mM of zinc nitrate, 40 mM of cobalt nitrate and 100 mM sodium nitrate. After electrodeposition, the samples were rinsed with water and ethanol and then dried in a vacuum oven for 12 hr. Further, dried samples were annealed at 300°C for 2 hr at a heating rate of 1°C/min in ambient air.

The $ZnCo_2O_4$ nanoflakes were successfully deposited over porous 3D-Ni thin film current collector and Ni foam for comparison by simple electrochemical deposition. The electrodeposition was carried out in an electrolyte solution containing zinc, cobalt nitrate with potential sweep between –1.2 to 0.2 V. During the potential sweep, two reduction peaks were observed at 0.01 and –0.9 V. The first reduction peak was associated with the reduction of dissolved nitrate into hydroxyl species (OH^-) (Brownson and Lévy-Clément 2008, Therese and Kamath 2000),

$$NO_3^- + H_2O + 2e^- \rightarrow NO_2^- + 2OH^- \tag{xi}$$

The second anodic peaks at –0.9 V corresponds to the electrolysis of water and deposition of zinc, cobalt hydroxide on the working electrode. Finally, the prepared ZnCo-hydroxide coated current collector was annealed at 300°C for 2 h in ambient condition to convert into Zn-Co spinel. The overall reaction can be represented as follows,

$$xZn^{2+} + 2xCo^{2+} + 6x(OH) \rightarrow Zn_xCo_{2x}(OH)_{6x} \tag{xii}$$

$$Zn_xCo_{2x}(OH)_{6x} + \frac{1}{2}xO_2 \rightarrow xZnCo_2O_4 + 3xH_2O \tag{xiii}$$

The FE-SEM image of the $ZnCo_2O_4$ coated porous, 3D-Ni is shown in Fig. 7(a,b). After $ZnCo_2O_4$ deposition, there was no change observed in the porous and 3D structure, which confirms that the fabricated porous film is stable under electrodeposition and thermal treatment conditions. Further,

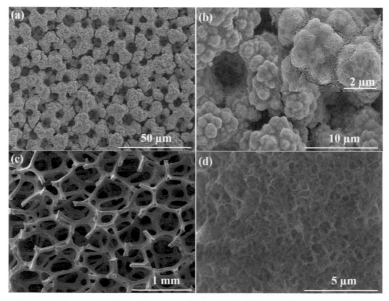

Fig. 7. FE-SEM image of, (a, b) $ZnCo_2O_4$ nanoflake coated porous 3D-Ni network at lower and higher magnifications; the right inset shows the higher magnified image (scale: 2 μm), (c, d) $ZnCo_2O_4$ nanoflake coated Ni foam (lower and higher magnifications).

FE-SEM image shows the uniform deposition of $ZnCo_2O_4$ nanoflakes over the 3D metal network. The higher magnified image (inset of Fig. 7b) indicates the deposition of wrinkled $ZnCo_2O_4$ nanoflakes with the size of few tens of nanometer. For comparison, we have deposited $ZnCo_2O_4$ nanoflakes over the commercial nickel foam using the same experimental condition. The surface morphology of the $ZnCo_2O_4$ coated nickel foam is shown in Fig. 7(c,d). The higher magnified image confirms the uniform deposition of $ZnCo_2O_4$ nanoflakes over the Ni foam.

X-ray diffraction (XRD) and EDX analysis of these results are displayed in Fig. 8. The EDX result (Fig. 8g) confirms the existence of Zn, Co, O and Ni elements in the sample. The XRD results clearly show three major peaks which are corresponding to 3D Ni metal. The diffraction peaks observed at (220), (311), (440), and (620) planes corresponds to the spinel cubic $ZnCo_2O_4$ (JCPDS card no. 23–1390) (Luo et al. 2012) and there is no impurity peaks related to ZnO and/or Co_2O_3.

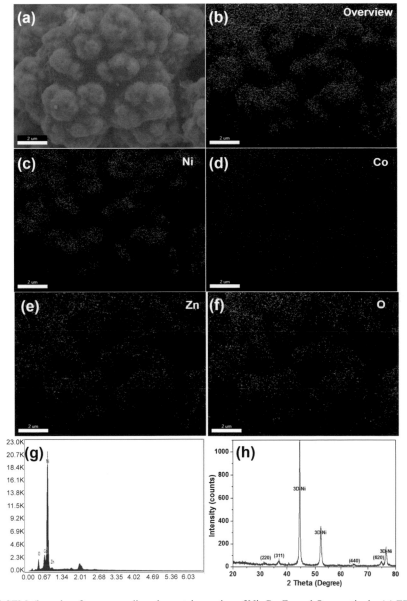

Fig. 8. (a) FE-SEM, (b, c, d, e, f) corresponding elemental mapping of Ni, Co, Zn and O respectively, (g) EDAX spectrum, and (h) X-ray diffraction pattern of $ZnCo_2O_4$ nanoflakes coated porous 3D-Ni film.

2.3 *Electrochemical Studies of ZnCo₂O₄ Nanoflakes Decorated Porous, 3D-Ni Electrode*

Cyclic voltammogram (CV), galvanostatic charge-discharge (GCD) and electrochemical impedance analysis of the fabricated $ZnCo_2O_4$ nanoflakes decorated porous, 3D structured electrode was evaluated in 1 M KOH electrolyte using three electrode set-up.

For comparison, the CV curve of $ZnCo_2O_4$ nanoflakes decorated 3D Ni film and Ni foam electrode is measured at a scan rate of 5 mV s⁻¹ in potential window of 0 to 0.7 V and the results are shown in Fig. 9a. The resultant CV curve of the both samples showed two distinct redox peaks at 0.567 V and 0.458 V respectively. The small reduction peak at 0.2 V corresponds to the conversion between $Co(OH)_2$ and CoOOH (Ott et al. 2011) and pair of oxidation and reduction peaks at 0.567 and 0.458 V correspond to the conversion between CoOOH and CoO_2 (Huang et al. 2014, Sun et al. 2013). However, Zn in the compound rarely participates in the electrochemical redox reaction, but it promotes electrochemical activities and improves the conductivity of the material (Xia et al. 2011). The overall reaction represented as,

$$Co(OH)_2 + OH \Leftrightarrow CoOOH + H_2O + e^-$$ (xiv)

$$CoOOH + OH \Leftrightarrow CoO_2 + H_2O + e^-$$ (xv)

Figure 9(b, c) presents the CV curve of $ZnCo_2O_4$ nanoflakes decorated porous 3D-Ni film and Ni foam in a potential range of 0 to 0.7 V at various scan rates (10–150 mV s⁻¹). The resultant CV curves of the both samples show pair of redox peaks and both cathodic and anodic peaks shifting towards positive and negative potential side with maximum peak shift of 36 mV (80 mV for Ni foam)

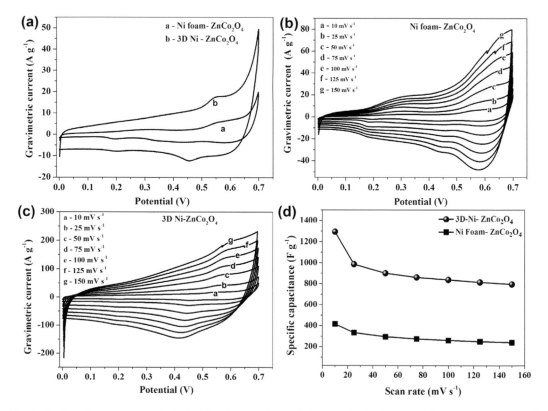

Fig. 9. Cyclic voltammogram (CV) of (a) $ZnCo_2O_4$ nanoflakes coated porous 3D-Ni film and Ni foam at scan rate of 5 mV s⁻¹ in 1 M KOH. (b) CV of $ZnCo_2O_4$ nanoflakes coated Ni foam at different scan rates. (c) CV of $ZnCo_2O_4$ nanoflakes coated porous 3D-Ni thin film current collector at different scan rates. (d) Specific capacitance of $ZnCo_2O_4$ nanoflakes coated porous 3D-Ni thin film current collector and Ni-foam at various scan rates.

at a higher scan rate of 150 mV s⁻¹. The minimum redox peak shift confirms the low polarisation and higher conductive nature of the porous 3D electrode during faradic reaction, which facilitates the higher rate capability. The specific capacitance of the porous 3D-Ni and Ni foam decorated $ZnCo_2O_4$ is calculated from the CV curves at different scan rates from 10 to 150 mV s⁻¹ using Eq. (2) (Chen et al. 2013),

$$C_{sp} = \int IdV \Big/ mvV \qquad (2)$$

where C_{sp} is a specific capacitance (F g⁻¹), I is response current (A), ΔV is working potential window (V), v is scan rate (mV s⁻¹) and m is the mass of the loaded electroactive material (g). The average mass loading of the electroactive materials is 0.1 and 0.3 mg cm⁻² for porous 3D-Ni and Ni-foam current collector respectively. The average specific capacitance of porous 3D-Ni and Ni-foam decorated $ZnCo_2O_4$ at various scan rates is shown in Fig. 9d. The porous 3D-Ni current collector sample shows an excellent specific capacitance of 1295 F g⁻¹ compare to Ni foam current collector (416.3 F g⁻¹) at scan rate of 10 mV s⁻¹. Further, the specific capacitance is gradually decreased with increasing scan rates and even at higher scan rate (150 mV s⁻¹) 3D-Ni current collector retains the specific capacitance of 795.1 F g⁻¹ with retention ratio of 61 per cent, which is higher than Ni foam current collector (Csp: 240 F g⁻¹).

The electrochemical capacitive properties of the fabricated 3D-Ni and Ni foam current collector based electrodes were tested using galvanostatic charge-discharge study in a potential range of 0–0.6 V with different current densities (Fig. 10). The specific capacitance of the electrode was calculated by using Eq. (3) from galvanostatic charge-discharge curve (GCD).

$$C_{sp} = I\Delta t \Big/ m \Delta V \qquad (3)$$

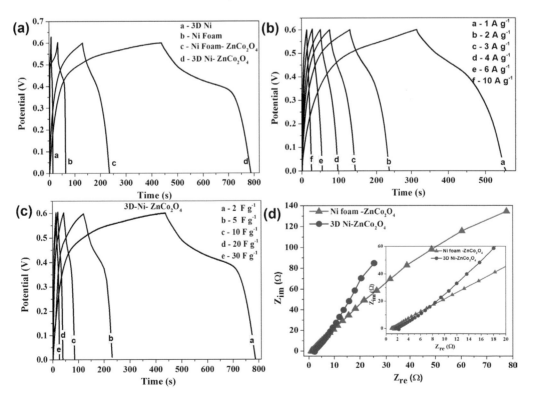

Fig. 10. Galvanostatic charge-discharge curves of (a) 3D-Ni, Ni foam, $ZnCo_2O_4$ nanoflakes coated porous 3D-Ni thin film current collector and Ni-foam at current density of 2 A g⁻¹, (b) $ZnCo_2O_4$ nanoflakes coated Ni-foam at different current densities, (c) $ZnCo_2O_4$ nanoflakes coated porous 3D-Ni thin film current collector at different current densities. (d) Nyquist plots of $ZnCo_2O_4$ nanoflakes coated porous 3D-Ni thin film current collector and Ni-foam collector in 1 M KOH.

where, C_{sp} is a specific capacitance (F g^{-1}), I, ΔV and Δt are discharge current (s), operating potential window (V) and discharge time (s), respectively, and m is mass of electroactive material (g). Figure 10a shows a typical GCD curve of the ZnCo$_2$O$_4$ nanoflakes decorated porous 3D-Ni, Ni foam and bare current collector at a current density of 2 A g^{-1}. The resultant GCD curve shows a pair of redox plateau in charge-discharging curve confirming the faradic reaction. A porous 3D-Ni decorated ZnCo$_2$CO$_4$ nanoflakes electrode shows a higher discharge time compare to other electrodes at the same current density, confirming the higher capacitance of the fabricated electrode. The charge-discharge curve of ZnCo$_2$O$_4$ nanoflakes decorated porous 3D-Ni and Ni foam electrode at various current density is shown in Fig. 10(b, c). The specific capacitance of the both electrodes is calculated at different current densities using Eq. (3) and shown in Fig. 11a. The fabricated porous 3D-Ni decorated ZnCo$_2$O$_4$ nanoflakes show a remarkably higher specific capacitance of 1170 F g^{-1} (at 2 A g^{-1}) compare to the Ni foam electrode (366.7 F g^{-1} at 2 A g^{-1}). The fabricated porous 3D-Ni based electrode retains the capacitance of 600 F g^{-1} at higher current density of 30 A g^{-1}, demonstrating the higher stability of electrode even at higher current density. The specific capacitance of the porous 3D-Ni based electrode is comparable to the previously reported values, which are tabulated in Table 1.

The higher specific capacitance of porous 3D-Ni based electrodes could be due to higher surface area, porous nature and shorter diffusion and conduction path for electrons and ions, which favors the higher ionic movement between electrode and electrolyte interface leading to faster redox kinetics. To understand better ion diffusion and electron transfer kinetics, we have measured electrochemical

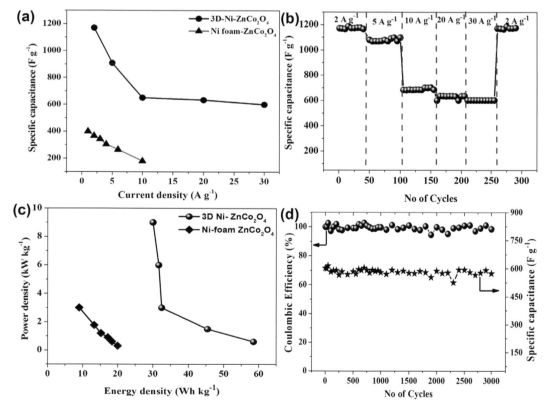

Fig. 11. (a) Specific capacitance of ZnCo$_2$O$_4$ nanoflakes coated porous 3D-Ni film and Ni foam at various current densities. (b) Rate capability of ZnCo$_2$O$_4$ nanoflakes coated porous 3D-Ni thin film current collector at different current densities. (c) Ragone plot of ZnCo$_2$O$_4$ nanoflakes decorated porous 3D-Ni and Ni foam. (d) The performance of cyclic stability and coulombic efficiency of ZnCo$_2$O$_4$ nanoflakes coated porous 3D-Ni thin film current collector at current density of 20 A g^{-1}.

Table 1. The reported specific capacitance of $ZnCo_2O_4$ at different morphology.

S. No.	$ZnCo_2O_4$ Morphology	Current Density	Specific Capacitance	Reference
1	Microsphere	$2 A g^{-1}$	$853.6 F g^{-1}$	Chen et al. 2015
2	Nanosheets	$2 A g^{-1}$	$1000.7 F g^{-1}$	Chen et al. 2015
3	Porous microspheres	$1 A g^{-1}$	$647.1 F g^{-1}$	Deng et al. 2013
4	Flower like microspheres	$1 A g^{-1}$	$689.4 F g^{-1}$	Jiang et al. 2015
5	Rod	$1 A g^{-1}$	$604.52 F g^{-1}$	Wang et al. 2015a/b
6	Mesoporous nanoflakes	$2 A g^{-1}$	$1220 F g^{-1}$	Huang et al. 2014
7	Nanorods	$2 A g^{-1}$	$1220 F g^{-1}$	Zhang et al. 2011
8	Porous nanoparticles	$1 A g^{-1}$	$457 F g^{-1}$	Liu et al. 2013
9	Porous nanotubes	$10 A g^{-1}$	$770 F g^{-1}$	Wang et al. 2014
10	Nanoneedle arrays @ porous carbon nanofiber	Scan rate: $2 mV s^{-1}$	$1384 F g^{-1}$	Chen et al. 2013

impedance (EIS) spectrum and the results are shown in Fig. 10d. In Nyquist plot, no distinct semicircle is observed at higher frequency region, confirming the higher ionic and electronic conductivity during redox reaction (Chen et al. 2013). In the lower frequency region, a straight line indicates the Warburg impedance which implies the ion diffusion resistance between the electrolyte and the electroactive material surface. The porous 3D-Ni based electrode shows more parallel behavior towards the imaginary Z axis, which is an indication of higher electrolyte ions diffusion on electrode surfaces. The higher diffusion of electrolyte ions in the porous 3D-Ni based electrode is due to the presence of micron size pores and interspace between the particles leading to faster ion movement between electrolyte and electrode.

The rate capability of the electrode was measured at different current densities (2, 5, 10, 20, and 30 A g^{-1}) from 50 continuous GCD cycles. Figure 11b shows the rate capability of the porous 3D-Ni based electrodes. The electrochemical stability of porous 3D-Ni decorated $ZnCo_2O_4$ nanoflakes is measured by multiple galvanostatic charge-discharge method at a current density of 20 A g^{-1} in stable potential windows from 0 to 0.6 V, as shown in Fig. 11d. The reversibility of the electrode is checked through Coulombic efficiency, which is calculated according to Eq. (4) (Guan et al. 2014),

$$\eta = (t_d/t_c) \times 100\% \tag{4}$$

where η is coulombic efficiency, t_d and t_c are the discharging and charging time (s) respectively. At a higher current density of 20 A g^{-1}, the porous 3D-Ni decorated $ZnCo_2O_4$ nanoflake electrode retains the specific capacitance of 574 F g^{-1} with retention ratio of 95 per cent after 3000 charge-discharge cycles indicating higher cyclic stability of the electrode. The higher coulombic efficiency (98 per cent) of the electrode is clear sign of higher reversibility of the fabricated electrode. The calculated power and energy densities are represented in the Ragone plot, as shown in Fig. 11c. The porous 3D-Ni decorated $ZnCo_2O_4$ nanoflakes electrode delivered a higher energy density of 58.5 Wh Kg^{-1} and a maximum power density of 9 kW Kg^{-1}, which is higher than Ni foam based electrode (energy density: 20 Wh Kg^{-1} at power density of 3 kW Kg^{-1}).

The higher specific capacitance, excellent rate capability and good cyclic stability of the porous 3D-Ni decorated $ZnCo_2O_4$ nanoflakes electrodes are due to the presence of micron sized pore and the nanospace between interconnected and interspaced Ni particles in 3D electrodes, providing the excellent path for electrolyte as well as enhancing the electroactive surfaces. Finally, the interconnected Ni nanostructure provides the excellent electrical conductivity which may reduce the contact resistance of the electrode and the current collector.

3. Fabrication of NiCo$_2$O$_4$ Nanoneedles Decorated Ni Seed Layer @ Ni Foam Electrodes

Various strategies have been employed to improve the performance of the supercapacitors by developing high capacitive electrodes with new architectures. NiCo$_2$O$_4$ appeals huge interest towards the energy storage application, because of multiple oxidative states and higher electrical conductivity. Here, NiCo$_2$O$_4$ nanoneedles were hierarchically grown on Ni seed layer deposited on 3D Ni foam as an electrode material. The introductions of Ni seed layer play a major role, for instance improving the adhesion between NiCo$_2$O$_4$ and Ni foam thereby reducing the contact resistance and improving the active surface area for NiCo$_2$O$_4$ growth and electrochemical reactions.

3.1 Fabrication of Binder Free NiCo$_2$O$_4$ Nanoneedles/Ni Seed Layer @ Ni Foam

Ni seed layer is deposited on commercial nickel foam (NF) by a dynamic hydrogen bubble template cathodic deposition (Li et al. 2007). Here, Ni seed layer is deposited at various time such as 10, 20 and 30s, which is denoted as Ni10@NF, Ni20@NF and Ni30@NF, respectively. Further, NiCo$_2$O$_4$ nanoneedles are grown on Ni seed layer deposited NF (Ni@NF) by a simple hydrothermal method at 120°C for 12 h in the solution mixture of nickel nitrate (0.582 g), cobalt nitrate (1.164 g) and urea (1.44 g) in 70 mL ethanol. Finally, the as grown samples are thermally annealed at high temperature of 300°C for 3 hr at a heating rate of 1°C/min. Here after, the NiCo$_2$O$_4$ grown on various Ni foams such as, bare NF, Ni10@NF, Ni20@NF and Ni30@NF are named as NiCo$_2$O$_4$@NF, NiCo$_2$O$_4$/Ni10@ NF, NiCo$_2$O$_4$/Ni20@NF and NiCo$_2$O$_4$/Ni30@NF respectively. The detailed preparation method is schematically presented in Fig. 12.

The surface morphology of the fabricated electrode is analysed through FE-SEM. Figure 13 shows the image of bare Ni foam and Ni seed layer deposited on Ni foam. Ni seed layer deposited foam shows a rough surface with uniformly distributed Ni particles, but bare Ni foam shows a smooth surface. The Ni seed layer plays an important role in NiCo$_2$O$_4$ growth such as it increases the surface area and acts as a nucleating site for further metal nanostructure growth. The surface morphology of the NiCo$_2$O$_4$ grown on different surface is analysed and shown in Fig. 14. The result clearly indicates

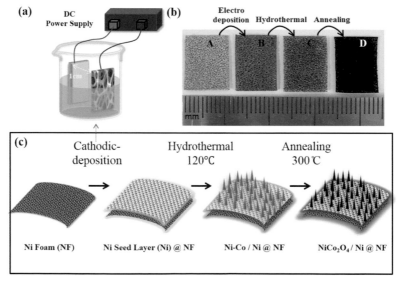

Fig. 12. (a) Schematic diagram of the experimental setup used for Ni electrodeposition. (b) Digital photographs of (A) bare Ni foam, (B) Ni seed layer electrodeposited Ni foam (Ni@NF) (black), (C) NiCo$_2$O$_4$ deposited by hydrothermal on Ni seed layer@Ni foam (NiCo$_2$O$_4$/Ni@NF) (pink) and (D) annealed NiCo$_2$O$_4$ deposited Ni seed layer@Ni foam (black). (c) Schematic illustration of the preparation of NiCo$_2$O$_4$/Ni seed layer@NF (NiCo$_2$O$_4$/Ni@NF).

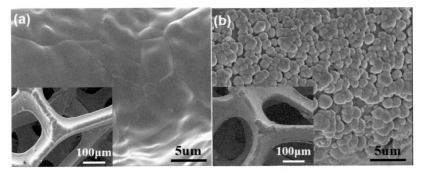

Fig. 13. FE-SEM image of (a) bare Ni foam, (b) Ni seed layer deposited Ni foam. Insets: low magnification image.

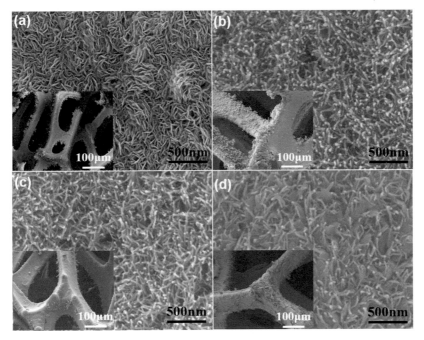

Fig. 14. FE-SEM images of $NiCo_2O_4$ nanostructure on (a) bare Ni foam, (b) Ni10@Ni foam, (c) Ni20@ Ni foam, (d) Ni30@ Ni foam. Insets: low magnification (100 μm).

that the $NiCo_2O_4$ grown on bare smooth Ni foam shows nanoplates-like structure, but Ni seed layer allows the growth of a needle-like structure due to availability of large number of nucleating (seed) sites. From this result, it is evident that the seed layer plays an important role in the growth of $NiCo_2O_4$ nanostructure. The density of the $NiCo_2O_4$ nanoneedles increases with increasing deposition time from 10 to 30 s and afterwards it starts to interconnect with nearest nanoneedles to form a mixed morphology due to higher density of nucleation sites, allowing the growth of denser nanoneedles.

Further, the $NiCo_2O_4$ nanoneedles were vertically grown to the substrate with the length of fewer nm and the diameter of ~ 25 nm at bottom and ~ 15 nm at top. The vertically grown homogeneous $NiCo_2O_4$ nanoneedles could be potential for high performance supercapacitor applications, due to higher electroactive surface area, good electrical conductivity of 1D nanostructure, shorter path length ions/electrons and nanospace between nanoneedles for easily accessible electrolyte ions over the nanoneedles surfaces. Moreover, Ni seed layer provides good interfacial contact between $NiCo_2O_4$ and Ni foam, which makes lower interfacial resistance allowing faster reversible redox reaction.

EDX spectrum (Fig. 15a) shows the presence of elemental Ni, Co, O and no impurities in the sample. The atomic percentage of O, Co and Ni was found to be 61.31 per cent, 22.83 per cent, 15.86 per cent respectively, confirming the $NiCo_2O_4$ stoichiometry. Figure 15b shows the XRD patterns of bare NF, $NiCo_2O_4$/NF ($NiCo_2O_4$@NF, without Ni seed layer) and $NiCo_2O_4$/Ni seed-layer @NF ($NiCo_2O_4$/Ni@NF), respectively. From the XRD patterns, three major diffraction peaks at 44°, 51°, and 76.3° corresponding to (111), (200) and (220) plane of Ni foam was observed. Diffraction peaks at 31.1°, 36.6°, 55.3°, 58.7° and 64.8° corresponding to (220), (311), (422), (511) and (440) planes of spinel $NiCo_2O_4$ structure (JCPDS no. 20–0781) was also noted (Yin et al. 2016). No secondary phases were observed in the pattern confirming the higher chemical purity and homogeneous crystal structure.

TEM images showed, the presence of needle like structure of interconnected $NiCo_2O_4$ nanoparticles with mesopores (Fig. 16(a,b)). From HR-TEM image (Fig. 16c), the interplanar distance of (220) plane is measured and it is ca. 0.288 nm, corresponds to the spinel $NiCo_2O_4$ (Wu et

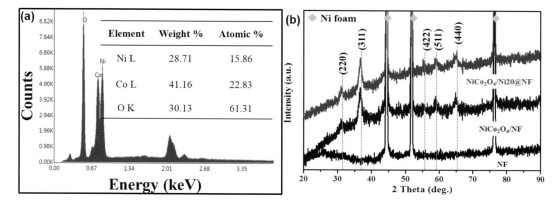

Fig. 15. (a) EDX spectra of $NiCo_2O_4$ nanoneedles. (b) XRD pattern of $NiCo_2O_4$ nanoneedles and nanoplates.

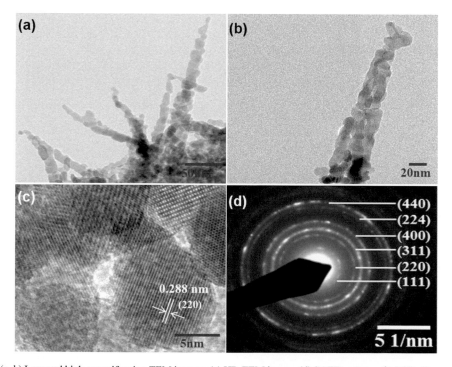

Fig. 16. (a, b) Low and high-magnification TEM images, (c) HR-TEM image, (d) SAED pattern of a $NiCo_2O_4$ nanoneedles.

al. 2015). The corresponding selected area of electron diffraction (SAED) pattern (Fig. 16d) showed well-defined diffraction rings, which corresponds to the plane of spinel $NiCo_2O_4$ crystal structure (JCPDS no. 20-0781) (Wu et al. 2015, Zhang et al. 2014).

3.2 Electrochemical Characterisation of $NiCo_2O_4$ Nanoneedles/Ni Seed Layer @ Ni Foam

In order to demonstrate the electrochemical performance of the fabricated electrode, cyclic voltagram (CV), galvanostatic charge-discharge (GCD) and electrochemical impedance spectrum in 1M KOH electrolyte at scan rate of 50 mV s^{-1} was performed. The resultant CV curves of all the samples (Fig. 17a) showed redox peaks at ~ 0.42 and ~ 0.24 V corresponding to the reversible faradic reaction of M−O/M−O−OH (M: Co, Ni ions) and OH$^-$ (reversible reactions of Ni^{2+}/Ni^{3+} and Co^{3+}/Co^{4+} with anions OH$^-$) (Zhang and Lou 2013). The whole redox reaction in alkaline electrolyte is expressed as (Zhu et al. 2014a/b),

$$NiCo_2O_4 + OH^- + H_2O \leftrightarrow NiOOH + 2CoOOH + e^- \tag{5}$$

$$CoOOH + OH^- \leftrightarrow CoO_2 + H_2O + e^- \tag{6}$$

The maximum current density and higher integrated CV area are observed for $NiCo_2O_4$/Ni20@NF samples when compared to other samples, because of improved interface, reduced contact resistance (Yoon et al. 2004) and higher active surface area.

Further, the effect of Ni seed layer with various deposition times (10, 20 and 30 s) were studied through galvanostatic charge-discharge measurements at a current density of 1 A g^{-1} and the results are shown in Fig. 17b. The measured charge–discharge curves clearly showed the discharge plateau, which is due to the faradic reaction. Moreover, the higher discharge time is observed for the $NiCo_2O_4$/

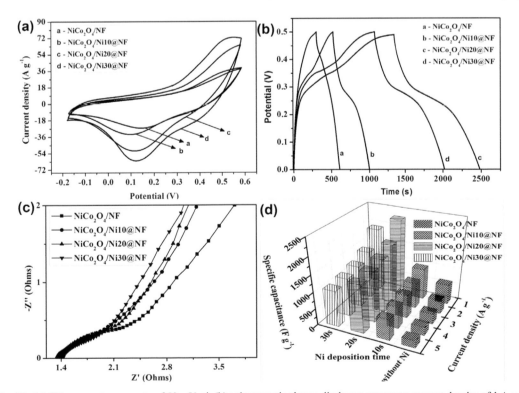

Fig. 17. (a) CV curves at a scan rate of 50 mV s^{-1}, (b) galvanostatic charge-discharge curves at a current density of 1 A g^{-1}, (c) electrochemical impedance spectrum, (d) specific capacitance at various current densities of $NiCo_2O_4$/Ni@NF electrodes with different seed layers (0, 10, 20, and 30s).

Ni20@NF, and then reduced again at longer deposition time (30 s) of Ni seed layer, which may be due to low accessible electroactive sites and therefore increased series resistance caused by loosely deposited Ni seed layer. Higher specific capacitance of around 2284 F g^{-1} was measured for NiCo$_2$O$_4$/Ni20@NF electrode at a current density of 1 A g^{-1}, which was higher than NiCo$_2$O$_4$/NF (610 F g^{-1}) and NiCo$_2$O$_4$/Ni10@NF (952 F g^{-1}), NiCo$_2$O$_4$/Ni30@NF (1898 F g^{-1}). The calculated specific capacitance of the NiCo$_2$O$_4$/Ni20@NF are 2284, 1788, 1608, 1504 and 1440 F g^{-1} at a current density of 1, 2, 3, 4 and 5 A g^{-1} respectively and these values are higher than those of NiCo$_2$O$_4$@NF (610, 412, 318, 256 and 220 F g^{-1}), NiCo$_2$O$_4$/Ni30@NF (1898, 1436, 1248, 1152 and 1080 F g^{-1}), NiCo$_2$O$_4$/Ni10@NF (952, 784, 702, 664 and 630 F g^{-1}) at the same current density (Fig. 17d). The higher electrochemical performance of NiCo$_2$O$_4$/Ni20@NF may be due to the enhanced contact between active materials and electrolytes, shorter diffusion path length for electrons and ions, which results in lower contact resistance, faster ions and electron transfer during the redox reaction and finally, large number of accessible electroactive sites.

The electron transfer kinetics of the NiCo$_2$O$_4$/Ni@NF sample were studied by the electrochemical impedance spectroscopy (EIS). The Nyquist plots of the NiCo$_2$O$_4$/Ni@NF electrode is shown in Fig. 17c. The serial resistance (R$_s$) of the samples decreases with the introduction of Ni seed layer and further decreases by increasing Ni deposition time (20 s). But longer deposition time (30 s) slightly increases the R$_s$, which may due to loosely connected Ni seed layer over the surface. The solution resistance values are 1.37, 1.3, 1.4, and 1.42 Ω for the NiCo$_2$O$_4$/Ni@NF samples with seed layer of different deposition times 20, 30, 10 and 0 s, respectively. All the samples showed small semi-circle and the vertical line at the low frequency region, which moves towards the imaginary axis after Ni seed layer deposition, indicating the higher capacitive behaviour.

3.3 *Electrochemical Characterisation of NiCo$_2$O$_4$//NiCo$_2$O$_4$ Symmetric Supercapacitor Device*

To test the feasibility of the fabricated NiCo$_2$O$_4$ decorated NF electrodes for real-world applications; we have assembled a simple symmetric device by sandwiching a two NiCo$_2$O$_4$/Ni@NF with filter paper as a separator. The electrochemical performance of the NiCo$_2$O$_4$//NiCo$_2$O$_4$ symmetric supercapacitor are tested in liquid electrolyte (1 M KOH). Figure 18a shows the CV curves of symmetric device at different working potential from 0.6 to 1.1 V at a scan rate of 50 mV s^{-1}. The resultant CV curves indicates a stable operating potential at 1.0 V. CV at different scan rates from 5 to 100 mV s^{-1} confirms the retention of rectangular shape in all scan rates and increased redox current of the device with increasing the scan rates (Fig. 18b).

The galvanostatic charge–discharge measurement is carried out at different current densities from 0.5 to 2.5 A g^{-1}. The calculated specific cell capacitance of the symmetric cell is about 131.5 F g^{-1} at a current density of 0.5 A g^{-1} (Fig. 18c). The cell capacitance decreases with increasing current density due to low ionic diffusion. The electrochemical impedance analysis was also measured in a frequency range from 100 kHz to 0.1 Hz and shown in the inset of Fig. 18d.

The equivalent series resistance (ESR) is deduced from EIS spectrum, which is 0.49 Ω. The lower ESR value of the device is due to the 3D nature of the Ni foam and additionally deposited Ni seed layer, which facilitates easy access of electrolyte ions to electroactive surface and shorter diffusion path length for ions and electron. The charge-discharge studies showed stability of 2000 cycles (with 92 per cent capacitance retention) at a current density of 2.5 A g^{-1} (Fig. 18d). The calculated maximum cell energy and power density of the symmetric supercapacitor was found to be 18.26 Wh kg^{-1} (at a power density of 250 W kg^{-1}) and 1250 W kg^{-1} (at energy density of 10.42 Wh kg^{-1}), respectively. The calculated full cell specific capacitance is comparable to the previously reported values of RuO$_2$//RuO$_2$ symmetric capacitor (52.66 F g^{-1} at a current density of 0.625 A g^{-1}) (Xia et al. 2012) and activated carbon//NaMnO$_2$ asymmetric device (38.9 F g^{-1}) (Qu et al. 2009).

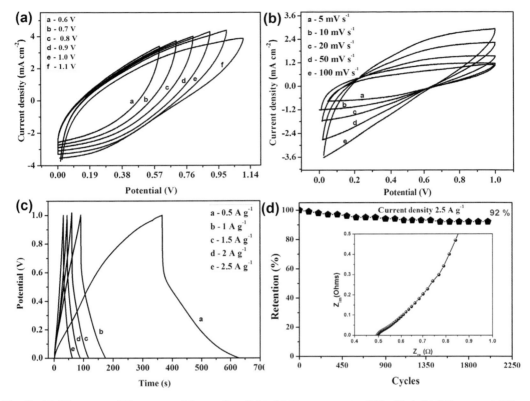

Fig. 18. (a) CV curves at different potential range from 0.6 to 1.1 V at a scan rate of 50 mV s^{-1}, (b) CV curves at different scan rates, (c) Galvanostatic charge-discharge curves at different current densities, (d) cyclic stability at a current density of 2.5 A g^{-1} (inset: electrochemical impedance spectrum) of the symmetric supercapacitor.

4. Fabrication of Mesoporous 3D NiCo$_2$O$_4$/MWCNT Nanocomposite and Carbon Aerogels

It is well known that the mesoporous materials (especially metal/metal oxide and their composite network morphologies) have been extensively utilised for various potential applications due to their high surface area and good mechanical properties (Yuan et al. 2013, Padmanathan and Selladurai 2014). Aerogels are one of the efficient materials to generate a porous solid that have exceptionally low density (0.003 g cm^{-3}) and large surface area (up to 1600 m^2 g^{-1}), making them useful in a wide range of applications from energy to bio-applications (Xu et al. 2010). Aerogels are derived from wet gel in which liquid components are replaced with air. Basically, aerogels are prepared by two step process. In first step, wet gel is prepared by traditional sol-gel method and in second step is dried at critical drying method. Based on the drying method it is classified as xerogel, aerogel and cryogel. The flow chart of general preparation (Pierre and Pajonk 2002) is given in Fig. 19.

Further, the drying process (CO$_2$ drying, ambient pressure drying, freeze-drying, microwave drying and vacuum drying) is a key role to achieve good porous aerogel structure with improved electrochemical performance. Among the various drying methods, CO$_2$ drying attracted greater attention due to their unique advantages, such as simple process, mechanically durable porous network structures and low processing temperature, etc.

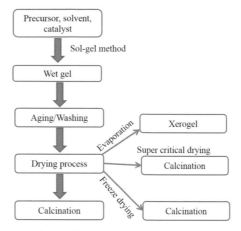

Fig. 19. General flow chart for aerogel synthesis process.

4.1 Fabrication of Mesoporous 3D NiCo$_2$O$_4$/MWCNT Nanocomposite Aerogels

The MWCNT was purified according to the reported procedure (Sivabalan et al. 2016). Briefly, 0.2 wt per cent of the purified MWCNT was dispersed in 5 mL of DMF by sonication for 15 min. The NiCl$_2$.6H$_2$O (3 wt per cent) and CoCl$_2$.6H$_2$O (3 wt per cent) was also dissolved in DMF (5 mL) in separate vial. These two solutions were mixed together followed by sonication for 60 min to get a homogenous solution. Then appropriate amount of epichlorohydrin is added into the homogenous solution and allowed for the gelation for 15 min. After gelation period, the solution turned into wet gel (bluish green). Subsequently, it is rinsed with fresh DMF and acetone for several times. Further, the washed wet gel is dried in supercritical CO$_2$ drying process for 8 hrs at 14°C under pressure of 1000 psi for evacuating the solvents to from the aerogels. Further, we have prepared different NiCo$_2$O$_4$-MWCNT nanocomposite aerogels by varying the MWCNT concentration (0.2, 0.7, 1.4, 2.1 wt per cent) under identical conditions.

The surface morphology of fabricated nanocomposite aerogel was analysed through FE-SEM and the results are displayed in Fig. 20. The prepared NiCo$_2$O$_4$ aerogels were highly porous in nature with average NiCo$_2$O$_4$ nanoparticle size of ~ 30 ± 5 nm. In case of NiCo$_2$O$_4$/2.1 wt per cent-MWCNT nanocomposite aerogel, it seems that the NiCo$_2$O$_4$ nanoparticles are randomly distribute over the aerogel and nanoparticles are interconnected through MWCNT to form a network-like structure, which is most important for reducing the aggregation of metal particles itself as well as enhancing the electrochemical performance of the materials. The surface area for the NiCo$_2$O$_4$ aerogel and NiCo$_2$O$_4$/2.1 wt per cent-MWCNT nanocomposite aerogel are listed in the inset of Fig. 21a. The result

Fig. 20. FE-SEM images of (a) NiCo$_2$O$_4$ aerogel, (b) NiCo$_2$O$_4$/MWCNT nanocomposite aerogel.

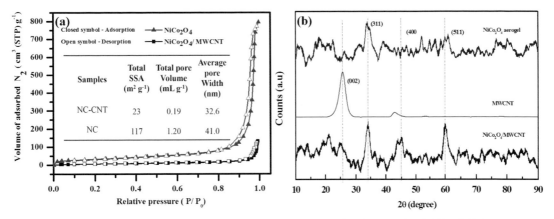

Fig. 21. (a) N_2-adsorption study and (b) XRD patterns of pure MWCNT, $NiCo_2O_4$ aerogel and $NiCo_2O_4$/2.1 wt per cent-MWCNT aerogel.

shows that the surface area of the $NiCo_2O_4$ aerogel is much higher when compared to $NiCo_2O_4$/2.1 wt per cent-MWCNT nanocomposite aerogel. The possible reason is due to the mesoporous 3D network structures formed by connecting the metal nanoparticles with MWCNT in the nanocomposite aerogels. Figure 21b shows the XRD patterns of pristine MWCNT, $NiCo_2O_4$ aerogel and $NiCo_2O_4$/2.1 wt per cent-MWCNT nanocomposite aerogels. The weak three peaks at $2\theta = 34°, 43°, 60°$, corresponding to (311), (400) and (511) planes, are observed, which are in good agreement with cubic crystal structure of spinel $NiCo_2O_4$ (JCPDS: 2-1074).

4.2 Electrochemical Characterisation of Mesoporous 3D NiCo₂O₄/MWCNT Nanocomposite Aerogel Electrode

The electrochemical performance of the fabricated electrode was tested in 2 M KOH electrolyte and the results are shown in Fig. 22. From CV results showed redox peaks clearly indicating the faradic reaction of electrode materials. The redox current and area under the curve was seen to be increased with the increasing MWCNT amount.

Figure 22(b) shows the GCD curves of aerogel electrodes at a current density of 1 A g^{-1}. The GCD curve shows a plateau arising due to faradic reaction of $NiCo_2O_4$ corroborating with the CV results. $NiCo_2O_4$/2.1 wt per cent-MWCNT nanocomposite aerogel electrodes showed higher discharge time. The calculated specific capacitance for the $NiCo_2O_4$/MWCNT aerogels with 0.2, 0.7, 1.4 and 2.1 wt per cent of MWNCT contents were found to be 344, 548, 628 and 1010 F/g, respectively. These capacitance values are comparable to the reported literatures, for instance $NiCo_2O_4$/MWNCT hybrids (678 F g^{-1}) and NiO (670 F g^{-1}), Co_3O_4 (705 F g^{-1}) composites (Zhang et al. 2013, Deng et al. 2015, Zhu et al. 2014a/b). The calculated specific capacitances of the various electrodes at different current densities are shown in Fig. 22c.

The $NiCo_2O_4$ with 2.1 wt per cent of MWCNT nanocomposite aerogel electrodes exhibited higher specific capacitance compare to pure $NiCo_2O_4$ and other $NiCo_2O_4$/MWCNT nanocomposite aerogels. The improved specific capacitance is mainly attributed due to the unique mesoporous structure of nanocomposite aerogel with a highly conductive MWCNT network. Even at high current density (5 A g^{-1}), the $NiCo_2O_4$/2.1 wt per cent-MWCNT electrode retains 30.7 per cent of its specific capacitance (i.e., from 1010 to 310 F g^{-1}), as shown in Fig. 22c. The result indicates that the $NiCo_2O_4$/MWCNT nanocomposite aerogel has an excellent rate capability, which is important to achieve high power and energy densities. The results showed that the $NiCo_2O_4$/2.1 wt per cent-MWCNT nanocomposite aerogel electrode retains 77 per cent of its initial specific capacitance at 1 A g^{-1} even after 1,000 cycles (Fig. 22d).

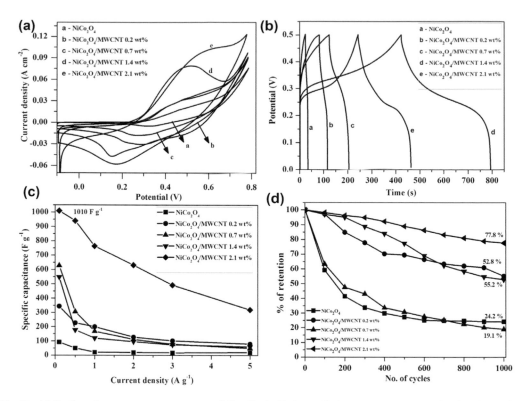

Fig. 22. (a) Cyclic voltammograms at a scan rate of 50 mV s^{-1}, (b) charge-discharge curves at current density of 1 A g^{-1}, (c) specific capacitance at various current densities, (d) cycling stability of NiCo$_2$O$_4$/MWCNT aerogels with different MWCNT contents.

Conclusion

In summary, this chapter described about the fabrication of porous, three dimensional (3D) electrode materials for supercapacitor applications. Here, two different strategies are followed to improve the energy density of supercapacitor device by designing a porous, 3D current collector and another way is to synthesis of mesoporous, 3D composite aerogel as an electrode materials. At first, this chapter described about the fabrication method and mechanism of a hierarchical structured porous film current collector through dynamic hydrogen bubble template cathodic deposition method. Further, electroactive materials decorated over the porous current collector through electrodeposition and hydrothermal method. Finally, this chapter discussed about the fabrication of highly porous 3D interconnected network aerogel based materials for supercapacitor applications.

First, a porous 3D-Ni current collector was fabricated on stainless steel using dynamic hydrogen bubble template method. Secondly, the electroactive material such as ZnCo$_2$O$_4$ nanoflakes was decorated over the porous, 3D-Ni current collector by simple low cost electrodeposition method. The fabricated porous, 3D electrode showed a higher specific capacitance of 1170 F g^{-1} current density of 2 A.g^{-1} with excellent rate capability and good cycle stability of 95 per cent after 3,000 cycles at a higher current density of 20 A g^{-1}. The high performance of the porous 3D-Ni based electrode may be due to the availability of micron size pore which facilitates a higher electrolyte and electrode interfaces, the interspace between Ni particles, porous Ni provides the higher surface active area for electrochemical redox reaction as well as to deposit thin electroactive materials and interconnected ramified Ni particles reduces the internal resistance which allows higher rate capability, thin layer of electroactive material ZnCo$_2$O$_4$ reduces the diffusion path for ion and electron.

Second part described about the fabrication of hierarchical $NiCo_2O_4$ nanoneedles on Ni seed layer deposited nickel foam by simple hydrothermal method and followed by thermal annealing. The fabricated hierarchal electrode delivered a higher specific capacitance of 2284 F g^{-1} at a current density of 1 A g^{-1} for $NiCo_2O_4$/Ni20@NF electrode than other electrodes $NiCo_2O_4$@NF (610 F g^{-1}), $NiCo_2O_4$/Ni10@NF (952 F g^{-1}) and $NiCo_2O_4$/Ni30@NF (1898 F g^{-1}).

Finally, symmetric supercapacitor device was fabricated using $NiCo_2O_4$/Ni@NF. The fabricated symmetric supercapacitor delivered a maximum specific cell capacitance of 131.5 F g^{-1} at a current density of 0.5 A g^{-1} with retention of 92 per cent in liquid electrolyte. The fabricated symmetric device showed a maximum cell energy and power density of around 18.26 Wh kg^{-1} and 1250 W kg^{-1} respectively. The deposited Ni seed layer provided the higher surface area to grow $NiCo_2O_4$ and strong adhesion between $NiCo_2O_4$ nanostructure and nickel foam reduces the contact resistance.

A simple and efficient approach has been introduced for synthesis of mesoporous 3D $NiCo_2O_4$/ MWCNT nanocomposite by aerogel method. The fabricated mesoporous, 3D electrode $NiCo_2O_4$/2.1 wt per cent-MWCNT delivered a higher specific capacitance of 1010 F g^{-1} at a current density of 0.1 A g^{-1} with higher stability of 77 per cent after 1,000 cycles. The higher capacitance and higher cyclic stability of the electrode due to porous 3D interconnected conducting network allows fast electrochemical kinetics and increases the electroactive sites.

In future, supercapacitor researches have moved towards the development of high performance device with high energy and power density with additional features like flexible, light weight. The development of new design in the prospective of current collector and electroactive materials are highly desirable. Further, action should be taken in the design and development of new porous current collector with various metal and metal alloys for supercapacitor devices with flexible nature. A good understanding required about the fabrication of high electroactive materials, new device architecture, tunable porosity and better understanding about electrochemical kinetics.

Acknowledgements

This work was supported by the Basic Research Laboratory Program (2014R1A4A1008140) through the National Research Foundation (NRF) funded by the Ministry of Science, ICT & Future Planning of Republic of Korea.

References

Arico, A.S., P. Bruce, B. Scrosati, J.M. Tarascon and W. van Schalkwijk. 2005. Nanostructured materials for advanced energy conversion and storage devices. Nat. Mater 4(5): 366–377.

Ashassi-Sorkhabi, H., P. La'le Badakhshan and E. Asghari. 2016. Electrodeposition of three dimensional-porous Ni/Ni(OH)$_2$ hierarchical nano composite via etching the Ni/Zn/Ni(OH)$_2$ precursor as a high performance Pseudocapacitor. Chem. Eng. J. 299: 282–291.

Attard, G.S., P.N. Bartlett, N.R.B. Coleman, J.M. Elliott, J.R. Owen and J.H. Wang. 1997. Mesoporous platinum film from lyotropic liquid crystalline phases. Science 278: 838–840.

Bartlett, P.N., J.J. Baumberg, P.R. Birkin, M.A. Ghanem and M.C. Netti. 2002. Highly ordered macroporous gold and platinum films formed by electrochemical deposition through templates assembled from submicron diameter monodisperse polystyrene spheres. Chem. Mater 14: 2199–2208.

Bartlett, P.N., B. Gollas, S. Guerin and J. Marwan. 2002. The preparation and characterisation of H1-e palladium films with a regular hexagonal nanostructure formed by electrochemical deposition from lyotropic liquid crystalline phases. PCCP 4: 3835–3842.

Brownson, J.R.S. and C. Lévy-Clément. 2008. Electrodeposition of α- and β-cobalt hydroxide thin films via dilute nitrate solution reduction. Phys. Status Solidi. B 245:1785–1791.

Chen, G.F., Y.Z. Su, P.Y. Kuang, Z.Q. Liu, D.Y. Chen, X. Wu et al. 2015. Polypyrrole shell@3D-Ni metal core structured electrodes for high-performance supercapacitors. Chem. Eur. J. 21: 4614–4621.

Chen, H., L. Hu, Y. Yan, R. Che, M. Chen and L. Wu. 2013. One-step fabrication of ultrathin porous nickel hydroxide-manganese dioxide hybrid nanosheets for supercapacitor electrodes with excellent capacitive performance. Adv. Energy Mater 3: 1636–1646.

Cheng, J., Y. Lu, K. Qiu, H. Yan, X. Hou, J. Xu et al. 2015. Mesoporous $ZnCo_2O_4$ nanoflakes grown on nickel foam as electrodes for high performance supercapacitors. Phys. Chem. Chem. Phys. 17(26): 17016–17022.

Cherevko, S. and C.H. Chung. 2011. Direct electrodeposition of nanoporous gold with controlled multimodal pore size distribution. Electrochem. Commun. 13: 16–19.

Cherevko, S., X. Xing and C.H. Chung. 2010. Electrodeposition of three-dimensional porous silver foams. Electrochem. Commun. 12: 467–470.

Conway, B.E. 1999. Electrochemical supercapacitors: Scientific fundamentals and technological applications, Kluwer Academic/Plenum, New York.

Deng, D., B.S. Kim, M. Gopiraman and I.S. Kim. 2015. Needle-like MnO_2/activated carbon nanocomposites derived from human hair as versatile electrode materials for supercapacitors. RSC Adv. 5: 81492–81498.

Deng, M.J., P.J. Ho, C.Z. Song, S.A. Chen, J.F. Lee, J.M. Chen and K.T. Lu. 2013. Fabrication of Mn/Mn oxide core–shell electrodes with three-dimensionally ordered macroporous structures for high-capacitance supercapacitors. Energy Environ. Sci. 6: 2178–2185.

Deng, Y., W. Huang, X. Chen and Z. Li. 2008. Facile fabrication of nanoporous gold film electrodes. Electrochem. Commun. 10: 810–813.

Devrayan, K., D. Lei, H.Y. Kim and B.S. Kim. 2015. Flexible transparent electrode based on PANi nanowire/nylon nanofiber reinforced cellulose acetate thin film as supercapacitor. Chem. Eng. J. 273: 603–609.

Dong, X.C., H. Xu, X.W. Wang, Y.X. Huang, M.B. Chan-Park, H. Zhang et al. 2012. 3D Graphene-cobalt oxide electrode for high performance supercapacitor and enzymeless glucose detection. ACS Nano 6: 3206–3213.

Dubal, D.P., O. Ayyad, V. Ruiz and P. Gómez-Romero. 2015. Hybrid energy storage: the merging of battery and supercapacitor chemistries. Chem. Soc. Rev. 44: 1777–1790.

Dunn, B., H. Kamath and J.M. Tarascon. 2011. Electrical energy storage for the grid: a battery of choices. Science 334: 928–935.

El-Kady, M.F., Y. Shao and R.B. Kaner. 2016. Graphene for batteries, supercapacitors and beyond. Nat. Rev. Mater 1: 16033.

Erlebacher, J., M.J. Aziz, A. Karma, N. Dimitrov and K. Sieradzki. 2001. Evolution of nanoporosity in dealloying. Nature 410: 450–453.

Ferna´ndez, D., P. Maurer, M. Martine, J.M.D. Coey and M.E. Möbius. 2014. Bubble formation at a gas-evolving microelectrode. Langmuir 30: 13065–13074.

Franceschini, E.A., G.A. Planes, F.J. Williams, G.J.A.A. Soler-Illia and H.R. Corti. 2011. Mesoporous Pt and Pt/Ru alloy electrocatalysts for methanol oxidation. J. Power Sources 196(4): 1723–1729.

Gao, Y., S. Chen, D. Cao, G. Wang and J. Yin. 2010. Electrochemical capacitance of Co_3O_4 nanowire arrays supported on nickel foam. J. Power Sources 195: 1757–1760.

Gao, Y., H. Jin, Q. Lin, X. Li, M.M. Tavakoli, S.F. Leung et al. 2015. Highly flexible and transferable supercapacitors with ordered three-dimensional MnO_2/Au/MnO_2 nanospike arrays. J. Mater Chem. A 3: 10199–10204.

Guan, B., D. Guo, L. Hu, G. Zhang, T. Fu, W. Ren et al. 2014. Facile synthesis of $ZnCo_2O_4$ nanowire cluster arrays on Ni foam for high-performance asymmetric supercapacitors. J. Mater Chem. A 2(38) 16116–16123.

Hall, P.J., M. Mirzaeian, S.I. Fletcher, F.B. Sillars, A.J.R. Rennie, G.O. Shitta-Bey et al. 2010. Energy storage in electrochemical capacitors: designing functional materials to improve performance. Energy Environ. Sci. 3(9): 1238–1251.

Huang, M., Y. Zhang, F. Li, L. Zhang, Z. Wen and Q. Liu. 2014. Facile synthesis of hierarchical Co_3O_4@MnO_2 core–shell arrays on Ni foam for asymmetric supercapacitors. J. Power Sources 252: 98–106.

Jiang, H., Y. Guo, T. Wang, P.L. Zhu, S. Yu, Y. Yu et al. 2015. Electrochemical fabrication of $Ni(OH)_2$/Ni 3D porous composite films as integrated capacitive electrodes, RSC Adv. 5: 12931

Jiang, J., Y. Li, J. Liu, X. Huang, C. Yuan and X.W. Lou. 2012. Recent advances in metal oxide-based electrode architecture design for electrochemical energy storage. Adv. Mater 24: 5166–5180.

Ko, T.H., S. Radhakrishnan, W.K. Choi, M.K. Seo and B.S. Kim. 2016. Core/Shell-like $NiCo_2O_4$-decorated MWCNT hybrids prepared by a dry synthesis technique and its supercapacitor applications. Mater Lett. 166: 105–109.

Ko, T.H., S. Radhakrishnan, M.K. Seo, M.S. Khil, H.Y. Kim and B.S. Kim. 2017. A green and scalable dry synthesis of $NiCo_2O_4$/graphene nanohybrids for high-performance supercapacitor and enzymeless glucose biosensor applications. J. Alloys Compd. 696: 193–200.

Kundu, M. and L. Liu. 2013. Direct growth of mesoporous MnO_2 nanosheet arrays on nickel foam current collectors for high-performance pseudocapacitors. J. Power Sources 243: 676–681.

Lei, D., K. Devarayan, M.K. Seo, Y.G. Kim and B.S. Kim. 2015. Flexible polyaniline-decorated carbon fiber nanocomposite mats as supercapacitors. Mater Lett. 154: 173–176.

Li, Y., Y. Song, Y.Y. Yang and X.H. Xia. 2007. Hydrogen bubble dynamic template synthesis of porous gold for nonenzymatic electrochemical detection of glucose. Electrochem. Commun. 9: 981–988.

Liu, B., B. Liu, Q. Wang, X. Wang, Q. Xiang, D. Chen et al. 2013. New energy storage option: Toward $ZnCo_2O_4$ nanorods/nickel foam architectures for high-performance supercapacitors. ACS Appl. Mater Interfaces 5(20): 10011–10017.

Lu, X., M. Yu, G. Wang, T. Zhai, S. Xie, Y. Ling et al. 2013. H-TiO_2@MnO_2//H-TiO_2@C core–shell nanowires for high performance and flexible asymmetric supercapacitors. Adv. Mater 25: 267–272.

Lukatskaya, M.R., B. Dunn and Y. Gogotsi. 2016. Multidimensional materials and device architectures for future hybrid energy storage. Nat. Commun. 7: 12647.

Luo, W., X. Hu, Y. Sun and Y. Huang. 2012. Electrospun porous $ZnCo_2O_4$ nanotubes as a high-performance anode material for lithium-ion batteries. J. Mater Chem. 22: 8916–8921.

Ma, W., H. Nan, Z. Gu, B. Geng and X. Zhang. 2015. Superior performance asymmetric supercapacitors based on $ZnCo_2O_4@$ MnO_2 core-shell electrode. J. Mater Chem. A 3(10): 5442–5448.

Miller, J.R. and P. Simon. 2008. Electrochemical capacitors for energy management. Science 321(5889): 651–652.

Nikolić, N.D., K.I. Popov, L.J. Pavlovic and M.G. Pavlovic. 2007. Phenomenology of a formation of a honeycomb-like structure during copper electrodeposition. J. Solid State Electrochem. 11: 667–675.

Niu, H., X. Yang, H. Jiang, D. Zhou, X. Li, T. Zhang et al. 2015. Hierarchical core-shell heterostructure of porous carbon nanofiber@$ZnCo_2O_4$ nanoneedle arrays: advanced binder-free electrodes for all-solid-state supercapacitors. J. Mater Chem. A 3(47): 24082–24094.

Norskov, J.K., T. Bligaard, A. Logadottir, J.R. Kitchin, J.G. Chen, S. Pandelov et al. 2005. Trends in the exchange current for hydrogen evolution. J. Electrochem. Soc. 152: J23.

Ott, A., L.A. Jones and S.K. Bhargava. 2011. Direct electrodeposition of porous platinum honeycomb structures. Electrochem. Commun. 13: 1248–1251.

Padmanathan, N. and S. Selladurai. 2014. Controlled growth of spinel $NiCo_2O_4$ nanostructures on carbon cloth as a superior electrode for supercapacitors. RSC Adv. 4: 8341–8349.

Pickering, H.W. and C. Wagner. 1967. Electrolytic dissolution of binary alloys containing a noble metal. J. Electrochem. Soc 114: 698–706.

Pierre, A.C. and G.M. Pajonk. 2002. Chemistry of aerogels and their applications. Chem. Rev. 102: 4243–4265.

Plowman, B.J., L.A. Jones and S.K. Bhargava. 2015. Building with bubbles: The formation of high surface area honeycomb-like films via hydrogen bubble templated electrodeposition. Chem. Commun. 51: 4331–4346.

Qiu, K., Y. Lu, D. Zhang, J. Cheng, H. Yan, J. Xu et al. 2015. Mesoporous, hierarchical core/shell structured $ZnCo_2O_4/MnO_2$ nanocone forests for high-performance supercapacitors. Nano Energy 11: 687–696.

Qiu, T., B. Luo, M. Giersig, E.M. Akinoglu, L. Hao, X. Wang et al. 2014. Au@MnO_2 core–shell nanomesh electrodes for transparent flexible supercapacitors. Small. 10: 4136–4141.

Qu, Q.T., Y. Shi, S. Tian, Y.H. Chen, Y.P. Wu and R. Holze. 2009. A new cheap asymmetric aqueous supercapacitor: Activated carbon//$NaMnO_2$. J. Power Sources 194: 1222–1225.

Radhakrishnan, S., H.Y. Kim and B.S. Kim. 2016. Expeditious and eco-friendly fabrication of highly uniform microflower superstructures and their applications in highly durable methanol oxidation and high-performance supercapacitors. J. Mater Chem. A 4: 12253–12262.

Shin, H.C., J. Dong and M. Liu. 2003. Nanoporous structures prepared by an electrochemical deposition process. Adv. Mater. 15(19): 1610–1614.

Shin, H.C. and M. Liu. 2004. Copper foam structures with highly porous nanostructured walls. Chem. Mater 16: 5460–5464.

Shin, H.C. and M. Liu. 2005. Three-dimensional porous copper–tin alloy electrodes for rechargeable lithium batteries. Adv. Funct. Mater 15: 582–586.

Simon, P. and Y. Gogotsi. 2008. Materials for electrochemical capacitors. Nat. Mater 7(11): 845–854.

Sivabalan, J.S., T.H. Ko, S. Radhakrishnan, C.M. Yang, H.Y. Kim and B.S. Kim. 2016. Novel MWCNT interconnected $NiCo_2O_4$ aerogels prepared by a supercritical CO_2 drying method for ethanol electrooxidation in alkaline media. Int. J. Hydrogen Energy 41: 13504–13512.

Sun, M.H., S.Z. Huang, L.H. Chen, Y. Li, X.Y. Yang, Z.Y. Yuan et al. 2016. Applications of hierarchically structured porous materials from energy storage and conversion, catalysis, photocatalysis, adsorption, separation, and sensing to biomedicine. Chem. Soc. Rev. 45: 3479–3563.

Sun, Z., S. Firdoz, E. Ying-Xuan Yap, L. Li and X. Lu. 2013. Hierarchically structured MnO_2 nanowires supported on hollow Ni dendrites for high-performance supercapacitors. Nanoscale 5: 4379–4387.

Therese, G.H.A. and P.V. Kamath. 2000. Electrochemical synthesis of metal oxides and hydroxides. Chem. Mater 12: 1195–1204.

Trasatti, S. 1972. Work function, electronegativity, and electrochemical behaviour of metals: III. Electrolytic hydrogen evolution in acid solutions. J. Electroanal. Chem. Interfacial Electrochem. 39: 163–184.

Van Damme, S., P. Maciel, H. Van Parys, J. Deconinck, A. Hubin and H. Deconinck. 2010. Bubble nucleation algorithm for the simulation of gas evolving electrodes. Electrochem. Commun. 12: 664–667.

Velev, O.D. and E.W. Kaler. 2000. Structured porous materials via colloidal crystal templating: From inorganic oxides to metals. Adv. Mater 12: 531–534.

Vogt, H., O. Aras and R.J. Balzer. 2004. The limits of the analogy between boiling and gas evolution at electrodes. Int. J. Heat Mass Transfer. 47(4): 787–795.

Wang, C., J. Xu, M.F. Yuen, J. Zhang, Y. Li, X. Chen et al. 2014. Hierarchical composite electrodes of nickel oxide nanoflake 3d graphene for high performance pseudocapacitors. Adv. Funct. Mater 24: 6372–6380.

Wang, G., L. Zhang and J. Zhang. 2012. A review of electrode materials for electrochemical supercapacitors. Chem. Soc. Rev. 41: 797–828.

Wang, Q., J. Yan and Z. Fan. 2016a. Carbon materials for high volumetric performance supercapacitors: Design, progress, challenges and opportunities. Energy Environ. Sci. 9: 729–762.

Wang, Q., L. Zhu, L. Sun, Y. Liu and L. Jiao. 2015a. Facile synthesis of hierarchical porous $ZnCo_2O_4$ microspheres for high-performance supercapacitors. J. Mater Chem. A 3(3): 982–985.

Wang, S., J. Pu, Y. Tong, Y. Cheng, Y. Gao and Z. Wang. 2014. $ZnCo_2O_4$ nanowire arrays grown on nickel foam for high-performance pseudocapacitors. J. Mater Chem. A 2(15): 5434–5440.

Wang, T., B. Zhao, H. Jiang, H.P. Yang, K. Zhang, M.M.F. Yuen et al. 2015b. Electro-deposition of CoNi2S4 flower-like nanosheets on 3D hierarchically porous nickel skeletons with high electrochemical capacitive performance, J. Mater. Chem. A 3: 23035.

Wang, Y., Y. Song and Y. Xia. 2016b. Electrochemical capacitors: mechanism, materials, systems, characterization and applications. Chem. Soc. Rev. 45: 5925–5950.

Wu, J., R. Mi, S. Li, P. Guo, J. Mei, H. Liu et al. 2015. Hierarchical three-dimensional $NiCo_2O_4$ nanoneedle arrays supported on Ni foam for high-performance supercapacitors. RSC Adv. 5: 25304–25311.

Xia, H., Y.S. Meng, G. Yuan, C. Cui and L. Lu. 2012. A symmetric RuO_2/RuO_2 supercapacitor operating at 1.6 V by using a neutral aqueous electrolyte. Electrochem. Solid-State Lett. 15:A60–A63.

Xia, X.H., J.P. Tu, Y.Q. Zhang, Y.J. Mai, X.L. Wang, C.D. Gu et al. 2011. Three-dimention porous Nano-Ni/Co(OH)$_2$ nanoflake composite film: A pseudocapacitive material with superior performance. J. Phys. Chem. C 115: 22662–22668.

Xu, J., L. Gao, J. Cao, W. Wang and Z. Chen. 2010. Preparation and electrochemical capacitance of cobalt oxide (Co_3O_4) nanotubes as supercapacitor material. Electrochem. Acta 56:732–736.

Xu, J., X. Wang, X. Wang, D. Chen, X. Chen, D. Li and G. Shen. 2014. Three-dimensional structural engineering for energy-storage devices: From microscope to macroscope. Chem. Electro. Chem. 1: 975–1002.

Yan, J., Q. Wang, T. Wei and Z. Fan. 2014. Recent advances in design and fabrication of electrochemical supercapacitors with high energy densities. Adv. Energy Mater 4: 1300816–1300859.

Yang, F., K. Cheng, X. Xue, J. Yin, G. Wang and D. Cao. 2013. Three-dimensional porous Ni film electrodeposited on Ni foam: high performance and low-cost catalytic electrode for H_2O_2 electrooxidation in KOH solution. Electrochim. Acta 107: 194–199.

Yang, G.M., X. Chen, J. Li, Z. Guo, J.H. Liu and X.J. Huang. 2011. Bubble dynamic templated deposition of three-dimensional palladium nanostructure catalysts: approach to oxygen reduction using macro, micro, and nano-architectures on electrode surfaces. Electrochim. Acta 56: 6771–6778.

Yin, X., C. Tang, L. Zhang, Z.G. Yu and H. Gong. 2016. Chemical insights into the roles of nanowire cores on the growth and supercapacitor performances of Ni-Co-O/Ni(OH)$_2$ core/shell electrodes. Sci. Rep. 6: 21566.

Yoon, B.J., S.H. Jeong, K.H. Lee, H.S. Kim, C.G. Park and J.H. Han. 2004. Electrical properties of electrical double layer capacitors with integrated carbon nanotube electrodes. Chem. Phys. Lett. 388: 170–174.

Yuan, C., J. Li, L. Hou, J. Lin, X. Zhang and S. Xiong. 2013. Polymer-assisted synthesis of a 3D hierarchical porous network-like spinel $NiCo_2O_4$ framework towards high-performance electrochemical capacitors. J. Mater Chem. A 1:.11145–11151.

Zach, M.P. and R.M. Penner. 2000. Nanocrystalline nickel nanoparticles. Adv. Mater 12: 878–883.

Zhang, D., H. Yan, H. Lu, K. Qiu, C. Wang, Y. Zhang et al. 2014. $NiCo_2O_4$ nanostructure materials: Morphology control and electrochemical energy storage. Dalton Trans 43: 15887–15897.

Zhang, F., T. Zhang, X. Yang, L. Zhang, K. Leng, Y. Huang et al. 2013. A high-performance supercapacitor-battery hybrid energy storage device based on graphene-enhanced electrode materials with ultrahigh energy density. Energy Environ. Sci. 6: 1623.

Zhang, G. and X.W. Lou. 2013. General solution growth of mesoporous $NiCo_2O_4$ nanosheets on various conductive substrates as high-performance electrodes for supercapacitors. Adv. Mater 25: 976–979.

Zhang, H., X. Yu and P.V. Braun. 2011. Three-dimensional bicontinuous ultrafast-charge and -discharge bulk battery electrodes. Nat. Nano 6: 277–281.

Zhang, L.L. and X.S. Zhao. 2009. Carbon-based materials as supercapacitor electrodes. Chem. Soc. Rev. 28: 2520–2531.

Zhao, C. and W. Zheng. 2015. A review for aqueous electrochemical supercapacitors. Front Energy Res. 10.3389/fenrg.2015.00023.

Zhong, C., Y. Deng, W. Hu, J. Qiao, L. Zhang and J. Zhang. 2015. A review of electrolyte materials and compositions for electrochemical supercapacitors. Chem. Soc. Rev. 44: 7484–7539.

Zhu, Y., X. Ji, Z. Wu, W. Song, H. Hou, H. Wu et al. 2014a. Spinel $NiCo_2O_4$ for use as a high-performance supercapacitor electrode material: Understanding of its electrochemical properties. J. Power Sources 267: 888–900.

Zhu, Y., E. Liu, Z. Luo, T. Hu, T. Liu, Z. Li et al. 2014b. A hydroquinone redox electrolyte for polyaniline/SnO_2 supercapacitors. Electrochem. Acta 118: 106–111.

7

Porous Carbon Materials for Fuel Cell Applications

N. Rajalakshmi,[1,*] *R. Imran Jafri*[2] *and T. Ramesh*[1]

1. Introduction

Carbon, a unique element in the periodic table, exists in various solid forms like graphite (sp^2), diamond (sp^3), fullerenes (60 carbon atoms made up of 20 six-membered rings and 12 five membered rings) and carbon nanotubes (a mixture of sp^3 and sp^2 bonds). Graphite exists in two structural forms viz., hexagonal and rhombic forms. Three of the valence electrons in the carbon atoms are involved in forming sp^2 hybrid bonds and leading to fourth electron forming a π bond. These π electrons are mobile, thereby making graphite a good electrical conductor. These electrons within graphite can also react with other elements without disrupting the layer structure. The other form is carbon black, which is obtained when a carbon compound is heated in air with a limited supply of oxygen. It is regarded as an amorphous form of carbon, usually consisting of near spherical particles of graphite, typically below 50 nm in diameter that may coalesce into particle aggregates and agglomerates of around 250 nm in diameter.

The morphology and particle size distribution of carbon black is dependent on the source material and the process of its thermal decomposition. Carbon blacks prepared by pyrolysis at temperatures greater than 2500°C are amorphous, and are called as 'hard carbons'. If a graphitised sample is obtained when heated to such high temperatures and the layered structure is easily visible, they are used for fuel cell applications. A few carbon blacks suitable for fuel-cell applications are being supplied by Cabot Corporation (Vulcan XC-72R, Black Pearls BP 2000), Ketjen Black International, Chevron (Shawinigan), Erachem and Denka. Particle size and distribution determine directly the surface area and are probably the most important properties of carbon black in terms of end use applications.

Wood, coal, lignite, coconut shells and peat are all important raw materials for producing activated carbon. The particle sizes of these activated carbons are relatively large typically 20–30 μm. They can be activated to increase the microporosity with the proportion of pores that are < 2 nm and also the BET surface area range from 800–1200 $m^2 g^{-1}$ depending on the degree of activation.

Carbon is being used for many electrochemical applications due to its high electrical conductivity, chemical stability and low cost (Zhang et al. 2016). Carbon can be used in fuel cell systems as part of the structure of the fuel cell and stack viz., bipolar plate, as gas-diffusion layer in a proton-exchange

---

[1] Center for Fuel Cell Technology (CFCT), International Advanced Research Centre for Powder Metallurgy and New Materials (ARCI), 2nd Floor, IIT-M Research Park, Phase-1, 6, Kanagam Road, Taramani, Chennai-600113.

[2] B.S. Abdur Rahman Crescent University, vandalur, Chennai 600048.

* Corresponding author: rajalakshmi@arci.res.in

membrane fuel cell (PEMFC) and as an electrocatalyst or as an electrocatalyst support. They can be used as a reacting species in hydrocarbon fueled systems, as a potential means of storing hydrogen and as a fuel in the direct carbon fuel cell (DCFC) systems. Activated carbon has also been used as a support for industrial precious metal catalysts for many years, and has been a natural choice for supporting the electrocatalysts in phosphoric acid (PAFC), and PEMFCs.

Carbon has also been utilised in solid oxide fuel cell (SOFC) where the incomplete oxidation of carbon formed by the pyrolysis of methane on SOFC cermet anodes (Ni/YSZ and Ni/Gd-CeO$_2$). In these studies, both CO$_2$ and CO were produced from the anode exhaust and the temperatures were as high as 900°C. The concept of direct carbon conversion in fuel cells can be traced back in the 19th century where the electrochemical oxidation of carbon was done in baked coal with hydroxides as electrolytes, at 400–500°C. As the carbon was consumed rather than continuously supplied, the system is considered as a battery rather than a fuel cell.

In the present chapter we aim to provide a summary on the role of carbon for fuel cell applications especially as a structural element of fuel cells, bipolar plate, electrode designs, as well as supports for catalysts. The chapter also discusses about the role of carbon in direct carbon fuel cell. The present chapter looks into the recent developments made in fuel cell areas since 2010.

2. Polymer Electrolyte Membrane Fuel Cells

Of the different types of fuel cells, proton exchange membrane fuel cells (PEMFC) with a proton conducting polymer as electrolyte is being used widely due to its low operating temperature, modular nature, no green house gas emission (when hydrogen is used as fuel) and design flexibility (Rajalakshmi et al. 2017). Figure 1 shows the block diagram of a typical PEMFC, showing current collectors, flow channels, electrocatalyst layer, gas diffusion layer, proton exchange membrane (PEM) and catalyst support.

The basic reactions involved in a fuel cell with hydrogen and oxygen as fuel and oxidant respectively is given in Table 1.

At the anode, fuel (hydrogen in case of PEMFC) oxidation occurs giving rise to electrons. At the cathode oxygen reduction reaction (ORR) takes place. Among these two reactions, oxygen reduction reaction is complex reaction and the fuel cell performance is determined by the ability of the catalyst to perform ORR. Hence, researchers are focused of developing efficient ORR catalyst for PEMFCs.

Carbon plays an important role in the development of PEMFC. Mentioned below are different components of a PEMFC where carbon allotropes are being used.

1. **Bipolar plate cum current collector:** The fuel and oxidant flow channels are generally made up of graphite which also acts as current collector. In fuel cell stacks, bipolar graphite plates are used in which both fuel and oxidant flow channels are grooved on opposite sides of a single graphite plate. Low density (1.8 to 2.0 g/cc), good electrical conductivity, ease of availability, low cost and ease of machining makes the graphite ideal candidate as bipolar plates cum current collector cum flow distributor in fuel cells.

Table 1. Basic reactions involved in fuel cells.

Proton Exchange Membrane Fuel cells		
Anode: $H_2 \rightarrow 2H^+ + 2e^-$	$E^+ = 0.00V$	→ (i)
Cathode: $O_2 + H^+ + e^- \rightarrow H_2O$	$E^- = 1.23V$	→ (ii)
Total reaction $H_2 + \frac{1}{2} O_2 \rightarrow H_2$	$E_{Total} = E^- - E^+ = 1.23$ V	→ (iii)
Alkaline Fuel Cells		
Anode: $2H_2 + 4OH^- \rightarrow 4 H_2O + 4e^-$	$E^+ = -0.83$ V	→ (iv)
Cathode: $O_2 + 2H_2O + 4e^- \rightarrow 4 OH^-$	$E^- = 0.40$ V	→ (v)
Total reaction $H_2 + \frac{1}{2} O_2 \rightarrow H_2$	$E_{Total} = E^- - E^+ = 1.23$ V	→ (vi)

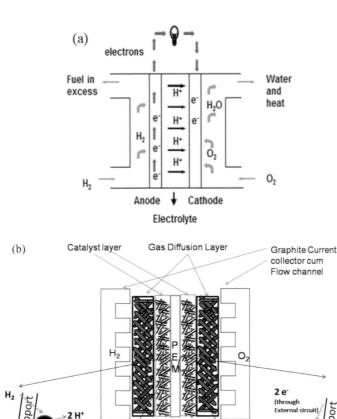

Fig. 1. (a) Block diagram of a hydrogen fuelled PEMFC and (b) figure showing current collectors, flow channels, electrocatalyst layer, gas diffusion layer, proton exchange membrane (PEM) and catalyst support.

2. **Gas diffusion layer (GDL):** Backing layer also called as macroporous layer is made up of carbon paper or carbon cloth over which a PTFE treated microporous carbon layer is coated. These two layers put together is called gas diffusion layer. The GDL helps in uniform distribution of fuel and oxidant gases, water management and achieving good adhesion between catalyst and backing layer. The GDL also helps in heat transfer and provide necessary mechanical strength to hold membrane electrode assembly (MEA) against the membrane swelling caused by water absorption. The degree of hydrophobicity can be altered using proper treatment on the GDLs. Carbon based GDLs are the favorite and widely used GDL in fuel cell stacks. The ease of fabrication, bulk production in the form of sheets, tunable pore size, tunable hydrophobicity and good electrical conductivity makes this carbon allotrope an ideal choice as GDL material in fuel cells.

3. **Catalyst layer:** Catalyst layer is the core region in the fuel cell where the oxidation and reduction reactions take place. Broadly, there are two types of catalysts (i) supported and (ii) un-supported catalysts. Supported catalysts are the one in which the catalyst particles (generally platinum and platinum alloys) are coated on carbon materials for effective utilisation of the catalyst and improved three phase boundary (interface formed between electrode, electrolyte and reactant gases). The catalyst supports should have good stability in acidic and basic environment, good

electrical conductivity, good adhesion with catalyst particles and high surface area. Again the candidate which fits the bill is carbon based materials such as amorphous carbon, activated carbon, vulcan carbon, carbon nanotubes, carbon nanorods, graphene, etc. Recently, nitrogen doped carbons without any Pt particles have been demonstrated as catalysts for oxygen reduction reactions in fuel cells.

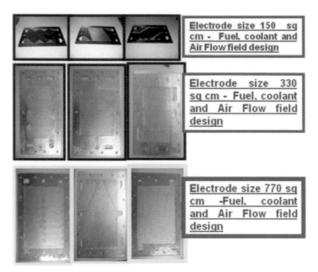

Fig. 2. Top two image shows the photograph of bipolar plates, SEM-Images of bare CC (a), one fibre of the bare CC (b) (Vivekananthan et al. 2015) and the bottom images show the TEM images of (a) Pt/C-01, (b) Pt/C-02 (Ye et al. 2014).

2.1 Bipolar Plates

Graphitic carbon impregnated with a resin, subject to pyrolytic process has been used as flow field plates for fuel cells. Although the resistivity increases due to addition of resin, the graphitic plates have however made gas impermeable (Rajalakshmi and Dhathathreyan 2008). POCO Graphite, USA and SGL Carbon, Germany are some of the companies offer graphite plates for fuel cells with high electronic and thermal conductivity, low contact resistance, corrosion resistant and can be easily machined. Pure graphite is an expensive material, brittle, porous and requires machining. However, machining the flow fields for the reactant supply in a carbon based plate, is a complicated and time-consuming process leading to high prices. In addition, graphite plates are brittle and porous; they have to be made impermeable to prevent the cross flow of the fuel and oxygen. The various materials based on carbon are classified in Fig. 3.

The alternative materials are graphite composite materials with polymer binders to achieve the desired properties and also to improve manufacturing technologies for bipolar plates including the flow fields and cooling channels. These composite materials are made with certain amount of conductive carbons like natural or synthetic graphite powder, carbon blacks, and carbon nanotubes to increase the conductivity with polymers as binders. The polymers are either thermoplastics such as poly vinylidene fluoride, polyethylene, polypropylene, liquid crystal polymers, and polyphenylene sulfide, or thermosets, such as vinyl ester, phenolic resins, and epoxy resins (Middelman 2003, Mehta and Cooper 2003, Wolf and Willert-Porada 2006).

The conducting particle forms a percolation network within the polymer matrix and retains the processability of the polymer. To obtain a mechanically stable bipolar plate with low gas permeability, a good quality of dispersion of the conductive component particles within the polymer binder is required. If pyrolytic impregnation is used they can operate at temperatures as high as 450°C, however, resin impregnated graphite is limited to 150°C. The specific volumetric and gravimetric power density of the stack gets reduced as the thickness of the flow field plates increases.

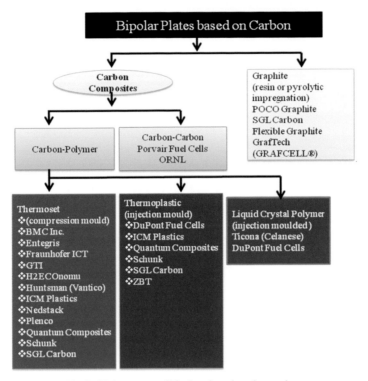

Fig. 3. Various types of bipolar plates based on carbon.

Robberg and Trapp have reviewed the subject of graphite plates and reported that the carbonaceous material could be in the form of 'natural' graphite, synthetic graphite, and non-graphitic carbon such as coke (Robberg and Trapp 2003).

The performance of carbon-polymer composites are lower compared to other materials mainly due to low electronic conductivity, which depends on the amount of graphite into the composite. Electrical conductivity ranges from 2.4 to 300 S cm^{-1} for carbon black and polyvinylidene fluoride (PVDF) and for composites with graphite loadings as high as 93 per cent, respectively. It has been observed that the graphite loading of 60–80 per cent and particle size of 50–200 μm have a pronounced effect on the electrical and thermal properties and also in the ease of processing (Del Rio et al. 2002).

Kuan et al. have systematically studied the effect of graphite content and powder size on the physical, thermal, mechanical, and electrical properties of a vinyl ester binder-based composite (Kuan et al. 2004). They have reported that the density increases with carbon content and decreases with carbon particle size while the porosity increases with graphite content and decreases with particle size. Electrical resistance decreases with increasing content and carbon particle size. Thermal expansion decreases with increased graphite content and decreased particle size and the flexural strength decreases with increased graphite content.

Flexible graphite is another carbon based material made by expansion of natural graphite by acid treatment with an intercalating agent along with heat treatment. This process is likely to increase the spacing between the planes of the hexagonal graphite structure in the 'c' axis, as much as 80–100 times. The expanded graphite in the form of powder is then compressed to the desired density and pressed to form the desired flow field plates. Flexible graphite are relatively inexpensive, meets the cost targets and has the advantage of very low contact resistance and density of 1 g cm^{-3} with good sealing characteristics. However, they suffer from the mechanical stability and have relatively high gas permeability (Mercuri et al. 2002).

Table 2. The representative values of various carbon plates used in fuel cells are given below (Brett and Brandon 2007).

Property	Bulk Moulding Compounds Inc. (BMC 940)	SGL Carbon SIGRACET® (BBP 4)	SGL Carbon SIGRACET® (PPG86)	Porvair Fuel Cells	POCO Graphite (AXF 5Q) (umimpregnated)	SS 316L
Type	Thermoset	Thermoset	Thermoplastic	Carbon-carbon composite	Graphite	Metallic
Density (g cm^{-3})	1.82	1.97	1.84	1.2–1.3	1.78 (20 per cent porous)	8
Electronic conductivity (S cm^{-1})	100[a] 50[b]	200[a] 41.6[b]	55.6[a] 18.2[b]	> 500[b]	680.3	13,513
Coefficient of thermal expansion X 10^{-6} (K^{-1})	30	5.8	49	2	7.9	16
Thermal conductivity (W m^{-1} K^{-1})	18.5[a] (at 85°C)	20.8	14	> 35[b]	95	16.3
Flexural strength (kg cm^{-2})	407.8	407.8	358.6	420–500	878.8	–
Permeability coefficient X 10^{-5} (cm^3 cm^{-2} s^{-1})	–	2.5	3.5	< 0.2	–	Gas tight
Max operating temperature	200°C Glass transition	≤ 180	≤ 80	> 400	> 400 (pyrolytic impregnation) 150°C (resign impregnation)	1400 m.p

[a]Through-plane property
[b]In-plane property

Grafoil™, an expanded graphite, has also been used for the manufacturing of flow field plates for use in fuel cells. Carbon-carbon composites, another carbon matrix reinforced with carbon fibers are now being used to make bipolar plates. These materials possess high strength, low density, chemical stability, high electrical and thermal conductivity, and the ability to operate at temperatures in excess of 400°C, which are suitable for fuel cell applications (Hwang and Hwang 2002).

It has been reported that the operating temperature of carbon-carbon plates are higher than carbon-polymer composites, and that also depends on the type of polymer. They have low density, 1.2–1.3 g cm^{-3}, and are mechanically stronger and are more electrically conductive than carbon-polymer composites. The major disadvantage is the dimensional tolerance due to shrinkage in the mold. Recently, M/S Porvair Fuel Cells has made improvements to the process, leading to higher accuracy in plate thickness, thereby avoiding seal leakage. The other disadvantage of carbon-carbon composites is the longer processing time than carbon-polymer molded plates (Besmann et al. 2000, Huang et al. 2005).

Liquid crystal polymer, with a carbon content of 85 per cent, has also been used to produce flow field plates by Ticona Engineering Polymers, Germany, molded by SGL Carbon with Vectra® LPC, which allows a high carbon content due to the low viscosity of the polymer. The higher carbon content improves the electrical and thermal conductivity and are also amenable for injection molding. Fuel cell stack cost has been considerably reduced by using engineered thermoplastics, manufactured by M/S. Ticona compared to metallic and graphite based plates. M/S DuPont is also manufacturing a LCP based composite plates (Zenite®) for fuel cell operation using low carbon (Wolf et al. 2006).

2.2 Gas Diffusion Layers

Gas diffusion layers (GDL) are porous media that serve as one of the electrode components for membrane-electrode assembly (MEA) in PEMFCs. The GDL play a key role for gas permeation to catalyst layer and water management during fuel cell operation. Although GDL does not take part in the electrochemical reaction, it functions significantly to provide the reactants access to the catalytic sites and effectively remove the reaction product, that is, water, from the electrode. In addition, it should have a high electronic conductivity to receive and accept electrons to and from the external circuit, physical, chemical stability and proper wetting characteristic for its use in PEMFC applications. The performance of the PEMFC is also dependent on effective transport properties of GDL. The GDL is typically hydrophobic in nature so that the surface and pores in the GDL are not blocked with product water that can possibly inhibit gas transport to the catalyst layer. The GDL consists of a macroporous substrate (MPS) (i.e., single layer GDL) or a thin carbon layer on a sheet of macroporous carbon cloth or carbon paper (i.e., dual-layer GDL). The typical range of GDL thickness is between 200 and 400 μm, with fiber diameter in the range of 7–10 μm (Park et al. 2012). The structure of dual-layer GDL is given in Fig. 4. As seen, the first layer in contact with the gas flow channel is the MPS serving as a gas distributor and a current collector. The second layer is a thin microporous layer (MPL) which contains carbon powder and hydrophobic and/or hydrophilic agent, primarily managing two-phase water flow.

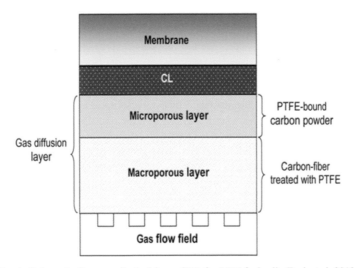

Fig. 4. Schematic diagram of a dual-layer GDL for PEM fuel cells (Park et al. 2012).

2.2.1 *Effect of a GDL Based on Carbon Paper or Carbon Cloth on PEM Fuel Cell Performance*

A commercially available GDL based on carbon paper or carbon cloth viz., SGL 10BB (carbon paper loaded with 5 wt. per cent PTFE and the MPL, SGL Carbon) and ELAT-LT-1400 W (carbon cloth loaded with no PTFE and the MPL, E-TEK) has been used as a macroporous substrate in fuel cell. The pore size distribution of SGL 10BB ranges between 0.01 to 0.1 μm in the MPL and from 6 to 300 um in the Carbon paper as shown in Fig. 5(a) (Park and Popov 2011). However, carbon cloth ELAT-LT-1400W exhibits a highly uniform pore sizes and has a maximum pore volume range from 0.1 to 10 μm. This indicates that the MPL in ELAT-LT-1400 W is significantly entrenched into carbon cloth, reducing large pores (dp > 6 μm) during the MPL deposition. As shown in Fig. 5b, SGL 10BB results in better fuel cell performance, which can be attributed to higher total pore volume. Also, the improved performance may be explained in terms of the water management, which results in better oxygen counter flow through the GDL. The fuel cell performance demonstrated that SGL 10BB resulted in better fuel cell performance using H_2/air. It has been observed that carbon paper provides more pore volume available for oxygen transport as well as prevent water accumulation in the GDL, resulting in higher oxygen concentration in the catalyst layer due to its more hydrophobic nature and dual pore size distribution.

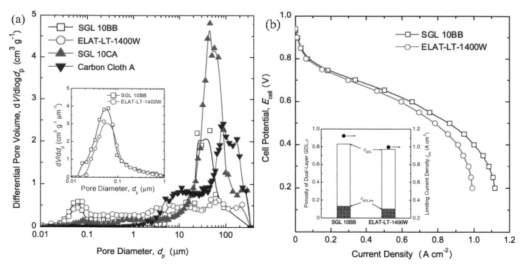

Fig. 5. (a) Pore Size distribution and (b) PEM Fuel cell performance of carbon paper and carbon cloth (Park and Popov 2011).

2.2.2 *Effect of PTFE Loading*

Poly Tetra Fluoro Ethylene (PTFE) commonly known as Teflon, is a fluorocarbon polymer having a high molecular weight. It is stable thermoplastic at room temperature with the density of about 2.2 g cm^{-3} and melting point of 327°C. PTFE is the electrical insulator with a thermal conductivity of approximately 0.25 W m^{-1} K^{-1}. GDLs are generally treated with different amounts of PTFE for water management in fuel cell. However, PTFE can alter the electrical and thermal resistance of a GDL Porosity also decreases with the increase in PTFE content, resulting in higher mass transport resistance. On the other hand, decrease in PTFE content would lower the capability for water removal (Tseng and Lo 2010). Velayutham et al. reported that the hydrophobic coating on the cathode GDL materials play an important role on the performance (Velayutham et al. 2007). It has been reported that the cell performance is good when the PTFE content in the GDL is 23 wt. per cent and the performance is very poor when the PTFE content is less than 10 wt. per cent, as shown in Fig. 6.

The cell voltages at 0.3 A cm^{-2} are 0.678 and 0.583 V for the GDL with 23 and 7 wt. per cent PTFE, respectively. But at 0.4 A cm^{-2} current density, the cell voltages are 0.611 and 0.422 V. Although the conductivity of the carbon paper is higher with low loading of PTFE, under experimental conditions the performance is affected due to poor removal of product water from the microporous layer. It has been observed that the product water removal rate is slow for GDL with low PTFE loading leading to higher stoichiometry of reactants.

The PTFE contents in GDL and ML which consists of carbon particles and binder, normally Teflon, have to be finely matched to get the best performance of the cells. The PTFE content on the microporous layer also play an important role in the performance. The product water that is formed in the active region during electrochemical reaction has to be removed continuously. The microporous carbon layer should have optimum hydrophobicity to remove the product water from the active layer.

In order to identify the suitable PTFE content in the MPL, the PTFE content in the GDM (substrate carbon paper) was fixed at 23 per cent and the PTFE content in the MPL was systematically changed from 10 to 32 wt. per cent. The carbon loading in the microporous layer was kept constant for all the experiments. The MEAs with 20 wt. per cent of PTFE in the microporous layer gave better performance than those with 10 and 32 wt. per cent PTFE. The performances of MEAs with other PTFE loadings were in the intermediate range. The PTFE content in the microporous layer does not have any influence on the performance up to the current density of 0.2 A cm^{-2}. This indicates that irrespective of the PTFE content, the MPL allows the gas, which is sufficient to draw the current

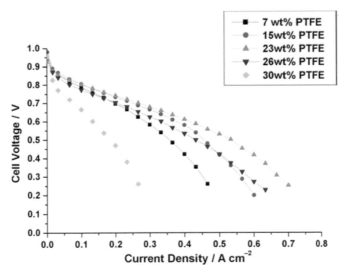

Fig. 6. Effect of PTFE on GDL (Velayutham et al. 2007).

density up to 0.2 A cm^{-2}. Although the resistance of the microporous layer is low due to PTFE loading, it does not help in improving the performance as it retains some product water, leading to flooding and restricts the reactant the flow of gas to the active layer. High loading of PTFE helps in removing the product water efficiently from the active layer, but the resistance of the MPL brings down the overall cell performance. The effect of PTFE variation in MPL is similar to that observed in the GDL. The effect of the PTFE content on the substrate materials has more influence on the performance at higher current density compared to MPL PTFE.

Tseng et al. (Tseng and Lo 2010) studied the effect of the PTFE loading of an MPL and is shown in Fig. 7(b). It can be observed that the MPL having 40 wt. per cent of PTFE has a better cell performance due to higher gas permeability and easier water removal. In other words, PTFE loading of 40 wt. per cent provides lower mass transport resistance in the high current density zone. Moreover, as shown in Fig. 7, the pore size distribution results also showed that the MPL with PTFE loading less than 25 wt. per cent had more small pores than the MPL with 25 and 40 wt. per cent PTFE. This may be due to the fact that during the baking process, part of PTFE vapourises, and the remaining part cures, producing larger pores. Therefore, the MPL with a greater PTFE loading has less small pores and higher gas permeability.

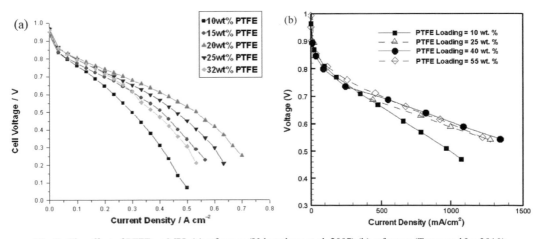

Fig. 7. The effect of PTFE on MPL (a) reference (Velayutham et al. 2007) (b) reference (Tseng and Lo 2010).

2.2.3 Effect of MPL Thickness on Cell Performance

The effect of the thickness of an MPL on the performance of a cell has been reported by Tseng et al. and it has been found that the averaged gas permeability increases with increase in the thickness of MPL, which may be due to the increase in the mean pore diameter (Tseng and Lo 2010). However, It has been reported that the 84 μm sample has better performance compared to 38 μm and 136 μm at high current density regions where mass transport dominates as shown in Fig. 8. It has been reported that the suitable MPL improves the reactant gas supply and facilitates the removal of the product water. A very thin MPL has smaller pores that can obstruct gas supply and increase mass transport resistance. However, an extremely thick MPL has a high electrical resistance because of the increased amount of PTFE. A thick layer also hinders the availability of the reactant gas due to the increased diffusion path.

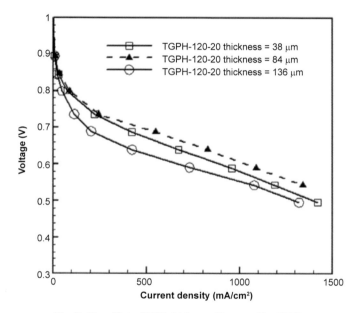

Fig. 8. The effect of MPL thickness (Tseng and Lo 2010).

2.2.4 Type of Carbon Materials

Various types of carbon-based materials have been used to effectively modify the pore structure, improved mass transfer, hydrophobicity and electronic conductivity of the GDL and MPL (Park et al. 2012, Lin and Chang 2015, Wang et al. 2006). Recently, *in situ*-grown CNT on GDL substrate has been used as PTFE-free MPL (Chen et al. 2012). Jordan et al. studied fuel performance of the GDLs with different carbon powders (Vulcan XC-72 and acetylene black) in the cathode MPL (10 wt per cent PTFE) with an O_2/Air at different operating conditions. The MPL which contains acetylene black led to higher power density than that with Vulcan XC-72 due to less porous structure in the MPL. Furthermore, when the MPL was heat-treated at 350°C for 30 min, the enhancement in performance was observed. They claimed that uniform distribution of the PTFE throughout the MPL by sintering makes the MPL more hydrophobic, resulting in better water management at the cathode.

The effects of *in situ* grown CNT layer on the mass transport and fuel cell performance by applying GDLs with different features has been studied (Xie et al. 2015), shown in Fig. 9. The *in situ* growth of CNT had an influence on the pore size distribution of both the carbon-fiber substrate and the MPL. Since the pore diameter of the carbon-fiber substrate is determined by the carbon fiber

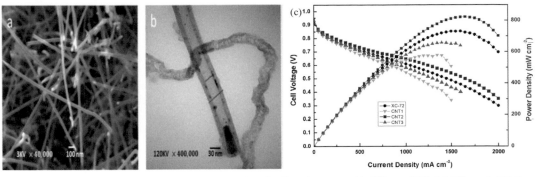

Fig. 9. SEM image of *in situ* grown CNT (a and b) and Fuel cell performance with different CNTs (c) (Xie et al. 2015).

spacing, the decreasing of macro pore volume might be attributed to the covering of CNTs on the carbon fiber which could enlarge their diameter and narrow the gap between them. Moreover, the CNTs twisted together results in the increasing of micro pore volume. The power density of the fuel cell with CNTs increased to 822 mW cm^{-2} using H$_2$/Air at 80 per cent RH. They reported that the GDLs with suitable hydrophobicity and proper pore structure were capable of reducing water flooding of the cathode side especially at high current density.

A composite micro-porous layer using carbon nano chain Pureblack + nano-fibrous carbon (VGCF-H) in the GDL for proton exchange membrane fuel cell was studied (Kannan and Munukutla 2007). From Fig. 10 the MPL with nano-chain Pureblack + nano-fibrous carbon (VGCF-H) based GDLs shows power density of as high as 0.55 Wcm^{-2} at 70°C using hydrogen/air, 100 per cent RH and ambient pressure with Nafion-112 membrane. The result exhibits an excellent gas diffusion (mass transport) characteristics of the GDLs fabricated with carbon nano-fibers along with Pureblack nano-chain carbon.

Bamboo was investigated for use as a GDL of PEMFC. The fibers were molded into bamboo fiber sheets (BFS) and were subjected to carbonisation to give carbonised bamboo fiber sheets (CBFS), in which the fibrous morphology remained. Both the in-plane and the through-plane electrical conductivities were measured after carbonisation (Kinumoto et al. 2015). Consequently, both the in-plane and through-plane electrical conductivities were measured and were found to be approximately 35 and 550 S cm^{-1}, respectively, after treatment at 3000°C. The electrical conductivities measured for carbon paper (CP) were approximately 150 and 1500 S cm^{-1}, respectively, and are significantly larger than those of CBFS. For the operation test, CBFS obtained via carbonisation at 3,000°C was employed because of its higher electrical conductivity and hydrophobicity. From Fig. 11, the open-circuit voltage (OCV) obtained for the cell based on CBFS was approximately 0.95 V and the maximum power density was 140 mW cm^{-1} at 0.45 V, demonstrating that the PEMFC could indeed be operated using CBFS as the GDL. On the other hand, the OCV obtained for the cell based on CP

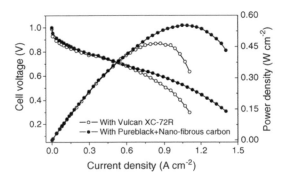

Fig. 10. Fuel cell performance of composite MPL (Pureblack + Nanofibrous carbon) (Kannan and Munukutla 2007).

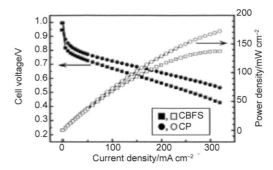

Fig. 11. Fuel cell performance of Bamboo based GDL (Kinumoto et al. 2015).

was slightly higher (50 mV) than that based on CBFS, and the maximum power density recorded was 170 mW cm^{-1} at 0.55 V. These OCV values were slightly lower than the theoretical OCV calculated for PEMFC under these conditions (1.18 V) (Kinumoto et al. 2015).

2.3 Catalyst Supports

2.3.1 Role of Carbon Nanostructures in Reducing the Pt Loading

Platinum is the extensively used catalyst for both hydrogen oxidation and oxygen reduction reaction in PEMFC because it outperforms all other catalyst in terms of activity, selectivity and stability as evidenced in the volcano plots (Holton and Stevenson 2013). At the current stage of technology, to increase the active surface and three phase boundary, it is dispersed on a support. Carbon materials are the suitable candidates for catalyst support due to their large specific surface area, strong corrosion resistance, good electrical conductivity and low cost. Commendable efforts have been made in the past two to three decades in bringing down the Pt loading (Pt content) in the fuel cell by the application of different carbon supports. The Pt loading has been reduced considerably from ~ 4 mg/cm^2 to ~ 0.6 mg/cm^2, as a result of the optimal dispersion of the Pt nanoparticles on carbon supports. Due to the Pt supply limitations, a further reduction in Pt loading is necessary for commercialisation of fuel cells. The US Department of Energy (DOE) has set 2017 technical target for total (anode + cathode) Pt group metals (PGM) loading to 0.125 mg Pt cm^{-2}, which is significantly lower than the current Pt loading of 0.4–1.0 mg Pt cm^{-2} used in PEMFCs. This target of 0.125 mg Pt cm^{-2} corresponds to the global target of 10 g of Pt per car by the year 2020 (Eberle et al. 2012). A review on the catalysts development for PEMFC and SOFC and the role of nanomaterials has been done by Dhathathreyan et al. which explains the details of all types of catalysts being used.

2.3.2 Types of Catalyst Supports

Porous carbon is an essential part of a catalyst layer in low temperature PEMFCs. Platinum catalyst nanoparticles are dispersed on high surface area carbon materials for effective utilisation of catalyst, improved three phase boundary and improved catalytic activity (Candelaria et al. 2012). The essential properties of a catalyst support should be (i) high specific surface area, (ii) high porosity (for the transfer of reactant and product species from the reaction sites), (iii) high corrosion resistance in acidic and basic environment in the fuel cell operating range and (iv) good electrical conductivity. Porous carbons very much satisfy these requirements due to their unique structure and properties. The commonly used catalyst support material are Vulcan XC72, Black Pearls, Ketjen EC300J, Ketjen EC600JD, Shawinigan, Denka black with specific surface areas ranging from 65–1800 m^2/g (Wang et al. 2015, Yu and Ye 2007). The bulk production of these carbons makes them as best choice for the industrial scale production of Pt supported carbons. However, these carbons suffer with the corrosion in the fuel cell environments thereby allowing the Pt particles to detach from the carbon support and aggregate into bigger clusters. This cluster formation reduces the effective surface area of Pt resulting in significant reduction in fuel cell performance (Zhang et al. 2010a/b). To address this issue, highly ordered nanostructures such as carbon nanotubes (Jha et al. 2013, Jafri et al. 2010a/b), carbon nanofibers (Andersen et al. 2013), graphene (Jafri et al 2010; Liu et al. 2014), mesoporous carbons have been demonstrated as catalyst support in PEMFC (Candelaria et al. 2012). CNTs with closed structures show excellent stability in fuel cell environments.

Multiwalled carbon nanotubes supported Pt showed excellent stability during the accelerated durability test compared to carbon black supported Pt (Shao et al. 2006). Using carbon nanotubes, ultra low loading of Pt can be achieved without any sacrifice in the performance of the PEMFC. Tang et al. have demonstrated ultra low loading (12 µg Pt cm^{-2}) multiwalled carbon nanotube supported Pt catalysts (Pt/MWNTs) at the cathode for PEMFC. A power density of 613 mW cm^{-2} and a mass activity of 250 A/mg Pt (based on cathode Pt loading) have been reported (Tang et al. 2007). Jha et al. (Jha et al. 2013) have recently demonstrated chemically modified single-walled carbon nanotubes (SWNTs) thin film catalyst support layers (CSLs) in PEMFCs, which were suitable for benchmarking against the US DOE 2017 targets. Use of the optimum level of SWNT-COOH functionality allowed the construction of a prototype SWNT-based PEMFC with total Pt loading of 0.06 mg Pt/cm^2—well below the value of 0.125 mg Pt/cm^2 set as per the US DOE 2017 technical target for total Pt group metals (PGM) loading (Jha et al. 2013). Hasche et al. have studied the activity, stability and degradation of multiwalled carbon nanotube (MWNT) supported Pt catalyst and compared the results with Vulcan XC 72R-supported Pt. Their results does not indicate any activity benefit of MWNT support for the kinetic rate of ORR but in the 'lifetime' regime, the MWCNT supported Pt catalyst showed clearly smaller electrochemically active surface area (ECSA) and mass activity losses compared to the Vulcan XC 72R supported Pt catalyst. In the 'start-up' regime, Pt on MWCNT exhibited a reduced relative ECSA loss compared to Pt on Vulcan XC 72R (Hasché et al 2010). The good stability can be due to enhanced adhesion between Pt atoms and the graphene sheets (Hasché et al. 2010).

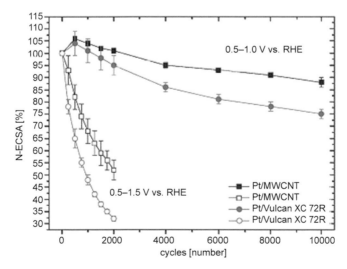

Fig. 12. Normalised ECSA loss due to voltage cycling from 0.5–1.0 V vs. RHE (solid symbols) and 0.5–1.5 V vs. RHE (hollow symbols) with 50 mV s^{-1} for Pt/MWCNT (square) and commercial Pt/Vulcan XC 72R (circular) (Hasché et al. 2010).

Since past decade, Graphene, a two-dimensional (2D) carbon material with single (or a few) atomic layer has attracted great attention from both fundamental science and applied research (Antolini 2012). Graphene and hetero atom doped graphene have been demonstrated as good catalyst support materials (Jafri et al. 2010a/b, Vinayan et al. 2012, Jafri et al. 2015). The combination of high specific surface area (theoretical value of 2600 m^2 g^{-1}), excellent electronic conductivity and potentially low manufacturing cost makes them a potential material for catalyst support in fuel cells. Since, it is highly difficult to obtain a single sheet of graphene in large quantities, researchers have focused their attention on few layered graphene nanosheets (tens of graphene sheets). Figure 13 shows TEM and high resolution TEM images of Pt on nitrogen doped graphene. Graphene sheets, tend to stack due to van der Waals force and π–π interaction thereby reducing the activity and utility of Pt particles supported on them (Jafri et al. 2015). To avoid this stacking and improve the porosity, CNTs and carbon black has been demonstrated as fillers (Jafri et al. 2015). Graphene also helps in reducing the Pt nanoparticle agglomeration. Functionalised graphene sheets supported Pt catalyst has been shown to have good catalytic activity and much better stability compared to commercial catalyst E-TEK owing to small Pt particle size and less agglomeration (Kou et al. 2009). Nitrogen doped graphene apart from being used as catalyst support, has also been demonstrated as Pt free ORR catalyst (Qu et al. 2010).

Since past two decades, there has been a surge in studies on another type of catalyst support material called mesoporous carbon which has controllable pore size, high surface area and large pore volumes (Ryoo et al. 2001). Template assisted synthesis like soft-templating and hard-templating strategies are proposed to synthesise high surface area carbons (Zhao et al. 2008). Silica based template assisted synthesis process involves infiltration of the pores of template by a carbon precursor followed by carbonisation and template etching. Silica templates like MCM-48, SBA-1 and SBA-15 having cubic and hexagonal frameworks have been successful in the synthesis of mesoporous carbon with large pore volumes and high specific surface areas up to 1800 $m^2 g^{-1}$. Zhang et al. have synthesised

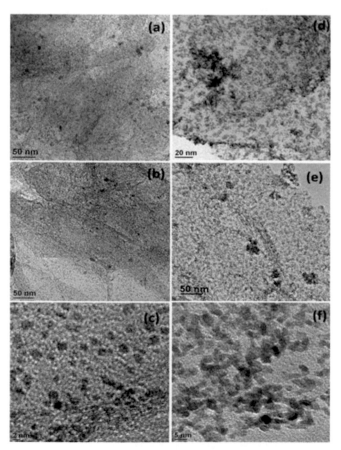

Fig. 13. TEM images of Pt/NG180, (d, e) TEM images of Pt/NGA800 and (c, f) High resolution-TEM images of Pt/NG180 and Pt/NGA800 respectively (Jafri et al. 2015).

three-dimensionally ordered mesoporous carbon sphere array (OMCS)-supported Pt nanoparticles (Pt/OMCS) and have demonstrated that smaller Pt particle size, greater Pt dispersion, larger specific electrochemically active surface area (ECSA) can be obtained with OMCS as catalyst support. A higher activity for methanol oxidation reaction (MOR) and ORR has been obtained compared to carbon black (Vulcan XC-72R)-supported Pt and commercial Pt/C catalysts (Zhang et al. 2014). Recently, the effect of hybridisation of Pt supported mesoporous CMK-3 into Pt-CB was studied with durability tests and proved to be a better stable support material (Park et al. 2016).

2.3.3 Carbon Support and Pt Interaction

The physical surface area of Pt nanoparticles is correlated with the its particle size using the relation $S = 6000/\rho d$, where ρ is the density of Pt and d is the particle diameter in Angstrom (Jafri et al. 2015). The reduction in Pt particle size greatly increases its surface area and hence more amount of Pt surface is available for fuel cell reactions. The Pt particle size and its morphology greatly depend up on the synthesis procedure used for Pt precursor reduction, Pt precursor and carbon interaction, the surface physicochemical properties and the structure of carbon support. Researchers are focusing on the interaction between Pt or Pt-alloy and carbon support which affects the structure and the dispersion of Pt or Pt alloy nanoparticles on the carbon support (Yu and Ye 2007, Wang et al. 2015). The mechanism of interaction which in turn affects the overall performance of Pt-carbon catalyst system needs to be understood clearly in order to develop highly active and efficient electrocatalysts. In the Pt or Pt-alloy carbon support system, carbon support is not a mere inert material but it can alter the systems galvanic potential, raise the electron density in the catalyst and lower the Fermi level (Yu and Ye 2007, Wang et al. 2015). Different researchers have reviewed the type of interaction between Pt and carbon support using different spectroscopic techniques (Yu and Ye 2007, Wang et al. 2015). The electron spin resonance and X-ray photoelectron spectroscopy (XPS) clearly demonstrate the electron donation by Pt to the carbon support. In the XPS studies, the Pt $4f_{7/2}$ signal shifted to higher values with respect to un-supported Pt due to the Pt support electronic effects (Yu and Ye 2007, Wang et al. 2015). Also, the surface treatment of carbon supports greatly influences the dispersion and the activity of the Pt particles. Different surface treatments like oxidative treatment (HNO_3, H_2SO_4, H_3PO_4, H_2O_2, O_2 or O_3, K_2ClO_3, $KMnO_4$, etc.) and thermal treatment have been performed to obtain uniform dispersion of Pt nanoparticles. The functionalisation also helps in good adhesion of Pt particles by acting as anchoring sites. Artyushkova et al. (Artyushkova et al. 2012) have studied the structure to property relationship of carbon black samples when used as catalyst support in PEMFC (Fig. 14).

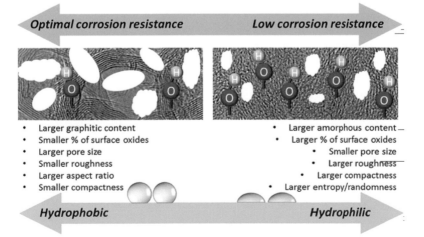

Fig. 14. Structure-to-property relationship for carbon black samples with optimal and low corrosion resistance (Artyushkova et al. 2012).

They have found that samples with the highest potential activity for oxygen reduction reaction (ORR) and highest corrosion were found to have high surface area, high roughness and large amounts of surface oxides. However, samples with moderate activity and high resistance towards corrosion have large amounts of graphitic carbons and elongated pores (Artyushkova et al. 2012).

Researchers have also demonstrated nitrogen doped carbons as catalysts support and also as Pt free catalysts in PEMFC. In the case of N doped carbons as catalyst support, it is possible that the oxophilic nature of the N-containing surface species could provide sites for surface groups that participate in the destruction of active reaction intermediates (Zhou et al. 2010). Nitrogen doping changes the electrical conductivity and density of states of the carbon support and is possible that these electronic effects can result in preferential growth of higher activity Pt facets (Sun et al. 2009, Zhou et al. 2010) (Fig. 15).

The adsorbed oxygen containing surface species facilitate reaction of strongly-absorbed intermediate reaction species that would otherwise block catalyst active sites, thereby increasing the net turn-over frequency of the electrochemical reaction (Zhou et al. 2010).

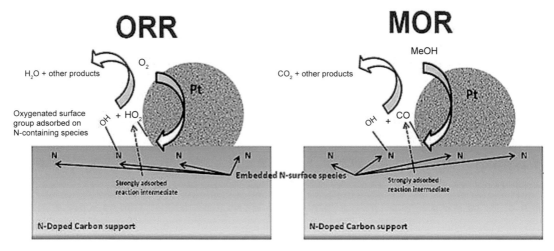

Fig. 15. Schematic of possible bi-functional ORR and MOR mechanisms involving oxophilic C–N defects near C/Pt catalyst particle interface in the case of low-density Pt catalyst (not to scale).

2.3.4 Carbon Based Metal Free Catalysts

As explained earlier, the role of nitrogen doped carbon supports is not only to enhance the catalytic activity of the Pt by a possible bi-functional mechanism, but they can also be used as Pt free catalysts in PEMFCs. Currently, scientists are intensely studying doped carbon catalysts for ORR application which can act as an alternative to the high cost Pt catalyst. With the emergence of N doped carbons, the dream of obtaining a Pt free catalyst with the activity and stability of pure Pt seems to be in the reach. Different types of N doped carbon materials such as N-doped graphene (Lai et al. 2012, Peng et al. 2013), N-doped CNT (Mo et al. 2012), N-doped mesoporous carbon (Niu et al. 2015, Liang et al. 2013), N-doped carbon nanocages (Chen et al. 2012), etc. have been studied. However, maintaining the stability of these catalysts remains to be a challenge. Peng et al. have reported Fe- and N- doped carbon catalyst Fe-PANI/C-Mela with graphene structure with a surface area up to 702 m^2 g^{-1} as an ORR catalyst in PEMFC. They have also done full cell studies (Fig. 16) using Fe-PANI/C-Mela as ORR catalyst and reported a maximum power density of 0.33 W/cm^2 at 0.47 V (Peng et al. 2013) but the stability degraded rapidly in 100 h long term test (Peng et al. 2013).

Mo et al. (Mo et al. 2012) synthesised a bamboo like Nitrogen-doped carbon nanotube (N-CNT) arrays by chemical vapour deposition using ferrocene as the catalyst precursor and imidazole as the

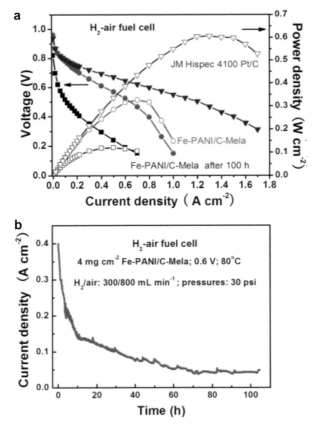

Fig. 16. (a) Polarisation plots of a single H$_2$-air PEMFC with Fe-PANI/C-Mela as the cathode before and after long-term stability test (loading: 4 mg cm^{-2}), and with JM Pt/C as cathode (Pt loading: 0.2 mg cm^{-2}). (b) The relationship of current density of single cell with Fe-PANI/C-Mela cathode and running time, at a constant fuel cell voltage of 0.60 V (Peng et al. 2013).

carbon and nitrogen precursor. Linear sweep voltammetry studies were conducted in both acidic and basic environment. They have concluded that the activity of N-CNTs is related to the pyridinic-type nitrogen content, and that N-CNTs show much better activity in an alkaline medium than in an acidic medium. Lai et al. (Lai et al. 2012) have studied N-graphene produced by annealing of graphene oxide (G-O) under ammonia or by annealing of N-containing polymer/reduced graphene oxide (RG-O) composite (polyaniline/RG-O or polypyrrole/RG-O). The effects of the N precursors and annealing temperature on the performance of the catalyst were investigated. They have demonstrated that the electrocatalytic activity of N-containing metal-free catalysts is highly dependent on the graphitic N content while pyridinic N species improve the onset potential for ORR.

However, the total atomic content of N in the metal-free, graphene-based catalyst did not play an important role in the ORR process. Graphitic N can greatly increase the limiting current density, while pyridinic N species might convert the ORR reaction mechanism from a 2e$^-$ dominated process to a 4e$^-$ dominated process (Lai et al. 2012). Recently, Niu et al., have demonstrated mesoporous N-doped carbons prepared by nanoparticle templates as an efficient electrocatalyst for ORR (Niu et al. 2015).

3. Direct Carbon Fuel Cells (DCFC)

The interest for carbon conversion in fuel cells is not new and has been pursued intermittently for about 150 years. Carbon fuel cells have emerged recently due to its two major advantages like higher conversion efficiency and concentrated CO$_2$ production stream, which can be directly collected for industrial use or sequestration. This technology promises significant opportunities for next generation coal power technology. Differing from other fuel cells, the DCFC is the only fuel cell type converting solid carbon into electricity without an intermediate reforming process (Gur 2013). There are inherent difficulties and technical challenges in carbon conversion in fuel cells. In a conventional power generation, coal is typically combusted in air or steam where irreversible mixing of at low efficiency. The efficiency is determined by the set of operating temperature and pressure, with a maximum efficiency as defined by the Carnot constraint. However in a electrochemical oxidation of carbon, the conversion occurs isothermally and isobarically, and the reactant and product streams are separate leading to minimum entropic losses due to mixing, thereby increasing the efficiency. The theoretical efficiency, η, for electrochemical conversion is defined by thermodynamic state functions as

$$\eta = \Delta G/\Delta H = \{1 - (T\Delta S/\Delta H)\} \tag{i}$$

Where T is the isothermal conversion temperature and ΔG, ΔH, and ΔS denote changes in Gibbs energy, enthalpy, and entropy, respectively.

Carbon fuel cells consists of an consists of an anode which houses the solid carbon fuel, where the oxidation occurs by the reaction of carbon with oxide ions through the electrolyte from the air compartment, cathode as shown in schematic Fig. 17.

$$C(s) + 2O^{2-}(electrolyte) \rightarrow CO_2(g) + 4e^-(electrode) \tag{ii}$$

$$O_2(g) + 4e^-(electrode) \rightarrow 2O^{2-}(electrolyte) \tag{iii}$$

The overall net reaction is

$$C(s) + O_2(g) = CO_2(g) \text{ --- } 1.02 \text{ V } @1100°C \tag{iv}$$

The challenge starts at the anode by partial oxidation of carbon leading to CO as shown in the reaction below.

$$C(s) + O^{2-}(electrolyte) = CO(g)+2e^- (electrode) \tag{v}$$

The direct electrochemical reaction of solid carbon oxide ion depends on the supply of oxygen in the form of OH^-, CO_3^{2-}, or a metal oxide. In some case, the carbon is provided as gaseous fuel such as CO and H_2 to the electrochemical reaction site in order to achieve conversion in fuel cells.

The desired anode reaction (ii) involves the transfer of four electrons for every carbon atom, and this occurs in multi reaction step, involves multi-step transfer of electrons sequentially by elementary reactions. The primary challenge in achieving carbon conversion is to effectively bring together the two reactants of reaction (iv), namely, the solid carbon fuel particles and the gaseous oxygen to react at the electrochemical reaction site (ERS) (Nabae et al. 2008, Nabae et al. 2009). Most of the research

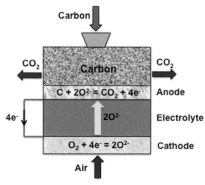

Fig. 17. Schematic of carbon fuel cells (Gur 2013).

has been focused on the effective ways of using solid fuel and the oxygen in an appropriate chemical form, to the electrolyte/electrode interface where electrochemical oxidation takes place. In addition, the equilibrium cell potential of 1.02 V is expected to be independent of temperature over a wide range, in accordance with complete oxidation reaction (iv) for carbon. Most studies also observed strong temperature dependence of the cell potential.

3.1 Electrolytes for DCFC

Three different concepts of a DCFC based on different electrolytes are been investigated. The performance data on the different concepts cannot be compared directly because they strongly depend on many factors such as the carbon fuel material, the operating temperature and the design. The various electrolytes, such as molten carbonates, molten hydroxides, and YSZ-based solid electrolytes (based on solid oxide fuel cells), are used for direct conversion of carbon into electricity. However there are recent reports about the use of mixed electrolytes, molten carbonate electrolytes (e.g., Li_2CO_3-Na_2CO_2-K_2CO_3) due to their high stability, electrical conductivity, low volatility, toxicity, and relatively low melting points. The hydroxide electrolyte based fuel cells has a low operating temperature of around 650°C due to its greater ionic conductivity than that of molten sodium carbonate. In addition, Molten NaOH also has lower over potentials and a greater oxidation rate compared to molten Na_2CO_3. Due to their lower operating temperatures, cheaper materials can be used for cell casing and cathodes, such as 300 series stainless steel and ultra-low carbon steel. However, carbonate formation has been an issue preventing the industrial application of molten NaOH as an electrolyte. A mitigating technique for the prevention of carbonate involves increasing the concentration of water in the electrolyte solution to drive the equilibrium thereby reducing the amount of O^{2-} in solution, which is an intermediate product in the formation of carbonate. This had several additional benefits like an increase in the ionic conductivity of the electrolyte/fuel solution and reduced corrosion rates of iron, nickel, and chromium, used in the cathode and cell casings.

The key elements of both the molten carbonate fuel cell (MCFC) and solid oxide fuel cell (SOFC) designs are combined in a single cell with the fuel dispersed within the molten carbonate electrolyte leading to hybrid direct carbon fuel cells (DCFC). In DCFC, carbon particles are suspended in the slurry containing the molten electrolyte, which is a mixture of Li_2CO_3 and K_2CO_3. This concept was conceived by Irvine and co-workers (Jiang and Irvine 2011) at the University of St Andrews, with the vision to alleviate the challenges with each cell type. Nabae and co-workers (Nabae et al. 2008) also studied the combination of the two technologies to solve some of these key issues like separation of the cathode and the molten carbonate via a solid oxide electrolyte, reducing the possibility of cathode corrosion, management of CO_2 in the electrolyte mix, etc. The primary motivation for incorporating molten carbonate in the SOFC anode chamber, is to improve the kinetics of electrochemical carbon oxidation in direct carbon as it is believed that the contact between the anode and carbon fuel is necessary. This effectively extends the anodic electrochemical reaction zone away from the triple phase boundary (TPB) of SOFC systems. The electrochemical oxidation of carbon also may occur in the slurry mixture because of the ionic conductivity of the carbonate and current carrying capacity of the carbon. Nabae and co-workers (Nabae et al. 2008), used carbon black (CB), and the cell was carried out as a physical contact-type direct carbon SOFC experiment, with no CO_2 feed. They also studied the effect of loading CB with 10, 30, and 50 per cent Ni on the cell performance and open circuit voltage (OCV) at various temperatures between 550 and 900°C. They found the most significant improvement to peak power occurred at 550°C by a factor of 7.6 for a Ni loading of 50 wt per cent.

Jain et al. investigated the performance of pyrolysed medium-density fiberboard (pMDF) by immersion the sticks in the eutectic molten carbonate mixture, in a saturated aqueous solution of the carbonates before drying, and milling of pMDF with carbonate powder (Jain et al. 2009). The best performance was achieved by immersion of pMDF in the molten carbonate solution at an operating temperature of 800°C followed by pMDF in aqueous solution and milled pMDF at 700°C. Jiang et al. (Jiang and Irvine 2011) studied the CB/carbonate mixture at varying carbonate loadings (0, 20, 50, and 80 mol per cent carbonate) and temperatures to determine the effect on OCV and achieved a peak power density of around 18 and 53 mW cm^{-2} at 700 and 800°C, respectively.

3.2 Anode Materials

Carbon in various forms are used for the conversion of fuel to electricity at the anode chamber, that includes activated carbons, synthetic carbons, amorphous carbon, carbon blacks, pyrolytic carbons, graphite, turbostratic carbon, etc. Various pretreatment methods are also adopted with the intent to improve the physical and chemical properties of carbons and increase their reactivity for electrochemical conversion. The early design is to use a solid carbon, which had various shortcomings, such as the leakage of impurities (e.g., ash) into the electrolyte and mechanical instability of coal-based electrodes, low reaction rates, economical and logistical impracticalities of carbon electrodes. Although various carbon fuels have been tested, a knowledge of the efficacy of carbon fuels is still unclear. It has been observed that the disordered carbon is more reactive due to a preponderance of edges sites and defects, but graphitic carbon with high electrical conductivity may also benefit the electrochemical reactivity due to their physical and chemical properties.

Galaghar et al. (Gallagher et al. 2010) studied a structure reactivity relationship for various nanostructured graphene-based carbons and also reported the important role of surface oxygenated species in the formation of CO_2. It has been observed that the nature and population of surface groups affect chemical reactivity as well as the physical properties of carbons. Although it has been reported that the pretreatment increases the activity by increasing the wettability. It is not clear how this pretreatment will be active at high temperatures. It has also been reported that coal and biomass chars are more active than activated carbons. There was also a discussion about the way carbon being supplied to the fuel cell in order to facilitate the formation of highly reactive char, which is desirable to obtain high power densities (Gür et al. 2010, Dong et al. 2009, Hackett et al. 2007, Li et al. 2009).

Weaver et al. found that carbon fuels with high surface areas, such as devolatilised coal, are more accessible to the anode reaction (Weaver et al. 1981). However, Cooper's group concluded that the effect of carbon's surface area alone on the discharge rate is relatively weak. The recent development is to utilise highly reactive carbon particulates dispersed in the molten electrolyte, between an anode and a cathode at high temperatures, thereby increasing the reaction rate. Although molten carbonates exhibit high ionic conductivity at elevated temperature, they are poor electrical conductors that limit the performance of these cells. One of the most promising methods employs molten metal as the anode electrode, which not only has certain oxygen conductivity but also has high electrical conductivity.

A new fuel cell with a fused metal/yttria-stabilised zirconia (YSZ)/platinum as electrolyte, and carbon as fuel fed to the fused iron in the anode compartment with air at the cathode, was reported by Yentekakis et al. which gave the best performance near the melting point of iron (~ 1536°C) (Yentekakis et al. 1989). CellTech Power LLC (Xu et al. 2013) used a layer of molten liquid tin as the anode to directly convert carbonaceous fuels including coal into electricity without gasification at 800–1000°C. This leads to a formation of solid SnO_2 layer formed at the electrolyte interface during fuel-cell operation at 700°C, which inhibits further the oxygen from the tin/electrolyte interface, leading to high anode over potential. Hence it is important to develop the fuel cell with a molten bath, and both the metal and metal oxide should have a low melting point. Based on that many metals like indium (In), lead (Pb), antimony (Sb), bismuth (Bi), and silver (Ag) were tested as anodes and has been reported that only $Sb-Sb_2O_3$ showed a resistance of 0.06 Ω cm^2 with a peak power density of 350 mW cm^{-2} at 700°C.

However, the electrochemical reactions and redox and transport processes within the liquid metal electrodes needs to be understood. Severe corrosion has been reported for scandium-stabilised

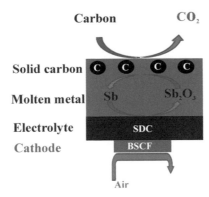

Fig. 18. DCFC with a molten antimony anode (Xu et al. 2013).

zirconia at high current densities. Watanabe et al. (Watanabe et al. 2016) studied the carbon/carbonate mixtures with various ternary carbonate composition (Li_2CO_3/Na_2CO_3/K_2CO_3) at 1073 K to form a slurry. The continuous power generation of DCFCs at 20 mA was achieved when the carbon was well wetted. It has been reported that a carbon wettability shows great impact on the performance.

In spite of many reports about the usage of various carbon fuels for direct carbon fuel cell applications, a knowledge of the suitability of carbon fuels is still unclear. Although the disordered carbon is more reactive due to edges sites and defects, the graphitic carbon with high electrical conductivity may also benefit the electrochemical reactivity. Hence the physical and chemical properties of carbon fuels play a significant role to enhance the electrochemical performance. There was a mixed opinion about the carbon fuels with high surface areas for efficient anode reaction and on the weak discharge rate. Hence, there is a need to develop approaches to improve the electrochemical reactivity of carbon fuels in the anode reaction. The other factors which are important apart from surface area are oxygen functional groups, particle size, pore distribution and their temperature dependent electrochemical activity.

A biomass-based activated carbon prepared from bamboo scraps under different activation temperatures, activation times, and impregnation ratios has been studied for use in direct carbon fuel cells (DCFCs) by Zhang et al. (Zhang et al. 2011). It was reported that high surface area is one of the critical parameter for high performance. Li et al. (Li et al. 2008) studied four commercial carbon fuels viz., activated carbon, two carbon blacks, and graphitic carbon and reported the relationship between carbon fuel surface complex and the concomitant electrochemical performance of the DCFC. They also reported the effects of operating conditions like carbon loading and particle size, stirring rate, and cell temperature to increase the performance of the DCFC. It has been observed that the electrochemical reactivity of four carbon fuels was in the order of AC > CB660 > CB220 > GC, which depends on the concentration of oxygen containing functional groups on the surfaces of the carbon fuels. They have also reported that carbon with high mesoporous surface area as well as rich surface oxygen-containing groups is most preferable.

Hao et al. (Hao et al. 2014) studied the effect of impurities present in solid carbonaceous fuel in a molten carbonate DCFC and found that the ash components, such as K, Ca, and Mg, involved fuel gasification as a catalyst leading to high performance compared to the expected performance of renewable biomass and wastes. Eom et al. (Eom et al. 2014, 2016) compared three different coals and their chars. The effect of gas on the performance DCFC system and reported that Adaro coal showed a maximum power density of 22.1 mW cm^{-2}, a 45 per cent performance improvement over the graphite particle at 15.3 mW cm^{-2}. It has been attributed to the fuel characteristics of the released gases, such as specific surface area, total pore volume, and amorphous structure.

In addition to reaction mechanisms, the carbonate ion transfer to the reaction site is important in enhancing the performance of a DCFC using carbon/carbonate slurry. Li et al. used carbon nanofibers

(CNFs) in a DCFC using molten carbonates and observed the changed morphology of short and sintered filamentous CNFs after discharge. However, finding the carbon reaction site is difficult in a DCFC using the carbon/electrolyte slurry because a large number of small carbon particles exist in the slurry. Watanabe et al. (Watanabe et al. 2016) used a different approach of making a single carbon pallet and studied the electrochemical oxidation in the molten carbonate electrolyte. It has been observed that TPB during discharge play a major role in the surface oxidation of carbon as well the particle morphology.

Conclusion

Carbon is a vital component for several types of fuel cell and also acts as a fuel in certain types of fuel cells. In the case of PEMFC, carbon in its various forms and types like particles, tubes, fibres, nanotubes, acts as a catalyst support. They are also used as catalysts in line with non-noble metal. All types of carbon materials are also used as support structure for gas diffusion for the reactants in the form of woven cloth and paper. The properties of the electrodes change, affecting the fuel cell performance, due to various modifications in either carbon or in the additives. Gas flow field plates for fuel cells were originally made of graphite on which the gas-flow channels were machined mechanically. However, many R & D institutions and manufacturing companies have developed variety of plates with carbon as major constituent and other additives like thermoset, thermo epoxy, liquid crystal polymers, etc., for cost and weight effectiveness. Several forms of carbon have been exploited for use in GDL and catalyst layer, for high dispersion of Pt and Pt-alloy catalysts, as a support structure. Although there are technical issues associated with the preparation and reactivity of the carbon, the successful development of direct carbon fuel cell could help to extend the use of fossil fuels for power generation as we move towards a more sustainable energy future.

Acknowledgements

The authors would like to thank Director, ARCI for his constant support and encouragement and also Technical Research Centre TRC (No. AI/l/65/ARCI/2014 (G)) Dept. of Science and Technology, Govt. of India for providing the financial support. One of the authors (R. Imran Jafri) would like to specifically acknowledge the financial assistance given by DST-INSPIRE program of Dept. of Science and Technology, Govt. of India.

References

Andersen, S.M., M. Borghei, P. Lund, Y.R. Elina, A. Pasanen, V. Kauppinen et al. 2013. Durability of carbon nanofiber (CNF) & carbon nanotube (CNT) as catalyst support for proton exchange membrane fuel cells. Solid State Ion. 231: 94–101.

Antolini, E. 2012. Graphene as a new carbon support for low-temperature fuel cell catalysts. Appl. Catal. B 123–124: 52–68.

Artyushkova, K., S. Pylypenko, M. Dowlapalli and P. Atanassov. 2012. Structure-to-property relationships in fuel cell catalyst supports: Correlation of surface chemistry and morphology with oxidation resistance of carbon blacks. J. Power Sources 214: 303–313.

Besmann, T.M., J.W. Klett, J.J. Henry and E. Lara-curzio. 2000. Carbon/Carbon composite bipolar plate for proton exchange membrane fuel cells. J. Electrochem. Soc. 147(11): 4083–4086.

Brett, D.J.L. and N.P. Brandon. 2007. Review of materials and characterization methods for polymer electrolyte fuel cell flow-field plates. J Fuel Cell Sci. Technol. 4(1): 29.

Candelaria, S.L., Y. Shao, W. Zhou, X. Li, J. Xiao, J.G. Zhang et al. 2012. Nanostructured carbon for energy storage and conversion. Nano Energy 1(2): 195–220.

Chen, S., J. Bi, Y. Zhao, L. Yang, C. Zhang, Y. Ma et al. 2012. Nitrogen-doped carbon nanocages as efficient metal-free electrocatalysts for oxygen reduction reaction. Adv. Mater. 24(41): 5593–5597.

Dhathathreyan, K.S., N. Rajalakshmi and R. Balaji. 2017. Nanomaterials for fuel cell technology. *In*: de Voorde, M.V., B. Raj and Y. Mahajan (eds.). Nanotechnology for Energy Sustainability. Wiley-VCH, ISBN: 978-3-527-34014-9.

Del Rio, C., M.C. Ojeda, J.L. Acosta, M.J. Escudero, E. Hontanon and L. Daza. 2002. New polymer bipolar plates for polymer electrolyte membrane fuel cells: Synthesis and characterization. J. Appl. Polym. Sci. 83(13): 2817–2822.

Dong, S., P. Alvarez, N. Paterson, D.R. Dugwell and R. Kandiyoti. 2009. Study on the effect of heat treatment and gasification on the carbon structure of coal chars and metallurgical cokes using fourier transform raman spectroscopy. Energy and Fuels 23(3): 1651–1661.

Eberle, U., B. Müller and R.V. Helmolt. 2012. Fuel cell electric vehicles and hydrogen infrastructure: status 2012. Energy Environ. Sci. 5(10): 8780.

Eom, S., S. Ahn, Y. Rhie, K. Kang, Y. Sung, C. Moon, G. Choi and D. Kim. 2014. Influence of devolatilized gases composition from raw coal fuel in the lab scale DCFC (direct carbon fuel cell) system. Energy 74(C): 734–740.

Eom, S., J. Cho, S. Ahn, Y. Sung, G. Choi and D. Kim. 2016. Comparison of the electrochemical reaction parameter of graphite and sub-bituminous coal in a direct carbon fuel cell. Energy and Fuels 30(4): 3502–3508.

Gallagher, K.G., G. Yushin and T.F. Fuller. 2010. The role of nanostructure in the electrochemical oxidation of model-carbon materials in acidic environments. J. Electrochem. Soc. 157: B820.

Gur, T.M. 2013. Critical review of carbon conversion in " Carbon Fuel Cells ." Chem. Rev. 113: 6179–6206.

Gür, T.M., M. Homel and A.V. Virkar. 2010. High performance solid oxide fuel cell operating on dry gasified coal. J. Power Sources 195(4): 1085–1090.

Hackett, G.A., J.W. Zondlo and R. Svensson. 2007. Evaluation of carbon materials for use in a direct carbon fuel cell. J. Power Sources 168: 111–118.

Hao, W., X. He and Y. Mi. 2014. Achieving high performance in intermediate temperature direct carbon fuel cells with renewable carbon as a fuel source. Appl. Energy 135: 174–181.

Hasché, F., M. Oezaslan and P. Strasser. 2010. Activity, stability and degradation of multi walled carbon nanotube (MWCNT) supported Pt fuel cell electrocatalysts. Phys. Chem. Chem. Phys 12(46): 15251.

Holton, O.T. and J.W. Stevenson. 2013. The role of platinum in proton exchange membrane fuel cells. Platinum Metals Rev. 57(4): 259–271.

Huang, J., D.G. Baird and J.E. McGrath. 2005. Development of fuel cell bipolar plates from graphite filled wet-lay thermoplastic composite materials. J. Power Sources 150(1-2): 110–119.

Hwang, J.J. and H.S. Hwang. 2002. Parametric studies of a double-cell stack of PEMFC using Grafoil™ flow-field plates. J. Power Sources 104(1): 24–32.

Jafri, R.I., T. Arockiados, N. Rajalakshmi and S. Ramaprabhu. 2010a. Nanostructured Pt dispersed on graphene-multiwalled carbon nanotube hybrid nanomaterials as electrocatalyst for PEMFC. J. Electrochem. Soc. 157(6): B874.

Jafri, R.I., N. Rajalakshmi, K.S. Dhathathreyan and S. Ramaprabhu. 2015. Nitrogen doped graphene prepared by hydrothermal and thermal solid state methods as catalyst supports for fuel cell. Int. J. Hydrogen Energy 40(12): 4337–4348.

Jafri, R.I., N. Rajalakshmi and S. Ramaprabhu. 2010b. Nitrogen doped graphene nanoplatelets as catalyst support for oxygen reduction reaction in proton exchange membrane fuel cell. J. Mater. Chem. 20: 7114–7117.

Jain, S.L., J. Barry Lakeman, K.D. Pointon, R. Marshall and J.T.S. Irvine. 2009. Electrochemical performance of a hybrid direct carbon fuel cell powered by pyrolysed MDF. Energy Environ. Sci. 2(6): 687.

Jha, N., P. Ramesh, E. Bekyarova, X. Tian, F. Wang, M.E. Itkis et al. 2013. Functionalized single-walled carbon nanotube-based fuel cell benchmarked against US DOE 2017 technical targets. Sci. Rep. 3: 2257.

Jiang, C. and J.T.S. Irvine. 2011. Catalysis and oxidation of carbon in a hybrid direct carbon fuel cell. J. Power Sources 196(17): 7318–7322.

Kannan, A.M. and L. Munukutla. 2007. Carbon nano-chain and carbon nano-fibers based gas diffusion layers for proton exchange membrane fuel cells. J. Power Sources 167: 330–335.

Kinumoto, T., T. Matsumura, K. Yamaguchi, M. Matsuoka, T. Tsumura and M. Toyoda. 2015. Material processing of bamboo for use as a gas diffusion layer in proton exchange membrane fuel cells. ACS Sustainable Chem. Eng. 3(7): 1374–1380.

Kou, R., Y. Shao, D. Wang, M.H. Engelhard, J.H. Kwak, J. Wan et al. 2009. Enhanced activity and stability of Pt catalysts on functionalized graphene sheets for electrocatalytic oxygen reduction. Electrochem. Commun. 11(5): 954–957.

Kuan, H.C., C.C. Ma, H.K. Chen and S.M. Chen. 2004. Preparation, electrical, mechanical and thermal properties of composite bipolar plate for a fuel cell. J. Power Sources 134(1): 7–17.

Lai, L., J.R. Potts, D. Zhan, L. Wang, C.K. Poh, C. Tang, H. Gong, Z. Shen, J. Lin and R.S. Ruoff. 2012. Exploration of the active center structure of nitrogen-doped graphene-based catalysts for oxygen reduction reaction. Energy Environ. Sci. 5(7): 7936.

Li, X., Z. Zhu, J. Chen, R. De Marco, A. Dicks, J. Bradley et al. 2009. Surface modification of carbon fuels for direct carbon fuel cells. J. Power Sources 186(1): 1–9.

Li, X., Z.H. Zhu, R. De Marco, A. Dicks, J. Bradley, S. Liu et al. 2008. Factors that determine the performance of carbon fuels in the direct carbon fuel cell. Ind. Eng. Chem. Res. 47(23): 9670–9677.

Liang, H.W., W. Wei, Z.S. Wu, X.L. Feng and K. Müllen. 2013. Mesoporous metal–nitrogen-doped carbon electrocatalysts for highly efficient oxygen reduction reaction. J. Amer. Chem. Soc. 135(43): 16002–16005.

Lin, S.Y. and M.H. Chang. 2015. Effect of microporous layer composed of carbon nanotube and acetylene black on polymer electrolyte membrane fuel cell performance. Int. J. Hydrogen Energy 40(24): 7879–7885.

Liu, J., D. Takeshi, K. Sasaki and S.M. Lyth. 2014. Defective graphene foam: A platinum catalyst support for PEMFCs. J. Electrochem. Soc. 161(9): F838–F844.

Mehta, V. and J.S. Cooper. 2003. Review and analysis of PEM fuel cell design and manufacturing. J. Power Sources 114(1): 32–53.

Mercuri, R.A., J.P. Capp, M.L. Warddrip and T.W. Weber. 2002. Flexible graphite article and method of manufacture. US 6,432,336 B1.

Middelman, E. 2003. Method for the production of conductive composite material. US 2003/0160352 A1.

Mo, Z., S. Liao, Y. Zheng and Z. Fu. 2012. Preparation of nitrogen-doped carbon nanotube arrays and their catalysis towards cathodic oxygen reduction in acidic and alkaline media. Carbon 50(7): 2620–2627.

Nabae, Y., K.D. Pointon and J.T.S. Irvine. 2009. Ni/C slurries based on molten carbonates as a fuel for hybrid direct carbon fuel cells. J. Electrochem. Soc. 156(6): B716.

Nabae, Y., K.D. Pointon and J.T.S. Irvine. 2008. Electrochemical oxidation of solid carbon in hybrid DCFC with solid oxide and molten carbonate binary electrolyte. Energy Environ. Sci. 1(1): 148–155.

Niu, W., L. Li, X. Liu, N. Wang, J. Liu, W. Zhou, Z. Tang and S. Chen. 2015. Mesoporous N-doped carbons prepared with thermally removable nanoparticle templates: An efficient electrocatalyst for oxygen reduction reaction. J. Amer. Chem. Soc. 137(16): 5555–5562.

Park, K.W., H.N. Yang, W.H. Lee, B.S. Choi and W.J. Kim. 2016. Effect of hybridization of Pt supported mesoporous-CMK-3 into Pt-CB as cathode catalyst on cell performance and durability in proton exchange membrane fuel cell. Microporous Mesoporous Mater. 220: 282–289.

Park, S., J.W. Lee and B.N. Popov. 2012. A review of gas diffusion layer in PEM fuel cells: Materials and designs. Int. J. Hydrogen Energy 37(7): 5850–5865.

Park, S. and B.N. Popov. 2011. Effect of a GDL based on carbon paper or carbon cloth on PEM fuel cell performance. Fuel. 90: 436–440.

Peng, H., Z. Mo, S. Liao, H. Liang, L. Yang, F. Luo et al. 2013. High performance Fe- and N-doped carbon catalyst with graphene structure for oxygen reduction. Sci. Rep. 3(10): 1765.

Qu, L., Y. Liu, J. B. Baek and L. Dai. 2010. Nitrogen-doped graphene as efficient metal-free electrocatalyst for oxygen reduction in fuel cells. ACS Nano de. 4(3): 1321–1326.

Rajalakshmi, N. and K.S. Dhathathreyan. 2008. Present trends in fuel cell technology development. Nova Science Publishers, Inc. New York.

Rajalakshmi, N., R.I. Jafri and K.S. Dhathathreyan. 2017. Research advancements in low-temperature fuel cells. pp. 35–65. *In*: Maiyalagan, T. and V.S. Saji (eds.). Electrocatalysts for Low Temperature Fuel Cells: Fundamentals and Recent Trends, John Wiley & Sons, Inc., New Jersey, United States.

Robberg, K. and V. Trapp. 2003. Graphite-based bipolar plates. pp. 308–314. *In*: Vielstich, W., H.A. Gasteiger and A. Lamm (ed.). Handbook of Fuel Cells–Fundamentals, Technology and Applications. John Wiley & Sons, West Sussex, England.

Ryoo, R., S.W. Joo, M. Kruk and M. Jaroniec. 2001. Meso-Tetra-alkynyl porphyrins for optical limiting—A survey of group III and IV metal complexes. Adv. Mater. 13(9): 652–656.

Shao, Y., G. Yin, Y. Gao and P. Shi. 2006. Durability study of Pt/C and Pt/CNTs catalysts under simulated PEM fuel cell conditions. J. Electrochem. Soc. 153(6): A1093.

Sun, S., G. Zhang, Y. Zhong, H. Liu, R. Li, X. Zhou et al. 2009. Ultrathin single crystal Pt nanowires grown on N-doped carbon nanotubes. ChemComm. (45): 7048–50.

Tang, J.M., K. Jensen, M. Waje, W. Li, P. Larsen, K. Pauley et al. 2007. High performance hydrogen fuel cells with ultralow Pt loading carbon nanotube thin film catalysts. J. Phys. Chem. C 111(48): 17901–17904.

Tseng, C.J. and S.K. Lo. 2010. Effects of microstructure characteristics of gas diffusion layer and microporous layer on the performance of PEMFC. Energy Convers. Manag. 51(4): 677–684.

Velayutham, G., J. Kaushik, N. Rajalakshmi and K.S. Dhathathreyan. 2007. Effect of PTFE content in gas diffusion media and microlayer on the performance of PEMFC tested under ambient pressure. Fuel Cells 7(4): 314–318.

Vinayan, B.P., R. Nagar, N. Rajalakshmi and S. Ramaprabhu. 2012. Novel platinum-cobalt alloy nanoparticles dispersed on nitrogen-doped graphene as a cathode electrocatalyst for PEMFC applications. Adv. Funct. Mater. 22(16): 3519–3526.

Vivekananthan, J., J. Masa, P. Chen, K. Xie, M. Muhler and W. Schuhmann. 2015. Nitrogen-doped carbon cloth as a stable self-supported cathode catalyst for air/H_2-breathing alkaline fuel cells. Electrochim. Acta 182: 312–319.

Wang, X., H. Zhang, J. Zhang, H. Xu, X. Zhu, J. Chen et al. 2006. A bi-functional micro-porous layer with composite carbon black for PEM fuel cells. J. Power Sources 162(1): 474–479.

Wang, Y., N. Zhao, B. Fang, H. Li, X. T. Bi and H. Wang. 2015. Carbon-supported Pt-based alloy electrocatalysts for the oxygen reduction reaction in polymer electrolyte membrane fuel cells: Particle size, shape, and composition manipulation and their impact to activity. Chem. Rev. 115: 3433–3467.

Watanabe, H., A. Kimura and K. Okazaki. 2016. Impact of ternary carbonate composition on the morphology of the carbon/carbonate slurry and continuous power generation by direct carbon fuel cells. Energy and Fuels 30(3): 1835–1840.

Weaver, R.D., S.C. Leach and L. Nanis. 1981. Proceedings of the 16th Intersociety Energy Conversion Engineering Conference 1: 717–721.

Wolf, H. and M. Willert-Porada. 2006. Electrically conductive LCP-carbon composite with low carbon content for bipolar plate application in polymer electrolyte membrane fuel cell. J. Power Sources 153(1): 41–46.

Xie, Z., G. Chen, X. Yu, M. Hou and Z. Shao. 2015. Carbon nanotubes grown *in situ* on carbon paper as a microporous layer for proton exchange membrane fuel cells. Int. J. Hydrogen Energy 40(29): 8958–8965.

Xu, X., W. Zhou and Z. Zhu. 2013. Samaria-doped ceria electrolyte supported direct carbon fuel cell with molten antimony as the anode. Ind. Eng. Chem. Res. 52(50): 17927–17933.

Ye, F., H. Liu, Y. Feng, J. Li, X. Wang and J. Yang. 2014. A solvent approach to the size-controllable synthesis of ultrafine Pt catalysts for methanol oxidation in direct methanol fuel cells. Electrochim. Acta 117: 480–485.

Yentekakis, I.V., P. Debenedett and B. Costa. 1989. A novel fused metal anode solid electrolyte fuel-cell for direct coal-gasification a steady-state model. Ind. Eng. Chem. Res. 28: 1414–1424.

Yu, X. and S. Ye. 2007. Recent advances in activity and durability enhancement of Pt/C catalytic cathode in PEMFC. Part I. Physico-chemical and electronic interaction between Pt and carbon support, and activity enhancement of Pt/C catalyst. J. Power Sources 172(1): 133–144.

Zhang, X., X. Cheng and Q. Zhang. 2016. Nanostructured energy materials for electrochemical energy conversion and storage: A review. J. Energy Che. 25(6): 1–18.

Zhang, C., L. Xu, N. Shan, T. Sun, J. Chen and Y. Yan. 2014. Enhanced electrocatalytic activity and durability of Pt particles supported on ordered mesoporous carbon spheres. ACS Catalysis 4(6): 1926–1930.

Zhang, J., Z. Zhong, D. Shen, J. Zhao, H. Zhang, M. Yang et al. 2011. Preparation of bamboo-based activated carbon and its application in direct carbon fuel cells. Energy & Fuels 25(5): 2187–2193.

Zhang, W., P. Sherrell, A.I. Minett, J.M. Razal and J. Chen. 2010. Carbon nanotube architectures as catalyst supports for proton exchange membrane fuel cells. Energy Environ. Sci. 3(9): 1286–1293.

Zhao, G., J. He, C. Zhang, J. Zhou, X. Chen and T. Wang. 2008. Highly dispersed pt nanoparticles on mesoporous carbon nanofibers prepared by two templates. J. Phys. Chem. C 112(4): 1028–1033.

Zhou, Y., K. Neyerlin, T.S. Olson, S. Pylypenko, J. Bult, H.N. Dinh et al. 2010. Enhancement of Pt and Pt-alloy fuel cell catalyst activity and durability via nitrogen-modified carbon supports. Energy Environ. Sci. 3(10): 1437.

Biomass Carbon: Prospects as Electrode Material in Energy Systems

P. Kalyani[1], and *A. Anitha[2]*

1. Introduction

Energy is an indispensable commodity and undoubtedly, it is the key to measure the quality of our lives in all sectors of modern economies. It has been projected that the world energy usage is doubling every 14 years (Smil 2008) and the need is exponentially escalating faster these days. But continuous usage of fossil fuels will result in its depletion and human lives without fuels are unimaginable and unmanageable. As a consequence, the combination of increasing need and diminishing supply leads to the energy crisis ultimately. By the year 2030, world-wide energy demand is projected to be at least twice of today's level. So this will lead to a situation where we would be forced to depend on imported energy at an inflated price. Needless to mention that pollution-free green energy systems are sought as replacement for currently used petroleum based fuels to meet energy requirements of today's demand. Hence for this reason, we are in dire need of alternative, cheap, eco-friendly and zero-side products based energy sources (Jupe et al. 2007).

Needless to mention that in today's society we stand before a change in energy paradigm and to face this situation, we need to develop technology that is compatible with the resources provided by nature and eventually to realise sustainability. Thus various energy storage technologies have been developed and to mention specifically, supercapacitors, batteries, fuel cells and conventional capacitors are the typical smart energy technologies. On comparing the energy storage systems, supercapacitors are superior over all other devices because of high power density (discharge at high current density), short charging time, long cycle life (no chemical reactions), high coulombic efficiency (high reversibility), and environmental friendliness (no heavy metals used), even though their energy density is lower. The charge storage mechanism in supercapacitor is predominantly due to double layer charging effects. It is further known that the performance of the electrode material is one of the dominating factors influencing the overall performance of a supercapacitor.

Electrode materials used in the capacitors (Yong et al. 2009) are of three main categories: carbon-based, transition metal oxides, and conducting polymers. Carbon based electrodes are much cheaper than metal oxides and conducting polymers and they have specific surface area larger than

[1] Department of Chemistry, DDE, Madurai Kamaraj University, Madurai 625 021.
[2] PG Department of Chemistry, V.V. Vanniaperumal College for Women, Virudhunagar 626 001.
  E-mail: anitha.karuppasamy@gmail.com
* Corresponding author: mkuddechemist@gmail.com

the latter. Various carbon-based materials such as activated carbons (Wang et al. 2007, Aida et al. 2007), carbon aerogels (Fang and Binder 2007, Liu et al. 2007), graphites (Mitra and Sampath 2004, Wang and Yoshio 2006, Gomibuchi et al. 2006), carbon nanotubes (Honda et al. 2007, Katakabe et al. 2005, Merritt 2003), carbon nanofibers (Xu et al. 2007, Kim and Lee 2004), and nano sized carbons (Sivakkumar et al. 2007, Honda et al. 2004, Eikerling et al. 2005), have been investigated for use as electrode materials of supercapacitors because of their accessibility, easy processability, non-toxicity, high chemical stability, and wide temperature range. At present activated carbons is widely used for the capacitors because of its high porosity and high surface area that favors good charge accumulation at the interface with the electrolyte and therefore high capacitance can be tapped.

World demand for virgin activated carbon is forecast to expand by 9 per cent per annum through 2014 to 1.7 million metric tons (World Activated Carbon Industry 2014). Activated carbon demand will benefit from a continuing intensification of the global environmental movement as well as rapid industrialisation. In most developing and developed countries, the use of activated carbon in pharmaceutical sector offers the strongest growth prospect. However, the preparation processes for these carbons require expensive raw materials, enormous time, energy and tedious preparation procedures.

Activated carbon production costs can be reduced by either choosing a cheap raw material or by applying a proper production method (Lafi 2001). Nevertheless, it is still a challenge to prepare activated carbon with very specific characteristics, such as even pore size distribution and using low-cost raw materials processed at low temperature (less energy costs). Therefore, it is necessary to find suitable low-cost raw materials that are economically attractive and at the same time present similar or even better characteristics than the conventional ones. The use of waste materials for the preparation of activated carbons is very attractive from the point of view of their contribution to decrease the costs of waste disposal, therefore helping environmental protection (Dias et al. 2007). Any cheap material with a high carbon content, low in organics can be used as a raw material for the production of activated carbon (Bansal et al. 1998). This means that the production of materials, especially from cheap and natural bio-precursors, namely the biomass, would be a highly forthcoming theme in today's science and engineering of materials. One way to produce activated carbons is to utilise the biomass wastes, the type of value added product which is being utilised as the electrode materials in energy systems like supercapacitors and lithium ion batteries.

Biomasses are derived by products of dried vegetation, crop residues and even garbages. It often refers to plants or plant-based materials that are not used for food or feed or for medical applications. An interesting fact is that people have been using biomass longer than any other energy source. Added to the above, biomass is also a renewable energy source because we can grow more in a short period of time. Biomass does not add carbon dioxide to the atmosphere as it absorbs the same amount of carbon in growing as it releases when consumed as a fuel (Magrini-Bair et al. 2009). Literature gives the main sources of biomass as (a) forest crops and residues, (b) agricultural crops and residues, (c) industrial residues, (d) animal residues and (e) municipal solid wastes and sewage.

Biomass is presently the fourth largest energy source (next to coal, oil and natural gas) in the world and provides about 14 per cent of world's total energy and 38 per cent of energy in developing countries demand (Hall et al. 1999). Different types of biomass materials are being used as the starting materials in producing (bio) carbon powders. Indeed, the choice of biomass materials for producing carbon for specific applications is based on their availability, cost and the ability to be converted into highly porous carbon powders after carbonisation. In this context, abundant agricultural discards/ wastes or refuse offer a secondary, inexpensive and renewable source of carbon for multifarious applications. These bio waste materials might have little or no economic value and may often present a disposal problem by and large. Therefore, there is a need to consume these low cost biomaterials in such a way that their conversion into activated carbon would add economic value, help reduce the cost of waste disposal and most importantly provide a potentially inexpensive alternative to the existing high-cost commercial activated carbon powders.

Based on the foregoing content, it is proposed that biomass can be effectively used to prepare carbon powders with diversified physical and chemical features and functionalities and these biomass derived carbon samples can be utilised for the application as electrode material for capacitors. Now researchers are exploring the possibility of employing unconventional yet novel precursors namely waste biomass to produce activated carbon for the application as supercapacitor electrode material without compromising supercapacitor performance. Supercapacitors making use of biomass derived porous carbon have the advantages like production of low cost carbon electrode components, environmental friendliness and good capacitive performance. All the more, the supercapacitors with biomass carbon electrodes stand as an important class of technology where one of the 3R principles in waste management is followed. The most popular and frequently used mantra in the area of production and consumption is the 3R principles and has been expanded as Reduce, Reuse and Recycle of commodities. The same principle is being applied to environmental protection and utilisation of wastes.

The current chapter provides insight on how biomass (zero-cost) can be effectively used to prepare carbon powders with diversified physical and chemical features and functionalities and also how the carbon samples can be utilised for the possible application as electrode material for energy systems like supercapacitors and lithium ion batteries. Various methodologies involved in the biomass carbon preparation, classification and physical/electrochemical characterisation techniques and finally their prospects as electrode materials in energy systems are discussed in the following sections.

2. Preparation of Biomass Carbon

2.1 Preparation of Biomass Carbon by Physical Activation

a. By pyrolysis

Pyrolysis or gasification is a thermo-chemical process that has been widely used to prepare biomass carbon. Pyrolysis is one form of energy recovery process, which has the potential to generate char, oil and gas products (Putun et al. 2005). During thermal treatment, the moisture and volatile matter contents of the biomass are removed and the remaining solid char shows remarkable difference in properties like surface area and pore structure on compared with the parent biomass materials. Pyrolysis is largely influenced by process parameters like particle size, temperature and heating rate. The process conditions can be optimised to maximise the production of biomass carbon.

Pyrolysis temperature has the most significant effect—followed by pyrolysis heating rate, the nitrogen flow rate and then finally the pyrolysis residence time. In general, higher pyrolysis temperature leads to a decreased yield of biomass carbon. The decrease in yield with increasing temperature could either be due to greater primary decomposition of biomass at higher temperature or through secondary decomposition of char residue. Varieties of biomass carbon are prepared by pyrolysis and they have extensive applications as electrode materials for supercapacitors and lithium ion batteries. Table 1 lists the biomass carbon prepared by pyrolysis, and its application (either as electrode materials for supercapacitors/lithium ion batteries), morphology and the maximum specific capacitance/specific capacity offered by them.

b. By Steam and CO_2 Activation

Steam/CO_2 are two-step physical activation process. It involves carbonisation of biomass followed by the activation of the resulting char at elevated temperature in the presence of suitable oxidising gases like CO_2, steam, air or their mixtures. List of biomass carbon prepared by steam and CO_2 activation, its application and the maximum efficiency are tabulated in Tables 2 and 3 respectively.

Table 1. Various biomass materials used as electrode materials for supercapacitors and Lithium ion batteries.

S. No.	Biomass	Electrode Material (supercapacitor/ lithium ion batteries)	Max. sp. Capacitance ($F\ g^{-1}$)/ Sp. capacity ($mAh\ g^{-1}$)	Description of Morphology	Reference
1	Tea dust	supercapacitor	69	No definite morphology	(Kalyani and Anitha 2015)
2	Jack fruit seed	supercapacitor	203	Crushed	(Kalyani and Anitha 2014)
3	Onion peel	supercapacitor	206	Crumbled	(Anitha and Kalyani 2014)
4	Neem leaves	supercapacitor	0.22	–	(Shekhar et al. 2014)
5	Neem leaves	supercapacitor	400	Highly porous network	(Biswal et al. 2013)
6	Durian Shell	supercapacitor	104	Uneven surface with aggregated particles	(Ong et al. 2012)
7	Egg shell	supercapacitor	297	Fibrous	(Li et al. 2012a,b)
8	Sea weed (Lesonia nigrescens)	supercapacitor	255	–	(Bichat et al. 2010)
9	Sea weed (Lesonia nigrescens)	supercapacitor	198	Aggregates of multiwalled nanocapsules**	(Raymundo-Pinero et al. 2006)
10	Waste news paper	supercapacitor	300	Good network of interconnected pores	(Misra 2006)
11	Duck weed	lithium ion batteries	1071	Porous	(Zheng et al. 2017)
12	Ox horn	lithium ion battery	1181	3D porous nanostructure	(Ou et al. 2015)
13	Willow Leaves	lithium ion battery	230–260	Spongy network	(Sun et al. 2013a)
14	Pomelo peel	lithium ion battery	452	Featureless wrinkle morphology	(Sun et al. 2013b)
15	Rice husk	lithium ion battery	691	–	(Fey and Chen 2001, Fey et al. 2010)
16	Peanut shell	lithium ion battery	3504	Porous network	(Fey et al. 2011)
17	Potato starch	lithium ion battery	531	Rough surface with many small pores	(Li et al. 2011)
18	Mangrove Charcoal	lithium ion battery	798	Highly disordered with micropores	(Liu et al. 2010)
19	Sugarcane Baggase	lithium ion battery	310	Honeycomb, flake, and plate structure	(Matsubura et al. 2010)
20	Cherry stones	LIB	200	Uneven surface and spongy	(Arrebola et al. 2010)

Table 1 contd....

Table 1 contd....

S. No.	Biomass	Electrode Material (supercapacitors/ lithium ion batteries)	Max. sp. Capacitance (Fg⁻¹)/ Sp. capacity (mAh g⁻¹)	Description of Morphology	Reference
21	Tea leaves	lithium ion battery	63	Bead like with porous structure	(Bhardwaj et al. 2007, 2008)
22	Bamboo Stem	lithium ion battery	3123	Bundled structure	
23	Coconut fiber	lithium ion battery	38	Heavily porous nature	
24	Jack fruit seed	lithium ion battery	130	Cotton ball like with pores disorganised	
25	Date seed	lithium ion battery	16	Fossilised porous rock with large pores	
26	Neem seed	lithium ion battery	15	Rectangular block with cavities	
27	Soap-nut seed	lithium ion battery	11	Cotton ball like with pores disorganised	
28	Peanut shell	lithium ion battery	4765	Finely divided without ordering	(Fey et al. 2003)
29	Cotton wool	lithium ion battery	600	–	(Peled et al. 1998)
30	Sugar	lithium ion battery	170–650	–	(Xing et al. 1996)

**from TEM

Table 2. Biomass carbon prepared by steam activation as electrode materials for supercapacitors and Lithium ion batteries.

S. No.	Biomass	Electrode Material (supercapacitors/ lithium ion batteries)	Description of Morphology	Max. sp. Capacitance (F g^{-1})/sp. Capacity (mAh g^{-1})	Reference
1	Rice husk + Beet sugar	supercapacitors	–	114	(Kumagai et al. 2013)
2	Oil palm empty fruit bunch	supercapacitors	–	150	(Farma et al. 2013)
3	Chicken dropping	supercapacitors	–	25	(Sato et al. 2013)
4	Sucrose	supercapacitors	Irregular morphology	160	(Wei and Yushin 2011a,b)
5	Rubberwood saw dust	supercapacitors	–	33	(Taer et al. 2010)
6	Pinecone hull	lithium ion battery	–	357	(Zhang et al. 2007)
7	Cotton cloth	lithium ion battery	–	1400	(Isaeva et al. 2003)

Table 3. Biomass carbon prepared by CO_2 activation as electrode materials for supercapacitors and Lithium ion batteries.

S. No.	Biomass	Electrode Material (supercapacitors/ lithium ion batteries)	Description of Morphology	Max. sp. capacitance (F g^{-1})/Sp. Capacity (mAh g^{-1})	Reference
1	Apple pulp	supercapacitors	–	109–187	(Centeno et al. 2009)
2	Pistachio shells	supercapacitors	–	60–125	(Wu et al. 2006)
3	Bamboo	supercapacitors	–	5–60	(Kim et al. 2006)
4	Firwood	supercapacitors	Honey comb	120	(Wu et al. 2005)

2.2 Preparation of Biomass Carbon by Chemical Activation

In the chemical activation process the two steps are carried out simultaneously, with the precursor being mixed with chemical activating agents, as dehydrating agents and oxidants. Chemical activation offers several advantages since it is carried out in a single step, combining carbonisation and activation, performed at lower temperatures and therefore resulting in the development of a better porous structure. Generally, chemical activation is the preferred route as it achieves higher yields, larger surface areas, low operating temperatures and is cost effective mostly. The most commonly used chemical activating agents are K_2CO_3, H_3PO_4, $ZnCl_2$, H_3PO_4, NaOH, HNO_3 and KOH. List of biomass carbon prepared by activating agents like $ZnCl_2$, KOH and others, its application and the maximum efficiency are tabulated in Tables 4, 5 and 6 respectively.

2.3 Preparation of Biomass Carbon by Combined Physical and Chemical Activation

As the physical and electrochemical properties of the biomass carbon largely depends on the preparation procedure, both physical and chemical activation methods are employed to obtain biomass carbon with entirely different properties. Various biomass carbons prepared by combined physical and chemical activation methods are listed in Table 7.

It is indicative from the above tables that the supercapacitor electrode materials are mostly prepared by KOH and $ZnCl_2$ activation. The ratio of activating agent/biomass varied from 1:1 to 5:1 (Babel and Jurewicz 2004). The excess activating agent is removed by washing with 0.1 M HCl solution then washed with water until pH 7 is achieved. Samples are then oven dried ready to be used in electrochemical testing and other characterisations.

Biomass derived activated carbons exhibit a high surface area (BET) which exceed 2500 m^2 g^{-1} and pore volume close to 2 cm^3 g^{-1} (Simon et al. 2013). The pores in the activated carbon are mainly micropores and least number of mesopores, because at higher temperatures most of mesopores break with decrease in pore diameter (Pagketanang et al. 2015).

Table 4. Biomass carbon prepared by ZnCl$_2$ activation as electrode materials for supercapacitors and Lithium ion batteries.

S. No.	Biomass	Electrode material (supercapacitor/ lithium ion batteries)	Description of Morphology	Max. sp. Capacitance (F g^{-1})/sp. Capacity (mAh g^{-1})	Reference
1	Potato waste	supercapacitor	Gravel-like particles	255	(Ma et al. 2015)
2	Papaya seed	supercapacitor	Spongy	472	(Kalyani et al. 2015)
3	Tea dust	supercapacitor	No definite morphology	98	(Kalyani and Anitha 2015)
4	Poplarcatkins	supercapacitor	Tubular	206	(Wei 2014)
5	Peanut shell + Rice husk	supercapacitor	–	245	(He et al. 2013a)
6	Rice husk	supercapacitor	Mesoporous**	243	(He et al. 2013b)
7	Ramie fibres	supercapacitor	–	287	(Du et al. 2012)
8	Sugarcane bagasse	supercapacitor	Nanoporous	138	(Si et al. 2011)
9	Sugarcane bagasse	supercapacitor	–	300	(Rufford et al. 2010)
10	Coffee shells	supercapacitor	Loose, disjointed structure with no definite shape	150	(Jisha et al. 2009)
11	Cationic starch	supercapacitor	Smooth with a few pores	136	(Wang et al. 2008)
12	Waste coffeeground	supercapacitor	–	368	(Rufford et al. 2008)
13	Banana fibre	supercapacitor	Fibrous stacks	74	(Subramanian et al. 2007)

** from TEM

Table 5. Biomass carbon prepared by KOH activation as electrode materials for supercapacitors and Lithium ion batteries.

S. No.	Biomass	Electrode Material (supercapacitor/ lithium ion battery)	Description of Morphology	Max. sp. Capacitance (F g^{-1})/ sp. Capacity (mAh g^{-1})	Reference
1	Soya bean	supercapacitor	Plenty of cavities	248	(Zhou et al. 2016)
2	Corncob	supercapacitor	Vast irregular granules	328	(Wang et al. 2015)
3	Coconut shell	supercapacitor	porous structure	356	(Jain and Tripathi 2014)
4	Cow dung	supercapacitor	Rough surface and highly porous	173	(Bhattacharjya and Yu 2014)
5	Eucalyptus leaves	supercapacitor	Spherical	442	(Jain and Tripathi 2013)
6	Coconut shell	supercapacitor	–	220	(Galinski et al. 2013)
7	Coconut shell + Melamine precursor	supercapacitor	–	368	(Jurewicz and Babel 2010)
8	Human hair + glucose	supercapacitor	Irregular spheres	264	(Si et al. 2013)
9	Water hyacinth	supercapacitor	Porous	509	(Senthil Kumar et al. 2013)
10	Waste tea leaves	supercapacitor	Nanoporous	330	(Peng et al. 2013)
11	Argan seed shell	supercapacitor	–	355	(Elmouwahidi et al. 2012)
12	Sunflower seed shell	supercapacitor	–	311	(Xiao et al. 2011)
13	Wheat straw	supercapacitor	Honeycomb	251	(Xueliang et al. 2010)
14	Sago waste	supercapacitor	Surface with small pits	64	(Aripin et al. 2010)

Table 5 contd. ...

...Table 5 contd.

S. No.	Biomass	Electrode Material (supercapacitor/ lithium ion batteries)	Description of Morphology	Max. sp. Capacitance (F g⁻¹)/ sp. Capacity (mAh g⁻¹)	Reference
15	Potato starch	supercapacitor	Irregular spheres	335	(Zhao et al. 2009)
16	Cherry stone waste	supercapacitor	–	230	(Olivares-Marin et al. 2009)
17	Firwood	supercapacitor	Honey comb	165	(Wu et al. 2006)
18	Bamboo	supercapacitor	–	15–65	(Kim et al. 2006)
19	Cherry stones	supercapacitor	–	174–232	(Centeno et al. 2009)
20	Recycled waste paper	supercapacitor	Interconnected in mesoporous range	180	(Kalpana et al. 2009)
21	Corn grains	supercapacitor	–	257	(Balathanigaimani et al. 2008)
22	Cationic starch	supercapacitor	Shallow concave round pores	238	(Wang et al. 2008)
23	Banana fiber	supercapacitor	Loose disjoint structure	66	(Subramanian et al. 2007)
24	Pistachio shells	supercapacitor	–	60–125	(Hu et al. 2007)
25	Rice husk	supercapacitor	–	180	(Guo et al. 2003)

Table 6. Biomass carbon prepared by various activating agents as electrode materials for supercapacitors and Lithium ion batteries.

S. No.	Biomass	Activating Agent	Description of Morphology	Electrode Material (supercapacitor/ lithium ion batteries)	Max. sp. Capacitance (F g⁻¹)/ sp. Capacity (mAh g⁻¹)	Reference
1	Waste tea	H_3PO_4	Irregular grainy with many surface voids	supercapacitor	123	(Isil Gurten Inal et al. 2015)
		K_2CO_3	Rough surface with many craters		203	
2	Tea dust	H_3BO_3	Interconnected porous particles	supercapacitor	203	(Kalyani and Anitha 2013)
3	Waste paper	HNO_3	Loose packed wire-like	supercapacitor	232	(Liu et al. 2012)
4	Sorghum pith	NaOH	Fibrous network with cavities	supercapacitor	220–320	(Senthil Kumar et al. 2011)
5	Rice husk	NaOH	–	supercapacitor	210	(Guo et al. 2003)

It is to be mentioned that microwave assisted heat treatment is one of the modern methods of producing activated carbon for supercapacitor applications as it is a facile, controllable, fast, and energy saving technique. Carbon xerogel on chemical activation using microwave radiation, results in micropores and mesopores formation in a time range between 6–30 min (Calvo et al. 2013). Microwave radiation procedure produces well modified surface nature with significant reduction in micropore volume and size (Puligundla et al. 2016, Nabais et al. 2004).

Next, hydrothermal carbonisation (HTC) has been suggested to be an alternate way to produce porous carbon for supercapacitor electrodes. Mixture of water and carbon precursor are thermally treated at temperature ranges of 150–300°C and 300–800°C for low and high temperature HTC, respectively (Zhang et al. 2015a, Sevilla and Fuertes 2009). This method gives high solid carbon yield, reduces the oxygen and hydrogen content in biomass carbon (Zhang et al. 2015b, Kalderis et al. 2014)

Table 7. Biomass carbon prepared by combined physical and chemical activation as electrode materials for supercapacitors and Lithium ion batteries.

S. No.	Biomass	Activating Agent	Electrode Material (supercapacitor/ lithium ion batteries)	Description of Morphology	Max. sp. capacitance (F g^{-1})/ sp. Capacity (mAh g^{-1})	Reference
1	Cotton	Pyrolysis/ KOH	supercapacitor	Helical and tubular	224	(Cheng et al. 2016)
2	Rice straw	Pyrolysis/ KOH	lithium ion battery	Parallel macroporous channels	599	(Zhang et al. 2009)
3	Cationic starch	$ZnCl_2 + CO_2$	supercapacitor	Fissured surface	139	(Wang et al. 2008)
4	Walnut shell	Physical/ Chemical	supercapacitor	–	292	(Jing et al. 2008)
5	Pistachio shell	$KOH + CO_2$	supercapacitor	–	25–47	(Hu et al. 2007)
6	Coffee shells	Pyrolysis/ KOH	lithium ion battery	Flake like	1200	(Hwang et al. 2007)
7	Banana fiber	Pyrolysis/ KOH	lithium ion battery	Fibrous stack	3123	(Manuel Stephen et al. 2006)
8	Firwood	$KOH + CO_2$	supercapacitor	Honey comb	197	(Wu et al. 2005)

along with very high BET surface area. Jain et al. (2015) obtained BET and mesopore areas of up to 2440 and 1121 m^2 g^{-1}, respectively on hydrothermal treatment of coconut shells with $ZnCl_2$ and H_2O_2 at temperature of 275°C. On chemical activation of the same material using $ZnCl_2$ at 500°C for 3 h by Azevedo et al. (2007), BET surface area of 1266 m^{2}g^{-1} was obtained. The chemical and structural properties of carbon derived from *prosopis africana* (common name: African mesquite/iron tree) waste plant material prepared by conventional hydrothermal and microwave assisted hydrothermal carbonisation was compared by Elaigwu and Greenway (2015). The authors concluded that microwave assisted hydrothermal carbonisation was faster in decomposing the *prosopis africana* and the degree of structure alteration was achieved within a short time when compared to the conventional approach (Elaigwu and Greenway 2015).

Ionothermal carbonisation is an attractive method to synthesis of porous carbon from biomass in single step using ionic liquids with high chemical and thermal stability, low melting point, low electrical and ionic conductivity and negligible vapour pressure. This method intends to produce carbon from glucose with high surface area of 2160 m^2 g^{-1} and total pore volume of 1.74 cm^3 g^{-1} (Pampel et al. 2016, Chang et al. 2015). The gravimetric capacitance obtained was 206 F g^{-1}.

Microporous and mesoporous carbon from fructose was synthesised through one step ionothermal method using iron by Lin et al. (2016). The surface area of 1200 m^2 g^{-1}, pore volume of 0.8 cm^3 g^{-1} and specific capacitance of 245 F g^{-1} at current density of 1 A g^{-1} were obtained. In the study different masses of fructose were dispersed in 10 ml of 1-butyl-3-methylimidazolium tetrachloroferrate [Bmim] [$FeCl_4$] were used as template, solvent and catalyst with advantage that it can be reused after recovery. Many other biomass derived structured carbon was produced ionothermally from carbohydrates (Xie et al. 2011), sugarcane baggase (Zhang et al. 2014) and bamboo (Guo and Fang 2014). However these carbons were not tested for supercapacitor applications.

Porous carbon from biomass can also be prepared by molten salt carbonisation (MSC) method in which the molten salt cracks the large molecules of biomass. In MSC the salt involved is melted at its melting point then biomass is immersed into the molten salt and carbonised at temperature greater than 400°C in inert atmosphere. After carbonisation the furnace is cooled to room temperature and the product washed with HCl and distilled water in order to remove salts within the product. It is reported that the particle size of the biomass alters the yield (Liu et al. 2014).

The presence of surface functionalities and heteroatoms such as O and N on carbon also play an important role in the pseudocapacitance behaviour of the electrode (Zhao et al. 2015, Seredych et al. 2008) Oxygen comes from both the activation and biomass itself, while nitrogen can come from the biomass or introduced into the carbon through doping (Cheng and Teng 2003). It has been reported that the capacitance of O and N containing carbonised chicken egg-shell membrane is 297 F g^{-1} with cyclic efficiency of 97 per cent after 10,000 cycles (Li et al. 2012 a/b). The activated carbon from the same precursor has specific capacitance of 203 F g^{-1} despite the fact that the specific surface area was 7 times higher than carbonised chicken egg-shell membrane. It has been reported also that oxygen and nitrogen rich activated carbon enhances the specific capacitance differently. While oxygen rich activated carbon exhibits lower capacitance because the electrolyte diffusion into pores is hindered by carboxyl surface, the nitrogen rich one exhibits higher capacitance (Elmouwahidi et al. 2012, Li et al. 2013). Furthermore it is reported that the adsorption and transport of electrolyte ions is enhanced by doping activated carbon with heteroatom such as sulphur (Zhao et al. 2015, Zhang et al. 2014, Seredych and Bandosz 2013). Heteroatom increases the wettability of the electrode which in turn increases the capacitance thus studies on doping different heteroatoms on the activated carbon are emphasised.

Understanding the various preparation processes of the activated carbon prepared from various biomass, the structure of biomass derived activated carbon materials are important in order to achieve the maximum efficiency on using as electrodes in supercapacitors and lithium ion batteries and hence a short description has been provided for better understanding.

3. Structure of Activated Carbon

The three important structure of activated carbons are described below.

a. Porous structure

All activated carbons show porous characteristics such as surface area, pore volume and pore size distribution and contain up to 15 per cent of mineral matter in the form of ash content (Bansal et al. 1988). The porous structure of activated carbon may be presumed to have been developed during the carbonisation process and further developed during activation also whereby tar and other carbonaceous materials which might be present in the spaces between the elementary crystallites got cleared off. The structure of pores and pore size distribution largely depends on the nature of the raw material and activation process route. The activation process removes disorganised carbon by exposing the crystallites to the action of activating agent which leads to the development of porous structure. Specific surface area and porosity are determined by physical adsorption of N_2 on the surface of activated carbon.

The pore systems of activated carbon are of different kinds and the individual pores may vary greatly both in size and shape.

Active carbons are associated with pores starting from less than a nanometer to several thousand nanometers. A conventional classification of pores according to their average width, which represents the distance between the walls of slit shaped pore or the radius of a cylindrical pore, proposed by Dubinin (Dubinin 1960) and officially adopted by the International Union of Pure and Applied Chemistry (IUPAC).

The pores of activated carbons are classified in three groups:

i) Micropores (diameter (d) < 2 nm),
ii) Mesopores (2 nm <d< 50 nm)
iii) Macropores (d > 50 nm)

The micropores (Moreno-Castilla 2004) constitute the largest part of the internal surface and are accessible to the adsorptive molecules. Generally, micropores contribute at least 90 per cent of the total surface area of an activated carbon, whereas the surface area of mesopores does not constitute more than

5 per cent of total surface area and the mesopore volume varies in between 0.1 and 0.2 cm³ g⁻¹. The contribution of macropores to the total surface area and pore volume is very small and does not exceed 0.5 m² g⁻¹ and 0.2 to 0.4 cm³ g⁻¹, respectively.

b. Crystalline structure

Microcrystalline structure of activated carbon starts to develop during the carbonisation process. The crystalline structure of activated carbons differed from the graphite with respect to the interlayer spacing. The interlayer spacing ranges between 0.34 and 0.35 in active carbons, and is 0.335 in the case of graphite. The basic structural unit of activated carbon is closely approximated by the structure of graphite. Based on the graphitising ability, activated carbons are classified into two types namely,

  i) graphitising carbon
 ii) non-graphitising carbon

Graphitising carbon had a large number of graphite layers oriented parallel to each other. The carbon obtained is delicate due to the weak cross linking between the neighbour micro crystallites and has a less developed porous structure. The non-graphitising carbons are hard due to strong crosslinking between crystallites and show a well-developed microporous structure (Franklin 1951). The formation of non-graphitising structure with strong cross links is promoted by the presence of associated oxygen or by an insufficiency of hydrogen in the original raw material. The schematic representations of the structures of graphitising and non-graphitising carbons are shown in Fig. 1.

c. Chemical structure

Besides the porous and crystalline structure, an active carbon surface has a chemical structure as well. It is known that the adsorption capacity of air (or water) on activated carbon is determined by its porous structure and is strongly influenced by a relatively small amount of chemically bonded heteroatoms like oxygen and hydrogen (Bansal et al. 1988). Moreover, activated carbons are invariably associated with significant amounts of oxygen, hydrogen (Rodriguez-Reinoso and Molina-Sabio 1992) and other heteroatoms like sulphur, nitrogen and halogens (Valix et al. 2006). These heteroatoms derived from the raw material involve in the structure of activated carbon during carbonisation process or they may be chemically bonded to the surface during activation (Rodriguez-Reinoso 1998). Literature show that the heteroatoms are bonded to carbon atoms of the edges and corners of the aromatic sheets or to the carbon atoms at defect positions to form carbon-oxygen, carbon-hydrogen, carbon-sulphur, carbon-nitrogen and carbon-halogen surface compounds, known as surface groups or surface complexes (Valix et al. 2006, Castro-Muniz et al. 2011).

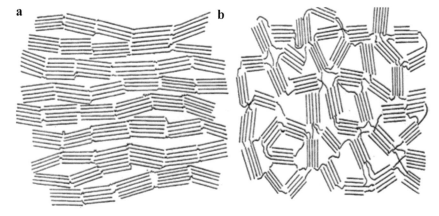

Fig. 1. Schematic illustration of structure of activated carbon: (a) graphitised carbon and (b) non-graphitised carbon (Adapted with modification from Franklin 1951).

4. Physico-Chemical Characterisation Techniques for Activated Carbon

Four techniques have been used to characterise the activated carbons to know about its structure, surface functional groups, morphology and porosity. To illustrate these techniques, carbon derived from apple pulp (Hesas et al. 2013) has been considered as a typical example.

a) X-ray diffractometry

The crystalline nature of activated carbon is identified by qualitative X-ray powder diffractometry (XRD). The identification of compounds is performed through comparison with standards in the Joint Committee of Powder Diffraction Standards (JCPDS). As an example, XRD pattern of activated carbon prepared from apple pulp is given in Fig. 2.

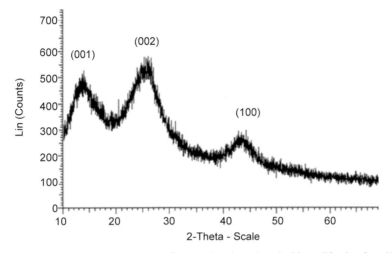

Fig. 2. Typical XRD pattern of the activated carbon from apple pulp (Adapted with modification from Hesas et al. 2013).

b) Scanning electron microscopy

Scanning electron microscopy (SEM) was used to identify the surface morphology. SEM image of activated carbon from papaya seed is given in Fig. 3.

Fig. 3. SEM image of the activated carbon from papaya seed (Adapted with modification from Kalyani et al. 2015).

c) N_2 adsorption/desorption studies

Specific surfaces area and porosity are determined by physical adsorption of nitrogen (N_2) gas molecules on the surface of the activated carbon. The N_2 adsorption-desorption isotherms of activated carbons derived by Hesas et al. (2013) from apple pulp is shown in Fig. 4. The isotherm of activated carbon from apple pulp has been basically classified as type II isotherm indicating unrestricted monolayer-multilayer adsorption. The filling of the micropores with nitrogen molecules occurs in the initial part of the type II isotherm, whereas the slope of the plateau at high relative pressure represents multilayer adsorption through macropores and external surface area. The changes in the slope of the isotherm at point $p/p_o > 0.1$ indicate that the monolayer coverage stage is completed and multilayer adsorption is about to begin.

d) Fourier transform infrared spectroscopy

The chemical functionality on the surface activated carbons (Fig. 5) has been qualitatively identified by Fourier transform infrared spectroscopy (FTIR).

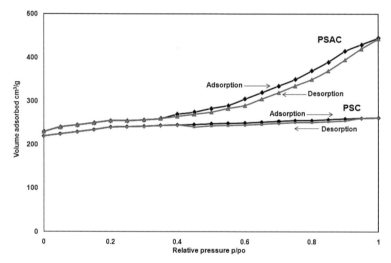

Fig. 4. N_2 adsorption isotherm of activated and virgin carbon derived from papaya seed (Adapted with modification from Kalyani et al. 2015).

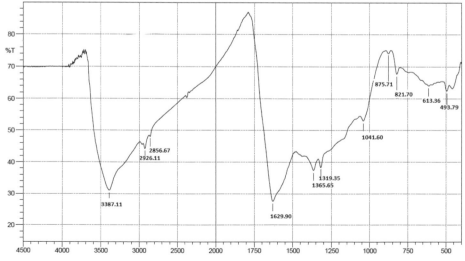

Fig. 5. Typical FTIR spectra of the activated carbon from onion peel (Adapted with modification from Anitha and Kalyani 2014).

5. Galvanostatic Charge-Discharge Studies

One of the main advantages of the EDLC when compared to the battery is the excellent cycle stability. Cycle stability is determined by galvanostatically charging/discharging the cell at constant currents between defined limiting voltages. The specific capacitance of the supercapacitor electrodes was also determined by galvanostatic charge-discharge technique. A short description on galvanostatic charge-discharge cycling studies has been provided here for better understanding of the technique.

Galvanostatic charge-discharge cycling studies (Chronopotentiometry) is a basic electrochemical technique to determine the capacitance of supercapacitors. In this technique constant current is applied to the electrodes. The potential of the working electrode against the reference electrode is measured as a function of time.

In non-faradaic processes, on applying constant current, the potential varies linearly with time due to interfacial charging at the interface between the electrode and the electrolyte. As soon as the current is applied, the potential jump appears instantaneously. In pseudocapacitors or hybrid capacitors, the variation of potential is not absolutely linear with time due to the existence of faradaic reactions. In the typical chronopotentiogram of the charging process, the voltage increases linearly at the beginning of the charging as the double layer capacitance at the electrode is charged until the voltage at which the non-faradaic reaction begins is reached. Then, the increase of voltage slows down due to the pseudocapacitive chemical process until the reactants in electrolyte near the surface are exhausted and the diffusion of reactants is no longer sufficient to maintain the reaction. The chronopotentiogram in this range also presents linearity with smaller slope if the pseudocapacitive process occurs continuously. Otherwise, the chronopotentiogram shows nonlinearity in this range. After the exhaustion of the reactant near the surface of the electrode, the voltage increases more quickly again. The slope change would not be obvious as the non-faradaic reactions occurs continuously over the whole voltage window.

Turning to the side of an ideal capacitor, the discharge curves are linear in the total range of potential with constant slopes, showing perfect capacitive behaviour. Galvanostatic charge/discharge of commercial activated carbon in organic medium (Fig. 6) presents a perfect performance of capacitor. The discharge capacitance of the electrodes (C) was calculated from the slope of the discharge curve,

$$C = I/(dV/dt)$$

where C is the cell capacitance in Farad (F), I the discharge current in Ampere (A) and dV/dt is the slope of the discharge curve in volts per second (V s^{-1}).

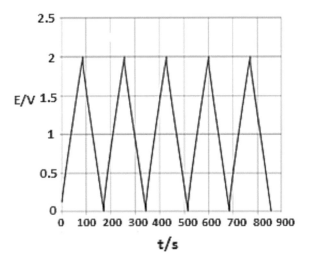

Fig. 6. Galvanostatic charge/discharge of commercial activated carbon (Adapted with modification from Fuertes et al. 2005).

$$C(F^{-1}) = i\Delta t/m\Delta V,$$

where i is the current used for charge-discharge; Δt is the time elapsed for the charge or discharge; m is the mass of the active electrode and ΔV is the voltage interval of the charge or discharge (Lee et al. 2004). The above equation was used to calculate the specific capacitance of the electrode in the three electrode configuration. On using two electrode cell assembly, the specific capacitances of the materials are calculated according to the following equation (Zhao et al. 2009):

$$C = 2It/m\Delta V$$

where C is the single electrode specific capacitance of electrode material, I is discharge current, t is discharge time, ΔV is voltage drop during discharge and m is the mass of electrode material in one electrode.

Supercapacitors store energy based on two capacitive behaviours viz., Electrical Double Layer (EDL) capacitance and pseudocapacitance. While the former is due to electrostatic interaction, pseudocapacitance is due to faradic phenomenon involving fast and reversible electrochemical reactions. In pseudocapacitance, the redox reactions are between the electrolyte and the electrode materials. The product of redox reactions are therefore electrons which are then transferred through electrode/electrolyte interfaces.

The number of charges stored in the electrode is proportional to the surface of the electrode, and energy stored is proportional to the amount of charge stored. Therefore materials with high surface area is important because the energy in electrochemical storage systems is stored on the surface. Generally the higher the specific surface area of activated carbon is, the higher the active surface area. However despite high specific surface areas attained for activated carbons that is around 2500 to 3000 $m^2 g^{-1}$, some activated carbons exhibit low specific capacitance. This might be due to poor pore size, type of the electrolyte used, scan rate, mass or surface area of the electrode. The relationship between surface area and capacitance is not always obvious due to the fact that capacitance is contributed by other factors such as pore size, electrical conductivity, and pore distribution and interconnectivity.

For supercapacitors application of both micropores and mesopores play an important role. Micropores are important in storing charges while mesopores store and facilitate charge transfer. Small pore size (0.68 nm) limits the electrolyte from accessing the entire active surface area of activated carbon, while the large pores size facilitates the storage and transport of charges. Moreover, the increase in the fraction of pore size which cannot be accessed by the electrolyte (when the average pore size is below 0.68 nm) in organic electrolyte is expected to decrease the capacitance.

Though the physico-chemical characterisation of activated carbon remains same for supercapacitor and lithium ion battery applications, electrochemical characterisation was done by galvanostatic charge–discharge profiles recorded between 3.00 and V at a 0.1 C rate on a multi-channel battery tester. From the galvanostatic charge–discharge (Fig. 7) studies, the first lithium insertion capacity of biomass carbon samples can be evaluated. From the cycling studies (Fig. 8), the coulombic efficiencies of the samples can also be calculated.

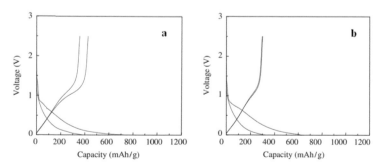

Fig. 7. Charge-discharge characteristics of carbon prepared from $ZnCl_2$ treated coffee shells at (a) 800°C (b) 900°C (Adapted with modification from Hwang et al. 2007).

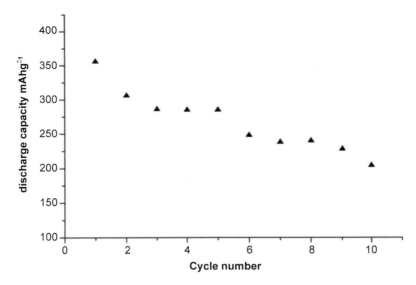

Fig. 8. Capacity as a function of cycle number for carbon derived from banana fibers treated with KOH (Adapted with modification from Manuel Stephan et al. 2006).

The size of the anode particles in the lithium ion battery greatly influences its performance. Small anodic particles contain short diffusion paths between them, which facilitate fast charge and discharge. Similarly, the larger surface area of the smaller anodic particles are prone to higher internal heat generation and lithium ion are consumed during the exothermic reaction at high temperatures compared to larger particles size, this leads to an increase in the irreversible capacity.

There is a direct correlation between the porosity of the anodic particle and its reversible capacity. An increase in porosity decreases the active surface area, reduces the electrical path into the anodic particles and reduces the accessibility of the lithium ions into the current collector. Although the pores will accommodate a large volume of electrolyte, they serve as a reaction point during the electrolyte decomposition process.

Anode materials, which have high surface area morphologies that provide large discharge capacity and high charge rate performance. During battery degradation, the ordered and radial structures of the carbon electrode may become less ordered, but this structural change is not the main contributor to battery degradation. Degradation can be either in the form of the lithium plating or the formation of the surface film.

Cycling the lithium ion batteries at high C-rate and high state of charge induces mechanical strain on the graphite lattice of the anode electrode due the steep gradient of lithium ions, and thus lattice parameter, in the particle. This mechanical strain caused by the insertion and de-insertion of the lithium ions cracks, fissures and splits the graphite particles thus making these particles less oriented as compared to the original platelets. Pressed graphite particles improve the ionic conductivity with a trade off in a decreased ohmic resistance and irreversible capacity loss. The nature and orientation of the graphite particles influences the reversible capacity of the anode. For instance, less-oriented graphite particles have a low reversible capacity due to more difficult lithium intercalation kinetics and to the formation of new boundaries between crystallite at which irreversible lithium ions/electrolyte interaction can occur (Berg 2015). While flake-like graphite particles have higher gravimetric capacity at higher C-rate compared to spherical particles.

6. Influence of Electrolyte on the Capacitive Performance of the Biomass Carbon

The electrolyte plays an important role in the capacitive performance, safety and the life time of a supercapacitor. It is a critical component for charge transport between the positive and negative electrodes. The ability to store charge is dependent on the accessibility of the ions to the porous surface area of the electrode material. Moreover, the attainable cell voltage of a supercapacitor depends upon the breakdown voltage of the electrolyte, and hence the possible energy density (which is dependent on voltage) is limited by the electrolyte. The behaviour of electrolyte ions in the pore during charging and discharging process is shown in Fig. 9.

Three types of electrolyte commonly used in supercapacitors:

  i) aqueous
 ii) non aqueous or organic
iii) ionic liquid

Aqueous electrolyte has a high ionic conductivity but a small electrochemical window (up to 1.2 V). The electrochemical window is the potential below which the electrolyte is neither reduced nor oxidised at an electrode. Aqueous electrolytes are potentially beneficial for large installations to store surplus power and unsteady electricity generated by natural energy resources, because of low cost, high safety, long lifetime and low internal resistance. Examples of these aqueous electrolytes are commonly H_2SO_4, Na_2SO_4 and KOH.

Organic electrolytes are the most commonly used in commercial devices, due to their higher electrochemical window of 3 V, which can significantly enhance the electrical charge (or energy) accumulated in supercapacitors than aqueous electrolytes. The lower electrical conductivity of organic electrolytes ($10–60$ S cm^{-1}) leads to a lower power density and higher energy density, since the energy density increases with the square of the voltage. Examples of most commonly used non aqueous electrolytes are propylene carbonate and acetonitrile electrolytes.

Ionic liquids composed of organic cations and inorganic anions. Their liquid phase range is large, so is their electrochemical window (6 V). Ionic liquids have high thermal stability and negligible vapour pressure. The problem is their high viscosity which reduces the ions migration.

The variation of capacitive performance of biomass carbon electrode in different electrolytes is given in Table 8.

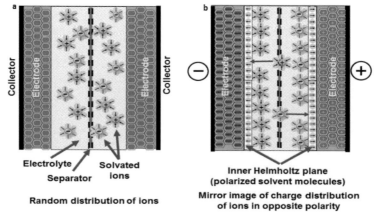

Fig. 9. Behaviour of electrolyte ion in the pore during (a) charge and (b) discharge (Adapted with modification Manaf et al. 2013).

Table 8. Influence of electrolyte on the capacitive performance of the biomass carbon electrode.

S. No.	Biomass Material	Activating Agent	Electrolyte	Sp. Capacitance (Fg^{-1})	Reference
1	Fir Wood	Steam	$NaNO_3$	89	Wu et al. 2005
			HNO_3	120	
			H_2SO_4	96	
2	Waste coffee ground	$ZnCl_2$	$MeEt_3NBF_4/AN$	100	Rufford et al. 2008, 2010
			H_2SO_4	368	
3	Seaweed (Lesonianigrescens)	Pyrolysis (unactivated)	KOH	201	Bichat et al. 2010
			H_2SO_4	255	
			Na_2SO_4	125	
4	Sucrose	CO_2	$EdMPNTf_2N$	170	Wei and Yushin 2011a,b
			$EMImBF_4$	148	
			H_2SO_4	160	
5	Jack fruit seed	Pyrolysis	H_2SO_4	316	Kalyani and Anitha 2014
			Na_2SO_4	203	

On comparing the electrochemical performance of biomass carbon samples in acidic and neutral electrolytes, CV measured in neutral electrolytes exhibits nearly rectangular shape and produces lower capacitance (on comparing with acid electrolyte), because in neutral electrolytes, the specific capacitance is mainly governed only by the non-faradic electrostatic sorption of ions at the double layer.

On the other hand, in acid electrolytes, the specific capacitance is governed both by the non-faradic electrostatic sorption of ions at the double layer and faradaic redox reactions due to surface functional groups on biomass carbon. The surface functional groups may play a hydrophilic role to enhance wettability of carbon electrode and quick charge propagation. If the wettability of the electrode is high, formation of electrostatic double layer on the surface of electrode will be promoted. Enhanced specific capacitance would be observed when pseudocapacitance couples with double layer capacitance.

Conclusion

Performance of the biomass carbon electrode is thus decided by its preparation procedure, surface area, total pore volume, volume percentage of mesopores coupled with functional groups which increases the wettability of the electrode surface and the nature of electrolyte. It will be now clear that the various sources listed in this chapter for producing biomass carbon is renewable and natural, obviously cost effective. Of late, vegetable sources are gaining significance as hi-end products like carbon and activated carbon could effectively be produced from these zero-cost raw materials. Thus the study opens up avenues for the utilisation of various zero-cost biomass for producing new carbon materials for energy device applications.

References

Aida, T., I. Murayama, K. Yamada and M. Morita. 2007. Analyses of capacity loss and improvement of cycle performance for a high-voltage hybrid electrochemical capacitor. J. Electrochem. Soc. 154: A798–A804.

Anitha, A. and P. Kalyani. 2014. Capacitive performance of onion peel derived carbon. IJCR 6: 8433–8438.

Aripin, H., L. Lestari, D. Ismail and S. Sabchevski. 2010. Sago waste based activated carbon film as an electrode material for electric double layer capacitor. Open Materials Science Journal 4: 117–124.

Arrebola, J.C., A. Caballero, L. Hernan, J. Morales, M. Olivares-Marin and V. Gomez-Serrano. 2010. Improving the performance of biomass-derived carbons in Li-ion batteries by controlling the lithium insertion process. J. Electrochem. Soc. 157: A791–A797.

Azevedo, D.C., J.C.S. Araujo, M. Bastos-Neto, A.E.B. Torres, E.F. Jaguaribe and C.L. Cavalcante. 2007. Microporous activated carbon prepared from coconut shells using chemical activation with zinc chloride. Microporous Mesoporous Mater. 100: 361–364.

Babel, K. and K. Jurewicz. 2004. KOH activated carbon fabrics as supercapacitor material. J. Phys. Chem. Solids 65: 275–280.

Berg, H. 2015. Batteries for electric vehicles. Cambridge University Press, Cambridge, UK.

Balathanigaimani, M.S., W.G. Shim, M.J. Lee, C. Kim, J.W. Lee and H. Moon. 2008. Highly porous electrodes from novel corn grains-based activated carbons for electrical double layer capacitors. Electrochem. Commun. 10: 868–871.

Bansal, R.C., J.B. Donnet and H.F. Stoeckli. 1988. Active carbon, Marcel Dekker, New York.

Bharadwaj, S., S.V. Jaybhaye, M. Sharon, D. Sathiyamoorthy, K. Dasgupta, P. Jagadale et al. 2008. Carbon nanomaterial from tea leaves as an anode in lithium secondary batteries. Asian J. Exp. Sci. 22: 89–93.

Bhardwaj, S., M. Sharon, T. Ishihara, S. Jayabhaye, R. Afre, T. Soga et al. 2007. Carbon material from natural sources as an anode in lithium secondary battery. Carbon Lett. 8: 285–291.

Bhattacharjya, D. and J.S. Yu. 2014. Activated carbon made from cow dung as electrode material for electrochemical double layer capacitor. J. Power Sources 262: 224–231.

Bichat, M.P., E. Raymundo-Pinero and F. Beguin. 2010. High voltage supercapacitor built with seaweed carbons in neutral aqueous electrolyte. Carbon 48: 4351–4361.

Biswal, M., A. Banerjee, M. Deo and S. Ogale. 2013. From dead leaves to high energy density supercapacitors. Energy Environ. Sci. 6: 1249–1259.

Calvo, E., N. Ferrera-Lorenzo, J. Menéndez and A. Arenillas. 2013. Microwave synthesis of micromesoporous activated carbon xerogels for high performance supercapacitors. Microporous Mesoporous Mater. 168: 206–212.

Castro-Muniz, A., F. Suarez-Garcia, A. Martinez-Alonso and J.M.D. Tascon. 2011. Activated carbon fibers with a high content of surface functional groups by phosphoric acid activation of PPTA. J. Colloid Interface Sci. 361: 307–315.

Centeno, T.A., F. Rubiera and F. Stoeckli. 2009. Recycling of residues as precursors of carbons for supercapacitors. Proceedings of the 1st Spanish National Conference on Advances in Materials Recycling and Eco-Energy. Madrid 04–12.

Chang, Y., M. Antonietti and T.P. Fellinger. 2015. Synthesis of nanostructured carbon through ionothermal carbonization of common organic solvents and solutions. Angew. Chem. Int. Ed. 54: 5507–5512.

Cheng, P.Z. and H. Teng. 2003. Electrochemical responses from surface oxides present on HNO_3-treated carbons. Carbon 41: 2057–2063.

Cheng, P., T. Li, H. Yu, L. Zhi, Z. Liu and Z. Lei. 2016. Biomass-derived carbon fiber aerogel as a binder-free electrode for high-rate supercapacitors. J. Phys. Chem. C 120: 2079–2086.

Dias, J.M., M.C.M. Alivim-Ferraz, M.F. Almeida, J. Rovera-Utrolla and M. Sanchez-Polo. 2007. Waste materials for activated carbon preparation and its use in aqueous-phase treatment: A review. J. Environ. Manage. 85: 833–846.

Du, Z., S. Zhang, T. Jiang, X. Wu, L. Zhang and H. Fang. 2012. Facile synthesis of SnO_2 nanocrystals coated conducting polymer nano wires for enhanced lithium storage. J. Power Sources 219: 199–203.

Dubinin, M.M. 1960. The potential theory of adsorption of gases and vapors for adsorbents with energetically non uniform surfaces. Chem. Rev. 60: 235–241.

Eikerling, M., A.A. Kornyshev and E. Lust. 2005. Optimized structure of nanoporous carbon-based double-layer capacitors. J. Electrochem. Soc. 152: E24–E33.

Elaigwu, S.E. and G.M. Greenway. 2015. Microwave-assisted and conventional hydrothermal carbonization of lignocellulosic waste material: Comparison of the chemical and structural properties of the hydrochars. J. Anal. Appl. Pyrol. 118: 1–8.

Elmouwahidi, A., Z. Zapata-Benabithe, F. Carrasco-Marin and C. Moreno-Castilla. 2012. Activated carbons from KOH-activation of argan (Arganiaspinosa) seed shells as supercapacitor electrodes. Bioresour. Technol. 111: 185–190.

Fang, B. and L. Binder. 2007. Enhanced surface hydrophobisation for improved performance of carbon aerogel electrochemical capacitor. Electrochim. Acta 52: 6916–6921.

Farma, R., M. Deraman, A. Awitdrus, I.A. Talib, E. Taer, N.H. Basri et al. 2013. Preparation of highly porous binderless activated carbon electrodes from fibres of oil palm empty fruit bunches for application in supercapacitors. Bioresour. Technol. 132: 254–261.

Fey, G.T.K. and C.L. Chen. 2001. High capacity carbons for lithium ion batteries prepared from rice husk. J. Power Sources 97-98: 47–51.

Fey, G.T.K., D.C. Lee, Y.Y. Lin and T. Prem Kumar. 2003. High-capacity disordered carbons derived from peanut shells as lithium intercalating anode materials. Synth. Met. 139: 71–80.

Fey, G.T.K., Y.D. Cho, C.L. Chen, Y.Y. Lin, T. Prem Kumar and S.H. Chan. 2010. Pyrolytic carbons from acid/base-treated rice husk as lithium-insertion anode materials. Pure Appl. Chem. 82: 2157–2165.

Fey, G.T.K., Y.Y. Lin, K.P. Huang, Y.C. Lin, T. Prem Kumar, Y.D. Cho et al. 2011. Green energy anode materials: Pyrolytic carbons derived from peanut shells for lithium ion batteries. Adv. Mat. Res. 415-417: 1572–1585.

Franklin, R.E. 1951. Crystalline growth in graphitizing and non-graphitizing carbons. Proc. R. Soc. A 209: 196–218.

Fuertes, A.B., G. Lota, T. Centeno and E. Frackowiak. 2005. Templated mesoporous carbons for supercapacitor application. Electrochim. Acta 50: 2799–2805.

Galinski, M., K. Babel and K. Jurewicz. 2013. Performance of an electrochemical double layer capacitor based on coconut shell active material and ionic liquid as an electrolyte. J. Power Sources 228: 83–88.

Gomibuchi, E., T. Ichikawa, K. Kimura, S. Isobe, K. Nabeta and H. Fujii. 2006. Electrode properties of a double layer capacitor of nano-structured graphite produced by ball milling under a hydrogen atmosphere. Carbon 44: 983–988.

Guo, F. and Z. Fang. 2014. Shape-controlled synthesis of activated bio-chars by surfactant-templated ionothermal carbonization in acidic ionic liquid and activation with carbon dioxide. BioResources 9: 3369–3383.

Guo, Y., J. Qi, Y. Jiang, S. Yang, Z. Wang and H. Xu. 2003. Performance of electrical double layer capacitors with porous carbons derived from rice husk. Mater. Chem. Phys. 80: 704–709.

Hall, D.O., J. House and I. Scrase. 1999. Introduction in industrial uses of biomass energy. Taylor & Francis. London.

He, X., P. Ling, J. Qiu, M. Yu, X. Zhang, C. Yu et al. 2013a. Efficient preparation of biomass-based mesoporous carbons for supercapacitors with both high energy density and high power density. J. Power Sources 240: 109–113.

He, X., P. Ling, M. Yu, X. Wang, X. Zhang and M. Zheng. 2013b. Rice husk-derived porous carbons with high capacitance by $ZnCl_2$ activation for supercapacitors. Electrochim. Acta 105: 635–641.

Hesas, R.H., A. Arami-Niya, W.M.A. Wan-Daud and J.N. Sahu. 2013. Preparation and characterization of activated carbon from apple waste by microwave-assisted phosphoric acid activation: Application in Methylene Blue Adsorption. BioResources 8: 2950–2966.

Honda, K., M. Yoshimura, K. Kawakita, A. Fujishima, Y. Sakamoto, K. Yasui et al. 2004. Electrochemical characterization of carbon nanotube/nanohoneycomb diamond composite electrodes for a hybrid anode of Li-ion battery and supercapacitor. J. Electrochem. Soc. 151: A532–A541.

Honda, Y., T. Haramoto, M. Takeshige, H. Shiozaki, T. Kitamura and M. Ishikawa. 2007. Aligned MWCNT sheet electrodes prepared by transfer methodology providing high-power capacitor performance. Electrochem. Solid-State Lett. 10: A106–A110.

Hu, C.C., C.C. Wang, F.C. Wu and R.L. Tseng. 2007. Characterization of pistachio shell-derived carbons activated by a combination of KOH and CO_2 for electric double-layer capacitors. Electrochim. Acta 52: 2498–2505.

Hwang, Y.J., S.K. Jeong, K.S. Nahm, J.S. Shin and A.M. Stephan. 2007. Pyrolytic carbon derived from coffee shells as anode materials for lithium batteries. J. Phys. Chem. Solids 68: 182–188.

Isaeva, I., G. Salitra, A. Soffer, Y.S. Cohen, D. Aurbach and J. Fischer. 2003. A new approach for the preparation of anodes for Li-ion batteries based on activated hard carbon cloth with pore design. J. Power Sources 119-121: 28–33.

Isil GurtenInal, I., S.M. Holmes, A. Banford and Z. Aktas. 2015. The performance of supercapacitor electrodes developed from chemically activated carbon produced from waste tea. Appl. Surf. Sci. 357: 696–703.

Jain, A. and S.K. Tripathi. 2014. Fabrication and characterization of energy storing supercapacitor devices using coconut shell based activated charcoal electrode. Mater. Sci. Eng. B 183: 54–60.

Jain, A. and S.K. Tripathi. 2013. Converting eucalyptus leaves into mesoporous carbon for its application in quasi solid-state supercapacitors. J. Solid State Electrochem. 17: 2545–2550.

Jain, A., C. Xu, S. Jayaraman, R. Balasubramanian, J. Lee and M. Srinivasan. 2015. Mesoporous activated carbons with enhanced porosity by optimal hydrothermal pre-treatment of biomass for supercapacitor applications. Microporous Mesoporous Mater. 218: 55–61.

Jing, Y., L. Yafei, C. Xiaomei, H. Zhonghua and Z. Guohua. 2008. Carbon electrode material with high densities of energy and power. ActaPhysico-ChimicaSinica 24: 13–19.

Jisha, M.R., Y.J. Hwang, J.S. Shin, K.S. Nahm, T. PremKumar, K. Karthikeyan et al. 2009. Electrochemical characterization of supercapacitors based on carbons derived from coffee shells. Mater. Chem. Phys. 115: 33–39.

Jupe, S.C.E., A. Michiorri and P.C. Taylor. 2007. Increasing the energy yield of generation from new and renewable energy sources. Renew. Energy 14: 37–62.

Jurewicz, K. and K. Babeł. 2010. Efficient capacitor materials from active carbons based on coconut shell/melamine precursors. Energy Fuels 24: 3429–3435.

Kalderis, D., M. Kotti, A. Mendez and G. Gasco. 2014. Characterization of hydrochars produced by hydrothermal carbonization of rice husk. Solid Earth 5: 477–483.

Kalpana, D., S.H., Cho, S.B. Lee, Y.S. Lee, R. Misra and N.G. Renganathan. 2009. Recycled waste paper—A new source of raw material for electric double-layer capacitors. J. Power Sources 190: 587–591.

Kalyani, P., A. Anitha and A. Darchen. 2015. Obtaining activated carbon from papaya seeds for energy storage devices. IJESRT 4: 110–122.

Kalyani, P. and A. Anitha. 2013. Refuse derived energy—Tea derived boric acid activated carbon as an electrode material for electrochemical capacitors. PortugaliaeElectrochimicaActa 31: 165–174.

Kalyani, P. and A. Anitha. 2014. On the (pseudo)capacitive performance of jack fruit seed carbon. IJRET 3: 225–238.

Kalyani, P. and A. Anitha. 2015. Capacitor behavior of activated carbon from used tea dust powder. Asian J. Chem. 26: 1365–1370.

Katakabe, T., T. Kaneko, M. Watanabe, T. Fukushima and T. Aida. 2005. Electric double-layer capacitors using bucky gels consisting of an ionic liquid and carbon nanotubes. J. Electrochem Soc. 152: A1913–A1916.

Kim, C., J.W. Lee, J.H. Ki and K.S. Yang. 2006. Feasibility of bamboo based activated carbons for an electrochemical supercapacitor electrode. Korean J. Chem. Eng. 23: 592–594.

Kim, S.U. and K.H. Lee. 2004. Carbon nanofiber composites for the electrodes of electrochemical capacitors. Chem. Phys. Lett. 400: 253–257.

Kumagai, S., M. Sato and D. Tashima. 2013. Electrical double-layer capacitance of micro- and mesoporous activated carbon prepared from rice husk and beet sugar. Electrochim. Acta 114: 617–626.

Lafi, W.K 2001. Production of activated carbon from acorns and olive Seeds. Biomass Bioenergy 20: 57–62.

Lee, Y.H., K.H. An, J.Y. Lee and S.C. Lim. 2004. Carbon nanotube-based supercapacitors. Encyclopedia of Nanoscience and Nanotechnology 1: 625–634.

Li, W., M. Chen and C. Wang. 2011. Spherical hard carbon prepared from potato starch using as anode material for Li-ion batteries. Mat. Lett. 65: 3368–3370.

Li, Z., L. Zhang, B.S. Amirkhiz, X. Tan, Z. Xu, H. Wang et al. 2012a. Eggshells for electronics: Portable supercapacitors. Mater. Views March 23.

Li, Z., L. Zhang, B.S. Amirkhiz, X. Tan, Z. Xu, H. Wang et al. 2012b. Carbonized chicken eggshell membranes with 3D architectures as high performance electrode materials for supercapacitors. Adv. Energ. Mater. 2: 431–437.

Li, Z., Z. Xu, X. Tan, H. Wang, C.M. Holt, T. Stephenson et al. 2013. Mesoporous nitrogen-rich carbons derived from protein for ultra-high capacity battery anodes and supercapacitors. Energy Environ. Sci. 6: 871–878.

Lin, X.X., B. Tan, L. Peng, Z.F. Wu and Z.L. Xie. 2016. Ionothermal synthesis of microporous and mesoporous carbon aerogels from fructose as electrode materials for supercapacitors. J. Mater. Chem. A 4: 4497–4505.

Liu, M.C., L.B. Kong, C. Lu, X.M. Li, Y.C. Luo and L. Kang. 2012. Waste paper based activated carbon monolith as electrode materials for high performance electric double-layer capacitors. RSC Adv. 2: 1890–1896.

Liu, T., R. Luo, W. Qiao, S.H. Yoon and I. Mochida. 2010. Microstructure of carbon derived from mangrove charcoal and its application in Li-ion batteries. Electrochim. Acta 55: 1696–1700.

Liu, X., C. Giordano and M. Antonietti. 2014. A facile molten-salt route to graphene synthesis. Small 10: 193–200.

Liu, X.M., R. Zhang, L. Zhan, D.H. Long, W.M. Qiao, J.H. Yang et al. 2007. Impedance of carbon aerogel/activated carbon composites as electrodes of electrochemical capacitors in aprotic electrolyte. New Carbon Mater. 22: 153–158.

Ma, G., Q. Yang, K. Sun, H. Peng, F. Ran, X. Zhao et al. 2015. Nitrogen-doped porous carbon derived from biomass waste for high-performance supercapacitor. Bioresour. Technol. 197: 137–142.

Magrini-Bair, K.A., S. Czernik, H.M. Pilath, R.J. Evans, P.C. Maness and J. Leventhal. 2009. Biomass derived carbon sequestering designed fertilizers. Ann. Env. Sci. 3: 217–225.

Manaf, N.S.A., M.S.A. Bistamam and M.A. Azam. 2013. Development of high performance electrochemical capacitor: A systematic review of electrode fabrication technique based on different carbon materials. ECS J. Solid State Sci. 2: M3101–M3119.

Manuel Stephan, A., T. Prem Kumar, R. Ramesh, S. Thomas, S.K. Jeong and K.S. Nahm. 2006. Pyrolitic carbon from biomass precursors as anode materials for lithium batteries. Mater. Sci. Eng. A 430: 132–137

Matsubara, E.Y., S.M. Lala and J.M. Rosolen. 2010. Lithium storage into carbonaceous materials obtained from sugarcane bagasse. J. Braz. Chem. Soc. 21: 1877–1884.

Merritt, R.P. 2003. CVD synthesis of carbon nanotubes as active materials for electrochemical capacitors. Florida Atlantic University.

Misra, R. 2006. Recycled waste paper—an inexpensive carbon material for supercapacitor applications. Masters thesis. Mahatma Gandhi ChitrakootGramodayaVishwavidyalaya. Chitrakoot.

Mitra, S. and S. Sampath. 2004. Electrochemical capacitors based on exfoliated graphite electrodes. Electrochem. Solid-State Lett. 7: A264–A268.

Moreno-Castilla, C. 2004. Adsorption of organic molecules from aqueous solutions on carbon material. Carbon 42: 83–94.

Nabais, J.V., P. Carrott, M.R. Carrott and J. Menendez. 2004. Preparation and modification of activated carbon fibres by microwave heating. Carbon 42: 1315–1320.

Olivares-Marin, M., J.A. Fernandez, M.J. Lazaro, C. Fernandez-Gonzalez, A. Macias-Garcia, V. Gomez-Serrano et al. 2009. Cherry stones as precursor of activated carbons for supercapacitors. Mater. Chem. Phys. 114: 323–327.

Ong, L.K., A. Kurniawana, A.C. Suwandia, C.X. Lin, X.S. Zhao and S. Ismadji. 2012. A facile and green preparation of durian shell-derived carbon electrodes for electrochemical double-layer capacitors. Prog. Nat. Sci. Mater. Int. 22: 624–630.

Ou, J., Y. Zhang, L. Chen, Q. Zhao, Y. Meng, Y. Guo et al. 2015. Nitrogen-rich porous carbon derived from biomass as a high performance anode material for lithium ion batteries. J. Mater. Chem. A 3: 6534–6541.

Pagketanang, T., A. Artnaseaw, P. Wongwicha and M. Thabuot. 2015. Microporous activated carbon from koh-activation of rubber seed-shells for application in capacitor electrode. Energy Procedia 79: 651–656.

Pampcl, J., C. Denton and T.P. Fellinger. 2016. Glucose derived ionothermal carbons with tailor made porosity. Carbon 107: 288–296.

Peled, E., V. Eshkenazi and Y. Rosenberg. 1998. Study of lithium insertion in hard carbon made from cotton wool. J. Power Sources 76: 153–158.

Peng, C., X.B. Yana, R.T. Wang, J.W. Lang, Y.J. Oub and Q.J. Xue. 2013. Promising activated carbons derived from waste tea-leaves and their application in high performance supercapacitors electrodes. Electrochim. Acta 87: 401–408.

Puligundla, P., S.E. Oh and C. Mok. 2016. Microwave-assisted pretreatment technologies for the conversion of lignocellulosic biomass to sugars and ethanol: A review. Carbon Lett. 17: 1–10.

Putun, A.E., N. Ozbay, E.P. Onal and E. Putun. 2005. Fixed-bed pyrolysis of cotton stalk for liquid and solid products. Fuel. Process Technol. 86: 1207–1219.

Raymundo-Pinero, E., F. Leroux and F. Beguin. 2006. High-performance carbon for supercapacitors obtained by carbonization of a seaweed biopolymer. Adv. Mater. 18: 1877–1882.

Rodriguez-Reinoso, F. 1998. The role of carbon materials in heterogeneous catalysis. Carbon 36: 159–175.

Rodriguez-Reinoso, F. and M. Molina-Sabio. 1992. Activated carbons from lignocellulosic materials by chemical and/or physical activation: An overview. Carbon 30: 1111–1118.

Rufford, T.E., D. Hulicova-Jurcakova, K. Khosla, Z. Zhu and G.Q. Lu. 2010. Microstructure and electrochemical double-layer capacitance of carbon electrodes prepared by zinc chloride activation of sugarcane bagasse. J. Power Sources 195: 912–918.

Rufford, T.E., D. Hulicova-Jurcakova, Z. Zhu and G.Q. Lu. 2008. Nanoporous carbon electrode from waste coffee beans for high performance supercapacitors. Electrochem. Commun. 10: 1594–1597.

Sato, K., S. Suemune, K. Nitta, C. Nakayama, S. Inokuma, S. Tonooka et al. 2013. A simple fabrication route of activated carbons from chicken droppings. J. Anal. Appl. Pyrol. 101: 86–89.

Senthilkumar, S.T., B. Senthilkumar, S. Balaji, C. Sanjeeviraja and R. Kalaiselvan. 2011. Preparation of activated carbon from sorghum pith and its structural and electrochemical properties. Mater. Res. Bull. 46: 413–419.

Senthilkumar, S.T., R. Kalaiselvan and J.S. Melo. 2013. The biomass derived activated carbon for supercapacitor. Proceedings of AIP Conference 124–127.

Seredych, M. and T.J. Bandosz. 2013. S-doped micro/mesoporous carbon-grapheme composites as efficient supercapacitors in alkaline media. J. Mater. Chem. A 1: 11717–11727.

Seredych, M., D. Hulicova-Jurcakova, G.Q. Lu and T.J. Bandosz. 2008. Surface functional groups of carbons and the effects of their chemical character, density and accessibility to ions on electrochemical performance. Carbon 46: 1475–1488.

Sevilla, M. and A.B. Fuertes. 2009. The production of carbon materials by hydrothermal carbonization of cellulose. Carbon 47: 2281–2289.

Shekhar, G., P.B. Karandikar and Mukesh Rai. 2014. Investigation of carbon material derived from leaves of tree for the electrodes of supercapacitor. IJEERT 2: 127–131.

Si, W., J. Zhou, S. Zhang, S. Li, W. Xing and S. Zhuo. 2013. Tunable N doped or dual N, S-doped activated hydrothermal carbons derived from human hair and glucose for supercapacitor applications. Electrochim. Acta 107: 397–405.

Si, W.J., X.Z. Wu, W. Xing, J. Zhou and S.P. Zhuo. 2011. Bagasse-based nanoporous carbon for supercapacitor application. J. Inorg. Mater. 26: 107–112.

Simon, P., P.L. Taberna. and F. Béguin. 2013. Electrical double-layer capacitors and carbons for EDLCs. pp. 131–165. *In*: Béguin F. and E. Frąckowiak (eds.). Supercapacitors: Materials, Systems, and Applications. Wiley-VCH Verlag GmbH & Co. KGaA, Weinheim, Germany.

Sivakkumar, S.R., J.M. Ko, D.Y. Kim, B.C. Kim and G.G. Wallace. 2007. Performance evaluation of CNT/polypyrrole/MnO_2 composite electrodes for electrochemical capacitors. Electrochim. Acta 52: 7377–7385.

Smil, V. 2008. Energy in nature and society: General energetics of complex systems. Cambridge, USA.

Subramanian, V., L. Cheng, A.M. Stephan, K.S. Nahm, S. Thomas, B. Wei et al. 2007. Supercapacitors from activated carbon derived from banana fibers. J. Phys. Chem. C 111: 7527–7531.

Sun, S., C.Y. Wang, M.M. Chen and M.W. Li. 2013a. Hard carbon prepared from willow leaves using as anode materials for Li-ionbatteries. Adv. Mater. Res. 724-725: 834–837.

Sun, X., X. Wang, N. Feng, L. Qiao, X. Li and D. He. 2013b. A new carbonaceous material derived from biomass source peels as an improved anode for lithium ion batteries. J. Anal. Appl. Pyrol. 100: 181–185.

Taer, E., M. Deraman, I.A. Talib, A.A. Umar, M. Oyama and R.M. Yunus. 2010. Physical, electrochemical and supercapacitive properties of activated carbon pellets from pre-carbonized rubber wood sawdust by CO_2 activation. Curr. Appl. Phys. 10: 1071–1075.

Valix, M., W.H. Cheung and K. Zhang. 2006. Role of heteroatoms in activated carbon for removal of hexavalent chromium from wastewaters. J. Hazard Mater. 135: 395–405.

Wang, H. and M. Yoshio 2006.Graphite, a suitable positive electrode material for high-energy electrochemical capacitors. Electrochem. Commun. 8: 1481–1486.

Wang, H., M. Yoshio, A.K. Thapa and H. Nakamura. 2007. From symmetric AC/AC to asymmetric AC/graphite, a progress in electrochemical capacitors. J. Power Sources 169: 375–380.

Wang, H., Y. Zhong, Q. Li, J. Yang and Q. Dai. 2008. Cationic starch as a precursor to prepare porous activated carbon for application in supercapacitor electrodes. J. Phys. Chem. Solids 69: 2420–2425.

Wang, D., Z. Geng, B. Li, and C. Zhang. 2015. High performance electrode materials for electric double-layer capacitors based on biomass-derived activated carbons. Electrochim. Acta 173: 377–384.

Wei, L. and G. Yushin. 2011a. Electrical double layer capacitors with sucrose derived carbon. Carbon 49: 4830–4838.

Wei, L. and G. Yushin. 2011b. Electrical double layer capacitors with sucrose derived carbon electrodes in ionic liquid electrolytes. J. Power Sources 196: 4072–4079.

Wei, Y. 2014. Activated carbon microtubes prepared from plant biomass (Poplar Catkins) and their application for supercapacitors. Chem. Lett. 43: 216–218.

World activated Carbon Industry. 2014. Available from http://www.reportliner.com/p090365/ World-Activated-Carbon-Industry.html> [Accessed 17 February 2014].

Wu, F.C., R.L. Tseng, C.C. Hu and C.C. Wang. 2005. Effects of pore structure and electrolyte on the capacitive characteristics of steam- and KOH-activated carbons for supercapacitors. J. Power Sources 144: 302–309.

Wu, F.C., R.L. Tseng, C.C. Hu and C.C. Wang. 2006. The capacitive characteristics of activated carbons-comparisons of the activation methods on the pore structure and effects of the pore structure and electrolyte on the capacitive performance. J. Power Sources 159: 1532–1542.

Xiao, L., X. Wei, Z. Shuping, Z. Jin, L. Feng, S.Q. Zhang et al. 2011. Preparation of capacitors electrode from sunflower seed shell. Bioresour. Technol. 102: 1118–1123.

Xie, Z.L., R.J. White, J. Weber, A. Taubert and M.M. Titirici. 2011. Hierarchical porous carbonaceous materials via ionothermal carbonization of carbohydrates. J. Mater. Chem. 21: 7434–7442.

Xing, W., J.S. Xue and J.R. Dahn. 1996. Optimizing pyrolysis of sugar carbons for use as anode materials in lithium-ion batteries. J. Electrochem. Soc. 143: 3046–3052.

Xu, B., F. Wu, S. Chen, C. Zhang, G .Cao and Y. Yang. 2007. Activated carbon fiber cloths as electrodes for high performance electric double layer capacitors. Electrochim. Acta 52: 4595–4598.

Xueliang, L., H. Changlong, C. Xiangying and S. Chengwu. 2010. Preparation and performance of straw based activated carbon for supercapacitor in non-aqueous electrolytes. Micropor. Mesopor. Mat. 131: 303–309.

Yong, Z., F. Hui, W. Xingbing and W. Lizhen. 2009. Progress of electrochemical capacitor electrode materials: A review. Int. J. Hydrogen Energy 34: 4889–4899.

Zhang, D., L. Zheng, Y. Ma, L. Lei, Q. Li, Y. Li et al. 2014. Synthesis of nitrogen-and sulfur-codoped 3D cubic-ordered mesoporous carbon with superior performance in supercapacitors. ACS Appl. Mater. Interfaces 6: 2657–2665.

Zhang, F., K.X. Wang, G.D. Li and J.S. Chen. 2009. Hierarchical porous carbon derived from rice straw for lithium ion batteries with high-rate performance. Electrochem. Commun. 11: 130–133.

Zhang, L., Q. Wang, B. Wang, G. Yang, L.A. Lucia and J. Chen. 2015a. Hydrothermal carbonization of corncob residues for hydrochar production. Energy Fuels 29: 872–876.

Zhang, L., S. Liu, B. Wang, Q. Wang, G. Yang and J. Chen. 2015b. Effect of residence time on hydrothermal carbonization of corn cob residual. BioResources 10: 3979–3986.

Zhang, P., Y. Gong, Z. Wei, J. Wang, Z. Zhang, H. Li et al. 2014. Updating biomass into functional carbon material in ionothermal manner. ACS Appl. Mater. Interfaces 6: 12515–12522.

Zhang, Y., F. Zhang, G.D. Li and J.S. Chen. 2007. Microporous carbon derived from pinecone hull as anode material for lithium secondary batteries. Mater. Lett. 61: 5209–5212.

Zheng, F., D. Liu, G. Xia, Y. Yang, T. Liu, M. Wu et al. 2017. Biomass waste inspired nitrogen-doped porous carbon materials as high-performance anode for lithium-ion batteries. J. Alloys Compd. 693: 1197–1204.

Zhao, S., C.Y. Wang, M.M. Chen, J. Wang and Z.Q. Shi. 2009. Potato starch-based activated carbon spheres as electrode material for electrochemical capacitor. J. Phys. Chem. Solids 70: 1256–1260.

Zhao, Y., M. Liu, X. Deng, L. Miao, P.K. Tripathi, X. Ma et al. 2015. Nitrogen-functionalized microporous carbon nanoparticles for high performance supercapacitor electrode. Electrochim. Acta 153: 448–455.

Zhou, X., H. Li and J.Yang. 2016. Biomass-derived activated carbon materials with plentiful heteroatoms for high-performance electrochemical capacitor electrodes. J. Energy Chem. 25: 35–40.

Mesoporous Silica: The Next Generation Energy Material

Saika Ahmed,[1] M. Yousuf Ali Mollah,[2] M. Muhibur Rahman[1] and Md. Abu Bin Hasan Susan[1,*]

1. Introduction

Nanoporous materials are continuous and solid network materials filled through voids, or structural pores in the nanodimension, i.e., 1–100 nm. These small sized pores give the nanoporous materials high pore volume and surface area, which consequently allow various practical and potential applications of such materials in the modern world. Few typical examples from real-world applications of these materials include zeolites, aerogels, several polymers, activated carbon, metal-organic frameworks, inorganic metal oxide materials, etc. Materials of this kind have emerged to be enormously important in the past decades in several applications like synthesis of nanomaterials, catalysis, separation techniques, drug delivery, purification of materials, energy storage, solar and fuel cells, batteries, membranes, optical and magnetic devices and so on.

Depending on the average size of the structural pores, nanoporous materials have been frequently categorised into microporous (average pore size < 2 nm), mesoporous (average pore size between 2 and 50 nm) and macroporous (average pore size > 50 nm). Mesoporous materials fall in the category of nanoporous materials; the studies on which are rapidly developing as an interdisciplinary research focus. Ordered mesoporous materials, specifically having an average and pretty uniform pore diameter between 2 and 50 nm, are receiving an upsurge of interest and their applications are extending from traditional fields like catalysis, adsorption and separation to high technology fields including chips, biotechnology, optoelectronics, sensors, etc. The beginning of the major interest in mesoporous materials, formed from nanoscale building blocks, goes back to the century when several discoveries turned heads towards the fact that one could successfully alter the properties of the mesoporous assemblies by controlled tailoring of the size of the pores and morphology of the assemblies of the constituents and thereby, utilise them in several targeted fields of applications. The notably important reasons why ordered mesoporous materials grabbed the focus of interest of the researchers and scientists around the globe are their highly regular nanopores, large surface area, liquid-crystal templated mesostructure and the periodically arranged organic–inorganic nanoarrays.

[1] Department of Chemistry, University of Dhaka, Dhaka, Bangladesh.
[2] University Grants Commission of Bangladesh, 29/1 Agargaon, Sher-E-Bangla Nagar, Dhaka, Bangladesh.
* Corresponding author: susan@du.ac.bd

Works have therefore been extensively carried out on mesoporous materials to gain a deep insight into the strategies of synthesis, methods, mechanism and thus, to establish a direct relationship between synthesis-properties-applications.

Among the mesoporous materials, ordered mesoporous silica has probably stood out to be the most extensively studied and potential one, due to their well defined and regular pores and pore channels, narrow pore size distribution, high thermal stability, easily tunable parameters for synthesis, which allow them to exhibit even better performances in different applications, compared to conventional non-porous silica particles (Islam et al. 2013). There has been a remarkable increase in the versatility in their properties and applications as well in the recent years. As a consequence, numerous attempts have been made to study and optimise the synthesis conditions for task specific applications. This chapter aims to make a brief summary of the major parameters for synthesis that have been found to influence the pore architecture and properties of mesoporous silica and then, to discuss some of their important real life applications in energy technologies, with emphasis on their potential as next generation materials usable in energy generation and storage.

2. Synthesis and Properties of Mesoporous Silica

The synthesis of mesoporous silica has widely been an intriguing topic. Even after adopting similar approaches and conditions for synthesis, significant dissimilarity is marked between the structure and architecture of the mesoporous silica synthesised. This necessitates a complete understanding of the mechanistic route of the synthesis. Systematic knowledge regarding the formation of mesostructures is essential and by proper combination of the different criteria of synthesis, variation in structure or/ and properties of silica may be obtained. Numerous approaches have so far been reported regarding the synthesis of mesoporous silica and mechanism of pore creations; few selective and important ones will be addressed in the following Sections.

2.1 Routes of Synthesis

The usual method for creation of pores in mesoporous materials, or in specific, mesoporous silica is through the use of supramolecular aggregates like micellar array of surfactants as templates, known as 'soft-templates' while the method is termed 'soft-templating method'. The surfactants in this case are referred to as 'structure-directing agents', since their structure and nature are crucial factors in templating the mesoporous materials. Such surfactants, when present in low concentrations in water, produce micellar aggregates; while in high concentrations, produce liquid-crystal like structures. Hydrothermal conditions are generally employed for the synthesis. Both acidic and basic conditions have been successfully used till date. The silica source, usually inorganic silicates are efficiently adsorbed on to the hydrophilic outward portion of the surfactant aggregates due to electrostatic interactions. The surfactant template is then removed by calcination at high temperature, leaving regular, ordered pores in the silica structure.

A brief overview of the different parameters of synthesis that can affect the structure, morphology and porosity of the resultant silica is given below.

2.1.1 Surfactant

The most widely used surfactants till date for this purpose are cationic surfactants containing quaternary ammonium salts, i.e., $C_nH_{2n+1}N(CH_3)_3Br$ ($n = 8$–22) due to their good solubility and ease of use in acidic or basic media. Cationic cetyltrimethylammonium bromide (CTAB) is the most common one; other cationic surfactants employed include gemini surfactants, bolaform surfactants, multi headgroup surfactants and cationic fluorinated surfactants (Huo et al. 1994a, Sakamoto et al. 2000, Tan et al. 2004, Shen et al. 2005). However, due to the toxicity and high cost of cationic surfactants; use of anionic surfactants containing carboxylates, sulfates, sulfonates, phosphates, terminal carboxylic acids (Che et al. 2003) and non-ionic surfactants are considered good replacements of the cationic ones. Among them, non-ionic surfactants have been found to be more promising, owing to their low cost, nontoxicity, availability, biodegradability and most importantly, easy control of their structure to obtain silica with desired porosity and morphology. Highly ordered mesoporous silica could be synthesised by using poly(ethylene oxide)-b-poly(propylene oxide)-b-poly(ethylene oxide) (PEO-PPO-PEO) triblock copolymers as the structure directing templates in acidic media. Zhao et al. (1998a,b) reported the synthesis of a novel mesoporous silica, SBA-15 using this surfactant, with well-ordered hexagonal uniform pores of diameters 50 to 300 Å and thick silica walls of thicknesses 31 to 64 Å. Pores were created after removal of the surfactant template by solvent extraction or calcination at high temperature. The variation of EO/PO ratio of the copolymers was found to alter the formation of the silica mesophase: lamellar mesostructured silica was found at low ratio of the triblock copolymer moieties, while higher ratio favored cubic mesostructures. Since then, SBA-15 has found wide range of applications till date due to their extraordinary properties like regular pores, high thermal stability, easy synthetic route, tunability of the properties by tailoring the block copolymer structure, etc. Other nonionic classes of surfactants used are Brij, Tergitol, Triton, Tween, Span, etc.

According to the available literatures, synthesis of mesoporous silica is usually carried out by either hydrothermal method or by non-aqueous method. Among them, the hydrothermal method is the most common one, where the principle of sol-gel transformation is applied during synthesis. At first, the surfactant solution is prepared, usually in water, where the silicate precursor is added. The silicate precursor undergoes acid or base-catalysed hydrolysis, followed by cooperative assembly and aggregation as a result of the interaction between the silicate oligomers and surfactant micelles. The aggregation then precipitates out of the solution in the gel form; after which, microphase separation and condensation of the silicate oligomers continue to take place. Direct indication of this phase transition of the hexagonal mesostructure in MCM-41 can be obtained by using *in situ* X-ray diffraction technique. Landry et al. (2001) implied that the MCM-41 silica is disordered before the formation of gels, which gives well-ordered mesostructures after precipitation.

2.1.2 Acidic/Basic Medium

Acidic or basic conditions are used as media for this kind of synthesis, since ordered mesostructures cannot be found from neutral solutions because of the rapid polymerisation and cross-linking of the silicates at pH around 7 (Brinker and Schrer 1990). Sodium hydroxide, potassium hydroxide, tetramethylammonium hydroxide ($(CH_3)_4NOH$) and tetraethylammonium hydroxide ($(C_2H_5)_4NOH$) are the most commonly used bases. During synthesis, the pH of the system does not remain consistent and it decreases initially due to the hydrolysis of the silicate network and then increases again due to the polymerisation and cross-linking of silica. In basic medium, the reversible polymerisation and cross-linkage of silicate species makes the use of basic medium diversified and synthesis of MCM-41 has been reported in basic medium (Kresge et al. 1992). Huo et al. (1994a,b) first introduced the use of acidic medium for the synthesis of mesoporous silica. They used several cationic and anionic surfactants and varied the pH of the reaction media using HCl, HBr acids and NH_3, NaOH and $(CH_3)_4$ NOH bases.

Usually, a lower pH, i.e., higher acid concentration facilitates the synthesis by increasing the reaction and precipitation rate. At high concentrations of acid, the PEO groups of the block copolymers are protonised at a faster rate, which increases the hydrophilicity of the block copolymer. Generally, a faster rate of precipitation leads to low quality and homogeneity of the product, for which an optimum acid concentration, below 4 M is used. Highly ordered SBA-16 mesosilicate with three dimensional cubic pores is synthesised at low concentrations of HCl (0.5 M), where the use of an additive *n*-butanol even reduces the rate. Different types of triblock copolymers have also been utilised to synthesise mesoporous silica having different pore structures, like bicontinuous cubic and face-centered cubic (Kleitz et al. 2003a,b, Kim et al. 2005). HCl, HNO_3, HBr, H_2SO_4 and other strong acids are used for this purpose.

Since the basic synthesis of mesoporous silica is a much faster one compared to acidic synthesis, control of the morphology is much easier in the acidic one. Due to the faster polymerisation rate, basic synthesis usually leads to formation of three dimensional networks; while acidic synthesis can actually help synthesising linear silicate oligomers with controllable morphology.

2.1.3 Cosolvents/Cosurfactants/Additives

Desired morphology of mesoporous silica with hard sphere, fiber, doughnut, rope, egg-sausage, gyroid and discoidlike hexagonal structures can be prepared by varying the structure of block copolymers, cosurfactants, cosolvents, or by the addition of strong electrolytes in acidic medium. Fiberlike structure can be formed in SBA-15 using tetramethyl orthosilicate (TMOS) as the silica source and P123 triblock copolymer ($EO_{20}PO_{70}EO_{20}$) as the structure-directing agent, whereas doughnutlike powder is obtained when *N,N*-dimethylformamide (DMF) and tetraethyl orthosilicate (TEOS) are used as additional cosolvent and silica source (Zhao et al. 2000). On the other hand, when a cationic surfactant, CTAB is used, hard spherical product is obtained; while prehydrolysed TEOS as the silica source produces gyroid and discoidlike morphology (Fig. 1).

The fiberlike morphology of the SBA-15 exhibits extraordinary stability even after calcination, which is evident from the SEM, TEM and XRD studies of the uncalcined and calcined silica samples. The condensation rate of silicate is faster in TMOS than TEOS as the silica source, resulting in high local curvature energy and thereby formation of fiberlike SBA-15. Slower condensation rate in TEOS is responsible for the more curved ropelike morphology.

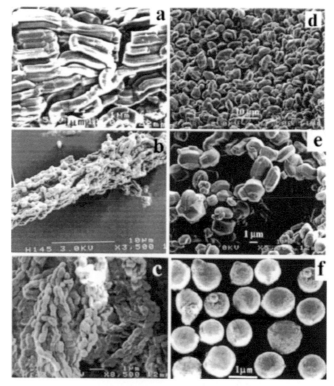

Fig. 1. SEM images of (a) as-synthesised SBA-15 prepared by using TEOS as a silica source; as-synthesised SBA-15 prepared by using small amounts of (b) DMF and (c) tetrahydofuran as the cosolvents; as-synthesised SBA-15 prepared in (d) Na_2SO_4 and (e) $MgSO_4$ solution; and (f) hard sphere SBA-15 prepared by using CTAB as a cosurfactant. (Reprinted with permission from ref. Zhao et al. 2000. Copyright 2000 American Chemical Society.)

When a cosurfactant and/or an organic additive are used along with a surfactant, the crystal structure and pore architecture of the resulting mesoporous silica material can be varied to a great deal. Figure 2 shows the sequential transition of the structure of micellar aggregates of triblock pluronic F127 surfactant with the addition of an anionic cosurfactant, sodium dioctyl sulfosuccinate (AOT) as well as a swelling agent, 1,3,5-trimethylbenzene (TMB) (Chen et al. 2005). The mesophase of the micellar assembly is cubic $Fm\bar{3}m$ at first, which changes to $Im\bar{3}m$ with added AOT and TMB. Further addition of these two converts the structure to two-dimensional $p6m$; at higher concentrations of both AOT and TMB, the structure finally consists of $Ia\bar{3}d$ mesostructure.

With the gradual addition of AOT and TMB, the hydrophobic volume of the micellar system increases accordingly, owing to the hydrogen bonding of the anionic AOT surfactants preferably with the more hydrophobic PPO blocks of F127 instead of the less hydrophobic PEO blocks. This leads to decreasing interface curvature of the micellar assembly and inversion of the characteristic truncated cone-shaped structures of AOT molecules. This swelling of the micellar volume is more pronounced due to the synergistic effect caused by the presence of a swelling agent like TMB in the system. This expansion in micellar volume brings out increase in the pore size of silica.

The studies of Zhao et al. 2000 revealed that when a polar cosolvent is added to the reaction mixture, they act as swelling agents to form microemulsions, resulting in the formation of large pores. Egg-sausagelike and ropelike SBA-15 particles have been obtained using ethanol and tetrahydrofuran (THF), respectively as the cosolvents. In addition, in the presence of highly charged and concentrated inorganic electrolytes like Na_2SO_4 or $MgSO_4$, SBA-15 is synthesised with gyroid and discoidlike morphology, respectively. Low-charge inorganic salts such as NaCl, LiCl, on the other hand, produce 3D continuous meso-macroporous membranes (Zhao et al. 1999). Polar cosolvent like DMF, cosurfactants and inorganic salts at high concentrations interact with the amphiphilic block copolymer species, lowering the local curvature energy and thus, the resultant more curved macroscopic structure gives rise to the more curved morphology. Their study implies that the morphology of SBA-15 is greatly controlled by the curvature energy at the interface of the inorganic silica/organic block copolymer species.

The nitrogen adsorption-desorption isotherms and the corresponding pore size distribution of silica synthesised using TMOS silica source with or without DMF cosolvent (Fig. 3) implies type IV isotherms with H1-type hysteresis, which is a characteristic of mesoporous materials with one dimensional hexagonal cylindrical channels.

Cubic $Fm\bar{3}m$ Cubic $Im\bar{3}m$ Hexagonal $p6m$ Cubic $Ia\bar{3}d$

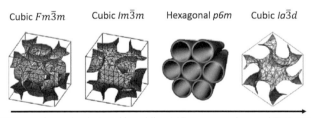

Mesophase transformation induced by adding AOT as co-template and TMB as swelling agent

F127   AOT   TMB

Interface curvature of the mixed micelles decreases with the augment of AOT and TMB amounts

Fig. 2. Schematic representation of the mesophasic transformation of F127 micellar assembly with addition of cosurfactant AOT and swelling agent TMB. The interface curvature is reduced by the expansion of the hydrophobic volume, arising from binding of AOT with the hydrophobic moieties and the synergistic effect from the additional presence of the swelling agent. (Reprinted with permission from ref. Chen et al. 2005. Copyright 2005 Royal Society of Chemistry.)

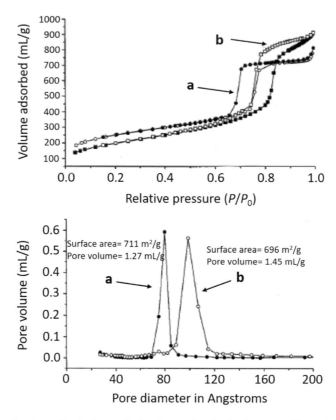

Fig. 3. Nitrogen adsorption-desorption isotherms (top) and pore size distribution curves (bottom) of (a) calcined fiberlike SBA-15 prepared using TMOS as the silica source and (b) calcined doughnutlike SBA-15 prepared by using DMF as a cosolvent. The pore size distribution curve was obtained from the analysis of the adsorption branch of the isotherm. (Reprinted with permission from ref. Zhao et al. 2000. Copyright 2000 American Chemical Society.)

The BET surface area and pore size of the silica synthesised from TMOS show comparable values with those from TEOS (Zhao et al. 1998a,b), which indicate that the surface area and pore size are not significantly altered by the silica source. Furthermore, the increased pore volume due to addition of the cosolvent (Fig. 3) suggests that the cosolvent results in swelling of the pores.

2.1.4 Temperature for Reaction and Hydrothermal Treatment

For synthesis, the usual approach is to choose a temperature higher than the individual critical micellar temperature (CMT) of the surfactant solution. Since the CMTs of cationic surfactants are relatively low, better control over the morphology can be achieved when synthesis is carried out at low temperatures, owing to the lower micellsation rate. Room temperature synthesis is thus carried out in the case of cationic surfactants.

However, in the case of nonionic surfactants, a higher temperature is common due to their higher CMT values, although low temperature has also been applied in some cases, resulting in highly ordered mesoporous silicates with large pores (Fan et al. 2005). Since a high temperature may also cause insolubility and precipitation of the nonionic surfactants, the temperature for synthesis must be lower than their cloud points. Lowering the temperature might avoid this problem, which at the same time may make the process extremely slow. Optimisation of the reaction temperature is therefore crucial for individual surfactant systems. For example, the optimum temperature for synthesis of SBA-15 is 35–40°C (Zhao et al. 1998a,b).

Temperature dependence of the pore size has been studied systematically by Martines et al. (2004) where the pore size was found to vary from microporous (below 2 nm) to mesoporous (close to 9 nm) range depending on the temperature for synthesis. They correlated this variation with two properties of the surfactant; at higher temperatures, the solubility limit of the nonionic surfactant prevents the formation of micelles because of high hydrophobicity and at lower temperatures, the CMT prevents the formation of micelles due to high hydrophilicity. MSU-3 type silica could be synthesised from nonionic P123 triblock surfactant template and TEOS silica source at different temperatures ranging from 5 to 65°C. At temperatures below 25°C, the particles exhibit spherical shape, similar to other MSU-X silica (Boissiére et al. 1999, 2000). When the temperature is raised above 35°C, the shape becomes stick-like and finally, above 45°C, the structure of the particles is lost and a fluffy powder is produced with a growing amorphous part of very small aggregates of silica, as observed under SEM (Fig. 4).

XRD studies of these samples also show this transition from an amorphous phase at temperature below 25°C to another amorphous phase above 45°C via a transition of phase showing three dimensional wormhole and hexagonal structures at the intermediate temperatures.

The nitrogen adsorption-desorption isotherms of such samples synthesised at different temperatures are shown in Fig. 5. With increasing temperature, the adsorption step shifts to higher pressure range, indicating an increase in pore size. At temperatures between 5 to 15°C, the isotherms

Fig. 4. SEM images of calcined MSU-3 silica synthesised at (a, b) 35; (c, d) 45; (e, f) 55; and (g, h) 65°C. (Reprinted with permission from ref. Martines et al. 2004. Copyright 2004 Elsevier.)

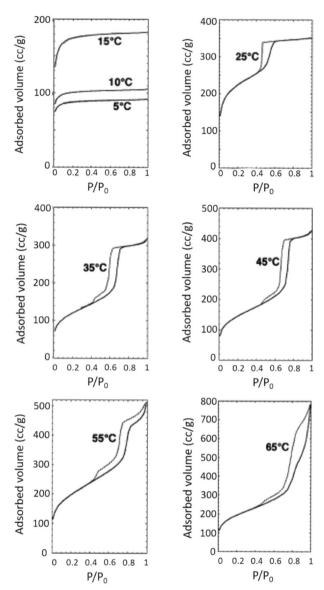

Fig. 5. Nitrogen adsorption-desorption isotherms of calcined MSU-3 silica synthesised at different temperatures from 5 to 65°C. (Reprinted with permission from ref. Martines et al. 2004. Copyright 2004 Elsevier.)

exhibit type I character, which indicates microporosity, while from 25 to 45°C, type IV isotherms are obtained showing H1-hysteresis loop, a very characteristic of mesoporosity; and, at temperature above 45°C, type II shaped isotherm is produced, which implies macroporosity. The increase in pore size was also evident from the pore size distribution curves.

After the solution reaction, hydrothermal treatment is applied to produce mesoscale regularity and order in the product. Reorganisation, growth and crystallisation take place during the hydrothermal treatment which is usually carried out at lower temperatures (80–130°C) in order to avoid the non-regularity and disorder arising from higher temperature (Chen et al. 1997). For cationic surfactants, the coulombic interaction of their ordered micro-domains with inorganic silicate is much stronger than that for nonionic surfactants to justify the choice of a higher temperature during hydrothermal treatment in the case of cationic surfactants.

2.1.5 Template Removal

As the silica sample crystallises and precipitates out of the solution as the gel phase after the synthesis and hydrothermal treatment, filtration and washing are applied to obtain the solid product. To obtain porosity in this sample, removal of the structure-directing surfactant template becomes necessary. This is done most commonly by calcination or solvent extraction. Calcination at high temperature in presence of air leads to destruction and subsequent removal of the surfactant aggregates; however, it can also damage the ordered silica network; thus, an optimum temperature needs to be maintained. For SBA-15, heating at a rate of 1–2°C/min up to 550°C is followed by maintenance of the temperature for 4–6 h. Temperatures higher than 350°C are usually sufficient to remove P123 type triblock copolymer, although much higher temperature, ca. 550°C is necessary to remove surfactants with long alkyl chains owing to their high thermal stability arising from the high degree of cross-linking.

Keene et al. (1999) systematically studied the step-by-step removal of the surfactant template, gradual liberation of the pores and evolution of hydrophobic surface using 'sample controlled thermal analysis' (SCTA), nitrogen gas adsorption, XRD, etc. By gradual elimination of the cationic surfactant, CTAB from an ordered silicate mesophase, they proposed the presence of two kinds of species in the network; major portion of which is relatively loosely bound, along with a small quantity of more strongly bound species on the silica surface. Most of the loosely bound surfactants disrupt from 100 to 220°C, to form hexadecene and a trimethylamine species and the surface created at this temperature is hydrophobic. The strongly bound surfactant species are removed at 195 to 220°C, through thermal decomposition of the pure surfactant. At temperatures above 200°C, the surface becomes hydrophilic and this hydrophilicity increases with temperature. However, temperature above 700°C results in rupture of the silica network. The mechanism of the surfactant degradation during calcination has been proposed in terms of Hoffman degradation mechanism (Hitz and Prins 1997).

Extraction of the surfactants by using suitable solvents is another method of removal of template. Ethanol or THF is usually used for this, while the use of slight amount of HCl can improve the cross-linkage of the silica frameworks and minimise the effects of template removal on ordered mesostructures (Inagaki et al. 1996). Yang et al. (2005) reported the comparative effect of removal of surfactant from a three dimensional cubic silica by calcination at 550°C and by treatment by H_2SO_4 followed by calcination at 250°C (Fig. 6).

The calcined samples exhibit typical type IV isotherms with the capillary condensation at high pressure and H2-type hysteresis loops, indicative of large mesopores with ink-bottle or cagelike shapes (Ravikovitch et al. 2002, Kleitz et al. 2003a). The sample treated with H_2SO_4 and calcined at

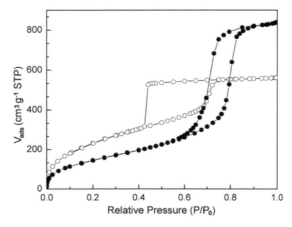

Fig. 6. Argon adsorption-desorption isotherms of cubic mesoporous silica with *bcc* structure after calcination at 550°C (open circles) or extraction by H_2SO_4 followed by calcination at 250°C (filled circles). (Reprinted with permission from ref. Yang et al. 2005. Copyright 2005 Royal Society of Chemistry.)

250°C, on the other hand, exhibits H1-type hysteresis indicating uniform mesopores with cylindrical geometry. Characterisation of the texture of the samples by non-local density functional theory (NLDFT) showed that the acid-treated materials have much larger pore volumes compared to the calcined samples, although the calcined samples have higher surface area, owing to the contribution of the micropores at the intrawall templated by polyalkylene oxide-based block copolymers (Ryoo et al. 2000, Ravikovitch et al. 2001).

Combination of solvent extraction and high temperature calcination was proposed by Grudzien et al. (2006) to obtain high pore volume and large pore size in the highly ordered mesoporous silica, SBA-16, synthesised using Pluronic F127 triblock copolymer as the structure directing agent and TEOS as the silica source. Figure 7(a,b) show comparative nitrogen adsorption-desorption isotherms of SBA-16 samples only calcined, or calcined + extracted by HCl/EtOH mixture at different temperatures.

The largest pore diameter is obtained for sample extracted at 250°C; however, the optimum temperature is around 350°C, since the sample prepared at 250°C contains a small residue of the polymer. Two distinguished peaks were observed in the pore size distribution curves Fig. 7(c,d); the first peak around 2 nm is due to the irregular micropores present in the mesopore walls and the small pores that interconnect ordered spherical cages (mesopores) as illustrated in Fig. 8; whereas the second peak reflects the distribution of ordered spherical mesopores.

Phase transition due to heating has been observed in MCM-41 mesoporous silica to give MCM-48 (Díaz et al. 2004). The small angle x-ray diffraction studies show that when MCM-41 is heated at 140°C for prolonged time, the initial predominant mesophase of hexagonal MCM-41 changes to the cubic *Ia3d* phase of MCM-48 after 20 h and eventually, after 48 h, the lamellar structure becomes predominant (Fig. 9).

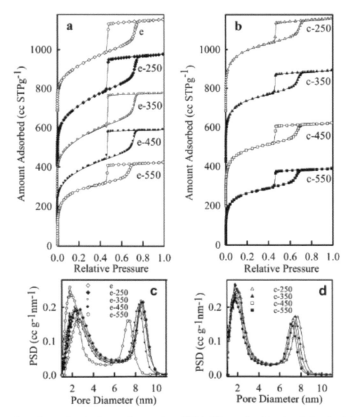

Fig. 7. Nitrogen adsorption–desorption isotherms for a series of SBA-16 samples (a) extracted and calcined; and (b) calcined at different temperatures. (c), (d) The corresponding pore size distributions. The extracted + calcined and calcined samples are expressed as e-x and c-x instead of SBA-16-e-x and SBA-16-c-x, respectively. (Reprinted with permission from ref. Grudzien et al. 2006. Copyright 2006 Royal Society of Chemistry.)

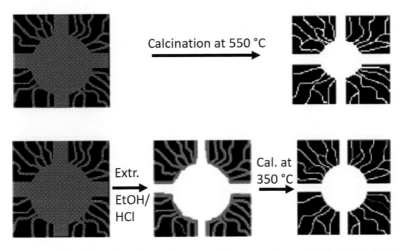

Fig. 8. Schematic representation of the polymeric template removal by commonly used calcination in air 550°C (top scheme) and combined extraction followed by controlled-temperature calcination at 300–350°C (bottom scheme). The large circles connected with straight channels represent interconnected ordered mesopores, whereas the curved thin ribbons denote irregular micropores at the walls of ordered mesopores. (Reprinted with permission from ref. Grudzien et al. 2006. Copyright 2006 Royal Society of Chemistry.)

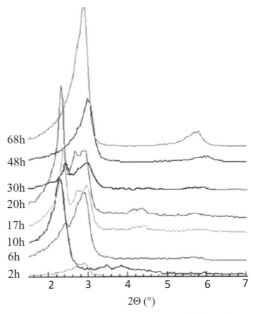

Fig. 9. XRD patterns of the as-made mesoporous silica synthesised at 140°C at different time showing phase transition. (Reprinted with permission from ref. Díaz et al. 2004. Copyright 2004 Royal Society of Chemistry.)

The mesostructures are long-range ordered, which makes them exhibit well-resolved diffraction peaks on the small angle XRD patterns ($2\theta = 2$–$10°$). This structural phase transition is also evident in the field emission SEM images (Fig. 10).

The conditions as preceded in the section give an insight to the criteria required for obtaining mesoporous silica with different structures. Various mesoporous silicates have so far been synthesised and reported with different symmetries (*p6mm, Ia3hd, Pm3hn, Im3hm, Fd3hm, Fm3hm*) (Fig. 13 of Wan and Zhao 2007), cell parameters and pore sizes to indicate the versatility and controllability of the synthetic process and hence, properties.

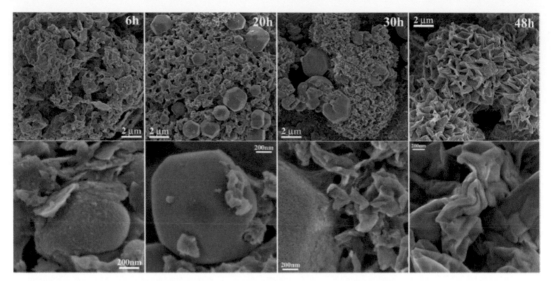

Fig. 10. FESEM images of the mesoporous solids synthesised at 140°C for 6, 20, 30 and 48 hrs. (Reprinted with permission from ref. Díaz et al. 2004. Copyright 2004 Royal Society of Chemistry.)

2.2 Mechanism of Pore Formation

The mechanism of pore formation in the MCM-41 silica, a member of the M41S family, was first proposed by Beck et al. 1992. The proposed mechanism was named as the liquid crystal templating (LCT) mechanism in which the liquid crystals phases of the surfactants were proposed to serve as the organic templates (Fig. 11).

The solution conditions like ionic strength, polarisability and charge of counter-ion, surfactant concentration, temperature and the addition of co-surfactants or additives define the structure of the liquid crystal formed in the surfactant solutions. According to their proposed mechanism, formation of the silica network takes place by two mechanistic pathways; one in which the liquid crystalline structure of the surfactant aggregates remain intact as the inorganic silicate source is added. Such as-synthesised MCM-41 silica does not exhibit porosity as the micelle arrays still remain undamaged in their liquid crystalline phase which is supported by the ^{13}C-NMR studies. The surfactant used here contains quaternary ammonium surfactant, which produces rodlike micelles (Tiddy 1980) in water with hexagonal arrangements. In the pathway 1, when the silicate precursor is added, the resulting silica structure also mimics this hexagonal pattern (Fig. 12). Pathway 2 depicts the approach in which

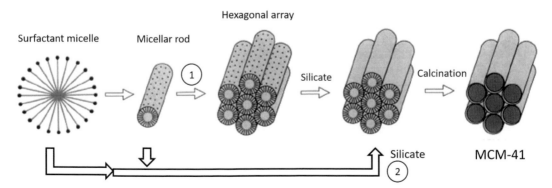

Fig. 11. The proposed possible mechanistic pathways by Beck et al. (1992) during the formation of MCM-41. (Reprinted with permission from ref. Beck et al. 1992. Copyright 1992 American Chemical Society.)

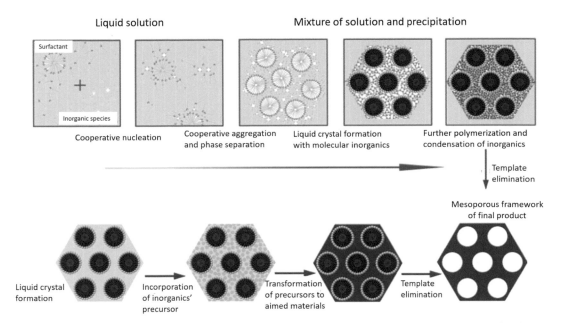

Liquid solution   Mixture of solution and precipitation

Cooperative nucleation   Cooperative aggregation   Liquid crystal formation   Further polymerization and
                          and phase separation   with molecular inorganics   condensation of inorganics

Template elimination

Mesoporous framework of final product

Liquid crystal formation   Incorporation of inorganics' precursor   Transformation of precursors to aimed materials   Template elimination

Fig. 12. Two synthetic pathways of mesoporous silica shown in a single diagram. (Wan and Zhao 2007) (Reprinted with permission from ref. Wan and Zhao 2007. Copyright 2007 American Chemical Society.)

the surfactant micelles re-order themselves after being entrapped in the added silicate (Fig. 12) and the inorganic silicate influences this arrangement.

After calcination, the desired porous MCM-41 is produced at the final stage of both the pathways. The two pathways are referred to as cooperative self-assembly and 'true' liquid-crystal templating processes, respectively. The length of the hydrophobic carbon chain of the surfactant determines the dimension of the micellar arrangement and thus, that of the pores. Addition of organic materials, however, can increase the size of the pores, as they become solubilised inside the hydrophobic cores of the surfactant aggregates. Silanol groups might also be present in the resulting silica structure depending on the calcination conditions, acting as charge balancing groups associated with the quaternary ammonium ions which roughly have similar spacing as that of the surfactant molecules in the micelles.

However, the LCT mechanism later appeared to have several shortcomings. The proposed Pathway 1 seems to be inappropriate in the case of hydrothermal synthesis of mesoporous silica and although high concentration of surfactants is required for establishing liquid crystal phases; MCM-41 mesoporous silica could still be synthesised using low concentration of CTAB.

Following the limitations of the LCT mechanism, several other mechanisms through extensive studies have been reported later on. Among them, Chen and co-workers (Chen et al. 1993) implied that rather than liquid crystal structures, MCM-41 is likely to form through the interaction of silicate with ordered rodlike micelles resulting in the formation of several monolayers of tubular silica deposited around the micellar external surface and subsequent spontaneous assembly of this complex structure giving hexagonally packed long-range ordered structure. Huo et al. (1994a,b), on the other hand, suggested a cooperative formation mechanism which has so far been the most popular and acceptable, depicting that the electrostatic charge balance in the molecular level between the head groups of the charged surfactants and the inorganic ions in solution from the silica source controls the formation of periodic biphase arrays in the liquid crystal form. The silicate ions polymerise through crosslinks at the interface which change the charge density at the inorganic layers, influencing the arrangements of surfactants. They prepared lamellar, hexagonal and *Pm3m* cubic phases by adding TEOS as the source of silica to solution of HCl and HBr and the x-ray diffraction studies of the solid recovered product showed that HCl favors the formation of all three phases, while HBr favors the hexagonal phase over the cubic one. The hexagonal phase was also clearly visible from the SEM and TEM photographs (Fig. 13).

Multinuclear magnetic resonance spectroscopy, SAXS and polarised optical microscopic measurements also supported the formation of hexagonal or lamellar packing of surfactants in highly alkaline micellar solutions in presence of anionic silicate oligomeric network (Firouzi et al. 1995, 1997).

SBA-15 is one of the other most popular forms of mesoporous silica in recent days and seemingly a promising candidate owing to its high degree of ordering, low cost, low toxicity, easy removal of template and good thermal stability. They are synthesised using nonionic Pluronics, $EO_y PO_x EO_y$ as the surfactant templates. Due to their high potential in practical applications, they are now-a-days being extensively studied with a view to establishing an accurate formation mechanism. Ruthstein et al. (2003, 2006) employed *in situ* x-band EPR spectroscopy in combination with electron spin-echo envelope modulation (ESEEM) and direct imaging and freeze-fracture replication cryo-TEM and directly correlated the structure of silica with the transformation of the microstructure in the micellar solution during the reaction time from spherical micelles into longer, stiffer threadlike micelles and then into bundle structure, producing a viscous phase, having similar dimension as the final silica material. The silicate precursor is then adsorbed followed by polymerisation at the hydrophilic micellar interface, which reduces the water content at the core/corona interface and thus, the curvature of the micellar structure. Subsequent removal of the surfactant by calcination or extraction creates pores in the structure and the hydrothermal temperature and amount of the silicate influence the size of the pores.

Extensive works have been reported on the pore formation mechanism in SBA-15 (Flodström et al. 2004, Khodakov et al. 2005). Yu et al. (2004) proposed a 'colloidal phase separation mechanism' (CPSM) for the formation of mesoporous materials in presence of inorganic salts by direct observation of the morphologies of the particles as a function of reaction time using SEM and TEM, which include three stages: (1) self-assembly of the surfactant block copolymers and formation of composites with the inorganic silicates, which is governed by the charge density matching between the surfactant head groups and hydrolysed inorganic oligomers; (2) formation of a liquid crystal phase by colloidal interaction of aggregates of block copolymer/silica species; and (3) phase separation of this new phase from water leading to further growth of this phase, which ends in developing the solid mesoporous material observed by the precipitation. The morphology of the final product is determined by the factor which is dominating between the free energy of mesophase formation (ΔG) and the surface free energy (F) of the liquid crystal aggregation phase. SEM images of the solid synthesised in 0.5 M KCl at 38°C and isolated at different periods after precipitation are shown in Fig. 14 which well supports the proposed CPSM mechanism.

Fig. 13. Scanning electron microscope (SEM) and transmission electron microscope (TEM) images of the hexagonal silica (left and right images respectively) (Huo et al. 1994a). (Reprinted with permission from ref. Huo et al. 1994a. Copyright Nature Publishing Group.)

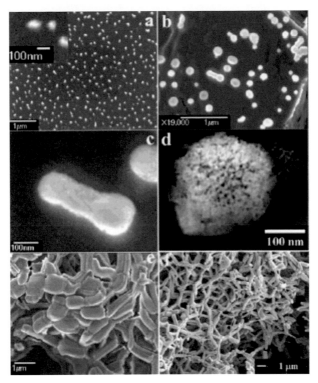

Fig. 14. Scanning microscopic images of the silica sample, isolated at different times after precipitation: (a) 2 min; (b) 5 min; (c) high magnification of (b); (e) 20 min; (f) 1 h; and (d) TEM image of sample separated after 5 min. Sample isolated 2 min after precipitation shows amorphous aggregates with diameters smaller than 100 nm (a), which becomes 50–100 nm after 5 min (b), TEM image of which shows wormlike mesoporous structure (d). Aggregation and directional elongation of the spheres is evident in a higher magnification in (c); while combination of rods and spheres are observed after 20 min (e) which eventually turns into a rodlike dominated one after 1 h (f) (Yu et al. 2004). (Reprinted with permission from ref. Yu et al. 2004. Copyright 2004 American Chemical Society.)

Evidence of this layer-by-layer growth was presented by Che et al. (2011) through direct images captured by cryo-SEM, based on which a mechanism of rod-unit repairing defects was proposed. Pore formation mechanism proposed for other forms of mesoporous silica was also reported by Yanagisawa et al. (1990), Inagaki et al. (1993) and Monnier et al. (1993).

2.3 Energy Considerations of the Interactions and Mechanistic Pathway

Monnier et al. (1993) and Huo et al. (1994b) proposed a synthesis model in terms of the free energy considerations. Since there are three types of assemblies: inorganic-organic, organic-organic and inorganic-inorganic found in the synthesis pathway of mesoporous silica, free energy is contributed by four processes and the overall free energy hence can be expressed as,

$$\Delta G = \Delta G_{inter}(A\rho_{inorg}, \rho_{org}, \ldots) + \Delta G_{org}(A\rho_{org}, \ldots) + \Delta G_{inorg}(A\rho_{inorg}, \ldots) + \Delta G_{sol} \qquad (1)$$

The four kinetic processes, to which the above free energy terms are related to, are the formation of interface of the inorganic/organic phases, organisation of organic array, interactions in the inorganic units and the chemical potential of the surrounding solution, as appeared in the equation respectively. A is the effective area occupied by the ionic surfactant head group in the organic phase and ρ is a variable representing the state of the phase related to the organisation and charge distribution of the various species within it, thus defining the charge and thickness of the interface wall. The term $\Delta G_{org}(A\rho_{org}, \ldots)$ results from the van der Waals forces and conformational energy of the hydrocarbon chains and the van der Waals and electrostatic interactions of the head groups within the organic phase, whereas ΔG_{inorg} ($A\rho_{\text{inorg}}$,...) is a total measure of all the free energy of the polymerised inorganic silicate network including the solvent, van der Waals and electrostatic interactions. The ΔG_{inter} ($A\rho_{\text{inorg}}$, ρ_{org},...) directly correlates with the van der Waals and electrostatic interactions in the inorganic/organic framework and ΔG_{sol} is contributed by the various species formed during nucleation and precipitation. The key factor is the interaction between the surfactant and inorganic species, i.e., the matching of charge density, so that the more negative ΔG_{inter}, the easier the assembly can occur.

The most important contributing factor in the total free energy is the arrangement of the organic array (Archibald et al. 1993, Dubois et al. 1993). Contributions from the other two factors, that is, from the inorganic/organic array and the inorganic network formations serve as perturbations of the organised array; which implies that ΔG_{org} ($A\rho_{\text{org}}$,...) < ΔG_{inter} ($A\rho_{\text{inorg}}$, ρ_{org},...) < ΔG_{inorg} ($A\rho_{\text{inorg}}$,...).

Huo et al. (1994a,b) proposed four pathways of interactions for the biphasic array formation, which are S^+I^-, S^-I^+, $S^+X^-I^+$ and $S^-X^+I^-$, where S^+ is surfactant cation, S^- is surfactant anion, I^+ is inorganic precursor cation, I^- is inorganic precursor anion, X^+ is cationic counterion and X^- is anionic counterion. The matching of the charge density is controlled by the surfactant head groups, which eventually governs the structure of the mesosilicate. In basic medium, Coulomb forces act between the silicate anions (I^-) and surfactant cations (S^+), while under acidic condition, the initial $S^+X^-I^+$ interaction gradually turns into the $(IX)^-S^+$ one via the double-layer hydrogen bonding interaction. Synthesis by the $S^+X^-I^+$, that is, interaction between surfactant and inorganic cations is also possible in the presence of counterions, X^-, which can be halides (Cl^-, Br^-, I^-), or other inorganic anions like SO_4^{2-}, NO_3^-, etc.

The repulsive interaction between anionic surfactants and silicate species, in comparison with that in the case of cationic surfactants, is the reason behind the poorly ordered mesostructures when anionic surfactants are used. Che et al. (2003) reported a synthetic route for a novel mesoporous silica using an anionic surfactant, organoalkoxysilane with quaternary ammonium organic group (*N*-trimethoxysilylpropyl-*N,N,N*-tributylammonium), in combination with an aminosilane or quaternised aminosilane as co-structure-directing agent (CSDA). The alkoxysilane site of the CSDA is co-condensed with inorganic precursors; while the ammonium site of the CSDA becomes attached to the silicon atoms and gets incorporated into the wall, producing well-ordered mesoporous silica via electrostatic interaction with the anionic surfactants.

Hydrogen bonds can also occur in such systems (Bagshaw et al. 1995, Tanev and Pinnavaia 1995) containing non-ionic surfactants where hydrogen bond type interactions take place between the hydrophilic surfaces of the flexible rod or wormlike micelles and the inorganic precursor, TEOS producing assembly of inorganic oxide framework. Instead of surfactants, amine derivatives-primary alkylamines like dodecylamine, hexadecylamine and *N,N*-dimethylalkylamines such as *N,N*-dimethyldodecylamine, *N,N*-dimethylhexadecylamine, etc. were used, which possess long hydrophobic carbon chains and hydrophilic amino groups, which have similar structure as surfactants. Hydrogen bonds have also been reported by Zhao et al. (1998b) in synthesis of SBA-15 where the assembly of the inorganic and organic periodic composite materials takes place by a hydrogen bonding $(S^0H^+)(X^-I^+)$ pathway.

Anderson et al. (2005) claimed that the structure-determining interface of the inorganic structure is the one between silica and water adsorbed at the micelle surface, not between silica and surfactant, which is in pretty ambiguity with the previous reports (Huo et al. 1994b). They synthesised SBA-1 mesoporous silica with cationic surfactant, hexadecyltriethylammonium bromide (HTEABr) of low concentrations, an order of magnitude below that normally required for mesophase formation and concluded that silica is not formed at the interface of the surfactant micelle, but at the interface with the adsorbed hydration layer surrounding the micelle.

3. Mesoporous Silica as a Next Generation Innovative Material

Over the past few decades, mesoporous silica and their functional derivatives have continuously found new dimensions of applications, like in catalysis, drug loading and delivery, nanosynthesis, purification and many more. Quite reasonably, it will not be extravagant to refer them as '*next generation innovative materials*'. However, research is still going on to explore more novel areas of applications in diversified fields. From an industrial stand point, we will focus on discussing some of their recent applications in energy sector and their great potential in energy technologies, energy production and storage, fuel cells and batteries.

Thermal Energy Storage

Mesoporous silica, either in pure form or modified, can be efficiently used to enhance the efficiency of shape-stabilised phase change materials (PCMs); which can serve as energy storage materials that can absorb or release huge amount of energy in the form of latent heat by going through sol-gel phase changing process. These PCMs can applied in solar energy storage, smart housing, insulation clothing, waste heat recovery, heat regulation of electronics and so on and thus, have a great potential to serve as an alternative energy source in the energy crisis in future. The most commonly used phase changing materials include paraffins, fatty acids, polyethylene glycols, hydrated salts, eutectic mixture, carbonates, etc. due to their high heat capacity, negligible tendency of supercooling, congruent melting, non-toxicity, low cost, small volume change and thermal stability (Baetens et al. 2010, Cao et al. 2014, Li et al. 2016). But the use of many of these materials is limited due to several advantages; however, propensity of leakage, low heat of fusion and low thermal conductivity are some of the disadvantages that restrict their wide-range applicability. Mesoporous silica, due to its uniform pores, ordered structure and high surface areas, can act as support matrix to immobilise and bind these materials via chemical interactions and thus, improve their thermal properties by overcoming the limitations. Polyethylene glycol (PEG), lauric acid and stearic acid are three of such materials whose thermal energy storage capability has been found to be greatly enhanced when prepared as composites along with mesoporous silica (He et al. 2014, Kadoono and Ogura 2014, Min et al. 2015, Qian et al. 2015, Wang et al. 2016, Chen et al. 2017). In most of the cases, the composite was homogeneous in structure, where the melting temperature of the PCM was greatly increased by the presence of silica network and the cycle life for heat storage/release was also enhanced manifold.

Energy storage can also be obtained by forcing a non-wetting liquid like water (e.g., mercury) to gain intrusion/extrusion in to/from a hydrophobic porous solid. The high pressure applied during the invasion of the liquid leads to high mechanical energy storage in the liquid, which can be converted in to interfacial energy caused by the forced development of an interface between the hydrophobic surface and the hydrophilic liquid. When the pressure is released, the pores are emptied during which the stored energy is again converted into mechanical energy. Mesoporous silica gels, grafted by alkyl groups of varying chain length, have been successfully used as a support solid to enclose such hydrophilic liquids (Gusev 1994, Eroshenko and Fadeev 1995, Fadeev and Eroshenko 1997, Gokulakrishnan et al. 2013, Mitran et al. 2015). The pressures of intrusion and extrusion of the liquid from the pore space and the contact angle at the wetting of surface are highly influenced by the length of the length of the grafted alkyl chain and the pore size of silica. The pressure and contact angle monotonically increase with increasing length of alkyl chain and decreasing mean radius of the pores.

Electrochemical Capacitor

Electrochemical capacitors are another form of energy storage devices that are widely being used to meet the current demands of energy and also a potential alternative energy source for the next generation. Various materials based on mesoporous silica have been employed in development of supercapacitors with improved specific capacitance, high battery density and superior cyclability. Supercapacitors can be classified based on either the formation of the electric double-layer at the interphase of electrode-electrolyte or the quick Faradic charge transfer reactions between the electrolyte and the electrode; referred to as 'electrical double layer capacitor' and 'pseudocapacitor', respectively.

Porous carbon has received enormous interest recently as electrodes for supercapacitors owing to their low-cost, versatility of structure/texture, good conductivity and high cycling life and the added advantage of tailor ability of the pore architecture and size by using variable templates for synthesis. Mesoporous silica is one of the materials that have been used as hard templates to efficiently control the texture of the porous carbon materials/films for obtaining good capacitive performance (Kwon et al. 2010, Zhi et al. 2014, Filonenko et al. 2015, Leyva-García et al. 2016).

Pseudocapacitive transition metal oxide nanorods (e.g., MnO_2, SnO_2, NiO) are generally considered as ideal materials to be used as supercapacitor electrodes despite their limitations like poor cycle life, low energy density, etc. (Cui et al. 2011, Huang et al. 2013, Singh et al. 2013, Wang et al. 2013, Zhang et al. 2013). However, such important energy storage materials, when synthesised inside mesoporous silica supported carbon nanomembranes as a matrix, are found to exhibit improved electrochemical performance in terms of specific capacitance, cycle life, electrical conductivity, chemical stability, mechanical strength and so on (Zhi et al. 2015).

Three dimensional interconnected macroporous graphene-based frameworks integrated with mesoporous silica shows enhanced supercapacitive performances, such as high specific capacitance, good rate capability and excellent cycling stability. Such porous graphene structures usually lack well-defined pore structure and size, which lead to low efficiency of mass transport and charge storage through the small pores; fortunately, when such materials are modified by integrating the small mesoporous channels of silica uniformly grown on the interior surface of the interconnected macroporous frameworks of graphene, hierarchical macro- and meso-porous architectures can be developed within the graphene 3D network, that leads to improvement of the storage capacities to a great deal (Wu et al. 2012).

Fuel Cells

Functionalised mesoporous silica is utilised as new, alternative approach toward overcoming the current limitations of polymer electrolyte membrane fuel cell technologies, like poor electrochemical properties at elevated temperatures (ca. 80°C), low relative humidity conditions and so on. The volatility of water in the polymer matrix at high temperatures is the crucial factor, which reduces the performance of the membrane and limits the operation and applicability of membrane fuel cells. Addition of mesoporous silica to the Nafion® membrane can not only inhibit the loss of water, but also significantly enhances the proton conductivity and electrochemical stability at high temperatures, owing to the high surface area of mesoporous silica that increases the amount of adsorbed water on the Nafion® membrane (Tominaga et al. 2007).

Introduction of imidazole-functionalised mesoporous silica (> 5 wt %) into the Nafion® membrane increases the glass transition temperature of the membrane due to enhanced intermolecular rigidity and improves the conductivity and thermochemical properties, in comparison with the pure Nafion® membrane-conductivity of which depends on the external humidity condition, due to incorporation of the imidazole moiety having high intrinsic proton conduction capacity (Amiinu et al. 2015). Similar enhancements in the proton conductivity (almost 8 times higher than that of pure Nafion®) can also be achieved by functionalising the mesoporous silica fillers with acid-base pairs such as sulfonic acid-amino groups, phosphoric acid-amino groups and carboxyl acid-amino groups (Yin et al. 2014,

2015), which also increases the water retention ability. The enhanced proton transfer capability arises from the reduced distance between the hydrophilic ionic domains inside the membranes, where the amino acids act as proton donor-acceptors (amino group as the proton acceptor and acid group as the proton donor) and the retained water as transfer bridges.

Apart from functionalised, non-functionalised mesoporous silica nanoparticles have also been used as additive to increase the proton conductivity of Nafion®. The efficiency of silica filled Nafion® composite significantly depends on the template during silica synthesis and the nature of silanol group of the silica network. According to the extensive works of Yang et al. (2016, 2017), the pore-directing agents (i.e., P123 and CTMA) can resist the methanol penetration from anode to cathode and this resistance increases with the loading amount of silica. The enhancement of the conductivity indicates the development of new ionic channels in the interphase region of silica fillers/polymer for proton transfer.

Batteries

With the rising interest in silicon based materials as anode materials in lithium ion batteries due to their significantly high specific capacity, retention of volume during cycling, low-cost and higher safety, works are being carried out to explore the prospect of mesoporous silica in this aspect. The controllable nanostructure of mesosilica offers great potential to be used for such purpose.

SnO_2 is a promising material for using in lithium ion batteries as anodes; however, if its limitations can be overcome, which include the significant volume expansion arising from the electrochemical reactions between Sn and Li^+ leading to the loss of capacity and rechargeability (Inagaki et al. 2002). Composite from SnO_2 incorporated into the periodic nanoholes or coated on the surface of silica network exhibits much higher capacity and lower initial irreversible capacity compared to pristine SnO_2, leading to the concept of using mesoporous silica for development of efficient and improved anode materials.

Carbon materials with three layers (carbon-silica-carbon) including mesoporous silica exhibit excellent specific capacity along with good number of cycle life (Cao et al. 2015). As the SiO_2 particles are confined between the carbon shell and the carbon core, the two carbon parts maintain the structural integrity and conductivity and keep the electrode highly conductive and active during cycling, thus ensuring the high capacity and cycle stability; while the carbon at core supports the SiO_2 particles, where the interior pores and voids create free space to accommodate the volume change and neutralise the volume change of the electrode during lithiation/delithiation.

Hydrogen Gas Generation and Storage

Hydrogen, the ultimate ideal energy carrier to solve the long-term environmental and energy issues, can effectively contribute to the reduction of atmospheric pollution and global dependence on fossil fuels. For this, optimisation of the catalyst condition is being carried out on a regular basis in order to obtain catalysts with the best performance and cyclability. Metal nanoparticles supported on mesoporous silica can act as highly efficient heterogeneous catalysts in different H_2 generation reactions, due to the higher surface area, homogeneous and smaller particle size of the nanoparticles. This has been evident for numerous H_2 generation and storage processes, such as hollow mesoporous silica supported ruthenium nanoparticles for hydrolysis of $NaBH_4$ (Peng et al. 2015), MCM-41 supported Au, Pd and Au-Pd metal catalysts for oxidation of benzene, toluene and *o*-xylene (da Silva et al. 2015), phenylamine functionalised SBA-15 supported Pd-Ag catalyst in the formic acid and carbon dioxide mediated reaction (Mori et al. 2017).

The metal catalysts, when present in their pure form, possess two problems; deactivation caused by coke formation and low selectivity to hydrogen (main reaction products being unwanted products). Improved performance in terms of the catalytic activity and selectivity to hydrogen gas has been observed while they are supported on mesostructured silica due to the ability of mesoporous silica to prevent agglomeration of metal particles owing to their porous structure. Mesoporous MCM-48 supported Co catalyst is found to exhibit an ethanol conversion of 100 per cent with a hydrogen selectivity higher than 80 per cent (Gac et al. 2011).

Conclusion

Mesoporous silica is an excellent material, thanks to its extraordinary properties, such as highly ordered structure, controllable porosity, crystal structure and morphology, high surface area, high thermal and mechanical stability, which make it a promising candidate to meet the demands of the next generation. Innovative applications of this material is being explored on a regular basis due to its great potential in diversified fields of applications involving energy storage, hydrogen gas production, synthesis using waste materials, separation and purification, controlled and targeted drug delivery, catalysis, as support in preparing nanomaterials and so on. Application of mesoporous silica in energy technologies is still in a primitive stage and requires further exploration of the possibilities. For targeted applications, tailoring of the morphology or porosity is necessary, which requires firm knowledge on the synthesis mechanism and structure of these types of silica. Works are still being carried out to optimise the synthesis conditions and to establish a flawless mechanism of the pore formation in order to have more control on the structural parameters. The wait for this day will be worthy considering the excellent prospective of mesoporous silica as a much better replacement of many synthetic materials.

References

Amiinu, I.S., W. Li, G. Wang, Z. Tu, H. Tang, M. Pan and H. Zhang. 2015. Toward anhydrous proton conductivity based on imidazole functionalized mesoporous silica/nafion composite membranes. Electrochim. Acta 160: 185–194.

Anderson, M.W., C.C. Egger, G.J.T. Tiddy, J.L. Casci and K.A. Brakke. 2005. A new minimal surface and the structure of mesoporous silicas. Angew. Chem. Int. Ed. 44: 3243–3248.

Archibald, D.D. and S. Mann. 1993. Template mineralization of self-assembled anisotropic lipid microstructures. Nature 364: 430–438.

Baetens, R., B.P. Jelle and A. Gustavsen, 2010. Phase change materials for building applications: A state-of-the-art review. Energ. Build 42: 1361–1368.

Bagshaw, S.A., E. Prouzet and T.J. Pinnavaia. 1995. Templating of mesoporous molecular sieves by nonionic polyethylene oxide surfactants. Science 269: 1242–1244.

Beck, J.S., J.C. Vartuli, W.J. Roth, M.E. Leonowicz, C.T. Kresge, K.D. Schmitt et al. 1992. A new family of mesoporous molecular sieves prepared with liquid crystal templates. J. Am. Chem. Soc. 114: 10834–10843.

Boissiére, C., A. van der Lee, A.E. Mansouri, A. Larbot and E. Prouzet. 1999. A double step synthesis of mesoporous micrometric spherical MSU-X silica particles. Chem. Commun. 20: 2047–2048.

Boissiére, C., A. Larbot, A. van der Lee, P.J. Kooyman and E. Prouzet. 2000. A new synthesis of mesoporous MSU-X silica controlled by a two-step pathway. Chem. Mater 12: 2902–2913.

Brinker, C.J. and G.W. Screr. 1990. Sol-Gel Science: The Physics and Chemistry of Sol-Gel Processing. Academic Press: New York.

Cao, L., F. Tang and G. Fang. 2014. Synthesis and characterization of microencapsulated paraffin with titanium dioxide shell as shape-stabilized thermal energy storage materials in buildings. Energ. Build 72: 31–37.

Cao, Xi., X. Chuan, R.C. Massé, D. Huang, S. Li and G. Cao. 2015. A three layer design with mesoporous silica encapsulated by a carbon core and shell for high energy lithium ion battery anodes. J. Mater Chem. A 3: 22739–22749.

Che, S., A.E. Garcia-Bennett, T. Yokoi, K. Sakamoto, H. Kunieda, O. Terasaki et al. 2003. A novel anionic surfactant templating route for synthesizing mesoporous silica with unique structure. Nat. Mater 2: 801–805.

Che, R., D. Gu, L. Shi and D. Zhao. 2011. Direct imaging of the layer-by-layer growth and rod-unit repairing defects of mesoporous silica SBA-15 by cryo-SEM. J. Mater Chem. 21: 17371–17381.

Chen, C.-Y., S.L. Burkett, H.-X. Li and M.E. Davis. 1993. Studies on mesoporous materials II. Synthesis mechanism of MCM-41. Microporous Mater 2: 21–34.

Chen, X.Y., L.M. Huang and Q.Z. Li. 1997. Hydrothermal transformation and characterization of porous silica templated by surfactants. J. Phys. Chem. B 101: 8460–8467.

Chen, D., Z. Li, Y. Wan, X. Tu, Y. Shi, Z. Chen et al. 2005. Anionic surfactant induced mesophase transformation to synthesize highly ordered large-pore mesoporous silica structures. J. Mater Chem. 16: 1511–1519.

Chen, Y., X. Zhang, B. Wang, M. Lv, Y. Zhua and J. Gao. 2017. Fabrication and characterization of novel shape-stabilized stearic acid composite phase change materials with tannic-acid-templated mesoporous silica nanoparticles for thermal energy storage. RSC Adv. 7: 15625–15631.

Cui, Y., G.H. Yu, L.B. Hu, M. Vosgueritchian, H.L. Wang, X. Xie et al. 2011. Solution-processed graphene/MnO$_2$ nanostructured textiles for high-performance electrochemical capacitors. Nano Lett. 11: 2905–2911.

da Silva, A.G.M., H.V. Fajardo, R. Balzer, L.F.D. Probst, A.S.P. Lovónc, J.J. Lovón-Quintanac et al. 2015. Versatile and efficient catalysts for energy and environmental processes: Mesoporous silica containing Au, Pd and Au-Pd. J. Power Sources 285: 460–468.

Díaz, I., J. Perez-Pariente and O. Terasaki. 2004. Structural study by transmission and scanning electron microscopy of the time-dependent structural change in M41S mesoporous silica (MCM-41 to MCM-48 and MCM-50). J. Mater Chem. 14: 48–53.

Dubois, M., T. Gulik-Krzywicki and B. Cabane. 1993. Growth of silica polymers in a lamellar mesophase. Langmuir 9: 673–677.

Eroshenko, V. and A.Y. Fadeev. 1995. Intrusion and extrusion of water in hydrophobized porous silica. Colloid J+ 57: 446–449.

Fadeev, A.Y. and V.A. Eroshenko. 1997. Study of penetration of water into hydrophobized porous silicas. J. Colloid Interf. Sci. 187: 275–282.

Fan, J., C.Z. Yu, J. Lei, Q. Zhang, T.C. Li, B.W.Z. Zhou et al. 2005. Low-temperature strategy to synthesize highly ordered mesoporous silicas with very large pores. J. Am. Chem. Soc. 127: 10794–10795.

Filonenko, S.M., N.D. Shcherban, P.S. Yaremov and V.S. Dyadyun. 2015. Sorption and electrochemical properties of carbon–silica composites and carbons from 2,3-dihydroxynaphthalene. J. Porous Mat. 22: 21–28.

Firouzi, A., D. Kumar, L.M. Bull, T. Besier, P. Sieger, Q. Huo et al. 1995. Cooperative organization of inorganic-surfactant and biomimetic assemblies. Science 267: 1138–1143.

Firouzi, A., F. Atef, A.G. Oertli, G.D. Stucky and B.F. Chmelka. 1997. Alkaline lyotropic silicate-surfactant liquid crystals. J. Am. Chem. Soc. 119: 3596–3610.

Flodström, K., H. Wennerström and V. Alfredsson. 2004. Mechanism of mesoporous silica formation. A time-resolved NMR and TEM study of silica-block copolymer aggregation. Langmuir 20: 680–688.

Gac, W., W. Zawadzki and B. Tomaszewska. 2011. Ethanol conversion in the presence of cobalt nanostructured oxides. Catal. Today 176: 97–102.

Gokulakrishnan, N., J. Parmentier, M. Trzpit, L. Vonna, J.L. Paillaud and M. Soulard. 2013. Intrusion/extrusion of water into organic grafted SBA-15 silica materials for energy storage. J. Nanosci. Nanotechno. 13: 2847–2852.

Grudzien, R.M., B.E. Grabicka and M. Jaroniec. 2006. Effective method for removal of polymeric template from SBA-16 silica combining extraction and temperature-controlled calcination. J. Mater Chem. 16: 819–823.

Gusev, V.Y. 1994. On thermodynamics of permanent hysteresis in capillary lyophobic systems and interface characterization. Langmuir 10: 235–240.

He, L., J. Li, C. Zhou, H. Zhu, X. Cao and B. Tang. 2014. Phase change characteristics of shape-stabilized PEG/SiO$_2$ composites using calcium chloride-assisted and temperature-assisted sol gel methods. Sol. Energy 103: 448–455.

Hitz, S. and R. Prins. 1997. Influence of template extraction on structure, activity and stability of MCM-41 catalysts. J. Catal. 168: 194–206.

Huang, Y., Y. Li, Z. Hu, G. Wei, J. Guo and J. Liu. 2013. A carbon modified MnO$_2$ nanosheet array as a stable high-capacitance supercapacitor electrode. J. Mater Chem. A 1: 9809–9813.

Huo, Q.S., D.I. Margolese, U. Ciesla, P.Y. Feng, T.E. Gier, P. Sieger et al. 1994a. Generalized synthesis of periodic surfactant/inorganic composite materials. Nature 368: 317–321.

Huo, Q.S., D.I. Margolese, U. Ciesla, D.G. Demuth, P.Y. Feng, T.E. Gier et al. 1994b. Organization of organic molecules with inorganic molecular species into nanocomposite biphase arrays. Chem. Mater 6: 1176–1191.

Inagaki, S., Y. Fukushima and K. Kuroda. 1993. Synthesis of highly ordered mesoporous materials from a layered polysilicate. J. Chem. Soc. Chem. Commun. 680–682.

Inagaki, S., Y. Sakamoto, Y. Fukushima and O. Terasaki. 1996. Pore wall of a mesoporous molecular sieve derived from kanemite. Chem Mater 8: 2089–2095.

Inagaki, S., S. Guan, T. Ohsuna and O. Terasaki. 2002. An ordered mesoporous organosilica hybrid material with a crystal-like wall structure. Nature 416: 304–307.

Islam, M.S., M.S. Miran, M.M. Rahman, M.Y.A. Mollah and M.A.B.H. Susan. 2013. Polyaniline-silica composite materials: Influence of silica content on the thermal and thermodynamic properties. J. Nanostructured Polym. Nanocomposites 9: 85–91.

Kadoono, T. and M. Ogura. 2014. Heat storage properties of organic phase-change materials confined in the nanospace of mesoporous SBA-15 and CMK-3. Phys. Chem. Chem. Phys. 16: 5495–5498.

Keene, M.T.J., R.D.M. Gougeon, R. Denoyel, R.K. Harris, J. Rouquerol and P.L. Llewellyn. 1999. Calcination of the MCM-41 mesophase: Mechanism of surfactant thermal degradation and evolution of the porosity. J. Mater Chem. 9: 2843–2850.

Khodakov, A.Y., V.L. Zholobenko, M. Impéror-Clerc and D. Durand. 2005. Characterization of the initial stages of SBA-15 synthesis by *in situ* time-resolved small-angle X-ray scattering. J. Phys. Chem. B 109: 22780–22790.

Kim, T.W., F. Kleitz, B. Paul, R. Ryoo. 2005. MCM-48-like large mesoporous silicas with tailored pore structure: Fabcile synthesis domain in a ternary triblock copolymer–butanol–water system. J. Am. Chem. Soc. 127: 7601–7610.

Kleitz, F., D.N. Liu, G.M. Anilkumar, I.S. Park, L.A. Solovyov, A.N. Shmakov and R. Ryoo. 2003a. Large cage face-centered-cubic *Fm3m* mesoporous silica: Synthesis and structure. J. Phys. Chem. B 107: 14296–14300.

Kleitz, F., S.H. Choi and R. Ryoo. 2003b. Cubic Ia3d large mesoporous silica: Synthesis and replication to platinum nanowires, carbon nanorods and carbon nanotubes. Chem. Commun. 2136–2137.

Kresge, C.T., M.E. Leonowicz, W.J. Roth, J.C. Vartuli and J.S. Beck. 1992. Ordered mesoporous molecular sieves synthesized by a liquid-crystal template mechanism. Nature 359: 710–712.

Kwon, T., H. Nishihara, Y. Fukura, K. Inde, N. Setoyama, Y. Fukushima et al. 2010. Carbon-coated mesoporous silica as an electrode material. Microporous Mesoporous Mat. 132: 421–427.

Landry, C.C., S.H. Tolbert, K.W. Gallis, A. Monnier, G.D. Stucky, P. Norby et al. 2001. Phase Transformations in mesostructured silica/surfactant composites. Mechanisms for change and applications to materials synthesis. Chem. Mater 13: 1600–1608.

Leyva-García, S., D. Lozano-Castelló, E. Morallón and D. Cazorla-Amorós. 2016. Silica-templated ordered mesoporous carbon thin films as electrodes for micro-capacitors. J. Mater Chem. A 4: 4570–4579.

Li, L., H. Yu, X. Wang and S. Zheng. 2016. Thermal analysis of melting and freezing processes of phase change materials (PCMs) based on dynamic DSC test. Energ. Build 130: 388–396.

Martines, M.A.U., E. Yeong, A. Larbot and E. Prouzet. 2004. Temperature dependence in the synthesis of hexagonal MSU-3 type mesoporous silica synthesized with pluronic P123 block copolymer. Micropor. Mesopor. Mat. 74: 213–220.

Mitran, R., D. Berger, C. Munteanu and C. Matei. 2015. Evaluation of different mesoporous silica supports for energy storage in shape-stabilized phase change materials with dual thermal responses. J. Phys. Chem. C 119(27): 15177–15184.

Min, X., M. Fang, Z. Huang, Y. Liu, Y. Huang, R. Wen et al. 2015. Enhanced thermal properties of novel shape-stabilized PEG composite phase change materials with radial mesoporous silica sphere for thermal energy storage. Sci. Rep. 5: Article no. 12964.

Mitran, R.-A., D. Berger, C. Munteanu and C. Mate. 2015. Evaluation of different mesoporous silica supports for energy storage in shape-stabilized phase change materials with dual thermal responses. J. Phys. Chem. C 119: 15177–15184.

Monnier, A., F. Schuth, Q. Huo, D. Kumar, D. Margolese, R.S. Maxwell et al. 1993. Cooperative formation of inorganic-organic interfaces in the synthesis of silicate mesostructures. Science 261: 1299–1303.

Mori, K., S. Masuda, H. Tanaka, K. Yoshizawa, M. Che and H. Yamashita. 2017. Phenylamine-functionalized mesoporous silica supported PdAg nanoparticle: A dual heterogeneous catalyst for the formic acid/CO_2-mediated chemical hydrogen delivery/storage. Chem. Commun. 53: 4677–4680.

Peng, S., B. Pan, H. Hao and J. Zhang. 2015. Hollow mesoporous silica supported ruthenium nanoparticles: A highly active and reusable catalyst for H_2 generation from the hydrolysis of $NaBH_4$. J. Nanomater 2015: 1–11.

Qian, T., J. Li, H. Ma and J. Yang. 2015. The preparation of a green shape-stabilized composite phase change material of polyethylene glycol/SiO_2 with enhanced thermal performance based on oil shale ash via temperature-assisted sol–gel method. Energy Mater. Sol. Energ. Mat. Sol. C 132: 29–39.

Ravikovitch, P.I. and A.V. Neimark. 2001. Characterization of micro- and mesoporosity in SBA-15 materials from adsorption data by the NLDFT method. J. Phys. Chem. B 105: 6817–6823.

Ravikovitch, P.I. and A.V. Neimark. 2002. Density functional theory of adsorption in spherical cavities and pore size characterization of templated nanoporous silicas with cubic and three-dimensional hexagonal structures. Langmuir 18: 1550–1560.

Ruthstein, S., V. Frydman, S. Kababya, M. Landau and D. Goldfarb. 2003. Study of the formation of the mesoporous material SBA-15 by EPR spectroscopy. J. Phys. Chem. B 107: 1739–1748.

Ruthstein, S., J. Schmidt, E. Kesselman, Y. Talmon and D. Goldfarb. 2006. Resolving intermediate solution structures during the formation of mesoporous SBA-15. J. Am. Chem. Soc. 128: 3366–3374.

Ryoo, R., C.H. Ko, M. Kruk, V. Antochshuk and M. Jaroniec. 2000. Block-copolymer-templated ordered mesoporous silica: Array of uniform mesopores or mesopore–micropore network? J. Phys. Chem. B 104: 11465–11471.

Sakamoto, Y., M. Kaneda, O. Terasaki, D.Y. Zhao, J.M. Kim, G. Stucky et al. 2000. Direct imaging of the pores and cages of three-dimensional mesoporous materials. Nature 408: 449–453.

Shen, S., A.E. Garcia-Bennett, Z. Liu, Q. Lu, Y. Shi, Yan et al. 2005. Three-dimensional low symmetry mesoporous silica structures templated from tetra-headgroup rigid bolaform quaternary ammonium surfactant. J. Am. Chem. Soc. 127(18): 6780–6787.

Singh, A.K., D. Sarkar, G.G. Khan and K. Mandal. 2013. Unique hydrogenated Ni/NiO core/shell 1D nano-heterostructures with superior electrochemical performance as supercapacitors. J. Mater Chem. A 1: 12759–12767.

Tan, B., A. Dozier, H.J. Lehmler, B.L. Knutson and S.E. Rankin. 2004. Elongated silica nanoparticles with a mesh phase mesopore structure by fluorosurfactant templating. Langmuir 20(17): 6981–6984.

Tanev, P.T. and T.J. Pinnavaia. 1995. A neutral templating route to mesoporous molecular sieves. Science 267: 865–867.

Tiddy, G.J.T. 1980. Surfactant-water liquid crystal phases. Phys. Rep. 57: 1–46.

Tominaga, Y., I.-C. Hong, S. Asai and M. Sumita. 2007. Proton conduction in nafion composite membranes filled with mesoporous silica. J. Power Sources 171: 530–534.

Wan, Y. and D. Zhao. 2007. On the controllable soft-templating approach to mesoporous silicates. Chem. Rev. 107(7): 2821–2860.

Wang, F., S. Xiao, Y. Hou, C. Hu, L. Liu and Y. Wu. 2013. Electrode materials for aqueous asymmetric supercapacitors. RSC Adv. 3: 13059–13084.

Wang, J., M. Yang, Y. Lu, Z. Jin, L. Tan, H. Gao et al. 2016. Surface functionalization engineering driven crystallization behavior of polyethylene glycol confined in mesoporous silica for shape-stabilized phase change materials. Nano Energy 19: 78–87.

Wu, Z.-S., Y. Sun, Y.-Z. Tan, S. Yang, X. Feng and K. Müllen. 2012. Three-dimensional graphene-based macro- and mesoporous frameworks for high-performance electrochemical capacitive energy storage. J. Am. Chem. Soc. 134: 19532–19535.

Yanagisawa, T., T. Shimizu, K. Kuroda and C. Kato. 1990. The preparation of alkyltriinethylaininonium–kaneinite complexes and their conversion to microporous materials. Bull. Chem. Soc. Jpn. 63: 988–992.

Yang, C.M., W. Schmidt and F. Kleitz. 2005. Pore topology control of three-dimensional large pore cubic silica mesophases. J. Mater Chem. 15: 5112–5114.

Yang, C.-W., K.-H. Chen and S. Cheng. 2016. Effect of pore-directing agents and silanol groups in mesoporous silica nanoparticles as nafion fillers on the performance of DMFCs. RSC Adv. 6: 111666–111680.

Yang, C.-W., C.-C. Chen, K.-H. Chen and S. Cheng. 2017. Effect of pore-directing agents in SBA-15 nanoparticles on the performance of Nafion®/SBA-15n composite membranes for DMFC. J. Membrane Sci. 526: 106–117.

Yin, Y., W. Deng, H. Wang, A. Li, C. Wang, Z. Jiangab et al. 2015. Fabrication of hybrid membranes by incorporating acid–base pair functionalized hollow mesoporous silica for enhanced proton conductivity. J. Mater Chem. A 3: 16079–16088.

Yin, Y., T. Xu, X. Shen, H. Wu and Z. Jiang. 2014. Fabrication of chitosan/zwitterion functionalized titania–silica hybrid membranes with improved proton conductivity. J. Membr. Sci. 469: 355–363.

Yu, C., J. Fan, B. Tian and D. Zhao. 2004. Morphology development of mesoporous materials: A colloidal phase separation mechanism. Chem. Mater 16: 889–898.

Zhang, H.N., Y.J. Chen, W.W. Wang, G.H. Zhang, M. Zhuo, H.M. Zhang et al. 2013. Hierarchical Mo-decorated Co_3O_4 nanowire arrays on Ni foam substrates for advanced electrochemical capacitors. J. Mater Chem. A 1: 8593–8600.

Zhao, D., J. Feng, Q. Huo, N. Melosh, G.H. Fredrickson, B.F. Chmelka et al. 1998a. Triblock copolymer syntheses of mesoporous silica with periodic 50 to 300 angstrom pores. Science 279(5350): 548–552.

Zhao, D., Q. Huo, J. Feng, B.F. Chmelka and G.D. Stucky. 1998b. Nonionic triblock and star diblock copolymer and oligomeric surfactant syntheses of highly ordered, hydrothermally stable, mesoporous silica structures. J. Am. Chem. Soc. 120: 6024–6036.

Zhao, D., P. Yang, B.F. Chmelka and G.D. Stucky. 1999. Multiphase assembly of mesoporous-macroporous membranes. Chem. Mater 11: 1174–1178.

Zhao, D.Y., J.Y. Sun, Q.Z. Li and G.D. Stucky. 2000. Morphological control of highly ordered mesoporous silica SBA-15. Chem. Mater 12: 275–279.

Zhi, J., Y. Wang, S. Deng and A. Hu. 2014. Study on the relation between pore size and supercapacitance in mesoporous carbon electrodes with silica-supported carbon nanomembranes. RSC Adv. 4: 40296–40300.

Zhi, J., S. Deng, Y. Wang and A. Hu. 2015. Highly ordered metal oxide nanorods inside mesoporous silica supported carbon nanomembranes: High performance electrode materials for symmetrical supercapacitor devices. J. Phys. Chem. C 119: 8530–8536.

POROUS MATERIALS IN ENERGY GENERATION

10

3*d* Block Transition Metal-Based Catalysts for Electrochemical Water Splitting

*Md. Mominul Islam** and *Muhammed Shah Miran*

1. Introduction

The industrial revolution has introduced an exponential increase of the world's population, which is accompanied by an increasing demand of energy. The world has to face a great challenge of energy of about 30 trillion watt for its estimated 9 billion people by 2050 (Energy Report 2011, Liu et al. 2012, Ewan and Allen 2005, Conway 1999). At present, more than 80 per cent of energy needs are met by fossil fuels that are finite natural resources, and are depleting at a rapid rate currently. For achieving the milestone of the use of '100 per cent Renewable Energy by 2050' set to reduce pollutant emission, environmental damage and ensure energy security, the search for clean, renewable and reliable alternatives has become increasingly more urgent than ever. The production of energy on a large scale can offset between demand and supply of energy, indeed, but the efficient storage and distribution protocol of energy would assist to save the energy through the minimisation of system loss arisen during its supply to the end-user. The storage of energy especially the renewable one, i.e., solar radiation, is a must to ensure the supply of energy when sunlight disappears. Moreover, the long-distance or remote distribution of electrical energy also urges the storage in some ways such as decentralised and centralised stationary energy storages, e.g., renewable fuels, compressed air, flywheel, batteries, supercapacitors, pumped hydroelectric, superconducting magnet and solar thermal (Cook et al. 2010). In this regard, the water splitting that offers the production of H_2 gas, a renewable green fuel, can be considered as an excellent energy storage and distribution module.

In this chapter, the roles of 3*d* block transition metals that are highly abundant in the Earth crust and hence are low-cost on electrochemical water splitting reactions are addressed. The recent research works on preparation, processing and modification of these metals-based structured electrocatalysts and their application in water splitting are comprehensively reviewed and the catalytic properties are compared by highlighting relevant thermodynamics and kinetics involved in electrochemical water splitting. With the essences of experimental and theoretical works, the future directions in catalytic electrochemical water splitting are also pointed out.

Department of Chemistry, Faculty of Science, University of Dhaka, Dhaka 1000, Bangladesh.
 E-mail: shahmiran@du.ac.bd
* Corresponding author: mominul@du.ac.bd

2. Energy Storage: $H_2O \leftrightarrow H_2$ Cycle

As a continuation of the discoveries of 'animal electricity' by Galvani in 1776 and 'voltaic electricity' by Volta in 1800, redox flow batteries, lithium-ion batteries, capacitors, fuel cells, solar cells, and water splitting in electrolysers are nowadays considered as contemporary energy conversion and storage protocols (Conway 1999, Ewan and Allen 2005). The storage systems such as batteries and capacitors have been considered as a secondary source of energy, while fossil fuels, fuel cells and solar cells are the primary sources. Now the question is whether H_2O is a source of energy like coal, gasoline, solar light and so on or an energy reservoir to be answered.

$$H_2O(l) \rightarrow H_2(g) + \frac{1}{2} O_2(g) \qquad \Delta E = 1.23 \text{ V} \qquad (1)$$

$$H_2(g) \quad + \frac{1}{2} O_2(g) \rightarrow H_2O(l) \qquad \Delta H = -286 \text{ kJ} \qquad (2)$$

H_2O decomposes into H_2 and O_2 gases called splitting of water of which H_2 is a green fuel with a calorific value of 286 kJ mol^{-1} (Pagliaro et al. 2010). On the contrary, the splitting reaction is very endothermic and the thermodynamic energy of 1.23 V requires for splitting of one mole H_2O (Eq. 1). The energy needed for water splitting is much higher than that would generate during the combustion of one mole of H_2 produced. This rationally justifies that H_2O is not a primary source of energy. Upon combustion or use as a fuel (i.e., as a proton-source) in H_2-O_2 fuel cell, H_2 produces only H_2O. Thus, the consumption and formation as by-product of H_2O occurs in a cycle, i.e., $H_2O \leftrightarrow H_2$ cycle (Fig. 1). In the $H_2O \leftrightarrow H_2$ cycle, H_2 gas is truly an energy carrier and hence $H_2O \leftrightarrow H_2$ cycle is rather a secondary source of energy like capacitors and batteries.

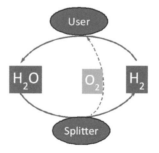

Fig. 1. $H_2O \leftrightarrow H_2$ cycle.

3. Importance of Water Splitting

The water splitting in producing H_2 gas is justified above as an energy storage cycle where the energy is stored in the smallest possible configuration: chemical bonds (Energy Report 2011, Liu et al. 2012, Ewan and Allen 2005). In addition, the necessity of a large-scale production of H_2 gas in the near future is obvious since H_2 is a green fuel in power generating industries as well as a feed in H_2-O_2 fuel cell, a promising and green electricity production device. Truly, the demand of H_2 gas has even nowadays been increasing day by day due its consumption in fuel cells. Furthermore, a large scale harvesting of solar energy requires a huge electrical energy storage devices such as batteries and capacitors. The $H_2O \leftrightarrow H_2$ cycle is superior to mechanical-based storage systems and capacitors or batteries, since H_2 fuels provide the most effective way of storing energy with 2–3 orders of magnitude higher energy density (Ni et al. 2006). Moreover, the fuels have superior functions over capacitors and batteries concerning power quality, stability, regulation, spinning reserve, load levelling and a long term reserves (Cook et al. 2010, Pagliaro et al. 2010). Therefore, establishing a method for catalytic water splitting is a challenge to researchers worldwide.

4. Methods of Water Splitting

The most recent literature survey using digital source showed that about thirteen thousand articles has been published on water splitting. Several methods have been employed for splitting of water as represented in Fig. 2. These include photocatalytic, solar thermochemical cycles, biomass gasification, photoelectrochemical, electrochemical, pyrolysis and biological conversion (Khan et al. 2016, Zou and Zhang 2015, Wang et al. 2012). Development of photochemical route for splitting of water by consuming solar energy to produce a large scale H_2 fuel is an attractive scientific and technological goal to address the increasing global energy demand and to reduce the impact on climate change from energy production (Domenech et al. 2008, Hou et al. 2016, Rudnik and Wloch 2013, Karkas et al. 2015). Photoelectrochemistry with semiconductor electrodes is a promising strategy for using photon energy to drive endergonic reactions, thus storing it in a 'solar fuel'. One of the important requirements for a photocatalyst to be useful for water splitting is to possess the minimum potential difference of 1.23 V. Direct approach with visible light driven semiconductor material could really be an alternative to the electrochemical water splitting run by consuming electricity. TiO_2 photocatalyst (bandgap ~ 2.4 eV) advantageously absorbs visible light and valence band edge sufficiently positive relative to the water oxidation potential (Liu et al. 2013). Mixed metal oxides, such as $BiVO_4$ is recently used as photocatalyst (Guo and Speidel 2014) and photoanode for photoelectrochemical water decomposition (Burstein 2005, Viswanathan 2013). Unfortunately, most semiconductors with suitable band structures for water splitting absorb mostly ultraviolet light. Indirect water splitting with hybrid photoelectrochemical system in which visible light is harvested with solar cell (e.g., dye-sensitised cell) to electricity that finally consumes by an electrochemical water splitting system to produce H_2 gas is effective in this regard. It is thus remarkably noted that, for converting freely available solar energy into usable form, electrochemical water splitting is important, i.e., electrochemical water splitting serves as storage system of solar energy.

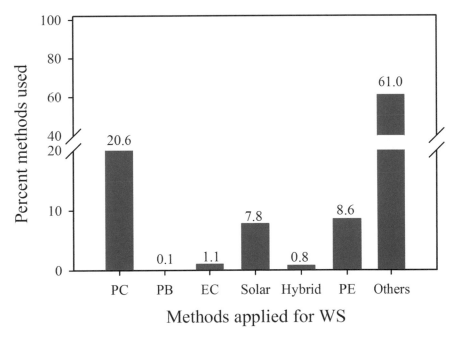

Fig. 2. Recent works on water splitting carried out by Photochemical (PC), Photobiological (PB), Electrochemical (EC), Solar, Hybrid, Photoelectrochemical (PE). The other methods include thermal, chemical, hydroelectric, etc. Source: SciFinder® Scholar.

Among all these technologies, the electrochemical water splitting has been believed to be only the suitable method for a large-scale renewable H_2 production in the near future (Liu et al. 2012, Khan et al. 2016), although currently ca. 95 per cent of H_2 is produced from non-renewable source, i.e., natural gas steam reforming (Zou and Zhang 2015). Despite it has been known to the scientists since 1800, the electrochemical water splitting producing H_2 fuel has to be the focal points of research in the energy sector especially where a continuous production and storing of energy are necessary (Liu et al. 2012, Khan et al. 2016). Thus it may be mentioned that, even though, the operational cost seems to be higher, the technology-based on direct water splitting for H_2 production, i.e., non-catalytic, is still important as energy storage system, like capacitors and batteries.

5. 3d Block Transition Metal Catalysts and Water Splitting

The chart shows $3d$ block transition metals starting from scandium to zinc. It is known to us that nowadays zinc has been excluded from the family of transition metals. Therefore, scandium to copper is the prime interest of this chapter. It is noted that use of any materials for application purposes depends on many factors including abundance, prices, processability, and environmental friendliness and so on. The abundance of $3d$ block transition elements has been known to be higher than those of other metals (see in Fig. 3).

Figure 3 represents that $3d$ block transition metals exist with a relatively high abundance and hence are relatively cheap. Among the metals under consideration iron is low pricy and highly abundant. A survey of recent research on water splitting with $3d$ block transition metals based catalysts has been executed and the results of the survey are illustrated in Fig. 4. Researches with nickel and cobalt based catalysts are found to be maximum although their abundances are relatively low. Later we will be acquainted with the reason for prime uses of these two elements where their activities towards water splitting are compared.

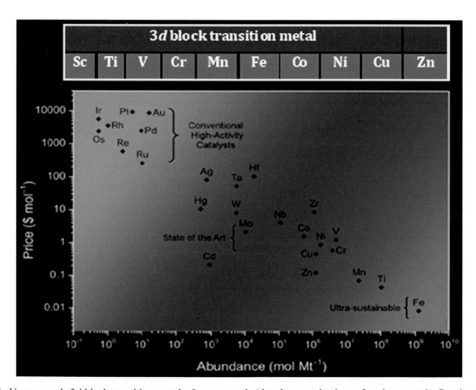

Fig. 3. Upper panel: $3d$ block transition metals. Lower panel: Abundance and prices of various metals. Reprinted with permission from Wiley-VCH (Martindale and Reisner 2016).

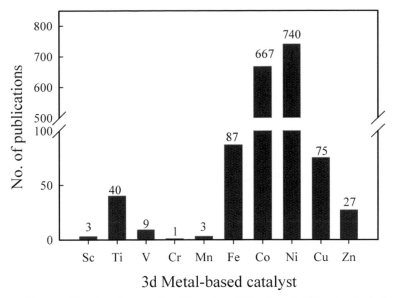

Fig. 4. *3d* block transition metal based catalysts employed for water splitting through different methods. Source: SciFinder® Scholar.

6. Electrochemical Water Splitting

Water splitting is a very endothermic reaction and the simplest way of providing the energy required for water splitting is by running the reaction electrochemically. Direct electrolysis of water at the electrode surface can be performed in which the electrical energy the driving force of chemical reactions is leading to the decomposition of water to H_2 and O_2 gases (Fig. 5). The electrical energy can be supplied from renewable source such as wind, hydro and solar energy harvesting system for the splitting of water molecule at the electrode surface, then this method can be considered as renewable, indeed. The formation of H_2 occurred at cathode is known as hydrogen evolution reaction (HER), while that takes place at anode is called oxygen evolution reaction (OER) (Eqs. 3 and 4).

$$\text{Anodic: } 2H_2O(l) \rightarrow O_2(g) + 4H^+(aq) + 4e^- \tag{3}$$

$$\text{Cathodic: } 4H_2O (l) + 4e^- \rightarrow 2H_2(g) + 4OH^- (aq) \tag{4}$$

Practically, the electrical energy (i.e., 1.23 V) required for direct electrochemical water splitting is much higher than that would generate for the combustion of equivalent amount of H_2 produced (Koseki et al. 2007, Liu et al. 2012). This is because the complexity of OER associated to a $4e^-/4H^+$ removal and to the formation of a new oxygen-oxygen bond (Eq. 3) that implies severe kinetic hurdles and high activation energy, which can be overcome by the employment of suitable catalysts (Trasatti 1972, Trasatti 1984, Bloor et al. 2014, Martindale and Reisner 2016). The catalytic routes that would make surplus of the energy stored in H_2 fuel produced minus the consumption of electricity (energy) required for water splitting would thus be explored for establishing the $H_2O \leftrightarrow H_2$ cycle as a sustainable source of renewable energy.

6.1 Energetics of Electrochemical Water Splitting

The thermodynamic voltage of water splitting is known to be 1.23 V (25°C and 1 atm) regardless of the reaction media. In practice, to perform water splitting electrochemically at a practical rate, one must apply extra potential on the cell (Fig. 5) and the overall operational potential (η_{op}) for water splitting in volt can be described as:

$$\eta_{op} = 1.23 + \eta_a + |\eta_c| + \eta_\Omega \qquad (5)$$

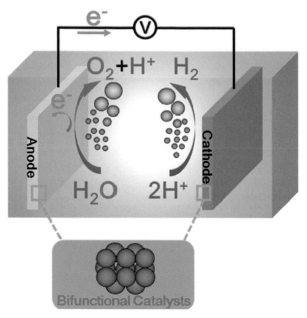

Fig. 5. Electrochemical cell for catalytic overall water splitting. Reprinted with Permission from Royal Society of Chemistry (Yan et al. 2016).

where, η_Ω represents the excess potential require to overcome internal resistance of the system such as solution resistance and contact resistance. η_a and η_c are the over potentials required to surmount the intrinsic activation barriers of the anode and cathode, respectively. The intrinsic activation barriers can be minimised by using highly active OER and HER catalysts.

The most difficult task of electrocatalytic studies is to figure out experimental kinetic data in a rational fashion. So far the well-practiced way is to compare experimental η_a or η_c measured using different electrode materials at a particular current density (j). The catalytic performance of a particular electrode material can also be revealed by analysing the voltammetric response using Tafel equation (Bard and Faulkner 2001) where the dependence of steady-state j on η_a or η_c for water splitting is analysed. η logarithmically relates to j and its linear portion is given as Tafel equation ($\eta > 0.05$ V).

$$\eta = a + b \log j \qquad (6)$$

where b is Tafel slope and is related to the mechanism of the electrode reaction. From this equation, when $\eta = 0$, the obtained j is called exchange current density (j_0). This represents the intrinsic activity of the catalysts under equilibrium states. Thus, a material possessing a high j_0 and a small b is considered as a good catalyst for both the HER and OER. It may be remarkably mentioned that when comparing j_0, it is necessary to ensure that the comparison is carried out based on same number of active sites. In practice, a more convenient way characterising the activities of different electrodes is to compare the relevant η at a given value of j (current/geometric area), by maintaining similar mass loading of the active material. In particular, the η needed to yield a j value of 10 mA cm^{-2} is widely adopted to evaluate the performance of different catalysts. In fact, j value of 10 mA cm^{-2} is expected for a 12.3 per cent efficiency of a solar water splitting device.

Turn over frequency (TOF) is the important characteristics of a catalyst for a particular reaction. The value of TOF in case of electrochemical water splitting can be estimated from the experimental j according to the following relationship (Browne et al. 2017):

$$TOF = \frac{j_\eta}{Q \times 4} \qquad (7)$$

where, j_η is the current density at the applied potential η, Q is the charge and 4 is number of electron involved with HER and OER shown in reactions (3 and 4). TOF number of a catalyst or turnover number per time unit characterises its level of activity. So the TOF is the total number of moles transformed into the desired product by one mole of active site per h. The larger the value of TOF is, the more active the catalyst is.

6.2 Mechanism of Electrochemical Water Splitting

Electrochemical water splitting has been suitably executed in aqueous acidic, neutral, basic, and buffer solutions. Depending on the type of electrode material and its catalytic activity towards water splitting the electrolytic media have generally been designed. Some special purposes, other medium such as surfactant-based organised media and ionic liquid-based media (Islam et al. 2008) have been employed for electrochemical water splitting.

The modes of adsorption of water molecule on electrode surface are depicted in Fig. 6, where adsorption of water occurs via the central oxygen atom on positively charged surface while two hydrogen atoms approach towards negatively charged surface. Sergio Trasatti has revealed that the energy difference arisen due to the reorientation of water molecule on positively *sp* block metal surface (Al) and *d* block transition metal (Au) is about 0.4 V (Trasatti 1972). The reduction on cathode surface and oxidation on anode surface of water molecules lead to ultimately the formation of molecular H_2 and O_2, respectively. The formation reactions of H_2 and O_2 through water splitting at the electrode surface are not straightforward. Various steps involving HER and OER in aqueous solution are described below.

Fig. 6. Modes of adsorption of water molecules on positively and negatively charged electrode surface (Trasatti 1972).

6.2.1 Hydrogen Evolution Reaction

HER has been explored extensively for almost all transition metals (M). HER process generally involves three major steps of which two of them are electrochemical concerned with electron transfer (Eqs. 8–11) and the rest one is simply desorption of product from electrode surface (Eq. 12).

Step 1. Electrochemical adsorption (*Volmer reaction*)

$$\text{In acidic solution: } H^+ + M + e^- \rightarrow MH_{ads} \qquad (8)$$

$$\text{In basic solution: } H_2O + M + e^- \rightarrow MH_{ads} + OH^- \qquad (9)$$

Step 2. Electrochemical desorption (*Heyrovsky reaction*)

$$\text{In acidic solution: } MH_{ads} + H^+ + e^- \rightarrow M + H_2 \tag{10}$$

$$\text{In basic solution: } MH_{ads} + H_2O + e^- \rightarrow M + OH^- + H_2 \tag{11}$$

Step 3. Chemical desorption (*Tafel reaction*)

$$2MH_{ads} \rightarrow 2M + H_2 \tag{12}$$

6.2.2 Oxygen Evolution Reaction

The mechanism of OER is more complicated compared to that of HER. Generally, the evolution of O_2 occurs at metal oxide surface, but not from the clean/bare metal surface, i.e., the formation of O_2 by the cleavage of O-H is actually followed by the formation of metal oxide under applied potential. Therefore, the mechanism of OER is thus different for metal oxides with different surface structures. Generally, the following steps are considered to take place during OER (Trasatti 1984, Yan et al. 2016).

$$M + H_2O \rightarrow MOH + H^+ + e^- \tag{13}$$

$$MOH \rightarrow MO + H^+ + e^- \tag{14}$$

or

$$2MOH \rightarrow MO + M + H_2O \tag{15}$$

$$2MO \rightarrow 2M + O_2 \tag{16}$$

or

$$MO + H_2O \rightarrow M + O_2 + 2H^+ + 2e^- \tag{17}$$

The removal of oxygenated species from the metal (Eqs. 16 or 17) is the rate-determining step of OER (Trasatti 1984).

6.3 Catalysis in Electrochemical Water Splitting

The International Energy Agency has recognised a key obstacle to the widespread implementation of electrochemical water splitting technologies to be the development of cheap and active materials that catalyse both the HER and OER with high efficiency and stability. The prime challenge for the researchers is to develop efficient water splitting electrocatalysts that can minimise η_a and $|\eta_c|$ and improve the efficiency of water splitting. Other factors should also be considered in designing the catalysts: the catalysts should have (a) tolerance for a wide pH range, (b) long-term stability, (c) composition with inexpensive earth-abundant materials, and (d) simple and economical preparation and fabrication methods.

6.3.1 Effect of Media on Electrochemical Water Splitting

Strongly acidic or alkaline conditions of the electrolytic solution is suitable for efficient water splitting because such electrolytes have high ionic conductivity that results in low ohmic resistance (i.e., low η_Ω in Eq. 5) losses and that allow for the necessary transfer of protons or OH^- ions between the compartments of the electrolyser (Fig. 5). In alkaline conditions, the HER is comparatively more sluggish as it directly depends on the anodic OER which supplies the protons to cathode by the deprotonation of OH^- ions (Eqs. 13 and 14) and affects the HER kinetics. The OER on the other hand has a different story. The kinetics of OER in acidic and alkaline media varies depending on the material by which it is being catalysed. The noble metal catalysts such as Ir and Ru and their compounds catalyse OER in acidic medium with ease than in alkaline conditions. On the other hand, the catalysts derived out of $3d$ transition metals (Fe, Co, Ni and Ti) catalyse the OER in alkaline medium more favourably than in acidic medium. In alkaline conditions, all the proposed mechanisms begin with an essential elementary step of hydroxide coordination to the active site and proceeds via different proposed other elementary steps (Bockris and Otagawa 1983). The kinetic barriers associated with each elementary step raise the overall η_{op} required.

As is described above, the total kinetic η of an electrolyser is the sum of η at each of the electrodes of the electrolysis cell. Kinetic η for the half reactions of water splitting generally are lower for electrocatalysts in contact with strongly acidic or strongly alkaline electrolytes than for electrocatalysts in contact with neutral electrolytes. Reduction of a positively charged proton (Eq. 8) is easier than reduction of a neutral species, water (Eq. 9), at cathode. Thus, the electrocatalysts for the HER exhibit much lower η in acid than at neutral pH (Tilak et al. 1981, Bockris and Reddy 1970). Conversely, the oxidation of a negatively charged OH^- ion is easier than oxidation of a neutral species, water. So the electrocatalysts for the OER exhibit much lower η in alkaline media than at neutral pH (Bockris and Reddy 1970).

Ionic liquids (ILs) have also been successfully used as electrolytes for water electrolysis and in other applications (Islam et al. 2008, Fiegenbaum et al. 2013). Water splitting using ILs medium are also attempted due to their unique properties such as high thermal stability and high ionic conductivity. Imidazolium based ILs have been used in the direct efficient electrolysis of water (Souza et al. 2007, Yue et al. 2011). Triethylammoniumpropanesulfonic acid tetrafluoroborate ([TEA.PS]BF$_4$) IL has been used as an electrolyte in the water electrolysis. The electrolysis of water with this ionic conductor produces high current densities with high efficiencies even at room temperature. A system using [TEA.PS]BF$_4$ in an electrochemical cell with platinum electrodes has current densities up to 1.77 Acm^2 and efficiencies between 93 and 99 per cent at temperatures ranging from 25 to 80°C, respectively (Fiegenbaum et al. 2013).

6.3.2 Effects of Catalysts on Electrochemical Water Splitting

As described above, there are several thermodynamic and kinetic barriers in electrochemical water splitting. These factors cause huge energy loss while doing water splitting and ultimately increase in the values of η of HER and OER at cathode and anode, respectively. Therefore, the electrocatalysts become the key components of commercial water splitting system in increasing the efficiency by reducing the kinetic η for the HER and OER. In previous days, for HER Pt, Ir or Ru and their compounds have been employed as the state-of-the-art electrocatalyst. Figure 7 represents the volcano curve of various metals catalysts employed for HER. The reason for using the noble Pt, Ir or Ru for HER can be easily understood from this curve where value of j_0 is on the top for these metals.

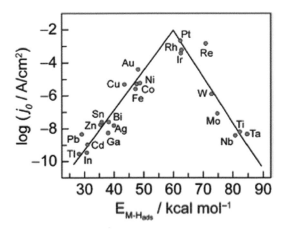

Fig. 7. Volcano curve for the HER on metal electrodes in acidic media. The log of j_0 is plotted versus the M–H bond energy for each metal surface. Reprinted with permissions from American Chemical Society (Cook et al. 2010) and Elsevier (Trasatti 1972).

Similar trend of catalytic activity as of HER of these metals towards OER also prevails as can be seen in Fig. 8. It is clear that the most active catalysts of both HER and OER are precious Pt and RuO_2 or IrO_2. These metals are noble, expensive and less abundant metals (see Fig. 3) and hence there is no real possibility to scale up these materials for global demand due to their scarcity. To avoid such drawbacks, the researchers have then started looking at the $3d$ transition metal-based catalysts such as Mn, Ni, Co and Fe for water splitting since their activities towards both HER and OER are reasonably high (see Figs. 7 and 8). In this regard, it may be mentioned that only three transition metals (Ti, Mn and Fe) are truly abundant in Earth's crust and within $3d$ block elements, Fe is by far the most plentiful (Fig. 3).

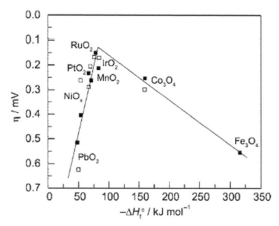

Fig. 8. Volcano plot showing activity for OER on metal oxide surfaces vs. the enthalpy of transition of the oxide in acidic (□) and basic (■) solutions. η measured relative to j of 0.1 mA cm^{-2}. Reprinted with permissions from American Chemical Society (Cook et al. 2010) and Elsevier (Trasatti 1984).

6.3.2.1 3d Block Transition Metal-Based Catalysts

Ni-based homogenous catalysts have been used as alkaline electrolysers on the commercial scale (Singh et al. 2000). A Ni-based oxide film has been synthesised which liberates O_2 in quasi-neutral conditions at η of ~ 425 mV and shows long lasting stability (Dinca et al. 2010). It has been demonstrated that Fe complexes catalyse water oxidation reactions to produce O_2 (Fillol et al. 2011). A water-soluble complex based on Fe has been reported to be a very efficient catalyst for water oxidation when an oxidant is added and hence liberated O_2 bubbles. The concentration of the catalyst and the oxidant has been found to be strongly affecting the oxidation process.

Nickel sulphide (Ni_3S_2) has been tested to show a potential activity towards the OER process and NiSOH has been proposed to be the active species though Ni chalcogenides have grown as attractive HER catalysts (Yan et al. 2016). Lately, Feng et al. have reported the synthesis of stable $\{\overline{2}10\}$ high-indexed faceted Ni_3S_2 nanosheet arrays supported on nickel foam (NF) as active bi-functional HER-OER catalysts (Feng et al. 2015). The obtained Ni_3S_2/NF material has been revealed to exhibit efficient and ultra-stable electrocatalytic activity towards the HER and OER for their nanocatalytic feature and high index facets. NF supported NiSe nanowire film has been synthesised by the hydrothermal reaction of NiF with NaHSe (Tang et al. 2015). Zhang et al. presented the interface engineering of MoS_2/Ni_3S_2 heterostructures on NF as an advanced electrocatalyst for both the OER and HER in alkaline solution (Zhang et al. 2016a,b,c). With the help of density functional theory (DFT) calculations, they argued that the interfaces established between Ni_3S_2 and MoS_2 and the *in situ* formed interfaces between NiO_2 and MoS_2 accelerated the simultaneous adsorption of hydrogen- and oxygen-containing intermediates, which consequently results in the improved efficiency of overall water splitting.

An extensive study of electrodeposited pure and mixed Ni/Fe oxides (those which are cheaper compared to the noble Pt materials) has been carried out in three different NaOH electrolytes with various Fe impurity concentrations (< 1 ppb, 5 ppb and 102 ppb) (Browne et al. 2016). A number of the mixed Ni/Fe catalysts containing Fe impurities at a concentration of 5 ppb have exhibited higher OER activities with higher TOF in NaOH than that of the same catalyst containing < 1 ppb. These Ni/Fe oxide materials have also been reported to be cheaper than Pt group materials, rendering these Ni/Fe catalysts more practical and economical. To determine the Ni and Fe species formed before and after OER *ex situ* Raman spectroscopy and XPS have been employed. The oxidation state of the Ni species in the pure Ni material does not change during OER in any of NaOH media. While for the pure Fe and mixed Ni/Fe 50/50 materials, the oxidation states of the species vary with the concentration of Fe impurities in the NaOH solution (Browne et al. 2016).

Cobalt salts can be used as water oxidation catalysts, but their capability of water splitting has been observed to decrease with time because of the precipitation of the salt from the homogenous solutions (Brunschwig et al. 1983). Co-based catalysts have been shown to work very efficiently at a pH 7.0 (Zidki et al. 2012, Wasylenko et al. 2011). Heterogeneous cobalt oxide (Co_3O_4) has been investigated to work on the same pattern as other Co salts (Harriman et al. 1988). A homogenous cobalt polyoxametalate complex $[Co_4(H_2O)_2(\alpha\text{-}PW_9O_{34})_2]^{10-}$ has been shown to be highly efficient as compared to the reported heterogeneous Co complexes (Yin et al. 2010). The complex undergoes a proton-coupled electron transfer to form a stable $[Co(III)\text{-}OH]^{2+}$ species which on further oxidation forms a Co(IV) intermediate. The intermediate formed reacts with water to liberate O_2 (Yin et al. 2010). A facile route to synthesise $\alpha\text{-}Co(OH)_2$ nanostructures by heating a solution containing $CoCl_2$, ethylene glycol and oleylamine has been developed (Cho et al. 2016). These nanostructures exhibited high activity for electrochemical water splitting with a charge density comparable to that obtained using Pt electrode.

Typical example of electrochemical preparation of electrode materials and their application for electrochemical HER and OER with relevant analysis are represented in Fig. 9. The activation of the catalyst on the electrode surface can be simply performed with cyclic voltammetric scan. The catalytic activities of cobalt-based complexes are compared with Pt metals with respect to characteristics value of η values and Tafel slopes. Moreover, the stability of the electrode materials with the time of operation in splitting of water is also clarified by measuring potential with time.

El-Deab et al. have investigated OER on MnO_2 nanorods modified gold, Pt and glassy carbon electrode deposited electrochemically. It has been clarified that the underlaying electrode substrate can affect η of OER occurred on MnO_2 nanorods (El-Deab et al. 2007). Umena et al. have investigated O_2 evolving photosystem of Mn_4CaO_5 cluster. They have shown from the electron density mapping that all of the metal atoms in Mn_4CaO_5 cluster are located together with all of their ligands (Umena et al. 2011). They have also found that five oxygen atoms serve as oxo bridges linking the five metal atoms, and that four water molecules are bound to the Mn_4CaO_5 cluster and some of them may therefore serve as substrates for O_2 formation.

A low-cost and effective oxygen catalyst has been developed by studying $BaFeO_3$ (BFO)-based perovskites using DFT calculations for manifold applications including water splitting (Chen and Ciucci 2016). The A and B sites in the BFO perovskite structure of which Na, K, Rb, Ca, Sr, Y, La, and Pb have been substituted into the A site, and Sc, Ti, V, Cr, Mn, Co, Ni, Cu, Zn, Y, Zr, Nb, Ag, In, and Ce have been substituted into the B site. Specifically, the results of this calculation have shown that the A site substitution with Na and K lowers energy of the vacancy formation and the introduction of such elements might destabilise the cubic perovskite lattice. Regarding the B site substitution, Sc, Y, Nb and Ce have found to stabilise the cubic perovskite structure, but they sacrifice performance of water splitting. Ni, Cu, Zn, and Ag substitution have shown to increase the electronic conductivity and reduce the energy of vacancy formation at the expenses of stability.

Fe-based electrodes for either HER or OER have been reported. Tunable bi-functional Fe catalyst for both half reactions of water splitting in alkaline media with activity greater than formerly reported Co and Ni-based bi-functional catalysts has been examined (Martindale and Reisner 2016). In this

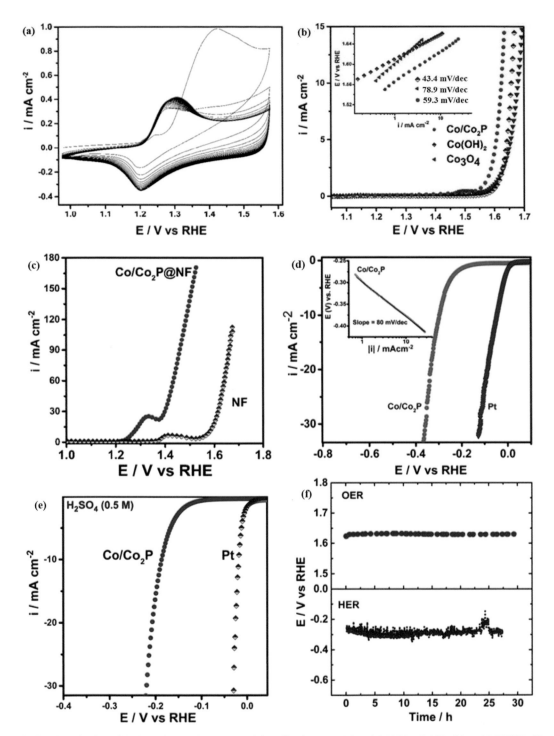

Fig. 9. (a) Activation of Co/Co$_2$P by continuous potential cycling between 1.0 and 1.58 V at 0.1 V s^{-1} in 1.0 M KOH. (b) Linear sweep voltammograms (LSVs) of Co/Co$_2$P in 1.0 MKOH juxtaposed with electrodeposited Co(OH)$_2$ and Co$_3$O$_4$, recorded under similar conditions. Inset represents the corresponding Tafel plot. (c) LSVs showing the OER activity of Co/Co$_2$P supported on nickel foam (NF) (Co/Co$_2$P@NF) by drop coating compared with the OER activity of pure NF. (d) LSVs showing the HER activity of Co/Co$_2$P and Pt in 1.0 M KOH (inset Tafel plot) and (e) in 0.5 M H$_2$SO$_4$. (f) Chronopotentiometric curves representing the stability of Co/Co$_2$P supported on glassy carbon during OER and HER at a *j* of 10 mA cm^{-2}. All the LSVs were recorded at a scan rate of 5 mV s^{-1} with electrode rotation at 1600 rpm to prevent the accumulation of gas bubbles. Reprinted with permission from American Chemical Society (Masa et al. 2016).

report, the mechanism of the electrochemically reversible bi-functional activity of iron-only electrodes towards overall water splitting has been proposed (discussed later).

Yan et al. have recently published an excellent review article on electrochemical water splitting with especially 3*d* block transition metal-based catalysts (Yan et al. 2016). Here, the reported data of η and Tafel slope related to electrocatalytic water splitting on mainly Ni, Co, Mo and Fe-based catalysts are compared. Figures 10 and 11 represent the comparison of catalytic activities of various Ni, Co, Mo and Fe-based compounds towards HER and OER, respectively. Generally Co, Ni and Mo-based catalysts show low values of η and Tafel slope for HER; the characteristics of an efficient catalyst as described above. On the other hand, the performances of the catalysts compared towards OER are almost the same and η values of OER are generally higher than those for HER. This indicates that OER is rather difficult than HER, while Tafel slopes of OER are generally lower than HER, indicating faster electrode kinetics.

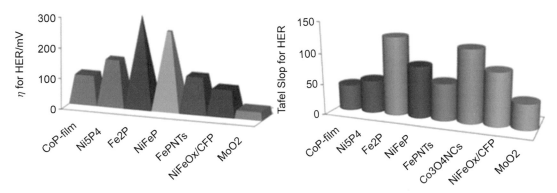

Fig. 10. η (Left panel) (vs. RHE) and Tafel slope (Right panel) for HER at corresponding *j* value of 10 mA cm^{-2} offered by different catalysts. Data were taken from the different references as cited in Yan et al. 2016.

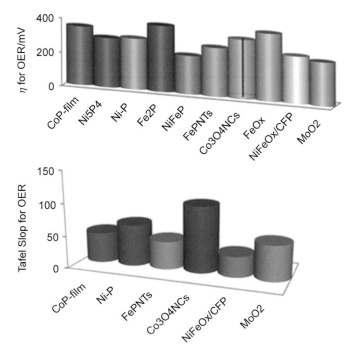

Fig. 11. η (Upper panel) (vs. RHE) and Tafel slope (Lower panel) for OER at corresponding *j* value of 10 mA cm^{-2} offered by different catalysts. Data were taken from the different references as cited in Yan et al. 2016.

Ti-based compounds have attracted the attention as effective water splitting electrocatalyst. Because of the proximity with the benchmark Pt, the reduced TiO_2 exhibits excellent HER catalytic activity in an acidic solution by exhibiting low onset potential of 75 mV and a Tafel slope of 88 mV dec^{-1}, and it demonstrates j value of 10 mA cm^{-2} at a potential of 198 mV (Swaminathan et al. 2016). The enhanced electro-conducting properties along with favourable surface energetic for H_{ads} of reduced $TiO_{1.23}$ ensure ultrafast HER kinetics. The defects that tailor the local atomic structure of nonconductive TiO_2 could be achieved via electrochemical cathodisation strategy.

Electrodeposited Ni diselenide nanoparticles-based film on conductive Ti plate has been prepared and found to be an efficient and robust electrode to catalyse both HER and OER in basic media (Pu et al. 2016). This electrode has been found to afford a j of 10 mA cm^{-2} at η of 96 mV for HER and j of 20 mA cm^{-2} at η of 295 mV for OER with strong durability in 1.0 M KOH.

6.3.2.2 *Effect of Porous Structure on Electrochemical Water Splitting*

The transition metals such as Ni, Fe, Co, Mn and Ti-based catalysts are so far found to be very promising for water splitting. In addition, they are highly abundant, cheap, efficient and easy to design them in a desired shape and structure. In this section, the effects of morphologies of electrocatalysts such as size, porosity and crystallinity on the efficiencies of HER and OER are discussed. During water splitting, two molecules of H_2 are produced for each molecule of O_2 (Eq. 1). So a pressure difference develops between the two sides of the cell (Fig. 5). This pushes liquid from the cathodic chamber to the anodic chamber, eventually resulting in a catastrophic failure of the electrolyser system. Porous separators made from asbestos, glass fibres, porous carbon, etc., prevent physical crossover of gas bubbles from one side of the cell to another, but do not prevent fluid flow. Porous separators indeed allow some product crossover, but can provide intrinsically safe operation, provided that the rate of crossover of the dissolved gases is small compared to the rate of gas evolution at the electrodes. Therefore, water splitting system that use porous separators require active pressure control (Renaud and Leroy 1982, Xiang et al. 2016).

Another important difficulty in water splitting is the high η of OER. To-date several effective strategies have been demonstrated for developing efficient materials including creating porous nanostructures, constructing a series of hybrids and optimising their structures and compositions. Among these favorable factors for enhancing OER performance, creating 3D porous morphology is a very promising approach to improve the OER performance since 3D porous morphology provides large surface area and abundant active sites, improves electrolyte permeability, and offers a better mass transport/diffusion and electron-transfer (Wang et al. 2013, Chen et al. 2013, Chaikittisilp et al. 2014, Ma et al. 2014). In fact, the mentioned matters decrease the value of η_{op} by decreasing especially the value of η_Ω as shown in Eq. 5.

Efficient synthetic approaches are significantly expected to create the OER catalysts with 3D porous nanostructures and to tailor their structure-related properties and OER applications. Specifically, freestanding porous nanostructures derived from electrodeposition can be directly utilised as electrode, and these efforts would assist to avoid binders (Zhou et al. 2013, Tang et al. 2015, Lu and Zhao 2015, Zhao et al. 2014).

Porous cobalt phosphide (CoP) and Co_3O_4-based glassy carbon electrodes have been fabricated by a facilely topological conversion method via a low-temperature multistep calcination reaction (Xu et al. 2015). The materials synthesised are of concave polyhedron (CPH) in morphology. To elucidate the transformation of porous structures during the topological conversion, the nitrogen adsorption-desorption isotherms of the samples have been performed and the samples have been found to be mesoporous. The Barrette-Joynere-Halenda pore-size distribution shows that the pore distribution of the as-prepared Co_3O_4 CPHs is mainly concentrate at approximately 5 nm. As for the as-prepared CoP CPHs, the pores become bigger during the low temperature phosphorisation with a predominant size of ca. 30 nm. Correspondingly, the Brunauer-Emmett-Teller (BET) special surface area of CoP CPHs is calculated to be 29.42 m^2 g^{-1}, which is lower than that of Co_3O_4 CPHs

(81.65 m^2 g^{-1}). The obtained porous nanostructural CoP electrocatalysts have improved electrocatalytic activity toward HER. In acid media, the CoP CPHs electrocatalysts have a low onset potential of ca. 30 mV, j_0 of 4.4×10^{-2} mA cm^{-2}, and a Tafel slope of 51 mV dec^{-1}. Meanwhile, it merely needs η of 133 mV to drive j of 10 mA cm^{-2}. Furthermore, the prepared material has good long term stability and its electrocatalytic activity can last at least 12 h in H$_2$SO$_4$.

Apart from the 3D porous features, the reactivity of active sites closely associated with the interplay between the catalyst surface and reagent also plays a significant role in the OER process. Integrating the favourable 3D porous nanostructures with accurately controlled active sites provides plentiful opportunities to create more functional nanomaterials and further enhance their OER performance. Recent experimental and theoretical studies revealed that the surface oxygen vacancies involved are beneficial for enhancing the electrochemical OER performance of spinel Co$_3$O$_4$ and NiCo$_2$O$_4$ (Wang et al. 2014, Bao et al. 2015). Morphology and composition play in the OER performance to rationally design freestanding 3D porous NiCo$_2$O$_4$ nanosheets with metal valence states alteration and abundant oxygen vacancies as robust electrocatalysts towards water splitting (Zhu et al. 2016).

Zhang et al. have successfully fabricated Co/Co$_9$S$_8$ core-shell structures anchored onto S,N co-doped porous graphene sheets (Co/Co$_9$S$_8$@SNGS) electrodes from S and N dual organic ligands assembled Co-metal-organic frameworks *in situ* grown on graphene oxide sheets (Co-MOFs@GO) by a simple pyrolysis approach (Zhang et al. 2016a,b,c). As electrocatalyst, Co/Co$_9$S$_8$@SNGS-1000 treated at 1000°C have been observed to exhibit superiorly bi-functional electrocatalytic activities toward both OER and HER in 0.1 M KOH solution (Fig. 12) with O$_2$ and H$_2$ generation efficiencies of 2.48 and 4.87 μmol min^{-1}, respectively, at an applied potential of 1.58 V (vs. RHE) under the given time range, affording about 100 per cent Faradaic yield (Zhang et al. 2016a,b,c).

Porous Ni-Fe bimetallic selenide nanosheets have been prepared on carbon fiber cloth by selenisation of the ultrathin NiFe-based nanosheet precursor (Wang et al. 2016). The outstanding catalytic performance and strong durability, in comparison to the advanced non-noble metal catalysts, have been derived from the porous nanostructure fabrication, Fe incorporation, and selenisation. The developed porosity of this material has been claimed to result in fast charge transportation and large electrochemically active surface area and enhance the release of O$_2$ bubbles from the electrode surface. Surface modification regarding the evolution of oxygen vacancies is facilely realised upon the sodium borohydride treatment, which is beneficial for the enhanced OER performance (Wang et al. 2016).

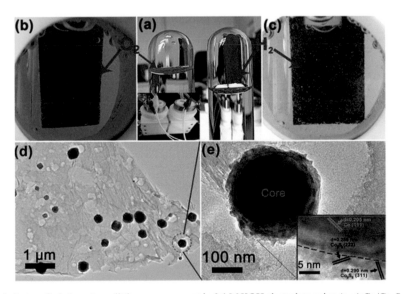

Fig. 12. Optical photos of whole water splitting measurement in 0.1 M KOH electrolyte using (a–c) Co/Co$_9$S$_8$@SNGS-1000 loaded Ni foam as cathode and anode, (d) TEM and (e) HRTEM images of Co/Co$_9$S$_8$@SNGS-1000. Reprinted with permission from Elsevier (Zhang et al. 2016c).

A mesoporous $LiNi_{1-x}Fe_xPO_4@C$ ($0 \leq x \leq 1$) nano-structure as a highly effective catalyst for electrochemical OER has been synthesised through a spray dry method (Ma et al. 2015). Another promising porous electrode has been developed by composed of porous CoP nanorod bundle arrays on Ti plate developed as a robust self-supported hydrogen-evolving cathode (Niu et al. 2015). This nano-array electrode has been found to exhibit excellent activity for HER in a wide pH range with low η, small Tafel slope, and long-term stability. Mesoporous molybdenum carbide nano-octahedrons have been prepared with ultrafine nanocrystallites which exhibit remarkable electrocatalytic performance for H_2 production from both acidic and basic solutions (Wu et al. 2015).

Feng et al. reported the synthesis of stable $\{\bar{2}10\}$ high-index faceted Ni_3S_2 nanosheet arrays supported on NF as active bifunctional HER–OER catalysts (Feng et al. 2015). The obtained Ni_3S_2/NF material have been reported to show efficient and ultra-stable electrocatalytic activity toward the HER and OER for its nanocatalytic feature and high index facets (Fig. 13). Theoretical calculations confirmed that the synergistic effect between the nanosheet array architecture and the $\{\bar{2}10\}$ high-indexed facets accounted for its uniquely efficient catalytic ability for overall water splitting. Calculated free-energy diagram of HER over $\{\bar{2}10\}$ and (001) surfaces at equilibrium potential are shown in Fig. 13c. Theoretical calculations confirmed that the synergistic effect between the nanosheet array architecture and the $\{\bar{2}10\}$ high-index facets accounted for its uniquely efficient catalytic ability for overall water splitting (Fig. 13c, d).

TOF is an important parameter to be considered for comparing catalytic effects especially for OER of water splitting. It has been found that porous state of a catalyst offers large value of TOF

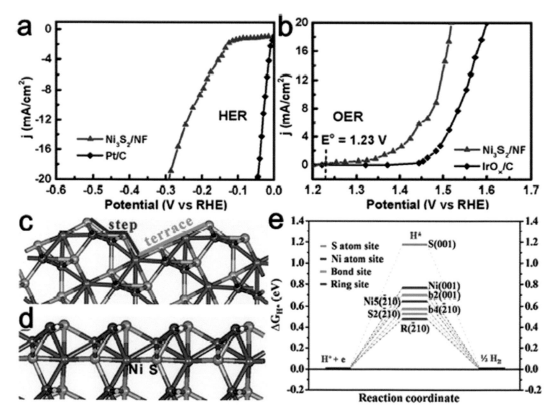

Fig. 13. Steady-state current density as a function of applied voltage in alkaline media (pH 14) over Ni_3S_2/NF in (a) HER and (b) OER. Most stable terminations of (c) $\{\bar{2}10\}$ and (d) (001) surfaces of Ni_3S_2. (e) Calculated free-energy diagram of HER over $\{\bar{2}10\}$ and (001) surfaces at equilibrium potential. The blue and yellow spheres represent Ni and S atoms, respectively, of Ni_3S_2. Reprinted with permission from American Chemical Society (Feng et al. 2015).

compared to that of packed film (Browne et al. 2017). TOF of electrochemical OER on Ni and Fe oxides electrodes prepared by electrodeposition method are 0.065–0.006 s^{-1} and 0.0015–0.0009 s^{-1}, respectively (Browne et al. 2017). Interestingly, the values of TOF for the same oxides prepared with hydrothermal method resulted in porous structure have been reported to be in the range of 0.28–0.10 s^{-1}. The formation of foam with hydrothermal method resulting in larger surface area for OER has been attributed for higher TOF than that obtained with the film prepared electrochemically.

6.3.2.3 Fabrication of Electrodes for Overall Water Splitting System

To search the efficient non-noble metal based and/or earth-abundant electrocatalysts for overall water splitting is critical to promote the clean-energy technologies for hydrogen economy. The expense of these types of metals can change the context in which this technology is important and viable. Industrial water electrolysis cells typically employ stainless steel or Ni-based electrodes. The overall energy efficiency of electrolysis is partly related to the HER. Ni shows a high activity towards the HER, however, it experiences extensive deactivation as a cathode during alkaline water electrolysis.

Deactivation of the cathode is manifested by either a current loss at a fixed electrode potential or an increase in η for HER at a constant current. To avoid such problems and enhance the overall efficiency of water electrolysis units, state-of-the-art electrodes have been developed by many researchers (Yan et al. 2016, Cook et al. 2010, Browne et al. 2017). According to the problem of Ni undergoing significant deactivation as a cathode during alkaline water electrolysis under certain conditions can be confronted by the addition of dissolved vanadium oxide (V_2O_5). It has been found that the dissolved V_2O_5 reactivates the Ni cathode and forms a V-bearing deposit (Abouatallah et al. 2001). In another study, Bocutti et al. investigated the HER on Ni-LaNi$_5$ and Ni-MmNi$_{3.4}$Co$_{0.8}$A$_{10.8}$ electrode materials in NaOH solution. The experimental data of steady-state polarisation curves and electrochemical impedance spectroscopy have showed a pronounced improvement in HER kinetics at these electrode materials (Bocutti et al. 2000). Suffredini et al. reported improvements and alternatives for the preparation of high area Ni and Ni-Co coatings as well as the deposition of a highly active Ni-Fe layer on mild steel substrates (Suffredini et al. 2000).

Li et al. have reported that Ni phosphide (Ni_xP_y) catalysts with the controllable phases as the efficient bi-functional catalysts for water electrolysis (Li et al. 2016). The phases of Ni_xP_y have been determined by the temperatures of the solid-phase reaction between the ultrathin Ni(OH)$_2$ plates and NaH$_2$PO$_2$·H$_2$O. The Ni_xP_y with the richest Ni_5P_4 phase synthesised at 325°C has been found to deliver efficient and robust catalytic performance for HER in the electrolytes with a wide pH range. NiFe nanosheets film on Ni foam act as an efficient bi-functional catalytic material with high activity and stability for overall water splitting in basic electrolytes (Lu and Zhao 2015). Co-Cophosphide (Co/Co$_2$P) NPs supported on Ni foam showed good performances in catalysing HER in an acidic electrolyte and OER in alkaline solution for overall water splitting on a glassy carbon electrode (Masa et al. 2016).

The overvoltage of full water splitting at *3d* block metal-based catalysts is compared in Fig. 14 and mechanism of bi-functional catalysts for HER and OER is represented in Fig. 15. Overvoltage with Fe-based catalyst is rather higher but it is profuse in the Earth and is non-toxic.

Bi-functional catalysts for both HER and OER have been designed with cheapest and highly abundant Fe-based compounds (Martindale and Reisner 2016). At anodic potential, the surface of electrode materials converts into an Fe(III) oxide-(hydroxide), FeO$_x$, species for OER catalysis. On the other hand, at a cathodic bias, FeO$_x$ are reduced to Fe(0) for HER catalysis when embedded in an Fe oxide matrix. Fe-based material regardless of their initial identity, the surface transforms under bias into the species involved in this cycle and shows very similar activity.

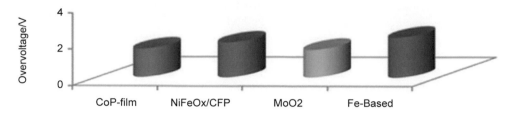

Fig. 14. Overall voltage (vs. RHE) for water splitting at corresponding j of 10 mA cm^{-2}. Data were taken from the different references as cited in Yan et al. 2016.

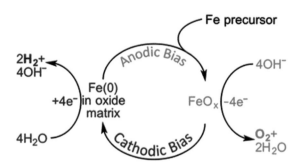

Fig. 15. Mechanism of overall water splitting in alkaline solution on bi-functional iron-based catalyst modified cathode and anode. An anodic potential generates the active FeO_x species for OER catalysis and a cathodic potential produces Fe(0) embedded in an iron oxide matrix as HER catalyst. Fe precursor represents the iron-based materials FeO_x, Fe, and FeS_x. Reprinted with permission from Wiley-VCH (Martindale and Reisner 2016).

Conclusion

The energy required for operating the water splitting system is η_{op} that is comprised of thermodynamic energy of 1.23 V plus η for OER (i.e., η_a) and for HER (i.e., η_c) and η arisen due to the internal resistance of the system itself (i.e., η_Ω). It is clear that all components except for thermodynamic energy can apparently be minimised to achieve the system with low operational η_{op}. The value of η_{op} for water splitting system made of noble metals such as Pd and Pt that catalysed both HER and OER are really low. However, it is still a great challenge to explore efficient, robust, cost-effective and environment friendly electrode materials for practical application in electrochemical water splitting technology.

In the electrochemical water splitting, the OER is more complicated compared to that of HER and the evolution of O_2 generally occurs at metal oxide surface, not from bare/clean metal surface. Under applied potential, O_2 evolution by the cleavage of O-H bond is in fact followed by the oxidation of metal surface to form metal oxide. Therefore, the mechanism of OER is thus different for oxides with different surface structures.

Theoretical and experimental studies clarify that the rate of OER linearly varies with the energy involved with the binding of oxygen with metal surface that occurs as the elementary step of water splitting (see Eq. 13 and Fig. 6). Bockris has clarified that $3d$ transition metal pervoskites have showed a strong linear relationship between catalysis and standard enthalpy of formation of $M(OH)_3$ (Bockris and Otagawa 1983). The results of DFT calculation (Rossmeisl et al. 2005) as shown in Fig. 16 suggest that for the oxidised metal surfaces such as Pt(111) and Au(111), the water layer is added over the adsorbed species and the hydrogen bonds between adsorbed O* and the water molecules are weak, whereas HO* and HOO* are polar and form stronger hydrogen bonds with water. HO* and HOO* species are stabilised relative to the O* species on the electrode surface. At low coverage, HO* and

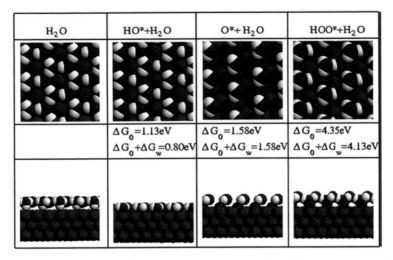

H_2O	HO^*+H_2O	O^*+H_2O	HOO^*+H_2O
	$\Delta G_0 = 1.13eV$	$\Delta G_0 = 1.58eV$	$\Delta G_0 = 4.35eV$
	$\Delta G_0 + \Delta G_w = 0.80eV$	$\Delta G_0 + \Delta G_w = 1.58eV$	$\Delta G_0 + \Delta G_w = 4.13eV$

Fig. 16. Structures from the left 2/3 of a monolayer of water, OH* and water, O* and water and OOH* and water. Top view (top) and side view (bottom). The free energies relative to water and H_2, for the intermediates without and with the water molecules are listed in the central row. Reprinted with permission from Elsevier (Rossmeisl et al. 2005).

HOO* are stabilised even further if the water molecules are adsorbed on the surface in plane with the dissociated species. At high η, the free energy of this step is downhill, indeed. At low η, H_2O dissociates to form O*, and after sufficient oxidation of electrode surface the O_2 evolution takes place. It has thus been concluded that the difficult step in water splitting is the formation of superoxy type (OOH) species on the surface by breaking of water molecule. New materials that can destabilise adsorbed O* relative to HOO* has been suggested to be developed for electrocatalytic water splitting. In this regard, the mixed 3*d* block transition metals or their oxides or composites with organic materials that may offer the expected surface properties suitable for water splitting would be the prior domain of research for developing cost-effective technology for electrochemical water splitting.

References

Abouatallah, R.M., D.W. Kirk, S.J. Thorpe and J.W. Graydon. 2001. Reactivation of nickel cathodes by dissolved vanadium species during hydrogen evolution in alkaline media. Electrochimica Acta 47: 613–621.

Bao, J., X. Zhang, B. Fan, J. Zhang, M. Zhou and W. Yang et al. 2015. Ultrathin spinel-structured nanosheets rich in oxygen deficiencies for enhanced electrocatalytic water oxidation. Angew. Chem. Int. Ed. 54: 7399–7404.

Bard, A.J. and L.R. Faulkner. 2001. Electrochemical methods: fundamentals and applications (2nd Ed) John Wiley & Sons. Inc, New York.

Bloor, L.G., P.I. Molina-Sanchez and M.D. Symes. 2014. Low pH electrolytic water splitting using earth abundant metastable catalysts that self-assemble *in situ*. J. Am. Chem. Soc. 136: 3304–3311.

Bockris, J.O.M. and A.K.N. Reddy. 1970. Modern electrochemistry, Perseus Publishing, Cambridge, MA.

Bockris, J.O. and T. Otagawa. 1983. Mechanism of oxygen evolution on perovskites. J. Phys. Chem. 87: 2960–2971.

Bocutti, R., M.J. Saeki, A.O. Florentino, C.L.F. Oliveira and A.C.D. Angelo. 2000. The hydrogen evolution reaction on codeposited Ni-hydrogen storage intermetallic particles in alkaline medium. Int. J. Hydrogen Energy 25: 1051–1058.

Browne, M.P., S. Stafford, M. O'Brien, H. Nolan, N.C. Berner and G.S. Duesberg et al. 2016.The goldilocks electrolyte: examining the performance of iron/nickel oxide thin films as catalysts for electrochemical water splitting in various aqueous NaOH solutions. Mater. Chem. A 4: 11397–11407.

Browne, M.P., J.M. Vasconcelos, J. Coelho, M. O'Brien, A.A. Rovetta and E.K. McCarthy. 2017. Improving the performance of porous nickel foam for water oxidation using hydrothermally prepared Ni and Fe metal oxides. Sustainable Energy Fuels 1: 207–216.

Burstein, G.T. 2005. A hundred years of Tafel's Equation: 1905–2005. Corros. Sci. 47: 2858–2870.

Brunschwig, B.S., M.H. Chou, C. Creutz, P. Ghosh and N. Sutin. 1983. Mechanisms of water oxidation to oxygen: cobalt(IV) as an intermediate in the aquocobalt(II)-catalyzed reaction. J. Am. Chem. Soc. 105: 832–4833.

Chaikittisilp, W., N.L.Torad, C. Li, M. Imura, N. Suzuki and S. Ishihara et al. 2014. Synthesis of nanoporous carbon–cobalt-oxide hybrid electrocatalysts by thermal conversion of metal–organic frameworks. Chem. Eur. J. 20: 4217–4221.

Chen, C. and F. Ciucci. 2016. Designing Fe-based oxygen catalysts by density functional theory calculations. Chem. Mater. 28: 7058–7065.

Chen, S., J. Duan, M. Jaroniec and S.Z. Qiao. 2013. Three-dimensional N-doped graphene hydrogel/NiCo double hydroxide electrocatalysts for highly efficient oxygen evolution. Angew. Chem. Int. Ed. 52: 13567–13570.

Cho, Y., J. Lee, T.T.-H. Nguyen, J.W. Bae, T. Yu and B. Lim. 2016. Facile synthesis of flower-like α-Co(OH)$_2$ nanostructures for electrochemical water splitting and pseudocapacitor applications. J. Ind. Eng. Chem. 37: 175–179.

Cook, T.R., D.K. Dogutan, S.Y. Reece, Y. Surendranath, T.S. Teets and D.G. Nocera. 2010. Solar energy supply and storage for the legacy and nonlegacyworlds. Chem. Rev. 110: 6474–6502.

Conway, B.E. 1999. Electrochemical supercapacitors: Scientific fundamentals and technological applications. Plenum Publishers, New York.

Dinca, M., Y. Surendranath and D.G. Nocera. 2010. Nickel-borate oxygen-evolving catalyst that functions under benign conditions. Proc. National. Acad. Sci. 107: 10337–10341.

Doménech, A., M.T. Doménech-Carbó and H.G.M. Edwards. 2008. Quantitation from Tafel analysis in solid-state voltammetry. Application to the study of cobalt and copper pigments in severely damaged frescoes. Anal. Chem. 80: 2704–2716.

El-Deab, M.S., M.I. Awad, A.M. Mohammad and T. Ohsaka. 2007. Enhanced water electrolysis: Electrocatalytic generation of oxygen gas at manganese oxide nanorods modified electrodes. Electro. Comm. 9: 2082–2087.

Energy Report. 2011. The energy report: 100% renewable energy by 2050. (https://www.worldwildlife.org/publications/the-energy-report.)

Ewan, B.C.R. and R.W.K. Allen. 2005. A figure of merit assessment of the routes to hydrogen. Int. J. Hydrogen Energy 30: 809–819.

Feng, L.L., G. Yu, Y. Wu, G.D. Li, H. Li and Y. Sun et al. 2015. High-index faceted Ni$_3$S$_2$ nanosheetarrays as highly active and ultrastable electrocatalysts for water splitting. J. Am. Chem. Soc. 137: 14023–14026.

Fiegenbaum, F., E.M. Martini, M.O. Souza, M.R. Becker and R.F. Souza. 2013. Hydrogen production by water electrolysis using tetra-alkyl-ammonium-sulfonic acid ionic liquid electrolytes. J. Power Sources 243: 822–825.

Fillol, J.L., Z. Codolà, I. Garcia-Bosch, L. Gómez, J.J. Pla and M. Costas. 2011. Efficient water oxidation catalysts based on readily available iron coordination complexes. Nat. Chem. 3: 807–813.

Guo, S.X. and A. Speidel. 2014. Facile electrochemical co-deposition of a graphene-cobalt nanocomposite for highly efficient water oxidation in alkaline media: Direct detection of underlying electron transfer reactions under catalytic turnover conditions. Phys. Chem. Chem. Phys. 16: 19035–19045.

Harriman, A., I.J. Pickering, J.M. Thomas and P.A. Christensen. 1988. Metal oxides as heterogeneous catalysts for oxygen evolution under photochemical conditions. J. Chem. Soc. Faraday Trans. 1. Phys. Chem. Cond. Phases. 84: 2795–2806.

Hou, C.C., W.F. Fu and Y. Chen. 2016. Self-supported Cu-based nanowire arrays as noble-metal-free electrocatalysts for oxygen evolution. Chem. Sus. Chem. 9: 20692073.

Islam, M.M., T. Okajima, S. Kojima, S. Kojima and T. Ohsaka. 2008. Water electrolysis: An excellent approach for the removal of water from ionic liquids. Chem. Commun. 2008: 5330–5332.

Karkas, M.D., O. Verho, E.V. Johnston and B. Akermark. 2015. Rapid water oxidation electrocatalysis by a ruthenium complex of the tripodal ligand tris (2-pyridyl) phosphine oxide. Chem. Sci. 6: 2405–2410.

Khan, S.A., S.B. Khan and A.M. Asiri. 2016. Electrocatalyst based on cerium doped cobalt oxide for oxygen evolution reaction in electrochemical water splitting. J. Matter. Sci. Electron. 27: 5294–5302.

Koseki, M., Y. Tanaka, H. Noguchl and T. Nishikawa. 2007. Effect of pH on the taste of alkaline electrolyzed water. J. Food Sci. 72: 298–301.

Li, J., L. Jing, X. Zhou, Z. Xia, W. Gao and Y. Ma et al. 2016. Highly efficient and robust nickel phosphides as bi-functional electrocatalysts for overall water-splitting. ACS Appl. Mater. Interfaces 8: 10826–10834.

Liu, N., K. Lee and P. Schmuki. 2013. Reliable metal deposition into TiO$_2$ nanotubes for leakage free interdigitated electrode structures and uses memristive electrode. Angew. Chem. Int. Ed. 52: 12381–12384.

Liu, R.-H., L. Zhang, X. Sun, H. Liu and J. Zhang. 2012. Electrochemical technologies for energy storage and conversion, Wiley VCH.: 2.

Lu, X. and C. Zhao. 2015. Electrodeposition of hierarchically structured three-dimensional nickel–iron electrodes for efficient oxygen evolution at high current densities. Nat. Commun. 6: 6616.

Ma, S., Q. Zhu, Z. Zheng, W. Wang and D. Chen. 2015. Nanosized LiNi$_{1-x}$Fe$_x$PO$_4$ embedded in a mesoporous carbon matrix for high-performance electrochemical water splitting. Chem. Commun. 51: 15815–15818.

Ma, T.Y., S. Dai, M. Jaroniec and S.Z. Qiao. 2014. Metal–organic framework derived hybrid Co$_3$O$_4$-carbon porous nanowire arrays as reversible oxygen evolution electrodes. J. Am. Chem. Soc. 136: 13925–13931.

Masa, J., S. Barwe, C. Andronescu, I. Sinev, A. Ruffand K. Jayaramulu et al. 2016. Low overpotential water splitting using cobalt-cobalt phosphide (Co/Co$_2$P) nanoparticles supported on nickel foam. ACS Energy Lett. 6: 1192–1198.

Martindale, B.C.M. and E. Reisner. 2016. Bi-functional iron-only electrodes for efficient water splitting with enhanced stability through in situ electrochemical regeneration. Adv. Energy Mater. 6: 1502095.

Ni, M., D.Y.C. Leung, M.K.H. Leung and K. Sumathy. 2006. An overview of hydrogen production from biomass. Fuel Processing Technol. 87: 461–472.

Niu, Z., J. Jiang and L. Ai. 2015. Porous cobalt phosphide nanorod bundle arrays as hydrogen-evolving cathodes for electrochemical water splitting. Electrochem. Commun. 56: 56–60.

Pagliaro, M., A.G. Konstandopoulos, R. Ciriminna and G. Palmisano. 2010. Solar hydrogen: Fuel of the near future. Energy Environ. Sci. 3: 279–287.

Pu, Z., Y. Luo, A.M. Asiri and X. Sun. 2016. Efficient electrochemical water splitting catalyzed by electrodeposited nickel diselenide nanoparticles based film. ACS Appl. Mater. Interfaces 8: 4718–4723.

Renaud, R. and R.L. Leroy. 1982. Separator materials for use in alkaline water electrolysers. Int. J. Hydrogen Energy 7: 155–166.

Rossmeisl, J., A. Logadottir and J.K. Nørskov. 2005. Electrolysis of water on (oxidized) metal surfaces. Chem. Phys. 319: 178–184.

Rudnik, E. and G. Wloch. 2013. Studies on the electrodeposition of tin from acidic-chloride-gluconate solutions. Appl. Surface. Sci. 265: 839–849.

Singh, R., J. Pandey, N. Singh, B. Lal, P. Chartier and J.-F., Koenig. 2000. Sol-gel derived spinel $MxCo_{3-x}O_4$ (M = Ni, Cu; $0 \leq x \leq 1$) films and oxygen evolution. Electrochim. Acta 45: 1911–1919.

Souza, R.F., J.C. Padilha, R.S. Gonçalves, M.O. Souza and J.R. Berthelot. 2007. Electrochemical hydrogen production from water electrolysis using ionic liquid as electrolytes: Towards the best device. J. Power Sources 164: 792–798.

Suffredini, H.B., J.L. Cerne, F.C. Crnkovic, S.A.S. Machado and L.A. Avaca. 2000. Recent developments in electrode materials for water electrolysis. Int. J. Hydrogen Energy 25: 415–423.

Swaminathan, J., R. Subbiah and V. Singaram. 2016. Defect-rich metallic titania ($TiO_{1.23}$)—an efficient hydrogen evolution catalyst for electrochemical water splitting. ACS Catal. 6: 222–2229.

Tang, C., N. Cheng, Z. Pu, W. Xing and X. Sun. 2015. NiSe nanowire film supported on nickel foam: An efficient and stable 3D bi-functional electrode for full water splitting. Angew. Chem. Int. Ed. 54: 9351–9355.

Tilak, B.V., P.W.T. Lu, J.E. Colman and S. Srinivasan. 1981. Electrolytic production of hydrogen. 2: 1–104. *In*: Bockris J. (ed.). Comprehensive Treatise of Electrochemistry, Springer.

Trasatti, S. 1972. Work function, electronegativity, and electrochemical behaviour of metals: III. Electrolytic hydrogen evolution in acid solutions. J. Electroanal. Chem. Interfacial Electrochem. 39: 163–184.

Trasatti, S. 1984. Electrocatalysis in the anodic evolution of oxygen and chlorine. Electrochim. Acta 29: 1503–1512.

Umena, Y., K. Kawakami, J.-R. Shen and N. Kamiya. 2011. Crystal structure of oxygen-evolving photosystem II at a resolution of 1.9Å. Nature 473: 55–60.

Viswanathan, V., A. Speidel, S. Gowda and A.C. Luntz. 2013. $Li-O_2$ kinetic overpotentials Tafel plots from experiment and first- principles theory. J. Phys. Chem. Lett. 4: 556–560.

Wang, J.H.-X. Zhong, Y.-L.Qin and X.-B. Zhang. 2013. An efficient three-dimensional oxygen evolution electrode. Angew. Chem. Int. Ed. 52: 5248 –5253.

Wang, W., J. Dong, X. Ye, Y. Li, Y. Ma and L. Qi. 2012. Heterostructured TiO_2 nanorod@nanobowl arrays for efficient photoelectrochemical water splitting. Small 11: 1469–1478.

Wang, Y., T. Zhou, K. Jiang, P. Da, Z. Peng and J. Tang et al. 2014. Electrocatalysis: reduced mesoporous Co_3O_4 nanowires as efficient water oxidation electrocatalysts and supercapacitor electrodes. Adv. Energy Mater. 4: 1400696.

Wang, Z., J. Li, X. Tian, X. Wang, Y. Yu and K. A. Owusu et al. 2016. Porous nickel–iron selenide nanosheets as highly efficient electrocatalysts for oxygen evolution reaction. ACS Appl. Mater. Interfaces 8: 19386–19392.

Wasylenko, D.J., C. Ganesamoorthy, J. Borau-Garcia and C.P. Berlinguette. 2011. Electrochemical evidence for catalyticwater oxidation mediated by a high-valent cobalt complex. Chem. Commun. 47: 4249–4251.

Wu, H.B., B.Y. Xia and X.-Y. Yu. 2015. Porous molybdenum carbide nano-octahedrons synthesized via confined carburization in metal-organic frameworks for efficient hydrogen production. Nat. Commun. 6: 6512.

Xiang, C., K.M. Papadantonakisab and N.S. Lewis. 2016. Principles and implementations of electrolysis systems for water splitting. Mater. Horiz. 3: 169–173.

Xu, M., L. Han, Y. Han, Y. Yu, J. Zhai and S. Donga. 2015. Porous CoP concave polyhedron electrocatalysts synthesized from metal-organic frameworks with enhanced electrochemical properties for hydrogen evolution. J. Mater. Chem. A 3: 21471–21477.

Yan, Y., B.Y. Xia, B. Zhao and X. Wang. 2016. A review on noble-metal-free bi-functional heterogeneous catalysts for overall electrochemical water splitting. J. Mater. Chem. A 4: 17587–17603.

Yin, Q.J., M. Tan, C. Besson, Y.V. Geletii, D.G. Musaev and A.E. Kuznetsov et al. 2010. A fast soluble carbon-free molecular water oxidation catalyst based on abundant metals. Science 328: 342–345.

Yue, C., D. Fang, L. Liu and T.-F. Yi. 2011. Synthesis and application of task-specific ionic liquids used as catalysts and/or solvents in organic unit reactions. J. Mol. Liq. 163: 99–121.

Zidki, T., L. Zhang, V. Shafirovich and S.V. Lymar. 2012. Water oxidation catalyzed by cobalt(II) adsorbed on silica nanoparticles. J. Am. Chem. Soc. 134: 14275–14278.

Zhang, B., C. Xiao, S. Xie, J. Liang, X. Chen and Y. Tang. 2016a. Iron–nickel nitride nanostructures *in situ* grown on surface-redox-etching nickel foam: efficient and ultra sustainable electrocatalysts for overall water splitting. Chem. Mater. 28: 6934–6941.

Zhang, J., T. Wang, D. Pohl, B. Rellinghaus, R. Dong and S. Liu et al. 2016b. Interface engineering of MoS_2/Ni_3S_2 heterostructures for highly enhanced electrochemical overall-water-splitting activity. Angew. Chem. Int. Ed. 55: 6702–6707.

Zhang, X., S. Liu, Y. Zang, R. Liu, G. Liu and G. Wang et al. 2016c. Co/Co$_9$S$_8$@S,N-doped porous graphene sheets derived from S, N dual organic ligands assembled Co-MOFs as superior electrocatalysts for full water splitting in alkaline media. Nano Energy 30: 93–102.

Zhao, Z., H. Wu, H. He, X. Xu and Y. Jin. 2014. A high performance binary Ni–Co hydroxide-based water oxidation electrode with three-dimensional coaxial nanotube array structure. Adv. Funct. Mater. 24: 4698–4705.

Zou, X. and Y. Zhang. 2015. Noble metal-free hydrogen evolution catalysts for water splitting. Chem. Soc. Rev. 44: 5148–5180.

Zhou, W., X.-J. Wu, X. Cao, X. Huang, C. Tan, J. Tian et al. 2013. Ni$_3$S$_2$ nanorods/Ni foam composite electrode with low overpotential for electrocatalytic oxygen evolution. Energy Environ. Sci. 6: 2921–2924.

Zhu, C., S. Fu, D. Du and Y. Lin. 2016. Facilely tuning porous NiCo$_2$O$_4$ nanosheets with metal valence-state alteration and abundant oxygen vacancies as robust electrocatalysts towards water splitting. Chem. Eur. J. 22: 4000–4007.

11

Wide Band Gap Nano-Semiconductors for Solar Driven Hydrogen Generation

Nur Azimah Abd Samad, Kung Shiuh Lau and *Chin Wei Lai**

1. Introduction

The rapid growth of gross domestic product (GDP) has played a critical role in global warming and climate change. The geophysical changes caused by the global warming and climate change have drawn an incredible amount of public attention (Chiroma et al. 2015, Cicea et al. 2014, Tang Qunwei et al. 2012). Rapid expansion of energy production and consumption has led to abnormal changes in the geographical and meteorological conditions making people more vulnerable to infectious diseases. Excessive energy consumption has dramatically increased carbon dioxide (CO_2) emission. Burning of fossil fuels such as coal causes the most CO_2 emission (Kaygusuz 2009, Ozyurt 2010).

Taking these facts into consideration, a clean and sustainable hydrogen energy has been introduced with the aim to ensure that environmental efficiency index is made achievable (Cicea et al. 2014). Hydrogen energy is one of the econometric models that can be used to promote the sustainable clean energy supply. There are variety of established techniques in literature for hydrogen generation. Pyrolysis and steam reforming of biomass, photo-reforming of organics, methanol steam reforming, ammonia decomposition through reactor technology, membrane reactor, borohydride hydrolysis reaction, and solar-driven water splitting system are few examples of potential methods of hydrogen production. Nevertheless, there is only a low degree of environmental efficiency (in green energy) being recognised and only countries such as United States, Japan and European Union countries have attained the desired environmental efficiency till-date (Chiuta et al. 2013, Clarizia et al. 2014, George et al. 2015, Hu et al. 2009, Iulianelli et al. 2014, Liu and Li 2009, Parthasarathy and Sheeba 2015).

Thermochemical techniques could be adopted for hydrogen generation. Fast pyrolysis is one of the modern thermochemical technologies to produce the bio-oil, followed by the bio-oil steam reforming, purification of water and steam gasification. Slow pyrolysis could be engaged with the steam gasification for hydrogen generation too (Parthasarathy and Sheeba 2015). On top of pyrolysis, ammonia decomposition for the hydrogen generation has also been well studied in recent years. There are basically two types of technology for ammonia decomposition. Ruthenium (Ru, as a catalyst) together with the carbon nanotubes (CNT) (as a supporting material) and potassium hydroxide (as a promoter) were used for ammonia decomposition in the aforementioned first technology.

Nanotechnology & Catalysis Research Centre (NANOCAT), 3rd Floor, Block A, Institute of Postgraduate Studies (IPS), University of Malaya, 50603 Kuala Lumpur, Malaysia.
* Corresponding author: cwlai@um.edu.my

Correspondingly, good dispersion of Ru promotes an excellent catalytic activity, while the basicity and conductivity of CNTs enhance the efficiency of the Ru catalyst (Yin et al. 2004). Second technology relates to the use of reactor technology for ammonia decomposition. It is noteworthy that lots of efforts have been paid to produce the portable and distributed power generation by adopting this technology (load-shedding technology). Over the past decade, reactor designs including operability, capacity of the power generation and efficiency were highly focused in the research studies. In a more advanced research works, microreactors and monolithic reactors that offer many more advantages have been widely investigated (Chiuta et al. 2013).

Apart from that, polymer electrolyte membrane (PEM) assisted by the methanol steam reforming is used to produce hydrogen with high purity. This technology is commonly known as inorganic membrane reactor. Specifically, MRs can be further categorised as photo catalytic MRs, zeolite MRs, polymeric MRs, enzyme MRs, dense and porous inorganic MRs, electro-chemical MRs and bio medical MRs. Among them, palladium-based membrane has been given special attention as it has the lowest permeability to hydrogen as compared to that of tantalum, vanadium and niobium. The last part for the methanol steam reforming conversion to hydrogen is the proton exchange membrane fuel cells. This particular process involves the chemical energy conversion (Peighambardoust et al. 2010). Correspondingly, nafion (sulfonated perfluorinated polymer) that possess excellent proton conductivity and high applied potential at low-medium temperature was commonly used as membrane (Iulianelli et al. 2014).

Organic materials photoreforming is also one of the techniques used for the hydrogen generation. This technology is almost similar to that of water splitting system. The only difference is that this method involves the use of organic materials as sacrificial agents. Methanol, glycerol and formic acid are few examples of the most frequently used sacrificial agents. Another key point about this technology is the use of single-catalyst or second catalyst for the conversion of organic materials to hydrogen. Alkaline medium is needed for hydrogen generation to produce the negative shift of the bands' positions (Chiuta et al. 2013).

The present chapter focuses on PEC water splitting system with emphasis on porous wide bandgap semiconductors for clean hydrogen production. Porous and/or nanoporous wide bandgap semi-conductors show excellent photoelectrochemical response and photoconversion efficiency due to their unique physico-chemical and electronic properties. The effects of porosity in titanium dioxide (TiO_2) and zinc oxide (ZnO) nanostructure film in influencing the photo-electrochemical response and photoconversion efficiency is discussed in the following subtopics.

According to Erne et al., anodic etching, deposition on an ordered template and the assembly of as-prepared nanoparticles are some approaches that can be applied in producing porous semiconductors (Erne et al. 1996). In this particular chapter, the anodic etching approach used for the formation of porous TiO_2 nanotubes and deposition method for porous ZnO nanostructures is reported. Referring to Wehrspohn, porosity plays an important role in the evolution of new physical properties. The new physical properties could be recognised from Raman Scattering via surface-related vibrational mode. Taking into account that the incident radiation wavelength will be greater than that of the average crystallite size, the optoelectronic properties will thereby be altered by surface excitations among pores. The surface vibrations include the absorption-desorption of photon by the pores, the diffusion and the stability of the porous nanostructures. The development of porous structure contributed to the stability against the dissolution of distinct crystallographic planes (Wehrspohn 2005).

It is also noteworthy that porous structure produces a sharp increase in the PL intensity especially at the near-band-edge (NBE) to the deep-level-emission (DLE) under ultraviolet and visible illuminations (Wehrspohn 2005). In this respect, pores could help in enhancing the photoresponse of materials. It has been reported by Kochergin and Föll that pores with nanometer size will increase the PL intensity because three dimensional of porous structures connect the photon transmission between crystals in a fastest way. By modulating the pore diameter and pore depth, the photoresponse could be increased. In addition, light absorption could be enhanced by adjusting the pores geometry (Kochergin and Föll 2009).

2. PEC Water Splitting

2.1 Historical Overview of PEC Water Splitting

To alleviate the dependence on the usage of fossil fuel, there has been a significant increase in the exploitation of renewable energy sources. The renewable energy sources with high efficiency, low cost and ease of production are highly sought (Kreuter and Hofmann 1998). Solar energy is one of those sources and this particular renewable energy could be harvested through the excitation of semiconductor materials, e.g., semiconductor photocatalyst (Aroutiounian et al. 2005, Tong et al. 2012). Becquerel first reported the work on semiconductor photocatalysis in 1839. In his design, silver chloride electrode was used as the counter electrode. Under the illumination of sunlight, voltage and electric current would be generated. This discovery has inspired the researchers to capture the sunlight energy and convert it into electric energy as an alternative energy sources. Five decades later, Charles Fritts developed a selenium/gold *p-n* junction device. The resulting device had a conversion electrical efficiency of around 1 per cent. Low efficiency and material's high cost prevented the use of such cells for energy supply.

In 1954, Bell Laboratories successfully designed a viable commercial *p-n* junction silicon solar cell with 6 per cent efficiency. From there onwards, photovoltaic, especially the silicon made solid-state junction devices, has been further improved for the global energy market (Okamoto et al. 2011, Solanki 2015). Correspondingly, a new generation of PEC water splitting cell, in which this photovoltaic system is integrated with an electrolyser to produce clean and portable H_2 energy carrier has emerged to challenge this dominance.

H_2 fuel stored within a fuel cell has been widely utilised as clean energy in recent years. Specifically, this fuel cell is composed of nanocrystalline materials with excellent chemical stability in aqueous solution. In a PEC water splitting cell, these materials are expected to have high current response to solar light irradiation (Centi and Perathoner 2009). In 1972, Honda and Fujishima reported on the first photoelectrochemical (PEC) water-splitting based on the Titanium dioxide (TiO_2) electrode. This research developed subsequent interests in photocatalysis of TiO_2 and made TiO_2 as an important component in many practical applications.

2.2 Basic Principle of PEC Water Splitting

The process of PEC water splitting, which is also known as solar-driven water splitting reaction, is a kind of chemical reaction catalysed by a photocatalyst to separate water into O_2 and H_2. Figure 1 illustrated the general schematic diagram of the overall PEC water splitting reaction catalysed by a semiconductor photocatalyst.

$$2H_2O\ (l) \rightarrow O_2\ (g) + 2H_2\ (g) \tag{1}$$

The PEC water splitting reaction belongs to a thermodynamic uphill reaction with large Gibbs free energy change for the process ($\Delta G° = 237.2$ kJ mol^{-1}) and it requires a minimum energy of 1.23 eV or 2.46 eV per molecule of H_2O for reaction to take place. Therefore, higher photon energy is required to overcome the Gibbs free energy in order to perform the reaction (Moniz et al. 2015). The mechanism of photocatalytic water splitting process can be described with two half reactions: (1) water oxidised to form O_2, and (2) proton reduction to hydrogen fuel, i.e., H^+/H_2 and O_2/H_2O at 0 V and 1.23 V, respectively, which are in line to the normal hydrogen electrode (NHE) (Kudo and Miseki 2009).

$$TiO_2 + 2\ h\nu \rightarrow TiO_2 + 2h^+ + 2e^- \tag{2}$$

$$2H_2O + 4h^+ \rightarrow O_2 + 4H^+ \tag{3}$$

$$2H^+ + 2e^- \rightarrow H_2 \tag{4}$$

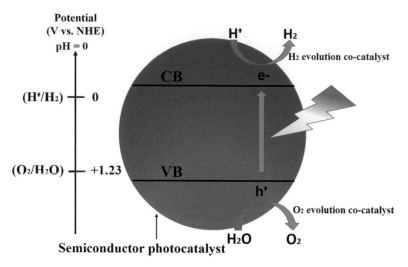

Fig. 1. Schematic diagram of the photocatalyse PEC water splitting reaction by a semiconductor (Ran et al. 2014).

Semiconductor-based photocatalyst (i.e., n-type TiO_2) will be photoexcited forming the electrons and holes pairs once being triggered by an incoming photon with energy equal or greater than that of its band gap. Accordingly, photocatalyst absorbs the photons, followed by promoting the electrons to the conduction band (CB) from the valence band (VB) leaving a hole. Within the picosecond timescale, these holes will diffuse to the surface-electrolyte interface forming the O_2 and H^+ ions by oxidising the water molecules (Eq. 3), while the electrons will move to the counter electrode through the external circuit. The reduction of H^+ ions to H_2 molecules (Eq. 4) at platinum electrode is basically induced by electric field or under external bias. The lifetime and pathway of the photogenerated charge carriers are critically important considering the fact that energy conversion efficiency relies on how fast the carriers recombine among themselves compared to the expected reaction (Moniz et al. 2015).

The rate of hydrogen generation through the light driven photocatalytic water splitting is basically affected by three main key factors. First, by the photons that are adsorbed by the material to provide sufficient energy to form the excited electron-hole pair. Second, the activities of photoexcited charge carriers like separation, recombination, trapping and migration and third, by the surface chemical reactions (utilising the surface active sites for the evolution of H_2 and O_2). When these conditions are favourable, photoexcited semiconductor (catalyst) tends to generate the electrons and holes pairs (Babu et al. 2015).

However, solar photocatalytic processes involve unavoidable energy loss. Specifically, electrons and holes pairs will be generated once the solar light illuminates towards the photocatalyst. These electron and hole pairs might experience rapid recombination and dissipate the absorbed energy as undesirable heat or phonon (Grimes 2007). Apart from electrons and holes recombination, energy loss also takes place via electrons transportation through lattice defects acting as the charge carrier traps. Different crystallinity of crystal lattice structures (polymorphs) (Jitputti et al. 2008) and material nanostructures will also affect the migration of the photo-generated charge carriers (Yan et al. 2011, Yang Min et al. 2014). Another possible reason for the poor photocatalytic activity may be attributed to the Joule heating resistant build up in the external circuit when the electrons flow from the photoanode to the counter electrode. Taking these facts into account, the efficient catalytic activity can be promoted by lowering the density of defects by increasing the crystallinity. Furthermore, modification of the size of the photocatalyst to nanoscale with the aim to provide a better diffusion path is also essential for the movement of photogenerated electrons and holes pairs (Leung et al. 2010, Martinson et al. 2006).

2.3 Material Selection for PEC Water Splitting

Taking into account the energy losses within the solar light driven photocatalytic activities for water splitting hydrogen generation, the materials used as the photocatalyst have to possess several properties as follows:

  i) Band edge position: Material with conduction band (CB) level more electronegative than that of the reduction of H⁺/H₂ level ($E_{H2/H2O}$); Its valence band (VB) level should be more electropositive than that of H₂O oxidation level ($E_{O2/H2O}$) in order to generate H₂ efficiently (Ni et al. 2007).
 ii) Band-gap: Maximise the solar light spectrum for the photoexcitation on the materials and to lower the electronic band gap (Misra and Raja 2010).
iii) Transportation of charge carriers: Minimise the recombination of the photogenerated charge carriers within materials to ensure a minimum losses of charge carriers during transportation from working electrode to counter electrode for high efficient H₂ generation (Grimes 2007).
 iv) Stability: The photocatalyst subjected under prolonged irradiation must be stable against the photocorrosion in electrolyte (Misra and Raja 2010, Moniz et al. 2015).

It is important to note that the electronic structure of a photocatalyst is important in determining its light-harvesting ability. Thus, the energetic levels (VB and CB) of each material play a critical part in PEC water splitting considering the ionisation potential and electron affinity are corresponded to their band edges. Figure 2 shows various types of semiconductors with their CB and VB edges related to vacuum and the normal hydrogen electrode (NHE) as standard for zero potential. It is clear that some materials show relatively wide band gap implying the poor harvest of the visible light spectrum. However, at the current moment, altering the band edges of the semiconductors in order to effectively increase the photocatalytic activity still remains as a big challenge. It is also clearly shown in Fig. 2 that the VB edges of most of the semiconductors are deeper than that of the O₂/H₂O oxidation potential. This gives us a hint that the evolution of the O₂ is still possible under the illumination. In contrast, the evolution of H₂ is not possible for the semiconductors in Fig. 2(b).

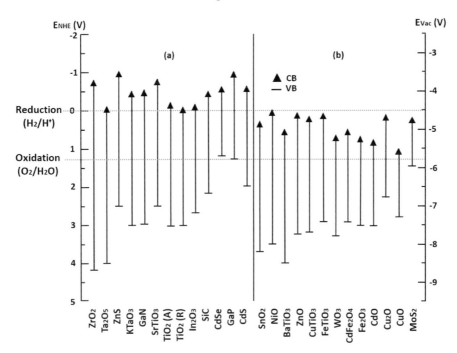

Fig. 2. Thermodynamically (a) suitable and (b) unsuitable materials. Electronic structure of different semiconductors and the relative position of their band edge with respect to the NHE and vacuum (Vac) (Babu et al. 2015, Castelli et al. 2013).

This is attributed to the fact that the H_2/H^+ reduction potential is higher than that of its conduction band energy and thereby, the excited electrons at the conduction band possess insufficient energy to reduce the H^+ to form the H_2 (Izumi et al. 2009). Although the band gaps of the semiconductors like MoS_2, Fe_2O_3 and WO_3 cover the whole visible region of solar spectrum, they are not considered as active materials due to their unsuitable band energies as compared to that of NHE. Furthermore, the semiconductors shown in Fig. 2(b) are classified as photocorrosive materials since their minimum CB does not reach the requirement of water splitting thermodynamically (Kudo and Miseki 2009). It is also important to note that the accessible of the flat band potential is an essential criterion to fulfil the redox potentials of water splitting hydrogen production.

2.4 One-Dimensional (1D) Nanostructure for PEC Water Splittting

In 1991, Iijima's successfully discovered CNT and since then generated unprecedented interest in the area of one-dimensional (1D) nanostructure in the field of nanotechnology due to its supreme molecular geometry properties with high surface area, distinct electrical properties like, high electron mobility, quantum confinement effect and high mechanical strength (Lieber 1998, Rao et al. 2006). This breakthrough has also triggered subsequent interests in the photocatalysis and made 1D nanostructure research in many practical applications, e.g., solar cells, supercapacitors, biomedical devices and lithium-ion batteries (Chen Jun Song et al. 2010a, Ji et al. 2012, Li Hui et al. 2014, Li Huaqiong et al. 2015).

3. The Engineering Behind TiO_2 Nanotubes Film

Recently, TiO_2 photocatalyst has been widely used as a photoelectrode in water splitting system due to its excellent promising features for the water splitting applications for hydrogen generation. TiO_2 are eco-friendly, have strong oxidation ability, possess good photocatalytic property and are highly photocorrosion resistant (Fujishima et al. 2000, Kudo and Miseki 2009). However, TiO_2 tends to suffer from shortcomings in terms of its large band gap (3.0 to 3.2 eV), which will subsequently result in poor visible light response and the rapid recombination of electrons and holes pairs. These limitations hinder the widespread use of TiO_2 as a photoelectrode in water splitting system (Lee Doh C et al. 2010, Lee Sooho et al. 2014b, Yang Min et al. 2010). In order to tackle these setbacks, some improvements on the visible light absorption of TiO_2 have been done. One of the methods is to couple TiO_2 with another small band gap semiconductor. This will eventually lead to an additional electronic state in the band-gap which in turn affects the optical, electronic and functionality of TiO_2. However, multi-component semiconductors are complex in nature because of the challenging task of predicting the material's properties.

Initially, research works on photocatalytic activities were only focused on the zero dimension (0D) TiO_2 nanoparticles since they showed excellent photocatalytic performances in hydrogen production, solar cells, adsorbents and sensors attributed to their large surface area and broadened band gap (Han et al. 2014, Liu Baoshun et al. 2012, Tang Yuxin et al. 2010). However, they tend to suffer from shortcomings in terms of fast recombination of electrons and holes, slow charge carrier transfer and high recycling cost (Ge et al. 2015, Wu Zhi et al. 2014). Also, the synthesis of 2D nanostructured materials is comparably complex and requires harsh experimental conditions (Lee Chang Soo et al. 2014a, Yao et al. 2014).

From then onwards, 1D self-organised TiO_2 nanotubes with porous structures have been extensively studied due to their distinctive advantages like large surface area and short lateral diffusion length. These strengths will provide a unidirectional electrical channel for the photo-induced charge carrier to transfer. In other words, the TiO_2 grains are stretched in the tube's growth direction. Thus, vertical transportation of charge carriers could be enhanced and the PEC water splitting performance could be further improved due to the lower recombination losses at the grain boundaries (Roy et al. 2011, Spadavecchia et al. 2013).

In order to get the right dimensions and morphologies, synthesis procedure for the production of self-organised nanotube arrays should be well controlled. Also, the length, wall thickness, pore diameter and inter-tube spacing of the nanotubes should be fine-tuned through the anodisation process. Maximising the specific surface area of porous TiO_2 nanostructure film is also very crucial. Highly ordered self-organising porous TiO_2 nanotubes film is generally considered to be an ideal photoelectrode in the water splitting system since their inner and outer wall surface area of the nanotubes increases the density of the available active sites for the photon absorption significantly. As a result, the use of porous TiO_2 nanotubes film in the present study is found to be better and is more effective to improve the PEC water splitting performance (Albu et al. 2008, Grimes 2007).

Over the past few years, various synthesis methods to form the 1D porous TiO_2 nanotubes have been explored. Hydro/solvothermal, template-assisted/sol-gel and electrochemical anodisation techniques are three main synthesis techniques used to prepare the 1D porous TiO_2 nanotubes (Kasuga et al. 1998, 1999, Lakshmi et al. 1997, Sander et al. 2004, Suzuki and Yoshikawa 2004). In the following section, these approaches to synthesis TiO_2 nanostructures is reviewed.

3.1 Synthesis of TiO_2 Nanostructures from Hydrothermal/Solvothermal Method

Hydrothermal method has been found to be the most widely used technique for the fabrication of 1D porous TiO_2 nanotubes. Hydrothermal method involves utilising a stainless steel vessel under high temperature and pressure. The main advantages adopting this method include simple procedure and low production cost. Besides, many literatures reported that nearly 100 per cent of the precursors could be converted into 1D porous TiO_2 nanostructured within one single procedure under hydrothermal process. In 1998, Kasuga et al. first reported the preparation of porous TiO_2-based nanotubular materials using the hydrothermal method (Kasuga et al. 1998, 1999). They claimed that porous TiO_2 nanotubes could be prepared by mixing the TiO_2 powder in a highly concentrated NaOH solution (10 M), followed by the heat treatment process at 110°C for 20 h. No sacrificial template is required for this approach. Since then, many investigations have been carried out by controlling the processing parameters, such as temperature, treatment duration and caustic concentration, in order to synthesise the nanotubes with desired dimension (Armstrong et al. 2005, Bavykin et al. 2004).

In fact, this hydrothermal synthesis could be performed either through acid-hydrothermal or alkali-hydrothermal depending on the nature of the reactants (Tian Jian et al. 2014). At the early stage, the reactants were composed of titanium salts and hydrochloric acid. These reactants usually form TiO_2 nanorods structure rather than nanotubular structure. In this respect, the reactants were altered to alkaline-based (high concentration of sodium hydroxide) in order to prepare the porous TiO_2 nanotubes via the dissolution–recrystallisation mechanism (Tian Jian et al. 2014). In 2005–2006, Tanaka and Peng et al. reported the effect of alkaline hydrothermal condition on the nanostructure formation. Ti substrate was used as the precursor (Peng and Chen 2006, Tanaka et al. 2005). Apart from that, Morgan and co-coworkers reported that different morphologies and structures of TiO_2 such as nanoparticles, nanotubes, and nanoribbons could be prepared by optimising the caustic concentration and temperature of Degussa P25 through the alkaline hydrothermal treatment (Kasuga method) (Morgan et al. 2008). These observations are particularly in line with the finding reported by Tanaka and Peng et al. This implies that TiO_2 with different phases and morphologies could be transformed to nanotubes under a specific hydrothermal condition (Lim et al. 2005, Nakahira et al. 2010).

Solvothermal method, which is identical to that of the hydrothermal method, is also one of the common synthesis approaches used to fabricate the 1D porous TiO_2 nanotubes (Nam et al. 2014, Wang Peifang et al. 2013). $TiCl_4$ or tetrabutyl titanate are the commonly used precursors. The only difference between hydrothermal and solvothermal methods is that solvothermal is usually conducted in an organic solvent (ethanol, ethylene glycol, n-hexane, etc.), while the hydrothermal is performed in the water solution (Hoa and Huyen 2013). Specifically, Wang et al. successfully synthesised a bundle of both nanowires and open-ended TiO_2 nanotubes using solvothermal method in the presence ethanol and glycerol as solvents (Wang Qiang et al. 2006). Besides, Zhao's group also successfully fabricated the vertically aligned TiO_2 nanorods using the solvothermal method and used it as a photoanodes for dye-sensitised solar cells (Zhao et al. 2014).

There are few drawbacks using either hydrothermal or solvothermal method. The reaction kinetics are slower which leads to longer reaction time. Also, various lengths of porous TiO_2 nanotubes were formed and the diameters of the large-scale fabricated nanotubes are non-uniform. These drawbacks limit most of their wide range of applications. In order to produce smaller range of high aspect ratio nanotubes, it is a must to understand the kinetic mechanism of each solvents towards the Ti precursors (Ge et al. 2016).

3.2 Synthesis of TiO₂ Nanostructures from Template-Assisted/Sol-Gel Method

The template-assisted method is one of the most prominent and well-established processes to produce porous TiO_2 nanotubes in the presence of TiO_2 precursor. In this case, an anodic aluminium oxide (AAO) nanoporous membrane with the arrays of parallel nanopores (with controllable diameter and length) are used as a potential template. This template is removed by chemical etching after the deposition of TiO_2 (Sander et al. 2004). However, some precursors (e.g., tetrabutyl titanate or titanium isopropoxide) mixed in acetic acid might need further treatment (e.g., purging or hydrolysis process) in order to form 1D porous TiO_2 nanotubes. In the last stage, the template was removed by chemical etching to obtain the porous TiO_2 nanotubes. The fabrication of porous TiO_2 nanotubes with controlled dimension using the template-assisted method has been reported in some literatures (Hoyer 1996, Lee Jiwon et al. 2011, Sander et al. 2004).

Sol-gel synthesis is also widely employed due to its capability in controlling the textural and surface properties of the resulting porous TiO_2 nanotubes. Specifically, this method is used primarily to fabricate the mix-metal oxides starting from a colloidal solution (sol). The sol basically acts as the precursor for the formation of the integrated network (gel) or discrete particles via various forms of hydrolysis and polycondensation reactions. These precursors are typically metal alkoxides, inorganic metal salts and metal-organic compound. Titanium tetrachloride and alkoxide Ti solutions are two main examples of the precursors. Generally, the precursor becoming a sol will chemically bond the TiO_2 into a template. The sol will then undergo the heat treatment at 150°C and/or vigorous stirring in acetone to remove the template (Tan et al. 2012, Xia et al. 2003, Zhang Dongbai et al. 2002). Sol-gel method is commonly constructed by the combination of template-assisted method to fabricate 1D porous TiO_2-based nanostructures. Joo et al. successfully synthesised the TiO_2 nanorods with a diameter of about 5 nm by the sol-gel method (Joo et al. 2005). Qiu and Attar et al. also reported the fabrication of the 1D well-aligned TiO_2 nanotube arrays, nanorods and nanowires using the modified template-assisted sol-gel method (Attar et al. 2009, Qiu et al. 2006).

In summary, porous TiO_2 nanotubes could be prepared via the template-assisted/sol-gel approaches which involves high production cost. Also, the difficulties in template removal stage might also damage the nanostructure (Hagen et al. 2014, Shin et al. 2004). In addition, varied lengths of porous TiO_2 nanotubes were produced using the titanium sol-gel precursor and/or template-assisted process. The porous TiO_2 nanotubes were either heaping on together or loosen tubes dispersed in solution. Considering these drawbacks, many advantages of the 1D orientation is lost (Roy et al. 2011).

3.3 Synthesis of TiO₂ Nanostructures from Electrochemical Anodisation Method

Electrochemical anodisation method has been widely studied in recent years to prepare the 1D porous TiO_2 nanotube arrays due to its facile synthesis procedure. In anodising cell, Ti is used as an anode connecting to the positive terminal of power source, while the platinum is used as cathode connecting to the negative terminal. The first literature about the use of electrochemical anodisation technique to form the porous TiO_2 was reported by Assefpour-Dezfuly and co-researches in 1984. Accordingly, they reported that TiO_2 porous structure could be synthesised via chemical etching in the presence of alkaline peroxide, followed by electrochemical anodisation in chromic acid solution (Assefpour-Dezfuly et al. 1984). In 1999, Zwilling and co-workers reported that self-organised porous TiO_2 could be obtained by anodising a Ti-based alloy in an acidic fluoride-based electrolyte (Zwilling et al. 1999). From then onwards, the electrochemical condition such as the voltage, time, concentration of electrolyte, reaction temperatures, as well as pH of electrolyte have been studied extensively to control the morphology and structure of the porous TiO_2 nanotubes.

Generally, the morphology of the self-ordered nanostructure is strongly affected by the type of the electrolyte, pH, applied potential, time and temperature. In the first generation, aqueous hydrofluoric acid based electrolytes were most commonly used in the fabrication of porous TiO_2 nanotubes. However, the length of the nanotubes obtained was merely a few hundreds of nanometer (200–500 nm) which is insufficient to generate favourable photocurrent. In this respect, further modifications have been done by using F-based inorganic and organic neutral electrolytes to produce longer and smoother nanotubes walls (Paulose et al. 2007). Accordingly, it is worth to notice that the anodisation voltage alters the morphology of the formed nanostructures while anodisation time mainly affects the length of porous TiO_2 nanotubes and temperature of the anodisation condition affects the dissolution rate. Taking these facts into account, porous TiO_2 nanotube arrays are normally grown into the optimum dimension under ambient conditions (20–25°C) (Ge et al. 2016). In short, appropriate selection of the anodisation parameters for the formation of porous TiO_2 nanotubes is the key to grow high quality nanotubes.

As a comparison, electrochemical anodisation method would lead to the formation of self-organised nanotubes array which is aligned perpendicularly with its metal (substrate) surface. All of the tubes are attached with the metal, herein is electrically conductive. The ability for the formation of dense nanotubes layer with controlled geometry makes the electrochemical anodisation an extremely versatile parallel structuring process.

3.4 Mechanism of Formation of TiO_2 Nanotubes

Generally, the porous TiO_2 nanotubes formation mechanism through the electrochemical anodisation method can be divided into three generations. Choi et al. was the first group reporting the mechanistic formation model (Fig. 3) of porous TiO_2 nanotubes based on the phenomena of the electrical breakdown of the TiO_2 during the porous TiO_2 formation (Choi et al. 2004). At the very beginning of the process, anodisation voltage was applied forming a barrier layer of TiO_2 on the substrate which grows thicker with increased voltage. The barrier layer will then undergo the crystal structure phase transformation into a denser structure like anatase or rutile inducing a high compressive stress on the oxide. When the electrical breakdown occurs, new pore will be created with the occurrence of sparking followed by the immediate passivation of these pores by the oxide. Repetition of the breakdown and re-passivation at both pore tips and inside the pores occurs resulting in the formation of the nanotube arrays.

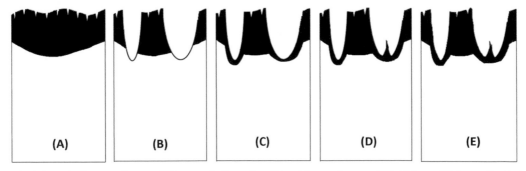

Fig. 3. Schematic diagram of porous TiO_2 nanotube formation: (A) oxide growth to maximal thickness (B) burst of oxide by the formation of crystallites (pore formation), (C) immediate repassivation of pore tips, (D) burst of repassivated oxide, and (E) dissolution of the formed oxide and second repassivation (Choi et al. 2004).

In 2006, another mechanistic model was reported by Mor et al. where the porous TiO_2 was grown on the Ti substrate. This is mainly due to the reaction of Ti metal with O^{2-} or OH^- ions from the electrolyte. A thin layer of oxide will then be formed on the surface as a barrier layer. Further oxidation of the Ti on the substrate has occurred thereafter to form a thicker oxide layer. The formation of thick oxide layer is mainly associated with the migration of these anions through the existing oxide layer. When these anions reach the metal/oxide interface, they would react with the metal. Ti^{4+} cations on the metal/oxide surface, followed by being ejected into the oxide/electrolyte interface under an electric field. The formation of the small pits in the oxide layer on the oxide/electrolyte interface originates from the localised dissolution of the oxide layer making this barrier layer at the bottom of the pits. The barrier layer becomes relatively thinner as the voltage was increased. The free O^{2-} anions in electrolyte then migrate to the metal/oxide interface to interact with the Ti metal to regain the oxide layer. Localised dissolution and oxidation will continuously lead to the growth of these pits into pores with various sizes and depths in the presence of the electric field (Mor et al. 2006). Figure 4 shows the schematic diagram of evolution of nanotubes proposed by Mor et al.

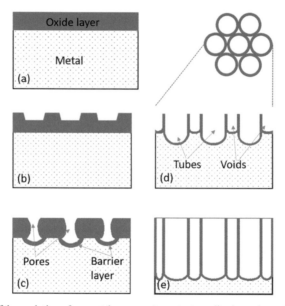

Fig. 4. Schematic diagram of the evolution of a nanotube arrays at constant anodisation voltage: (a) oxide layer formation, (b) pit formation on the oxide layer, (c) growth of the pit into scallop shaped pores, (d) metallic part between the pores undergoes oxidation and field assisted dissolution, and (e) fully developed nanotube array with a corresponding top view (Mor et al. 2006).

Another more systematic and constructive mechanistic model involving chemical reaction equations was proposed and reported by Patrick Schmuki (Taveira et al. 2005). Accordingly, a competition between the formation of the anodic oxide and the chemical dissolution of the oxide layer was discussed based on the formation of the nanotubes layer. As reported by Mor et al. in 2006, the initial layer of oxide was formed ascribed to the reaction of Ti metal with O^{2-} or OH^- ions from the electrolyte under an applied electrical current (Eq. 5).

$$Ti + H_2O \rightarrow TiO_2 + 2H^+ + 4e^- \tag{5}$$

$$TiO_2 + 6F^- + 4H^+ \rightarrow [TiF_6]^{2-} + 2H_2O \tag{6}$$

Random pits formed at the surface of the oxide layer were then continuously dissolved by the fluoride ions forming a soluble fluoride complex of $[TiF_6]^{2-}$ under applied electric field (Eq. 6). At the beginning of the anodisation process, the whole process is dominated by the field-assisted dissolution and the chemical dissolution is comparatively slow due to the large electric field across the thin oxide layer. Small random pits were also found on the surface of the oxide as a result of localised dissolutions. These pits will also act as the pore forming centres, which would eventually grow into pores. The field-assisted dissolution at the oxide barrier layer provides high inwards driving force for the F$^-$ ions to form a soluble fluoride complex of $[TiF_6]^{2-}$ ions. The fluoride complex formation tendency not only leads to a permanent chemical dissolution (attack) at the pores, but also prevents and/or minimise the migration of $Ti(OH)_xO_y$ precipitation layer from the Ti^{4+} ions at the oxide/solution interface. During anodisation, both continual growth of the oxide at the metal/oxide interface and the chemical dissolution of the oxide layer at the oxide/electrolyte interface occur simultaneously. When steady state was established, the growth rate of the porous nanotubes in length was similar to that of the thickness reducing dissolution rate of the outer interface of the oxide layer. Under this situation, the porous nanotube oxide layer will just continuously "eat" through the titanium substrate without thickening of the oxide layer (Macak et al. 2007).

3.5 Limitation of Porous TiO$_2$ Nanotubes Film

In recent years, TiO_2 has immerged as one of the promising photoelectrodes in water splitting system for hydrogen generation. However, the large bandgap energy of TiO_2 that can only function effectively under the UV region ($\lambda < 400$ nm) hinders the widespread use of TiO_2 as a photoelectrode in the water splitting cell (Khare et al. 2013). In short, the energy conversion efficiency from solar to hydrogen by TiO_2 photocatalytic water-splitting is still low due to the following reasons:

1. Recombination of photogenerated charge carriers: CB electrons can recombine with VB holes in a very fast manner and the energy was released in the form of unproductive heat or phonons (Martin et al. 2013).
2. Fast backward reaction: Backward reaction (i.e., recombination of hydrogen and oxygen into water) occurred easily since the decomposition of water into hydrogen and oxygen is an energy increasing process (Martin et al. 2014).
3. Inability to utilise visible light: The band gap of TiO_2 is about 3.2e V, and thereby only UV light can be utilised for hydrogen production. Owing to UV light accounts for about 4 per cent of the solar radiation energy while the visible light contributes about 50 per cent, the inability to utilise the visible light limits the efficiency of solar photocatalytic hydrogen production (Kitano et al. 2007).

In general, these drawbacks cannot be overcome by just optimising the dimensions of porous TiO_2 nanotubes itself via controlling the processing parameter of electrochemical anodisation as described in the previous section. Thus, considerable efforts have been paid to minimise the recombination losses of charge carriers and to extend the spectral response of TiO_2 to the visible spectrum (by incorporating an optimum amount of small band gap semiconductor, such as CdSe with porous TiO_2 nanotubes film). By adopting these approaches, the band gap narrowing effects of the TiO_2 could expand the excitation light range to the visible region and provide the sites that can reduce the recombination loss of charge carriers (Fernandes et al. 2016, Lv et al. 2015, Su et al. 2014).

3.6 Enhance the Photoactivity of TiO$_2$ Across Entire UV-Visible Region

One-dimension (1D) porous TiO_2 nanotubular structure has been widely used in the application of PEC water splitting due to its excellent properties like high corrosion resistant, efficient charge transport and good photocurrent stability (Reyes-Gil and Robinson 2013). In order to extend the visible light absorption of TiO_2, various research works have been done either by modifying the methods or decorating the TiO_2 with few percent of other elements. Table 1 summarises the details of the foreign element doping on TiO_2 photocatalyst by several researchers and their findings.

Chu et al. performed the solar water splitting using the transferred porous TiO_2 nanotubes arrays on a transparent conductive oxide substrate. Two co-reduction processes of flame and chemical

Table 1. Summary of the works reported on the different element decorated TiO_2 photocatalyst.

Materials	Method	Observations	Citation
Transferred TiO_2 nanotubes arrays on a TCO substrate	Immersed in 33 wt per cent H_2O_2 solution to separate TiO_2 nanotubes layer	- Rapid flame reduce (> 1000°C) anatase TiO_2 by generating O^{2-} vacancies and Ti^{3+}. - Chemical reduction of TiO_2 nanostructure in $TiCl_3$ successfully introduce O^{2-} vacancies. - Highest photocurrent density reported was ~ 2.0 mA/cm².	(Cho et al. 2015)
Au	Seed-growth method	- Photocurrent density of Au-TiO_2 and TiO_2 recorded at 0 V are 1.49 and 0.82 mA/cm².	(Pu ct al. 2013)
	Chemical bath deposition	- Au nanorod decorated TiO_2 exhibited three times higher under visible light illumination than bare TiO_2.	(Pu et al. 2013)
	Photocatalytic reduction method	- Highest photocurrent density obtained as ~ 150 µA cm⁻² when under illumination of visible light.	(Zhang Zhonghai et al. 2012)
Ag	Microwave-assisted chemical reduction	- Best hydrogen generation in the medium contain of ethanol (10 vt per cent) and H_2O under solar illumination is 1.34 µmol/cm²h.	(Wu Feng et al. 2013)
ZnO	Chemical bath deposition	- High electron mobility of ZnO improve the process of electron transfer between CB and VB. - Photocurrent density and hydrogen generated by ZnO-TiO_2 nanotubes as 1.24 mA/cm² and 11 µL/cm²h respectively.	(Momeni and Ghayeb 2015)
Cr_2O_3	Chemical bath deposition	- Photocurrent density of > 150 µA/cm² at 0.2 V (vs. Ag/AgCl) as compare to bare TiO_2 with only < 8 µA/cm² that they synthesised.	(Momeni and Ghayeb 2016)
H	Heat treatment	- Hydrogen treated porous TiO_2 nanotubes exhibited higher photocurrent density (~ 6 mA/cm²) and photoconversion efficiency (2.96 per cent).	(Chen Bo et al. 2013)
N	Hydrothermal	- Nitrogen was introduced to combine with the 2p states of oxygen in TiO_2 increase the VB band hence reduce the band gap energy. - Hydrogen generation rate: - UV light irradiation –40 µmol/h and 38.6 µmol/h with respect to N-TiO_2 and TiO_2 nanotubes. - Visible light illumination-N doped porous TiO_2 nanotubes was ~ 4.3 µmol/h where TiO_2 nanotubes is ~ 1.5 µmol/h.	(Park Sang-Sun et al. 2011)

reduction were applied in order to enhance the water splitting performance. First, rapid flame reduces (> 1000°C) the anatase TiO_2 by generating the oxygen (O^{2-}) vacancies and Ti^{3+}. These oxygen vacancies in TiO_2 act as the new electron donors enhancing the photocatalytic capability and the solar illumination absorption (Hu Yun Hang 2012). Second, the chemical reduction of the TiO_2 nanostructure in $TiCl_3$ introduces the oxygen vacancies and thereby improves the photocurrent density. The highest photocurrent density of ~ 2.0 mA/cm², which is ~ 15 per cent higher than that of the mono-reduced sample, was reported from this synergistically engineered co-reduction processes (Cho et al. 2015).

On top of that, the improvement in the photoelectrochemical water splitting could be done through the modification of methods and also by metal doping. Several successful cases have been reported and gold (Au) nanoparticles doped TiO_2 by Pu et al. is one of the examples. The Au nanoparticle decorated by the TiO_2 could effectively enhance the absorption of whole UV-visible solar spectrum as compare to that of the bare TiO_2. The photocurrent density recorded at 0 V was found to be 1.49 and 0.82 mA/cm². Pu and his teams also found that the decoration of Au nanorods on the TiO_2

could significantly enhance the photocurrent density under visible light illumination (> 430 nm, 73.3 mW/cm^2) which was three times higher than that of the bare TiO$_2$ (2.67 and 0.87 µA/cm^2) (Pu et al. 2013). Zhang and his co-workers also reported that the Au doped porous TiO$_2$ nanotubes exhibited the highest photocurrent density of ~ 150 µA/cm^2 under the illumination of visible light (Zhang Zhonghai et al. 2012). Apart from that, Wu et al. used the metal Ag to dope on the porous TiO$_2$ nanotubes. The Ag nanoparticles were assembled by the microwave-assisted chemical reduction onto anodised porous TiO$_2$ nanotubes. The best hydrogen generation in the medium containing ethanol (10 vt per cent) and H$_2$O under solar illumination was found to be 1.34 µmol/cm^{2}h. It was obtained by soaking the sample in 2 mM AgNO$_3$ solution with 3 min microwave irradiation (300 W) at 100°C (Wu Feng et al. 2013).

ZnO is commonly used to couple with the TiO$_2$. High electron mobility of ZnO tends to improve the process of electron transfer between CB and VB. This will significantly reduce the recombination of photogenerated charge carrier as compared to that of the TiO$_2$. It has been reported by Momeni and Ghayeb that the photocurrent density and hydrogen generated by ZnO-TiO$_2$ nanotubes were found to be 1.24 mA/cm^2 and 11 µL/cm^{2}h, respectively (Momeni and Ghayeb 2015). Furthermore, the transition metal oxide was also adopted to synthesise the nanocomposite. One of the commonly used transition metal oxide is Chromium III oxide (Cr$_2$O$_3$). Through chemical bath deposition, Cr$_2$O$_3$ was deposited onto the porous TiO$_2$ nanotubes. The photocurrent density of the nanocomposites was > 150 µA/cm^2 at 0.2 V (vs. Ag/AgCl) as compared to that of the bare TiO$_2$ with only < 8 µA/cm^2. The hydrogen generation performance was found to be 10.67 µL/cm^{2}h, which is 12.8 times higher than that of the bare TiO$_2$ (Momeni and Ghayeb 2016).

Many nonmetallic elements, especially N and H, have also been incorporated with TiO$_2$ to enhance its absorption-ability of visible light. For instance, Chen et al. reported the synthesis of hydrogen treated porous TiO$_2$ nanotubes. It is noteworthy that the hydrogen treated porous TiO$_2$ nanotubes showed higher photocurrent density (6 mA/cm^2) and 10 times higher photoconversion efficiency (2.96 per cent) as compared to that of the as-grown TiO$_2$ nanotubes (Chen Bo et al. 2013). Moreover, nitrogen doped porous TiO$_2$ nanotubes in PEC water splitting has also been reported by Park and co-workers, in which the anodised TiO$_2$ nanotube array was hydrothermally treated in mixed gases of NH$_3$/Ar for 3 hour at 773 K. It was proposed that the nitrogen tends to combine with the 2p states of oxygen in TiO$_2$. This phenomenon increases the VB and thereby reduces the band gap energy. Two modes (UV and visible spectrum) have been tested to identify the performance of water splitting reaction. The hydrogen generated rate after UV light irradiation was found to be 40 µmol/h and 38.6 µmol/h with respect to N-TiO$_2$ and TiO$_2$ nanotubes, respectively. Under the visible light illumination, the hydrogen evolution rate of N-doped porous TiO$_2$ nanotubes was ~ 4.3 µmol/h, while the porous TiO$_2$ nanotubes was ~ 1.5 µmol/h (Park Sang-Sun et al. 2011). Based on these findings, it could be concluded that the N-doped porous TiO$_2$ nanotubes generates more hydrogen as compared to that of the pure porous TiO$_2$ nanotubes.

4. The Engineering Behind ZnO Nanostructures

Besides TiO$_2$, ZnO is also one of the promising photoelectrodes used to increase the efficiency of PEC water splitting performance. This is mainly attributed to its excellent electrical, piezoelectric and optical properties. ZnO is known to perform well under the blue UV region with a direct bandgap of 3.37 eV. ZnO also exhibits a large excitation binding energy of about 60 meV at room temperature and generally exists in three crystal structures, i.e., wurtzite, zinc blende, and rocksalt. Among them, wurtzite is more preferable for electronic applications as wurtzite is thermochemically stable at room temperature and belongs to space groups $P6_3mc$ (Hermann-Mauguin notation) and (Schoenflies notation) (Kim et al. 2011c, Lepot et al. 2007, Park et al. 2002).

It has also been reported that porous ZnO provides excellent ratio especially at the near-band-edge (NBE) up to the deep-level emission (DLE) of the PL intensity (Jeon et al. 2010, Kim Min Su et al.

2011b). According to Jeon et al., the pore density is inversely proportional to the average diameter of the porous ZnO. Increasing the average diameter of pores will subsequently increase the PL intensity. This is mainly due to the larger surface area and higher surface-to-volume ratio of porous ZnO (Jeon et al. 2010, Kim Min Su et al. 2011b). Field emission scanning electron microscope (FESEM) and High resolution transmission electron microscope (HRTEM) images of ZnO with porous structure are shown below (Fig. 5).

It is worth to mention that the porous ZnO crystals with specific size and shape could be successfully produced by adopting atomic layer deposition, sol-gel, hydrothermal, solvothermal, electrodeposition, and/or chemical vapour deposition. In addition, the effects of surfactant's types on the formation of ZnO nanocrystals with various shapes and structures have also been studied (Lévy-Clément et al. 2005, Ramírez et al. 2010, Wong et al. 2003, Yiamsawas et al. 2009). However, the use of surfactants leads to highly complicated reaction and the use of additives are not environmentally friendly especially during the large-scale industrial production. Taking these facts into account, an additive-free technique has been developed to prepare ZnO nanostructures. Correspondingly, additive-free ZnO nanostructures could be prepared by using simple solution phase, autoclave and microwave techniques. The annealing effects on the morphology and properties of ZnO at low temperature will be discussed hence-forth. ZnO has also been listed as safe (GRAS) material by the Food and Drug Administration and can be used as food additive (Espitia et al. 2012) making it highly safe to work with.

It is noteworthy that most of the fabrication techniques to prepare the porous wurtzite ZnO nano-structures share the common growth conditions. The growth behaviour of the porous wurtzite ZnO nano-structures could be altered by controlling the growth kinetics. The most crucial variable for the formation of porous wurtzite ZnO nanostructures is the pH of the electrolyte in solution-based synthesis technique (i.e., sol-gel, hydrothermal, solvothermal, electrodeposition or chemical bath deposition). pH 5–6 is highly preferable. However, it is interesting to note that ZnO nanostructures

Fig. 5. FESEM images of (a) nanodisk-dendritic ZnO, (b) porous structure of nanodisk-dendritic ZnO, (c) ZnO nanorod, and (d) HRTEM of ZnO nanorod.

can be produced in the basic solution (pH of 9–13). In addition, reaction temperature could also be considered as one of the variables affecting the formation of porous wurtzite ZnO nanostructures. Many research works claimed that, using the physical and chemical vapour deposition techniques, the porous wurtzite ZnO nanostructures formation starts between 500°C–800°C.

For solution-based synthesis, the appropriate deposition temperature for the formation of the porous wurtzite ZnO nanostructures is 70–80°C because the polycrystalline structure would start to evolve at this temperature with randomly oriented grains (Abd Samad et al. 2016). It has also been reported that 0.001–0.1 M of zinc ion was more than sufficient for the formation of porous wurtzite ZnO nanostructures using the solution-based synthesis technique. Other than that, the growth direction and the diameter of porous ZnO nanostructures could be assisted by the catalyst, such as Au, Fe, and Sn. However, the growth of porous nanostructures may be terminated once it reaches the eutectic point of catalyst alloys or reactant. Substrates is also one of the essential elements in the formation of wurtzite porous ZnO nanostructures as the process of crystal growth at a certain orientation (on top of another crystal) was influenced by the surface of the substrate (Kim Kwang-Sik and Kim 2003, Miao et al. 2013, Zheng et al. 2013). In short, the morphologies of the grown nanostructures could be affected by the crystallographic structure together with the surfaces to be used and the atomic termination and charge status of the substrate. Modifying the ageing or deposition time could also give a very different result of the synthesised ZnO nanostructures.

4.1 Synthesis of ZnO Nanostructures from Sol-Gel Technique

According to Bahadur and co-researchers, the selection of precursor materials is critical in sol-gel technique considering they could determine the future morphology of the ZnO nanostructures. Besides, it has also been reported that the nitrate ions could form an agglomeration or island such as the dendrite structure, while the acetate ions could produce a smooth character and uniform ZnO nanostructures. On top of that, nitrate ions could produce a smaller crystallite size of ZnO nanostructures compared to that of the acetate ions as precursor. This is mainly associated with the increase in the basicity in nitrate ions electrolyte which produces the random and rapid crystallisation of ZnO nanostructures (Bahadur et al. 2007). Moreover, sol-gel technique produces an enhancement of preferential growth for ZnO deposition process and thereby, a highly ZnO crystalline solid in c-axis growth could be produced (Bornand and Mezy 2011).

4.2 Synthesis of ZnO Nanostructures from Hydrothermal Technique

Hydrothermal technique is also recognised as one of the potential methods for the formation of ZnO nanostructures. This technique generally produces nanostructure with narrow size particle distribution, high-quality growth orientation and good crystallisation (Aziz et al. 2014, Dai et al. 2013, Kim et al. 2011c). However, a growth directing agent during the hydrothermal process is needed to nucleate the ZnO nanostructures (Lepot et al. 2007). Also, the presence of a capping agent (polyvinylpyrrolidone, PVP) is needed to achieve the anisotropic growth of nanocrystals. In this respect, the surface energy of the crystallographic surfaces of ZnO could be altered (Du et al. 2005, Park et al. 2002). Also, it has been reported in literatures that the reaction time required for hydrothermal is much longer than that of the electrodeposition process. However, prolonging the reaction time would not affect the morphology of ZnO nanostructures because surfactant plays the main role in tailoring the morphology compared to that of reaction time (Lepot et al. 2007).

4.3 Synthesis of ZnO Nanostructures from Solvothermal Technique

Solvothermal is also widely employed to synthesis the ZnO nanostructure as its working principle is identical to that of the hydrothermal. Solvothermal technique was found to be more advantageous than that of both the sol-gel (Oliveira et al. 2003) and hydrothermal technique (Andersson et al. 2002). The size, shape distribution and crystallinity of ZnO nanostructures formed by adopting the solvothermal approach are more tunable by varying the precursor, surfactant and solvent used.

Varghese et al. reported that ZnO nanostructures can be easily prepared under the temperature of 200–300°C. Also, the presence of surfactants in the reaction mixture help ZnO nanostructures attain more uniform diameter (Varghese et al. 2007). Identical to that of the hydrothermal technique, polyvinylpyrrolidone (PVP) plays an important role in controlling the size and shape of ZnO. Indeed, PVP acts as a template to form the chain structures of ZnO crystals and thereby, the ZnO could grow up along these chains to form the nanostructures with the use of polymer template. On the other hand, PVP could form a shell surrounding these particles and prevent them from aggregating into larger particles since its steric effect could control the grain growth (Yiamsawas et al. 2009). As a growth directing agent, the morphology of ZnO could be well controlled by PVP depending on the interaction between ZnO and PVP during the crystal growth (Yiamsawas et al. 2009).

4.4 Synthesis of ZnO Nanostructures from Electrodeposition Technique

Among those aforementioned preparation techniques, electrodeposition technique can be considered as one of the most commonly used techniques for the preparation of ZnO nanostructures. Generally, the main advantage of the electrodeposition technique is its easiness for the production of ZnO nanostructures. In addition, low equipment cost, scalability and facile and precise control of nanostructure and morphology are added advantages of electrodeposition technique (Abd-Ellah et al. 2013). It is also noteworthy that ZnO nanostructures could be produced via the electrodeposition technique without the need of any surfactant. Specifically, Zn (II) concentration highly influences the growth of ZnO and oxygen precursors by adopting electrodeposition technique. In fact, the effects of Zn (II) concentration on the self-assembly of 1D ZnO growth have been extensively reported in many literatures (Chen Qiu-Ping et al. 2006, Cui and Gibson 2005, Elias et al. 2007, Elias Jamil et al. 2008, Könenkamp et al. 2000, Lévy-Clément et al. 2005, Mollar et al. 2006, Ramírez et al. 2010, Tena-Zaera Ramon et al. 2007, Tena-Zaera et al. 2005, Wong et al. 2003). ZnO nanorod could be formed via electrodeposition technique under certain operating conditions.

It has also been reported that the structures of ZnO produced using the electrodeposition technique is highly dependent on the applied current density. Aziz and co-researchers claimed that a very thin layer of nanorod structures could be obtained by using a very low current density of –0.1 mA/cm². Further increasing the current density up to –0.5 mA/cm² tends to produce the nanoporous-like ZnO layer. The diameter (75–150 nm) of the nanorod was also found to increase drastically generating almost no space between the nanorods when the applied current was increased to –1.0 mA/cm². Also, the well-defined hexagonal structure of ZnO nanorod was no longer detected at high current density of –1.5 mA/cm². The only reason explaining this is associated with the high dissolution and deposition rate as a result of the increase in the rate of all chemical reactions. At high current density of –2.0 mA/cm², a ZnO layer of nanocluster structures were formed due to the higher dissolution and deposition rate produced a higher conductance value. Therefore, it affects the nucleation densities and morphology (Abd-Ellah et al. 2013, Aziz et al. 2014, Samad et al. 2016).

4.5 Synthesis of ZnO Nanostructures from Chemical Vapour Deposition (CVD)

In recent years, chemical vapour deposition (CVD) has been identified to be the most prominent and well established physical technique to form ZnO nanostructures, especially the porous nanorods, at high temperature (700–900°C). Accordingly, ZnO nanostructures could be produced via the plasma-enhanced chemical vapour deposition (PE-CVD) and metal organic chemical vapour deposition (MO-CVD). In fact, CVD technique has been classified as a high cost technique due to expensive equipment are commonly required (Aziz et al. 2014, Kim Yong-Jin et al. 2009, Lee Keun Young et al. 2012). It has also been reported that temperature plays an important role in influencing the formation of ZnO nanostructures through the MO-CVD technique. The growth of ZnO nanostructures, especially in the aspect of crystal planes, energy difference and growth kinetic, would be affected by temperature. Besides, the transformation of nanostructures from conventional polycrystalline to arranged clusters of ZnO nanostructures was also observed (Khranovskyy and Yakimova 2012, Kim Kwang-Sik and Kim 2003, Saitoh et al. 1999, Sbrockey and Ganesan 2004, Tompa et al. 2006). Chi et al. added that, at high temperature (> 900°C), MO-CVD produced a better crystal quality as compared to that of those prepared at lower temperature. This is due to the fact that, under high temperature, the samples generally have higher photon emission efficiencies as a result of higher emission quantum efficiency (Chi et al. 2005).

PE-CVD provides layers of deposition at relatively low substrate temperature (200–300°C) (Dobkin and Zuraw 2003, Kern 2012). Another variables like substrate temperature, power density, gas pressure, gas composition and frequency would also affect the crystal growth and properties of metal oxide (Grimes et al. 2007).

4.6 Synthesis of ZnO Nanostructures from Atomic Layer Deposition (ALD)

The ALD technique has also illustrated its versatility in the deposition of dielectric and luminescent films for the electroluminescent flat panel displays. However, the big challenge for ALD is to establish a position in microelectronics. The accuracy in thickness control in very thin films deposition and 100 per cent conformity even on high aspect ratio structures make ALD technique to be ideal for the synthesis of ZnO (Clavel et al. 2010, Leskelä and Ritala 2002).

The basic steps to adopt ALD technique are as follows: First, the precursors need to be located in the growth chamber. When precursors reach the substrate, it will be scattered by purging the inert gas (N_2). In the meanwhile, the opening and closing of valves will be controlled by computer. The pressure in the chamber, monitored by vacuum gauge, is normally controlled at about 1–2 Torr. The temperature of the substrate is maintained at ~ 200°C for the deposition to take place. In order to achieve different thickness and crystallographic, the reaction is repeated up to 1,800 cycles (depending on the requirements) (Solís-Pomar et al. 2011).

Nowadays, plasma activation technique has increased the usability of ALD technique. It is because various precursors can be implemented by adopting this technique. Plasma-ALD could also improve the current existing process further (Leskelä and Ritala 2002). On top of that, it is worth to state that nanoparticles from the colloidal medium could be used as the precursor in this technique (Clavel et al. 2010). However, plasma-ALD suffers from limitation when the process is related to the electropositive metals (i.e., alkaline earth metals and rare earth metals) due to the lack of volatile compounds for ZnO deposition to occur (Leskelä and Ritala 2002). Unlike chemical vapour deposition, ALD proceeds through reactions solely at the surface of the substrate contributing to a self-limiting and layer-by-layer growth (Wu Mong-Kai et al. 2010). Table 2 summarises the pros and cons of the synthesis techniques to obtain ZnO nanostructures.

4.7 Crystallisation of ZnO Nanostructures

It has been reported in literatures that the crystallisation of ZnO nanostructures might affect the photocatalytic and PEC performance of ZnO nanostructures. Chen et al. 2006, Goh et al. 2011 and Cheng et al. 2007, confirmed that the as-prepared ZnO is in the amorphous phase (Abd Samad et al. 2015, Chen Zhuo et al. 2010b, Cheng et al. 2007, Goh et al. 2011, Hwang and Wu 2004, Khrypunov et al. 2011, Kim Sung Joong et al. 2008, Mao et al. 2002, Miller et al. 2014, Musić et al. 2003, Sun et al. 2012, Xu and Wang 2011, Zeng Jing Hui et al. 2009). In this respect, the heat treatment process might alter the properties (e.g., semiconducting property, photoconductivity, tunable electric property, thermal conductivity, piezoelectricity, and luminescent property) of ZnO ascribed to the crystallisation of the ZnO nanostructures.

Table 2. The advantages and disadvantages between synthesis techniques to obtain ZnO nanostructure.

Synthesis Methods	Advantages	Disadvantages
Sol-gel method	1. Processing temperature is very low (below 100°C) (Wright and Sommerdijk 2000). 2. It is a convenient production technique for any possible shape for film and bulk materials (Wright and Sommerdijk 2000). 3. Can create very fine powders (Carter and Norton 2007).	1. High cost of raw materials (Carter and Norton 2007). 2. Shrinkage and cracking occur during drying with high volume (Carter and Norton 2007).
Hydrothermal method	1. The ability to create crystalline phases which are not stable at the melting point (Liu Li et al. 2008). 2. Can process materials that have a high vapour pressure near melting point (Liu Li et al. 2008). 3. Can produce high quality of crystals while maintaining good control over composition (O'Donoghue 2006).	1. The need of expensive autoclaves (O'Donoghue 2006). 2. The impossibility of observing the crystal as it grows (O'Donoghue 2006).
Solvothermal method	1. This method is benefit of both sol-gel method and hydrothermal method (Andersson et al. 2002). 2. Having good control over the size, shape distribution, and crystallinity of nanostructures (Andersson et al. 2002).	1. The need of expensive autoclaves (O'Donoghue 2006). 2. The use of non-aqueous solution (Oliveira et al. 2003).
Chemical bath deposition	1. The use of unexpansive and precise equipment (Yi et al. 2007). 2. Allow the large scale of nanostructures production at a relatively low cost (Yi et al. 2007). 3. The low temperature growth method makes it suitable for flexible polymers, and the arrays can be readily patterned using standard synthesis procedures (Yi et al. 2007).	1. Two steps process, which co-operate with the hydrothermal process for the formation of ZnO nanorods (Yi et al. 2007).
Electrodeposition method	1. Simple, quick, economic, and able to control the crystallisation of ZnO nanostructures (Li Gao-Ren et al. 2007). 2. Low temperature condition, low equipment cost, and precise control of nanostructures (Abd-Ellah et al. 2013).	1. The solution, after use, needs to be safely disposed and could become a cause of environmental concern.
Chemical vapour deposition method (CVD)	1. Uniform distribution over large areas (Dobkin and Zuraw 2003, Plummer and Griffin 2001, Smith 1995, Wolf and Tauber 1986). 2. No compositional gradients across substrate (Dobkin and Zuraw 2003, Plummer and Griffin 2001, Smith 1995, Wolf and Tauber 1986). 3. During the sources change, there is no need to stop the vacuum (Dobkin and Zuraw 2003, Plummer and Griffin 2001, Smith 1995, Wolf and Tauber 1986).	1. Mostly involve safety and contamination (Dobkin and Zuraw 2003, Plummer and Griffin 2001, Smith 1995, Wolf and Tauber 1986). 2. Metal organics are pyrophoric (ignite in contact with air) (Dobkin and Zuraw 2003, Plummer and Griffin 2001, Smith 1995, Wolf and Tauber 1986). 3. High cost for compounds with sufficient purity (Dobkin and Zuraw 2003, Plummer and Griffin 2001, Smith 1995, Wolf and Tauber 1986).

Table 2 contd. ...

...Table 2 contd.

Synthesis Methods	Advantages	Disadvantages
	4. Due to the higher activation energy for reaction, it can be more selective in deposition (Dobkin and Zuraw 2003, Plummer and Griffin 2001, Smith 1995, Wolf and Tauber 1986). 5. MOCVD is suitable synthesised uniform ZnO nanorods in bulk quantities (Kim Kwang-Sik and Kim 2003).	4. The combination of chemical reaction and gas kinetics; lead the CVD process more complicated (Dobkin and Zuraw 2003).
Atomic layer deposition method (ALD)	1. Low processing temperature (Demmin 2001, Leskelä and Ritala 2002). 2. Accurate and facile thickness control, excellent conformality, high uniformity over a large area, good reproducibility, dense and pinhole-free structures, and low deposition temperatures (Wu Mong-Kai et al. 2010).	1. Low deposition rate (Demmin 2001). 2. The use of electropositive metals (alkaline earth metals, rare earth metals) form a challenge for chemical thin film depositions as they do not have many volatile compounds (Leskelä and Ritala 2002).

In 2007, Callister stated that heat treatment acting as a strain hardening remover, is a phenomenon that works with atomic diffusion. Also, the strengthening effect, where strain hardening and reduction in grain size takes place, could be eliminated by heat treatment process at elevated temperature. However, restoration that affects the grain growth could take place at an elevated temperature via recovery and recrystallisation. Nevertheless, heat treatment process allows the recrystallisation process to take place through the nucleation of new defect-free grains until an optimum size is achieved. Other effect of heat treatment is coring. Coring which occurs at an elevated temperature enables homogenisation to occur below the solidus point for a certain alloy or composite. Compositionally, homogenous grains could be obtained due to atomic diffusion (Callister and Rethwisch 2007).

Heat treatment process is mainly influenced by three main factors, i.e., applied temperature, heating time at an applied temperature and cooling rate to the room temperature. Heat treatment is closely related to the solid state transformation (kinetic consideration) and thereby, it influences the nucleation, growth, and transformation rates. Kinetics of solid state transformation could be measured as a function of time while the temperature is maintained as a constant (Eq. 7).

$$y = 1 - exp\ (-kt^n) \qquad (7)$$

where, y is the fraction of transformation, t is time, k & n are time-independent constant for a particular reaction.

Super-cooling and super-heating during the heat treatment are the rate of temperature change. High super-cooling and superheating will produce internal stress and temperature gradients that could lead to warping and cracking of structure (Callister and Rethwisch 2007). Considering these facts, temperature, heating rate, cooling rate and heating time need to be in equilibrium condition to achieve the best phase transformation.

Musić et al. reported that the use of precursors has a great impact on the crystallisation of ZnO as well as its properties. Specifically, the decomposition of urea produces a basic zinc carbonate. Further treatment of this basic zinc carbonate under 300°C will form ZnO. Meanwhile, the second method is hydrothermal technique. The use of $Zn_5\ (OH)_8\ (NO_3)_3\ (H_2O))_{2-x}\ (NH_3)_x$ as a complex compound will undergo hydrothermal treatment prior to ZnO formation. Musić et al. 2003, Chin and Chao 2013 and Molefe et al. concluded that the crystallisation either in hydrothermal technique or chemical bath deposition as well as radio frequency (RF) sputtering increases with increasing temperature (Chin and Chao 2013, Molefe et al. 2015, Musić et al. 2003).

According to Jeon et al. and Kim et al. 2011a, heat treatment also plays a major role in the pore formation on ZnO nanostructures. There are basically two phenomena of formation of porous ZnO during the heat treatment, i.e., desorption of zinc and oxygen at the defect sites and the decomposition of volatile gas and unreacted components. Heat treatment also provides thermal energy to the surfaces of ZnO. The migration of volatile gas and unreacted material to the ZnO surface tends to increase with increasing the thermal energy. Correspondingly, the volatile gas and unreacted material would agglomerate producing pores on the ZnO nanostructures (Jeon et al. 2010).

4.8 Modification of ZnO Nanostructures

Among the different nano-architectures, ZnO nanorod is the most studied photoelectrode in PEC water splitting application. The aforementioned shortcomings of the ZnO especially its poor visible light absorption and rapid recombination charge carrier losses hinder most of its use in electronic application. It is important to note that the utilisation of visible light from the solar energy and the reduction of the recombination losses of charge carriers are essential considering these could subsequently give rise to higher photoconversion efficiency in water-splitting applications.

4.8.1 Metal-coated ZnO Nanostructures

Lately, a number of potential modification techniques have been studied to improve the performance of porous ZnO. Coupling of ZnO with various noble metals (e.g., palladium, silver, osmium, iridium, platinum, and gold) is one of the techniques and/or strategies. Accordingly, this technique helps to prevent the corrosion and oxidation under moist air. Another technique is the coupling of ZnO semiconductor with transition metal ions (e.g., Titanium, Nickel, Cobalt, Manganese, Ferum, Copper, Chromium, and Vanadium). Specifically, transition metal ions are the metals with incomplete d-orbitals (Huang et al. 2014, Ullah and Dutta 2008, Yang Yefeng et al. 2013, Zeng Haibo et al. 2008a, Zeng Haibo et al. 2008b).

The coupling of metal with ZnO works very differently in the Fermi levels. The performance could be improved by tuning the work function of the band structure of ZnO and metal (Kapałka et al. 2010). The transfer of charge carriers is rectified in the coupled semiconductors (Gao et al. 1991, Kapałka et al. 2010, Linsebigler et al. 1995, Schierbaum et al. 1991). The electrons in the CB of the semiconductor are generated by the photon irradiations leading its Fermi level to be more negative in value (Kapałka et al. 2010, Subramanian et al. 2004). At the semi-conductor/metal interface, the energetic difference drives the electrons from semiconductor CB to metal nanoparticles giving rise to the secondary electrons transfer during the redox coupling with surrounding electrolyte (Kapałka et al. 2010, Yu et al. 2000).

4.8.2 Polyaniline-modified ZnO Nanostructures

Polyaniline (PANI) is the other coupling agent commonly used for the performance improvement of the ZnO semiconductor (Tang Qunwei et al. 2012, Zhang Hao et al. 2009b). Accordingly, PANI could act as a photo-corrosion inhibitor and photocatalytic activity improver. The separation efficiency is thereby high for photo-generated charge carriers on the hydride PANI/semiconductor interface. In the meantime, the inhibition of photo-corrosion of ZnO can take place as the PANI monolayer manages to transfer the photo-generated holes rapidly. Additionally, PANI also improves the performance of ZnO under the visible irradiation where the excited electron is delivered to CB of ZnO and the electrons are transferred to adsorbed electron acceptor. Hydroxyl radical produced will then catalyse the pollutant degradation (Zhang Hao et al. 2009b).

4.8.3 Carbon-modified ZnO Nanostructures

Outstanding performance of ZnO semiconductor could be achieved by graphite-like carbon doping or coupling (Huang et al. 2014, Zhang Liwu et al. 2009c). This is attributed to the fact that the graphite-like carbon coupled with ZnO could suppress the photo-corrosion by developing surface hybridisation. These carbon materials inhibit the coalescence and crystal growth of the ZnO at high temperature annealing as well. In addition, these carbon materials could improve the adsorption-ability and crystallinity as well as increase the reusability of this hybrid graphite-like carbon/ZnO for the photocatalysis reaction. Other advantage of this coupled ZnO is that it can perform well under an extreme pH condition (Huang et al. 2014, Mishra et al. 2013, Zhang Xinyu et al. 2015).

4.8.4 Semiconductor-modified ZnO Nanostructures

Coupling of ZnO semiconductor with other semiconductors types could also improve the photo-catalytic and PEC performance. It has been proven that the performance of ZnO (visible light absorption and recombination of charge carrier losses) could be enhanced by coupling with another semiconductor (Kim Young Kwang and Park 2011, Navarro Yerga et al. 2009). Many semiconductors such as Cadmium Selenide (CdSe), Gallium Arsenide (GaAs), Silicon Carbide (SiC), Boron Arsenide (BAs), Cadmium Sulfide (CdS), and metal oxide semiconductors could be coupled with ZnO. Charge carrier recombinations could be reduced by appropriate coupling of CB and VB of two different semiconductors. The reason is that the charge carriers will be transferred from one semiconductor to another semiconductor. This will thereby prolong the movement of charge carriers and prevent the occurrence of recombination process. Also, the composite of semiconductors can be activated in the visible region depending on the band-gap energy of the semiconductor used (Zhang Huanjun et al. 2009a). The impurity level existed due to the replacement of cationic ions within the crystal lattices facilitates better absorption in the visible region. By modifying the core semiconductor, the interfacial potential gradient will exist between materials. This corresponds to the energetic position that plays its role by providing a better charge carrier transportation and charge carrier separation (Bessegato et al. 2014, Lai et al. 2014). The electrons at a more negative CB would be injected to the more positive band when the semiconductor composites are illuminated by the photon. At the same time, holes in the more positive valence band will be moved to a more negative band. The separation of charge carrier could thereby be achieved and the lifetime of charge carriers could be increased. This will also contribute to a significant increase of interfacial charge transferring to the water (Kim et al. 2011a). It is also important to realise that an optimum content of second oxide semiconductor will enhance the performance by promoting an impurity level in the binary type semiconductors. However, an excessive second semiconductor may lead to disadvantageous scale of impurity levels and exhibit poor performance (Ghicov and Schmuki 2009, Lai et al. 2014).

4.8.5 Metal Oxide-modified ZnO Nanostructure Film as Photoelectrode

PEC water splitting solar energy conversion efficiency and photocatalytic performance are dependent on the light adsorption capability, charge separation competency, the movement of charge migration, charge recombination process and electrocatalytic activity on the surface of photoelectrode (Li Jiangtian and Wu 2015). By taking TiO_2-ZnO hybridisation for metal oxide-modified ZnO nanostructure film as an example, the TiO_2-ZnO hybridisation increases the adsorption efficiency of solar irradiation and overcomes the PEC photo-electrode inhibition of recombination by building a heterojunction. An electrons sink would then be produced from the heterojuction. Taking into account that TiO_2 and ZnO are good semiconductors and thereby, the quantum confinement effect would be generated from the nanostructures giving rise to high electron mobility. The binary oxide arises from the enrichment of second oxide on primary oxide which improves the transfer of photon energy. This is due to the fact that the photon energy is absorbed by the second oxide (Anpo et al. 1986). Anpo et al. also proved that low Ti content would enhance the photocatalytic activity in the primary oxide semiconductor (Anpo et al. 1986). Also Nakamura et al., Ihara et al., and Rehman et al. claimed that the semiconductor coupling may modify the band-gap via oxygen vacancies modification and oxygen sub-stoichiometry introduction. This particular modification helps improve the workability of the semi-conductor under the visible light (Ihara et al. 2003, Nakamura et al. 2000, Rehman et al. 2009). Grain boundaries of polycrystalline semiconductor are the place for the oxygen vacancies occur. A discrete state of about 0.75 eV and 1.18 eV could be generated by the CB of core semiconductor and thereby, the adsorption towards visible light can be improved (Nakamura et al. 2000, Rehman et al. 2009). Oxygen vacancies are well-known electron trappers. The existence of oxygen vacancies lying close to the CB of core semiconductor help increase the electron capturing and promote visible light adsorption at the surface of the semiconductor (Nakamura et al. 2000, Rehman et al. 2009). Justicia et al. also claimed that oxygen sub-stoichiometry can generate an overlap of defect states and reduce the catalyst band gap (Justicia et al. 2005).

Jlassi et al. also found that coupling a photoelectrode with another semiconductor would give better photocatalytic activity. Accordingly, TiO_2-ZnO based film illustrated a better photocatalytic efficiency than that of the pure P25 (particle size 25 nm) TiO_2. This could be explained by the increase in the efficiency of photocatalytic activity via charge-carrier lifespan improvement. Additionally, the increase in the efficiency of electron transfer from ZnO to TiO_2 could increase the photocatalytic activity. The improvement of photocatalytic activity is due to the occurrence of perturbation on the surface of composite structure (Jlassi et al. 2013).

Tian et al. revealed that crystallisation behaviour of the hybrid TiO_2-ZnO film is influenced by atomic ratio of Ti. Sol-gel technique was chosen for the formation of hybrid TiO_2-ZnO film in which the TiO_2-ZnO was directly mixed, followed by heat treatment at 500°C for 2 hour (Tian Jintao et al. 2009). Hernandez et al. studied and reported the formation of ZnO-TiO_2 nanoarrays and TiO_2-covered ZnO by using hydrothermal technique and non-acid sol-gel technique, respectively. The impregnation method was varied by about 3–10 min. From the PEC tests, the ZnO-TiO_2 materials exhibited a photocurrent density of 0.7 mA/cm^2 under solar light (AM 1.5G, 100 mW/cm^2) (Hernández et al. 2014).

Dao et al. prepared TiO_2-ZnO core shell via chemical synthesis. Specifically, Dao's team formed the TiO_2 and ZnO via the sol-gel and hydrothermal techniques, respectively. The heterostructures generated produced an enhancement in light scattering and gave better charge carriers' separation. Moreover, the switch current ratio of 140 was produced under the UV ray with reverse bias of –5 V and 250 A/W (Dao et al. 2013). Shao et al. has also discovered that high-quality ZnO-TiO_2 nanowires (NWs) could be fabricated via a facile-two steps method: hydrothermal technique for ZnO NWs and atomic layer deposition (ALD) technique for TiO_2 coating. It is noted that efficient charge carriers' separation of TiO_2-ZnO could quench the UV emission intensity and minimise/reduce the band-to-band recombination. However, it is interesting to figure out that the UV and visible light adsorption ability were improved due to the high refractive index of TiO_2. A maximum value of 495 A/W at 373 nm photoresponsitivity was detected, which was ~ 8 times higher than that of the bare ZnO NWs. The transient response was also improved by ~ 6 times compared to that of bare ZnO NWs (Shao et al. 2014).

Cheng et al. discovered that hybrid TiO_2-ZnO film could exhibit strong quenching of green emission under photoluminescence testing. This is associated with the enhancement of charge carriers' separation resulting from type II heterojunction existing near to the TiO_2-ZnO hybrid interface. On top of that, the change in E_g, high specific area and the increase in surface hydroxyl groups also provide a significant improvement in catalytic activity (Cheng et al. 2014). An optimum amount of incorporated TiO_2-based ZnO formation not only extends the lifetimes of charge carriers, but also suppresses the recombination losses effectively. The modification of ZnO could also lead to a higher photocatalytic activity and PEC performance than that of pure ZnO. Besides, the light absorption from the UV region to visible region could be enhanced as well.

Theoretically, three different types of semiconductor heterojunction could be organised by the band alignment: straddling gap (type I), staggered gap (type II), and broken gap (type III). The hybrid

TiO_2-ZnO film had a staggered gap (type II). The proposed mechanism (Cheng et al. 2014, Shao et al. 2014) is as follow. In an early stage, electrons and holes are at their lowest energy states. Those electrons and holes will then be spatially separated by the energy gradient existing at the interfaces. This phenomenon took place due to the UV ray/visible light illumination on different sides of the heterojunction. Under illumination, the electrons were transferred from the conduction band (CB) of ZnO to CB of TiO_2. Meanwhile, the holes were then transferred from the valence band (VB) of TiO_2 to VB of ZnO simultaneously. This process isolated the active electrons and holes leading to a decrease in the electron-hole pair recombination and an increase in lifespan.

Conclusion

Wide-bandgap porous nanostructured semiconductors are promising candidates for photoelectrochemical water splitting application for hydrogen generation. Design and development of porous nanostructured semiconductors by loading optimum content of secondary phases can improve the transportation of charge carriers and minimise the recombination losses thereby improving its water splitting performance. In addition, these porous nanocomposites also can be used for various optoelectronic applications such as solar cells, optical fibers, supercapacitor, and other energy generation and storage applications.

References

Abd-Ellah, M., N. Moghimi, L. Zhang, N.F. Heinig, L. Zhao, J.P. Thomas et al. 2013. Effect of electrolyte conductivity on controlled electrochemical synthesis of zinc oxide nanotubes and nanorods. J. Phys. Chem. C 117: 6794–6799.

Abd Samad, N.A., C.W. Lai and S.B. Abd Hamid. 2015. Easy formation of nanodisk-dendritic ZnO film via controlled electrodeposition process. J. Nanomater. 2015.

Abd Samad, N.A., C.W. Lai, K.S. Lau and S.B. Abd Hamid. 2016. Efficient solar-induced photoelectrochemical response using coupling semiconductor TiO_2-ZnO nanorod film. Materials 9: 937.

Albu, S.P., A. Ghicov, S. Aldabergenova, P. Drechsel, D. LeClere, G.E. Thompson et al. 2008. Formation of double-walled TiO_2 nanotubes and robust anatase membranes. Adv. Mater. 20: 4135–4139.

Andersson, M., L. Österlund, S. Ljungstroem and A. Palmqvist. 2002. Preparation of nanosize anatase and rutile TiO_2 by hydrothermal treatment of microemulsions and their activity for photocatalytic wet oxidation of phenol. J. Phys. Chem. B 106: 10674–10679.

Anpo, M., H. Nakaya, S. Kodama, Y. Kubokawa, K. Domen and T. Onishi. 1986. Photocatalysis over binary metal oxides. Enhancement of the photocatalytic activity of titanium dioxide in titanium-silicon oxides. J. Phys. Chem. A 90: 1633–1636.

Armstrong, G., A.R. Armstrong, J. Canales and P.G. Bruce. 2005. Nanotubes with the TiO_2-B structure. ChemComm. 2454–2456.

Aroutiounian, V., V. Arakelyan and G. Shahnazaryan. 2005. Metal oxide photoelectrodes for hydrogen generation using solar radiation-driven water splitting. Solar Energy 78: 581–592.

Assefpour-Dezfuly, M., C. Vlachos and E. Andrews. 1984. Oxide morphology and adhesive bonding on titanium surfaces. J. Mater. Sci. 19: 3626–3639.

Attar, A.S., M.S. Ghamsari, F. Hajiesmaeilbaigi, S. Mirdamadi, K. Katagiri and K. Koumoto. 2009. Sol–gel template synthesis and characterization of aligned anatase-TiO_2 nanorod arrays with different diameter. Mater. Chem. Phys. 113: 856–860.

Aziz, N.S.A., M.R. Mahmood, K. Yasui and A.M. Hashim. 2014. Seed/catalyst-free vertical growth of high-density electrodeposited zinc oxide nanostructures on a single-layer graphene. Nanoscale Res. Lett. 9: 1–7.

Babu, V.J., S. Vempati, T. Uyar and S. Ramakrishna. 2015. Review of one-dimensional and two-dimensional nanostructured materials for hydrogen generation. Phys. Chem. Chem. Phys. 17: 2960–2986.

Bahadur, H., A.K. Srivastava, R.K. Sharma and S. Chandra. 2007. Morphologies of sol-gel derived thin films of ZnO using different precursor materials and their nanostructures. Nanoscale Res. Lett. 2: 469–475.

Bavykin, D.V., V.N. Parmon, A.A. Lapkin and F.C. Walsh. 2004. The effect of hydrothermal conditions on the mesoporous structure of TiO_2 nanotubes. J. Mater. Chem. 14: 3370–3377.

Bessegato, G.G., T.T. Guaraldo and M.V.B. Zanoni. 2014. Enhancement of photoelectrocatalysis efficiency by using nanostructured electrodes. Modern Electrochemical Methods in Nano, Surface and Corrosion Science 271–319.

Bornand, V. and A. Mezy. 2011. An alternative approach for the oriented growth of ZnO nanostructures. Mater. Lett. 65: 1363–1366.

Callister, W.D. and D.G. Rethwisch. 2007. Materials Science and Engineering: An Introduction. Wiley New York.

Carter, C.B. and M.G. Norton. 2007. Ceramic Materials: Science and Engineering. Springer Science & Business Media.

Castelli, I.E., J.M. García-Lastra, F. Hüser, K.S. Thygesen and K.W. Jacobsen. 2013. Stability and bandgaps of layered perovskites for one-and two-photon water splitting. New J. Phys. 15: 105026.

Centi, G. and S. Perathoner. 2009. The role of nanostructure in improving the performance of electrodes for energy storage and conversion. Eur. J. Inorg. Chem. 2009: 3851–3878.

Chen, B., J. Hou and K. Lu. 2013. Formation mechanism of TiO_2 nanotubes and their applications in photoelectrochemical water splitting and supercapacitors. Langmuir 29: 5911–5919.

Chen, J.S., Y.L. Tan, C.M. Li, Y.L. Cheah, D. Luan, S. Madhavi et al. 2010a. Constructing hierarchical spheres from large ultrathin anatase TiO_2 nanosheets with nearly 100% exposed (001) facets for fast reversible lithium storage. J. Am. Chem. Soc. 132: 6124–6130.

Chen, Q.-P., M.-Z. Xue, Q.-R. Sheng, Y.-G. Liu and Z.-F. Ma. 2006. Electrochemical growth of nanopillar zinc oxide films by applying a low concentration of zinc nitrate precursor. Electrochem. Solid State Lett. 9: C58–C61.

Chen, Z., K. Shum, T. Salagaj, W. Zhang and K. Strobl. 2010b. ZnO thin films synthesized by chemical vapor deposition. Energy 8: 8–9.

Cheng, C., A. Amini, C. Zhu, Z. Xu, H. Song and N. Wang. 2014. Enhanced photocatalytic performance of TiO_2-ZnO hybrid nanostructures. Sci. Rep. 4.

Cheng, C., K.F. Yu, Y. Cai, K.K. Fung and N. Wang. 2007. Site-specific deposition of titanium oxide on zinc oxide nanorods. J. Phys. Chem. C 111: 16712–16716.

Chi, C.Y., S.C. Chin, Y.C. Lu, L. Hong, Y.L. Lin, F.Y. Jen et al. 2005. Nanostructures and optical characteristics of ZnO thin-film-like samples grown on GaN. Nanotechnology 16: 3084–3091.

Chin, H.-S. and L.-S. Chao. 2013. The effect of thermal annealing processes on structural and photoluminescence of zinc oxide thin film. J. Nanomater. 2013: 4.

Chiroma, H., S. Abdul-kareem, A. Khan, N.M. Nawi, A.Y. Gital, L. Shuib et al. 2015. Global Warming: Predicting OPEC Carbon Dioxide Emissions from Petroleum Consumption Using Neural Network and Hybrid Cuckoo Search Algorithm. Plos One 10 (art. e0136140): 21.

Chiuta, S., R.C. Everson, H. Neomagus, P. van der Gryp and D.G. Bessarabov. 2013. Reactor technology options for distributed hydrogen generation via ammonia decomposition: A review. Int. J. Hydrog. Energy 38: 14968–14991.

Cho, I.S., J. Choi, K. Zhang, S.J. Kim, M.J. Jeong, L. Cai et al. 2015. Highly efficient solar water splitting from transferred TiO_2 nanotube arrays. Nano Lett. 15: 5709–5715.

Choi, J., R.B. Wehrspohn, J. Lee and U. Gösele. 2004. Anodization of nanoimprinted titanium: A comparison with formation of porous alumina. Electrochim. Acta 49: 2645–2652.

Cicea, C., C. Marinescu, I. Popa and C. Dobrin. 2014. Environmental efficiency of investments in renewable energy: Comparative analysis at macroeconomic level. Renew. Sust. Energ. Rev. 30: 555–564.

Clarizia, L., D. Spasiano, I. Di Somma, R. Marotta, R. Andreozzi and D.D. Dionysiou. 2014. Copper modified-TiO_2 catalysts for hydrogen generation through photoreforming of organics. A short review. Int. J. Hydrog. Energy 39: 16812–16831.

Clavel, G., C. Marichy, M.G. Willinger, S. Ravaine, D. Zitoun and N. Pinna. 2010. CoFe2O4-TiO2 and CoFe$_2$O$_4$-ZnO thin film nanostructures elaborated from colloidal chemistry and atomic layer deposition. Langmuir 26: 18400–18407.

Cui, J. and U.J. Gibson. 2005. Enhanced nucleation, growth rate, and dopant incorporation in ZnO nanowires. J. Phys. Chem. B 109: 22074–22077.

Dai, S., Y. Li, Z. Du and K.R. Carter. 2013. Electrochemical deposition of ZnO hierarchical nanostructures from hydrogel coated electrodes. J. Electrochem. Soc. 160: D156–D162.

Dao, T., C. Dang, G. Han, C. Hoang, W. Yi, V. Narayanamurti et al. 2013. Chemically synthesized nanowire TiO2/ZnO core-shell pn junction array for high sensitivity ultraviolet photodetector. Appl. Phys. Lett. 103: 193119.

Demmin, J.C. 2001. Special report-process technology update: Progress on all fronts-Ulra-shallow junctions. Solid State Technology 44: 68–69.

Dobkin, D.M. and M.K. Zuraw. 2003. Principles of chemical vapor deposition. Springer.

Du, J., Z. Liu, Y. Huang, Y. Gao, B. Han, W. Li et al. 2005. Control of ZnO morphologies via surfactants assisted route in the subcritical water. J. Cryst. Growth 280: 126–134.

Elias, J., R. Tena-Zaera and C. Lévy-Clément. 2007. Electrodeposition of ZnO nanowires with controlled dimensions for photovoltaic applications: Role of buffer layer. Thin Solid Films 515: 8553–8557.

Elias, J., R. Tena-Zaera and C. Lévy-Clément. 2008. Effect of the chemical nature of the anions on the electrodeposition of ZnO nanowire arrays. J. Phys. Chem. C 112: 5736–5741.

Erne, B., D. Vanmaekelbergh and J. Kelly. 1996. Morphology and strongly enhanced photoresponse of GaP electrodes made porous by anodic etching. J. Electrochem. Soc. 143: 305–314.

Espitia, P.J.P., N.dF.F. Soares, J.S. dos Reis Coimbra, N.J. de Andrade, R.S. Cruz and E.A.A. Medeiros. 2012. Zinc oxide nanoparticles: Synthesis, antimicrobial activity and food packaging applications. Food Bioproc. Tech. 5: 1447–1464.

Fernandes, J.A., S. Khan, F. Baum, E.C. Kohlrausch, J.A.L. dos Santos, D.L. Baptista et al. 2016. Synergizing nanocomposites of CdSe/TiO_2 nanotubes for improved photoelectrochemical activity via thermal treatment. Dalton Transactions.

Fujishima, A., T.N. Rao and D.A. Tryk. 2000. Titanium dioxide photocatalysis. J. Photochem. Photobiol. C: Photochemistry Reviews 1: 1–21.

Gao, X., A. Hamelin and M.J. Weaver. 1991. Atomic relaxation at ordered electrode surfaces probed by scanning tunneling microscopy: Au (111) in aqueous solution compared with ultrahigh-vacuum environments. J. Chem. Phys. 95: 6993–6996.

Ge, M., C. Cao, J. Huang, S. Li, Z. Chen, K.-Q. Zhang et al. 2016. A review of one-dimensional TiO_2 nanostructured materials for environmental and energy applications. J. Mater. Chem. A.

Ge, M., C. Cao, S. Li, S. Zhang, S. Deng, J. Huang et al. 2015. Enhanced photocatalytic performances of n-TiO_2 nanotubes by uniform creation of p–n heterojunctions with p-Bi_2O_3 quantum dots. Nanoscale 7: 11552–11560.

George, L., S. Sappati, P. Ghosh and R.N. Devi. 2015. Surface site modulations by conjugated organic molecules to enhance visible light activity of ZnO nanostructures in photocatalytic water splitting. J. Phys. Chem. C 119: 3060–3067.

Ghicov, A. and P. Schmuki. 2009. Self-ordering electrochemistry: A review on growth and functionality of TiO_2 nanotubes and other self-aligned MO x structures. ChemComm. 2791–2808.

Goh, H.-S., R. Adnan and M.A. Farrukh. 2011. ZnO nanoflake arrays prepared via anodization and their performance in the photodegradation of methyl orange. Turk. J. Chem. 35: 375–391.

Grimes, C.A. 2007. Synthesis and application of highly ordered arrays of TiO_2 nanotubes. J. Mater. Chem. 17: 1451–1457.

Grimes, C.A., O.K. Varghese and S. Ranjan. 2007. Light, water, hydrogen: The solar generation of hydrogen by water photoelectrolysis. Springer Science & Business Media.

Hagen, D.J., I.M. Povey, S. Rushworth, J.S. Wrench, L. Keeney, M. Schmidt et al. 2014. Atomic layer deposition of Cu with a carbene-stabilized Cu (i) silylamide. J. Mater. Chem. C. 2: 9205–9214.

Han, G.S., S. Lee, J.H. Noh, H.S. Chung, J.H. Park, B.S. Swain et al. 2014. 3-D TiO_2 nanoparticle/ITO nanowire nanocomposite antenna for efficient charge collection in solid state dye-sensitized solar cells. Nanoscale 6: 6127–6132.

Hernández, S., V. Cauda, D. Hidalgo, V.F. Rivera, D. Manfredi, A. Chiodoni et al. 2014. Fast and low-cost synthesis of 1D ZnO–TiO_2 core–shell nanoarrays: characterization and enhanced photo-electrochemical performance for water splitting. J. Alloys Compd. 615: S530–S537.

Hoa, N.T.Q. and D.N. Huyen. 2013. Comparative study of room temperature ferromagnetism in undoped and Ni-doped TiO_2 nanowires synthesized by solvothermal method. J. Mater. Sci. Mater. Electron. 24: 793–798.

Hoyer, P. 1996. Formation of a titanium dioxide nanotube array. Langmuir 12: 1411–1413.

Hu, J.A., F. Zhu, I. Matulionis, T. Deutsch, N. Gaillard, E. Miller et al. 2009. Development of a hybrid photoelectrochemical (PEC) device with amorphous silicon carbide as the photoelectrode for water splitting. pp. 121–126. *In*: ShahedipourSandvik, F., E.F. Schubert, L.D. Bell, V. Tilak and A.W. Bett (eds.). Compound Semiconductors for Energy Applications and Environmental Sustainability, vol. 1167. Warrendale: Materials Research Society.

Hu, Y.H. 2012. A highly efficient photocatalyst—hydrogenated black TiO_2 for the photocatalytic splitting of water. Angew. Chem. Int. Ed. 51: 12410–12412.

Huang, K., Y. Li, S. Lin, C. Liang, H. Wang, C. Ye et al. 2014. A facile route to reduced graphene oxide–zinc oxide nanorod composites with enhanced photocatalytic activity. Powder Technol. 257: 113–119.

Hwang, C.-C. and T.-Y. Wu. 2004. Synthesis and characterization of nanocrystalline ZnO powders by a novel combustion synthesis method. Mater. Sci. Eng., B 111: 197–206.

Ihara, T., M. Miyoshi, Y. Iriyama, O. Matsumoto and S. Sugihara. 2003. Visible-light-active titanium oxide photocatalyst realized by an oxygen-deficient structure and by nitrogen doping. Appl. Catal., B 42: 403–409.

Iulianelli, A., P. Ribeirinha, A. Mendes and A. Basile. 2014. Methanol steam reforming for hydrogen generation via conventional and membrane reactors: A review. Renew. Sust. Energ. Rev. 29: 355–368.

Izumi, Y., T. Itoi, S. Peng, K. Oka and Y. Shibata. 2009. Site structure and photocatalytic role of sulfur or nitrogen-doped titanium oxide with uniform mesopores under visible light. J. Phys. Chem. C 113: 6706–6718.

Jeon, S.M., M.S. Kim, M.Y. Cho, H.Y. Choi, K.G. Yim, G.S. Kim et al. 2010. Fabrication of porous ZnO nanorods with nano-sized pores and their properties. J. Korean Phys. Soc. 57: 1477–1481.

Ji, Y., M. Zhang, J. Cui, K.-C. Lin, H. Zheng, J.-J. Zhu et al. 2012. Highly-ordered TiO_2 nanotube arrays with double-walled and bamboo-type structures in dye-sensitized solar cells. Nano Energy 1: 796–804.

Jitputti, J., Y. Suzuki and S. Yoshikawa. 2008. Synthesis of TiO_2 nanowires and their photocatalytic activity for hydrogen evolution. Catal. Commun. 9: 1265–1271.

Jlassi, M., H. Chorfi, M. Saadoun and B. Bessaïs. 2013. ZnO ratio-induced photocatalytic behavior of TiO_2–ZnO nanocomposite. Superlattices Microstruct. 62: 192–199.

Joo, J., S.G. Kwon, T. Yu, M. Cho, J. Lee, J. Yoon et al. 2005. Large-scale synthesis of TiO_2 nanorods via nonhydrolytic sol-gel ester elimination reaction and their application to photocatalytic inactivation of *E. coli*. J. Phys. Chem. B 109: 15297–15302.

Justicia, I., G. García, G. Battiston, R. Gerbasi, F. Ager, M. Guerra et al. 2005. Photocatalysis in the visible range of sub-stoichiometric anatase films prepared by MOCVD. Electrochim. Acta 50: 4605–4608.

Kapałka, A., G. Fóti, C. Comninellis and G. Chen. 2010. Electrochemistry for the Environment.

Kasuga, T., M. Hiramatsu, A. Hoson, T. Sekino and K. Niihara. 1998. Formation of titanium oxide nanotube. Langmuir 14: 3160–3163.

Kasuga, T., M. Hiramatsu, A. Hoson, T. Sekino and K. Niihara. 1999. Titania nanotubes prepared by chemical processing. Adv. Mater. 11: 1307–1311.

Kaygusuz, K. 2009. Energy and environmental issues relating to greenhouse gas emissions for sustainable development in Turkey. Renew Sust. Energ. Rev. 13: 253–270.

Kern, W. 2012. Thin film processes II. Academic press.

Khare, C., K. Sliozberg, R. Meyer, A. Savan, W. Schuhmann and A. Ludwig. 2013. Layered WO_3/TiO_2 nanostructures with enhanced photocurrent densities. Int. J. Hydrog. Energy 38: 15954–15964.

Khranovskyy, V. and R. Yakimova. 2012. Morphology engineering of ZnO nanostructures. Physica B Condens. Matter. 407: 1533–1537.

Khrypunov, G., N. Klochko, N. Volkova, V. Kopach, V. Lyubov and K. Klepikova. 2011. Pulse and direct current electrodeposition of zinc oxide layers for solar cells with extra thin absorbers. pp. 8–13. World Renewable Energy Congress WREC-2011, Linköping, Sweden.

Kim, C.H., Y.C. Park, J. Lee, W.S. Shin, S.J. Moon, J. Park et al. 2011a. Hybrid nanostructures of titanium-decorated ZnO nanowires. Mater. Lett. 65: 1548–1551.

Kim, K.-S. and H.W. Kim. 2003. Synthesis of ZnO nanorod on bare Si substrate using metal organic chemical vapor deposition. Physica. B Condens Matter. 328: 368–371.

Kim, M.S., K.G. Yim, S.M. Jeon, D.-Y. Lee, J.S. Kim, J.S. Kim et al. 2011b. Photoluminescence studies of porous ZnO nanorods. Jpn. J. Appl. Phys. 50: 035003.

Kim, S.J., J. Lee and J. Choi. 2008. Understanding of anodization of zinc in an electrolyte containing fluoride ions. Electrochim. Acta 53: 7941–7945.

Kim, Y.-J., J.-H. Lee and G.-C. Yi. 2009. Vertically aligned ZnO nanostructures grown on graphene layers. Appl. Phys. Lett. 95: 213101.

Kim, Y.J., Hadiyawarman, A. Yoon, M. Kim, G.C. Yi and C. Liu. 2011c. Hydrothermally grown ZnO nanostructures on few-layer graphene sheets. Nanotechnology 22 (art. 245603): 8.

Kim, Y.K. and H. Park. 2011. Light-harvesting multi-walled carbon nanotubes and CdS hybrids: Application to photocatalytic hydrogen production from water. Energy Environ. Sci. 4: 685–694.

Kitano, M., M. Matsuoka, M. Ueshima and M. Anpo. 2007. Recent developments in titanium oxide-based photocatalysts. Appl. Catal., A 325: 1–14.

Kochergin, V. and H. Föll. 2009. Porous semiconductors: Optical properties and applications. Springer Science & Business Media.

Könenkamp, R., K. Boedecker, M.C. Lux-Steiner, M. Poschenrieder, F. Zenia, C. Levy-Clement et al. 2000. Thin film semiconductor deposition on free-standing ZnO columns. Appl. Phys. Lett. 77: 2575–2577.

Kreuter, W. and H. Hofmann. 1998. Electrolysis: The important energy transformer in a world of sustainable energy. Int. J. Hydrog. Energy 23: 661–666.

Kudo, A. and Y. Miseki. 2009. Heterogeneous photocatalyst materials for water splitting. Chem. Soc. Rev. 38: 253–278.

Lai, C.W., J.C. Juan, W.B. Ko and S. Bee Abd Hamid. 2014. An overview: Recent development of titanium oxide nanotubes as photocatalyst for dye degradation. Int. J. Photoenergy 2014.

Lakshmi, B.B., P.K. Dorhout and C.R. Martin. 1997. Sol-gel template synthesis of semiconductor nanostructures. Chem. Mater. 9: 857–862.

Lee, C.S., J.K. Kim, J.Y. Lim and J.H. Kim. 2014a. One-step process for the synthesis and deposition of anatase, two-dimensional, disk-shaped TiO$_2$ for dye-sensitized solar cells. ACS Appl. Mater. Interfaces 6: 20842–20850.

Lee, D.C., I. Robel, J.M. Pietryga and V.I. Klimov. 2010. Infrared-active heterostructured nanocrystals with ultralong carrier lifetimes. J. Am. Chem. Soc. 132: 9960–9962.

Lee, J., D.H. Kim, S.-H. Hong and J.Y. Jho. 2011. A hydrogen gas sensor employing vertically aligned TiO$_2$ nanotube arrays prepared by template-assisted method. Sens. Actuators, B 160: 1494–1498.

Lee, K.Y., B. Kumar, H.-K. Park, W.M. Choi, J.-Y. Choi and S.-W. Kim. 2012. Growth of high quality ZnO nanowires on graphene. J. Nanosci. Nanotechnol. 12: 1551–1554.

Lee, S., K. Lee, W.D. Kim, S. Lee, D.J. Shin and D.C. Lee. 2014b. Thin amorphous TiO$_2$ shell on CdSe nanocrystal quantum dots enhances photocatalysis of hydrogen evolution from water. J. Phys. Chem. C 118: 23627–23634.

Lepot, N., M. Van Bael, H. Van den Rul, J. D'Haen, R. Peeters, D. Franco et al. 2007. Synthesis of ZnO nanorods from aqueous solution. Mater. Lett. 61: 2624–2627.

Leskelä, M. and M. Ritala. 2002. Atomic layer deposition (ALD): From precursors to thin film structures. Thin Solid Films 409: 138–146.

Leung, D.Y., X. Fu, C. Wang, M. Ni, M.K. Leung, X. Wang et al. 2010. Hydrogen production over titania-based photocatalysts. ChemSusChem. 3: 681–694.

Lévy-Clément, C., R. Tena-Zaera, M.A. Ryan, A. Katty and G. Hodes. 2005. CdSe-sensitized p-CuSCN/nanowire n-ZnO heterojunctions. Adv. Mater. 17: 1512–1515.

Li, G.-R., X.-H. Lu, D.-L. Qu, C.-Z. Yao, F.-l. Zheng, Q. Bu et al. 2007. Electrochemical growth and control of ZnO dendritic structures. J. Phys. Chem. C 111: 6678–6683.

Li, H., Z. Chen, C.K. Tsang, Z. Li, X. Ran, C. Lee et al. 2014. Electrochemical doping of anatase TiO$_2$ in organic electrolytes for high-performance supercapacitors and photocatalysts. J. Mater. Chem. A 2: 229–236.

Li, H., Y. Lai, J. Huang, Y. Tang, L. Yang, Z. Chen et al. 2015. Multifunctional wettability patterns prepared by laser processing on superhydrophobic TiO$_2$ nanostructured surfaces. J. Mater. Chem. B 3: 342–347.

Li, J. and N. Wu. 2015. Semiconductor-based photocatalysts and photoelectrochemical cells for solar fuel generation: A review. Catal. Sci. Technol. 5: 1360–1384.

Lieber, C.M. 1998. One-dimensional nanostructures: Chemistry, physics & applications. Solid State Commun. 107: 607–616.

Lim, S.H., J. Luo, Z. Zhong, W. Ji and J. Lin. 2005. Room-temperature hydrogen uptake by TiO$_2$ nanotubes. Inorg. Chem. 44: 4124–4126.

Linsebigler, A.L., G. Lu and J.T. Yates Jr. 1995. Photocatalysis on TiO$_2$ surfaces: principles, mechanisms, and selected results. Chem. Rev. 95: 735–758.

Liu, B., K. Nakata, M. Sakai, H. Saito, T. Ochiai, T. Murakami et al. 2012. Hierarchical TiO$_2$ spherical nanostructures with tunable pore size, pore volume, and specific surface area: Facile preparation and high-photocatalytic performance. Catal. Sci. Technol. 2: 1933–1939.

Liu, B.H. and Z.P. Li. 2009. A review: Hydrogen generation from borohydride hydrolysis reaction. J. Power Sources 187: 527–534.

Liu, L., S. Ryu, M.R. Tomasik, E. Stolyarova, N. Jung, M.S. Hybertsen et al. 2008. Graphene oxidation: Thickness-dependent etching and strong chemical doping. Nano Lett. 8: 1965–1970.

Lv, J., H. Wang, H. Gao, G. Xu, D. Wang, Z. Chen et al. 2015. A research on the visible light photocatalytic activity and kinetics of CdS/CdSe co-modified TiO_2 nanotube arrays. Surf. Coat. Technol. 261: 356–363.

Macak, J., H. Tsuchiya, A. Ghicov, K. Yasuda, R. Hahn, S. Bauer et al. 2007. TiO_2 nanotubes: Self-organized electrochemical formation, properties and applications. Curr. Opin. Solid State Mater. Sci. 11: 3–18.

Mao, D., X. Wang, W. Li, X. Liu, Q. Li and J. Xu. 2002. Electron field emission from hydrogen-free amorphous carbon-coated ZnO tip array. J. Vac. Sci. Technol. 20: 278–281.

Martin, D.J., K. Qiu, S.A. Shevlin, A.D. Handoko, X. Chen, Z. Guo et al. 2014. Highly efficient photocatalytic H_2 evolution from water using visible light and structure-controlled graphitic carbon nitride. Angew. Chem. Int. Ed. 53: 9240–9245.

Martin, D.J., N. Umezawa, X. Chen, J. Ye and J. Tang. 2013. Facet engineered Ag_3PO_4 for efficient water photooxidation. Energy Environ. Sci. 6: 3380–3386.

Martinson, A.B., J.E. McGarrah, M.O. Parpia and J.T. Hupp. 2006. Dynamics of charge transport and recombination in ZnO nanorod array dye-sensitized solar cells. Phys. Chem. Chem. Phys. 8: 4655–4659.

Miao, J.W., H.B. Yang, S.Y. Khoo and B. Liu. 2013. Electrochemical fabrication of ZnO-CdSe core-shell nanorod arrays for efficient photoelectrochemical water splitting. Nanoscale 5: 11118–11124.

Miller, D.R., S.A. Akbar and P.A. Morris. 2014. Nanoscale metal oxide-based heterojunctions for gas sensing: A review. Sens Actuators B Chem. 204: 250–272.

Mishra, D., J. Mohapatra, M. Sharma, R. Chattarjee, S. Singh, S. Varma et al. 2013. Carbon doped ZnO: Synthesis, characterization and interpretation. J. Magn. Magn. Mater. 329: 146–152.

Misra, M. and K. Raja. 2010. Ordered titanium dioxide nanotubular arrays as photoanodes for hydrogen generation. John Wiley & Sons: Chichester, UK.

Molefe, F.V., L.F. Koao, B.F. Dejene and H.C. Swart. 2015. Phase formation of hexagonal wurtzite ZnO through decomposition of $Zn(OH)_2$ at various growth temperatures using CBD method. Opt. Mater. 46: 292–298.

Mollar, M., J. Cembrero, M. Perales, M. Pascual and B. Marí. 2006. Obtención de columnas de znO.: Variables a controlar (y II). Boletín de la Sociedad Española de Cerámica y Vidrio 45: 278–282.

Momeni, M.M. and Y. Ghayeb. 2015. Visible light-driven photoelectrochemical water splitting on ZnO–TiO_2 heterogeneous nanotube photoanodes. J. Appl. Electrochem. 45: 557–566.

Momeni, M.M. and Y. Ghayeb. 2016. Fabrication, characterization and photoelectrochemical performance of chromium-sensitized titania nanotubes as efficient photoanodes for solar water splitting. J Solid State Electrochem. 20: 683–689.

Moniz, S.J., S.A. Shevlin, D.J. Martin, Z.-X. Guo and J. Tang. 2015. Visible-light driven heterojunction photocatalysts for water splitting—A critical review. Energy Environ. Sci. 8: 731–759.

Mor, G.K., O.K. Varghese, M. Paulose, K. Shankar and C.A. Grimes. 2006. A review on highly ordered, vertically oriented TiO_2 nanotube arrays: Fabrication, material properties, and solar energy applications. Sol. Energy Mater Sol. Cells 90: 2011–2075.

Morgan, D.L., H.-Y. Zhu, R.L. Frost and E.R. Waclawik. 2008. Determination of a morphological phase diagram of titania/titanate nanostructures from alkaline hydrothermal treatment of Degussa P25. Chem. Mater. 20: 3800–3802.

Musić, S., Đ. Dragčević, M. Maljković and S. Popović. 2003. Influence of chemical synthesis on the crystallization and properties of zinc oxide. Mater. Chem. Phys. 77: 521–530.

Nakahira, A., T. Kubo and C. Numako. 2010. Formation mechanism of TiO_2-derived titanate nanotubes prepared by the hydrothermal process. Inorg. Chem. 49: 5845–5852.

Nakamura, I., N. Negishi, S. Kutsuna, T. Ihara, S. Sugihara and K. Takeuchi. 2000. Role of oxygen vacancy in the plasma-treated TiO_2 photocatalyst with visible light activity for NO removal. J. Mol. Catal. Chem. 161: 205–212.

Nam, C.T., J.L. Falconer and W.-D. Yang. 2014. Morphology, structure and adsorption of titanate nanotubes prepared using a solvothermal method. Mater. Res. Bull. 51: 49–55.

Navarro Yerga, R.M., M.C. Álvarez Galván, F. Del Valle, J.A. Villoria de la Mano and J.L. Fierro. 2009. Water splitting on semiconductor catalysts under visible-light irradiation. ChemSusChem. 2: 471–485.

Ni, M., M.K. Leung, D.Y. Leung and K. Sumathy. 2007. A review and recent developments in photocatalytic water-splitting using TiO_2 for hydrogen production. Renew. Sust. Energ. Rev. 11: 401–425.

O'Donoghue, M. 2006. Gems: Their sources, descriptions and identification. Butterworth-Heinemann.

Okamoto, H., Y. Sugiyama and H. Nakano. 2011. Synthesis and modification of silicon nanosheets and other silicon nanomaterials. Chem. Eur. J 17: 9864–9887.

Oliveira, M.M., D.C. Schnitzler and A.J. Zarbin. 2003. (Ti, Sn) O_2 mixed oxides nanoparticles obtained by the sol-gel route. Chem. Mater. 15: 1903–1909.

Ozyurt, O. 2010. Energy issues and renewables for sustainable development in Turkey. Renew. Sust. Energ. Rev. 14: 2976–2985.

Park, S.-S., S.-M. Eom, M. Anpo, D.-H. Seo, Y. Jeon and Y.-G. Shul. 2011. N-doped anodic titania nanotube arrays for hydrogen production. Korean J Chem. Eng. 28: 1196–1199.

Park, W.I., D. Kim, S.-W. Jung and G.-C. Yi. 2002. Metalorganic vapor-phase epitaxial growth of vertically well-aligned ZnO nanorods. Appl. Phys. Lett. 80: 4232–4234.

Parthasarathy, P. and K.N. Sheeba. 2015. Combined slow pyrolysis and steam gasification of biomass for hydrogen generation—a review. Int. J. Energy Res. 39: 147–164.

Paulose, M., H.E. Prakasam, O.K. Varghese, L. Peng, K.C. Popat, G.K. Mor et al. 2007. TiO_2 nanotube arrays of 1000 μm length by anodization of titanium foil: Phenol red diffusion. J. Phys. Chem. C 111: 14992–14997.

Peighambardoust, S., S. Rowshanzamir and M. Amjadi. 2010. Review of the proton exchange membranes for fuel cell applications. Int. J. Hydrog. Energy 35: 9349–9384.

Peng, X. and A. Chen. 2006. Large-scale synthesis and characterization of TiO_2-based nanostructures on Ti substrates. Adv. Funct. Mater. 16: 1355–1362.

Plummer, J.D. and P.B. Griffin. 2001. Material and process limits *in silicon* VLSI technology. Proceedings of the IEEE 89: 240–258.

Pu, Y.-C., G. Wang, K.-D. Chang, Y. Ling, Y.-K. Lin, B.C. Fitzmorris et al. 2013. Au nanostructure-decorated TiO_2 nanowires exhibiting photoactivity across entire UV-visible region for photoelectrochemical water splitting. Nano Lett. 13: 3817–3823.

Qiu, J., W. Yu, X. Gao and X. Li. 2006. Sol–gel assisted ZnO nanorod array template to synthesize TiO_2 nanotube arrays. Nanotechnology 17: 4695.

Ramírez, D., H. Gómez, G. Riveros, R. Schrebler, R. Henríquez and D. Lincot. 2010. Effect of Zn (II) concentration on the morphology of zinc oxide nanorods during electrodeposition on very thin alumina membrane templates. J. Phys. Chem. C 114: 14854–14859.

Ran, J., J. Zhang, J. Yu, M. Jaroniec and S.Z. Qiao. 2014. Earth-abundant cocatalysts for semiconductor-based photocatalytic water splitting. Chem. Soc. Rev. 43:7787–7812.

Rao, C.N.R., A. Müller and A.K. Cheetham. 2006. The chemistry of nanomaterials: Synthesis, properties and applications. John Wiley & Sons.

Rehman, S., R. Ullah, A. Butt and N. Gohar. 2009. Strategies of making TiO_2 and ZnO visible light active. J. Hazard. Mater. 170: 560–569.

Reyes-Gil, K.R. and D.B. Robinson. 2013. WO_3-enhanced TiO_2 nanotube photoanodes for solar water splitting with simultaneous wastewater treatment. ACS Appl. Mater. Interfaces 5: 12400–12410.

Roy, P., S. Berger and P. Schmuki. 2011. TiO_2 nanotubes: Synthesis and applications. Angew. Chem. Int. Ed. 50: 2904–2939.

Saitoh, H., M. Satoh, N. Tanaka, Y. Ueda and S. Ohshio. 1999. Homogeneous growth of zinc oxide whiskers. Jpn. J. Appl. Phys. 38: 6873.

Samad, N.A.A., C.W. Lai and S.B.A. Hamid. 2016. Influence Applied Potential on the Formation of Self-Organized ZnO Nanorod Film and its Photoelectrochemical Response.

Sander, M.S., M.J. Cote, W. Gu, B.M. Kile and C.P. Tripp. 2004. Template-assisted fabrication of dense, aligned arrays of titania nanotubes with well-controlled dimensions on substrates. Adv. Mater. 16: 2052–2057.

Sbrockey, N.M. and S. Ganesan. 2004. ZnO thin films by MOCVD. III-Vs Review 17: 23–25.

Schierbaum, K., U. Kirner, J. Geiger and W. Göpel. 1991. Schottky-barrier and conductivity gas sensors based upon Pd/SnO 2 and Pt/TiO_2. Sens. Actuator B-Chem. 4: 87–94.

Shao, D., H. Sun, G. Xin, J. Lian and S. Sawyer. 2014. High quality ZnO–TiO_2 core–shell nanowires for efficient ultraviolet sensing. Appl. Surf. Sci. 314: 872–876.

Shin, H., D.K. Jeong, J. Lee, M.M. Sung and J. Kim. 2004. Formation of TiO_2 and ZrO_2 nanotubes using atomic layer deposition with ultraprecise control of the wall thickness. Adv. Mater. 16: 1197–1200.

Smith, D.L. 1995. Thin-Film Deposition: Principles and Practice. McGraw-hill New York, etc.

Solanki, C.S. 2015. Solar Photovoltaics: Fundamentals, Technologies and Applications. PHI Learning Pvt. Ltd.

Solís-Pomar, F., E. Martínez, M.F. Meléndrez and E. Pérez-Tijerina. 2011. Growth of vertically aligned ZnO nanorods using textured ZnO films. Nanoscale Res. Lett. 6: 1–11.

Spadavecchia, F., S. Ardizzone, G. Cappelletti, L. Falciola, M. Ceotto and D. Lotti. 2013. Investigation and optimization of photocurrent transient measurements on nano-TiO_2. J. Appl. Electrochem. 43: 217–225.

Su, L., J. Lv, H. Wang, L. Liu, G. Xu, D. Wang et al. 2014. Improved visible light photocatalytic activity of CdSe modified TiO_2 nanotube arrays with different intertube spaces. Catal. Lett. 144: 553–560.

Subramanian, V., E.E. Wolf and P.V. Kamat. 2004. Catalysis with TiO2/gold nanocomposites. Effect of metal particle size on the Fermi level equilibration. J. Am. Chem. Soc. 126: 4943–4950.

Sun, Q., Y. Lu, Y. Xia, D. Yang, J. Li and Y. Liu. 2012. Flame retardancy of wood treated by TiO_2/ZnO coating. Surf. Eng. 28: 555–559.

Suzuki, Y. and S. Yoshikawa. 2004. Synthesis and thermal analyses of TiO_2-derived nanotubes prepared by the hydrothermal method. J. Mater. Res. 19: 982–985.

Tan, A., B. Pingguan-Murphy, R. Ahmad and S. Akbar. 2012. Review of titania nanotubes: Fabrication and cellular response. Ceram Int. 38: 4421–4435.

Tanaka, S.-i., N. Hirose and T. Tanaki. 2005. Effect of the temperature and concentration of NaOH on the formation of porous TiO_2. J. Electrochem. Soc. 152: C789–C794.

Tang, Q., L. Lin, X. Zhao, K. Huang and J. Wu. 2012. p–n heterojunction on ordered ZnO nanowires/polyaniline microrods double array. Langmuir 28: 3972–3978.

Tang, Y., Y. Lai, D. Gong, K.H. Goh, T.T. Lim, Z. Dong et al. 2010. Ultrafast synthesis of layered titanate microspherulite particles by electrochemical spark discharge spallation. Chem. Eur. J. 16: 7704–7708.

Taveira, L., J. Macak, H. Tsuchiya, L. Dick and P. Schmuki. 2005. Initiation and growth of self-organized TiO_2 nanotubes anodically formed in $NH_4F/(NH_4)_2SO_4$ electrolytes. J. Electrochem. Soc. 152: B405–B410.

Tena-Zaera, R., J. Elias, G. Wang and C. Levy-Clement. 2007. Role of chloride ions on electrochemical deposition of ZnO nanowire arrays from O_2 reduction. J. Phys. Chem. C 111: 16706–16711.

Tena-Zaera, R., A. Katty, S. Bastide, C. Lévy-Clément, B. O'Regan and V. Muñoz-Sanjosé. 2005. ZnO/CdTe/CuSCN, a promising heterostructure to act as inorganic eta-solar cell. Thin Solid Films 483: 372–377.

Tian, J., L. Chen, Y. Yin, X. Wang, J. Dai, Z. Zhu et al. 2009. Photocatalyst of TiO_2/ZnO nanocomposite film: preparation, characterization, and photodegradation activity of methyl orange. Surf. Coat. Technol. 204: 205–214.

Tian, J., Z. Zhao, A. Kumar, R.I. Boughton and H. Liu. 2014. Recent progress in design, synthesis, and applications of one-dimensional TiO_2 nanostructured surface heterostructures: A review. Chem. Soc. Rev. 43: 6920–6937.

Tompa, G.S., S. Sun, L. Provost, D. Mentel, D. Sugrim, P. Chan et al. 2006. Large Area Multi-Wafer MOCVD of Transparent and Conducting ZnO Films. Pages 0957-K0909-0905. MRS Proceedings: Cambridge Univ Press.

Tong, H., S. Ouyang, Y. Bi, N. Umezawa, M. Oshikiri and J. Ye. 2012. Nano-photocatalytic materials: Possibilities and challenges. Adv. Mater. 24: 229–251.

Ullah, R. and J. Dutta. 2008. Photocatalytic degradation of organic dyes with manganese-doped ZnO nanoparticles. J. Hazard. Mater. 156: 194–200.

Varghese, N., L. Panchakarla, M. Hanapi, A. Govindaraj and C. Rao. 2007. Solvothermal synthesis of nanorods of ZnO, N-doped ZnO and CdO. Mater. Res. Bull. 42: 2117–2124.

Wang, P., Y. Ao, C. Wang, J. Hou and J. Qian. 2013. Preparation of graphene-modified TiO_2 nanorod arrays with enhanced photocatalytic activity by a solvothermal method. Mater. Lett. 101: 41–43.

Wang, Q., Z. Wen and J. Li. 2006. Solvent-controlled synthesis and electrochemical lithium storage of one-dimensional TiO_2 nanostructures. Inorg. Chem. 45: 6944–6949.

Wehrspohn, R.B. 2005. Ordered Porous Nanostructures and Applications. Springer.

Wolf, S. and R.N. Tauber. 1986. Silicon Processing for the VLSI Era vol. 1: Process technology. Lattice Press.

Wong, M., A. Berenov, X. Qi, M. Kappers, Z. Barber, B. Illy et al. 2003. Electrochemical growth of ZnO nano-rods on polycrystalline Zn foil. Nanotechnology 14: 968.

Wright, J.D. and N.A. Sommerdijk. 2000. Sol-Gel Materials: Chemistry and Applications. CRC press.

Wu, F., X. Hu, J. Fan, E. Liu, T. Sun, L. Kang et al. 2013. Photocatalytic activity of Ag/TiO_2 nanotube arrays enhanced by surface plasmon resonance and application in hydrogen evolution by water splitting. Plasmonics 8: 501–508.

Wu, M.-K., M.-J. Chen, F.-Y. Tsai, J.-R. Yang and M. Shiojiri. 2010. Fabrication of ZnO nanopillars by atomic layer deposition. Mater. Trans. 51: 253–255.

Wu, Z., Y. Wang, L. Sun, Y. Mao, M. Wang and C. Lin. 2014. An ultrasound-assisted deposition of NiO nanoparticles on TiO_2 nanotube arrays for enhanced photocatalytic activity. J. Mater. Chem. A 2: 8223–8229.

Xia, Y., P. Yang, Y. Sun, Y. Wu, B. Mayers, B. Gates et al. 2003. One-dimensional nanostructures: Synthesis, characterization, and applications. Adv. Mater. 15: 353–389.

Xu, S. and Z.L. Wang. 2011. One-dimensional ZnO nanostructures: Solution growth and functional properties. Nano Res. 4: 1013–1098.

Yan, S., L. Wan, Z. Li and Z. Zou. 2011. Facile temperature-controlled synthesis of hexagonal Zn_2GeO_4 nanorods with different aspect ratios toward improved photocatalytic activity for overall water splitting and photoreduction of CO_2. ChemComm. 47: 5632–5634.

Yang, M., Y. Ji, W. Liu, Y. Wang and X. Liu. 2014. Facile microwave-assisted synthesis and effective photocatalytic hydrogen generation of Zn_2GeO_4 with different morphology. RSC Adv. 4: 15048–15054.

Yang, M., N.K. Shrestha and P. Schmuki. 2010. Self-organized CdS microstructures by anodization of Cd in chloride containing Na_2S solution. Electrochim. Acta 55: 7766–7771.

Yang, Y., Y. Li, L. Zhu, H. He, L. Hu, J. Huang et al. 2013. Shape control of colloidal Mn doped ZnO nanocrystals and their visible light photocatalytic properties. Nanoscale 5: 10461–10471.

Yao, H., W. Fu, H. Yang, J. Ma, M. Sun, Y. Chen et al. 2014. Vertical growth of two-dimensional TiO_2 nanosheets array films and enhanced photoelectrochemical properties sensitized by CdS quantum dots. Electrochim. Acta 125: 258–265.

Yi, S.H., S.K. Choi, J.M. Jang, J.A. Kim and W.G. Jung. 2007. Low-temperature growth of ZnO nanorods by chemical bath deposition. J. Colloid Interface Sci. 313: 705–710.

Yiamsawas, D., K. Boonpavanitchakul and W. Kangwansupamonkon. 2009. Preparation of ZnO nanostructures by solvothermal method. J. Microscopy Society of Thailand 23: 75–78.

Yin, S.F., B.Q. Xu, X.P. Zhou and C.T. Au. 2004. A mini-review on ammonia decomposition catalysts for on-site generation of hydrogen for fuel cell applications. Appl. Catal., A 277: 1–9.

Yu, J., X. Zhao and Q. Zhao. 2000. Effect of surface structure on photocatalytic activity of TiO2 thin films prepared by sol-gel method. Thin Solid Films 379: 7–14.

Zeng, H., W. Cai, P. Liu, X. Xu, H. Zhou, C. Klingshirn et al. 2008a. ZnO-based hollow nanoparticles by selective etching: elimination and reconstruction of metal−semiconductor interface, improvement of blue emission and photocatalysis. ACS Nano 2: 1661–1670.

Zeng, H., P. Liu, W. Cai, S. Yang and X. Xu. 2008b. Controllable Pt/ZnO porous nanocages with improved photocatalytic activity. J. Phys. Chem. C 112: 19620–19624.

Zeng, J.H., B.B. Jin and Y.F. Wang. 2009. Facet enhanced photocatalytic effect with uniform single-crystalline zinc oxide nanodisks. Chem. Phys. Lett. 472: 90–95.

Zhang, D., L. Qi, J. Ma and H. Cheng. 2002. Formation of crystalline nanosized titania in reverse micelles at room temperature. J. Mater. Chem. 12: 3677–3680.

Zhang, H., G. Chen and D.W. Bahnemann. 2009a. Photoelectrocatalytic materials for environmental applications. J. Mater. Chem. 19: 5089–5121.

Zhang, H., R. Zong and Y. Zhu. 2009b. Photocorrosion inhibition and photoactivity enhancement for zinc oxide via hybridization with monolayer polyaniline. J. Phys. Chem. C 113: 4605–4611.

Zhang, L., H. Cheng, R. Zong and Y. Zhu. 2009c. Photocorrosion suppression of ZnO nanoparticles via hybridization with graphite-like carbon and enhanced photocatalytic activity. J. Phys. Chem. C 113: 2368–2374.

Zhang, X., J. Qin, R. Hao, L. Wang, X. Shen, R. Yu et al. 2015. Carbon-Doped ZnO Nanostructures: Facile Synthesis and Visible Light Photocatalytic Applications. J. Phys. Chem. C 119: 20544–20554.

Zhang, Z., L. Zhang, M.N. Hedhili, H. Zhang and P. Wang. 2012. Plasmonic gold nanocrystals coupled with photonic crystal seamlessly on TiO2 nanotube photoelectrodes for efficient visible light photoelectrochemical water splitting. Nano Lett. 13: 14–20.

Zhao, J., J. Yao, Y. Zhang, M. Guli and L. Xiao. 2014. Effect of thermal treatment on TiO$_2$ nanorod electrodes prepared by the solvothermal method for dye-sensitized solar cells: Surface reconfiguration and improved electron transport. J. Power Sources 255: 16–23.

Zheng, Z., Z.S. Lim, Y. Peng, L. You, L. Chen and J. Wang. 2013. General route to ZnO nanorod arrays on conducting substrates via galvanic-cell-based approach. Sci. Rep. 3.

Zwilling, V., E. Darque-Ceretti, A. Boutry-Forveille, D. David, M.-Y. Perrin and M. Aucouturier. 1999. Structure and physicochemistry of anodic oxide films on titanium and TA6V alloy. Surf. Interface Anal. 27: 629–637.

NEW PERSPECTIVES
AND TRENDS

12

Nature and Prospective Applications of Ultra-Smooth Anti-Ice Coatings in Wind Turbines

Hitesh Nanda,[1] *P.N.V. Harinath,*[2] *Sachin Bramhe,*[2] *Thanu Subramanian,*[1]
Deepu Surendran,[2] *Vinayak Sabane,*[2] *M.B. Nagaprakash,*[1] *Rishikesh Karande,*[1]
Alok Singh[2] and *Avinash Balakrishnan*[2,*]

1. Introduction

The ability to design new materials that can withstand environmental challenges has been critical for human survival. Issues, such as crop decomposition due to the excessive moisture and hypothermia as a result of wet clothing, has provided mankind the needed motivation to develop novel protective barriers that could help in effectively repelling water in form of condensed moisture, rain, snow and ice. Many species have undergone evolution that allows them to resist the detrimental effects of water. Many surfaces in nature, most notably the leaves of the lotus flower and duck feathers are known to be super hydrophobic. These are the desired features that is been aspired by many industries for repelling moisture in materials in several situations (Barthlott and Neinhuis 1997, Wang et al. 2012). Most of the products are derived directly from nature, such as animal furs or natural fibres, which are further improved by incorporating organic or synthetic oils and waxes to withstand harsh climatic conditions involving water (Holman and Jarrell 1923, Burney 1935). These water repelling strategies has been tuned for continuous improvement throughout the human history until the modern understanding of liquid-solid interactions which allowed products to be designed for more advance features. Practical applications for super hydrophobic surfaces are abundant and versatile, if they can be imparted into fabrics, water resistant garments, transparent surfaces, e.g., glass windows and wind shields, outdoor optical devices, such as solar panels or satellites, etc. The surfaces have craved to be dirt and dust free to maintain their functionality. This can be achieved if the material could remove impacting and condensed water droplets from its surface and thus possess self-cleaning ability. Super hydrophobic surfaces can also lower the drag between the liquid and the surfaces of boat hulls or pipes thereby improving its efficiency. It can also reduce the ice adhesion in cold regions on machineries related to logistics, construction, heat exchangers, aviation, refrigerator units, power lines, meteorological instruments antennas, radars, etc., and can improve functionality and safety significantly (Kim et al.

[1] Suzlon Blade Technology, Suzlon Energy Limited, One Earth, Hadapsar, Pune 411028, India.
[2] Suzlon Blade Technology-Materials Lab, Suzlon Energy Limited, Survey No. 588, Paddhar Village, Bhuj-Bhachau Highway, Bhuj, Kutch-370105 India.
* Corresponding author: avinash.balakrishnan@suzlon.com

2008, Shillingford et al. 2014, Wilson et al. 2013). Icing has been a major concern in colder regions which accounts for lethal accidents, major power losses, lost capital and other nuisances around the globe (Mara et al. 1999, National 1996). The techniques for dealing with accreted ice have so far been costly, laborious and challenging. Though, surface sciences combined with nanotechnology have given encouraging results and reinitiated interest towards icephobic materials, it is stated that there are yet no surface coatings or materials which could be identified as perfectly ice-phobic (Laforte 2005). This emphasises the fact that the demand for anti-ice or lowered ice adhesion surfaces truly exists as it hinders many industries and operators especially in colder regions of the earth such as Artic.

2. Wind Turbines in Cold Regions

Wind energy is a green renewable power source with no significant atmospheric pollution. Currently the global capacity of installed wind energy stands at more than 40 GW with a 30 per cent annual growth rate predicted over the next decade (Dalili et al. 2009). In order to sustain such a huge growth, it becomes imperative to necessitate research into the management of financial and environmental risks associated with the large-scale operation of commercial wind ventures. Fundamental to the efficient extraction of power from the wind, wind turbine technology has a unique technical, structural and aerodynamic blade designs which subsequently meets the power demands. With the increasing number of turbine installations, wind companies have pushed their way into icy Nordic environments with moisture-laden winds. These cold and windy environments create significant operational issues for a wind farm. Ice accretion on wind turbine blades can be detrimental to turbine performance, durability, and also to the safety of other operating turbines in the vicinity.

Prior to addressing the causes and mitigation measures, the significant problems that icing causes to commercial wind generation are discussed.

3. Effects of Icing

The following problems are directly related to icing and cold climate:

3.1 Complete Stall

Severe icing can lead to complete stall or stoppage of the turbine thereby resulting in significant energy loss. For instance, in the winter of 2002–2003, the Appelbo wind turbines in Sweden reported complete turbine stoppage for 7 weeks due to icing issue. As per the Swedish statistical incident database between 1998 and 2003, a total of 1337 such 'stoppages' were reported-resulting in a total downtime of 161,523 h. Seven percent of these incidents were related to the cold climate which resulted in 8022 h (5 per cent) of production loss. Finland reported 4230 h of down time between 1996 and 2001 due to icing (Jasinski et al. 1998).

Power losses ranging from 0.005 to 50 per cent of the annual production can occur depending on icing intensity and its duration on the site, wind turbine models and the evaluation methodology (Gillenwater 2008, Laakso and Peltola 2005, Tammelin 2005).

3.2 Measurement Errors

The anemometers, wind vanes and temperature sensors can be significantly affected by icing. Its been reported that wind speed measurement errors can be as high as 40 per cent for an ice-free anemometer and 60 per cent for a standard anemometer during icing events (Fortin et al. 2005). Higher air density at lower temperatures and air-foil modifications due to icing can lead to overproduction of the wind turbine. Overproduction of up to 16 per cent has been recorded (Jasinski et al. 1997).

3.3 Mechanical and Electrical Failures

Ice accretion increases the load on the blades and on the tower structure, causing high amplitude vibrations and/or resonance as well as mass imbalance between blades. Studies (Antikainen and Peuranen 2000) on the effects of icing time and shapes on the aerodynamic balance of turbine blades have suggested that both mass and aerodynamic imbalance can occur even in the early stages of icing. Jasinski et al. (1998) studies have shown that the onset of ice accretion could cause a slight increase in surface roughness that, in turn, increases the drag coefficient thereby reducing power production.

Although the extreme loads are considered at the design level, the fatigue loads will shorten the lifetime for the components (Makkonen 1994). These effects were demonstrated in the WECO (Wind Energy Production in Cold Climates) project (Tammelin 1996–1998) where they found the following:

a) Heavy ice masses can cause higher deterministic loads.
b) Asymmetric masses will cause unbalance during rotation.
c) Ice accretion will increase the excitation of edgewise vibrations.
d) Resonance can occur due to the changed natural frequencies of the blades, particularly for lightweight rotor blades.

Besides, low temperatures also affect the oil viscosity and changes the dimensions and mechanical properties of different components of the wind turbine. This results in possible overheating and higher fatigue charges on components; one of the most affected being gearboxes whose lifetime is considerably reduced (Seifert 2003). Snow infiltration in nacelle and extreme low temperature can lead to condensation in the electronics (Laakso et al. 2003).

3.4 Safety Hazard

Ice thrown from rotating blades can pose a serious safety threat, particularly when the wind farm borders public roads, housing settlements, power lines, and shipping routes. Heavy ice accumulation on blades can be thrown at a distance of up to 1.5 times the combined height of the turbine and the rotor diameter (Tammelin et al. 2000). Monte-Carlo simulations have shown that (Battisti et al. 2005) the odds to be hit by a piece of ice weighing between 0.18 to 0.36 kg under moderate icing conditions are 1 in 10. These studies have prompted a recommendation that for installation sites with a high probability of icing, the distance between the turbine and the nearest object should adhere to the following equation—with the effect of slopes also taken into account for mountainous sites:

$$d = 1.5\ (D + H) \tag{1}$$

where d is the projected distance ice can be thrown, D equals the diameter of the rotor and H is the height of the nacelle. This equation gives a rough estimate of the area at risk (Homola 2005).

4. Current Turbine Blade Icing Solutions

Over the past few years, significant effort has been put to determine and model ice prevention techniques. Most of these techniques are derived from the aviation industry and can be categorised in two classes: active and passive. Active (de-icing) techniques rely on an external systems applied to the blade while passive (anti-icing) methods depend on the physical properties of the blade to prevent ice accumulation. The former technique removes the ice from the surface after its formation, while the latter prevents the initiation of icing (Laakso et al. 2003).

The following section briefly describes the active and passives methods employed for de-icing and anti-icing.

4.1 Active Protection Techniques

Active protection methods employ thermal, chemical, and pneumatic techniques and act as de-icing or anti-icing systems.

4.1.1 Active De-icing Systems

Small aircrafts often use mechanical de-icing systems for example, inflatable rubber boots on leading edge of the wings which can expand and contract on ice-prone areas. These systems have not proven to be practical for wind turbine application for many reasons. The obvious reason is the increased aerodynamic interference and noise that is generated by an inflated boot. Further, this additional mechanical complexity associated with such a system significantly adds to the maintenance burden during the 20-year life span of the turbine (Talhaug et al. 2005).

4.1.2 Active Anti-icing Systems

Most of the active methods that have been developed till date have been on thermal systems that can remove the ice by applying heat to the blade. This approach has been restricted to prototype stages. Heating the blade's surface can be achieved through several techniques:

a) Direct heating: The electrical resistance heating system consists of a heating membrane or element that is applied to the blade surface. These simple thermal ice prevention systems have been successfully employed in the aerospace industry. Similar heating systems were developed for wind energy application in the mid-1990s (Talhaug et al. 2005). For instance, the Finnish blade heating system employs carbon fibre elements mounted to the blade surface and has the proven operating experience with installation in 18 turbines at various sites with a total of nearly 100 operating in winters (Jasinski et al. 1998). A direct resistance heating system developed by JE has shown to work effectively as well, but has not yet seen mass production (Homola 2005).

b) Indirect heating: This method heats the inside of the blade by blowing warm air or by using a radiator, and the heat is conducted to the outer surface. Such a system has been employed on an Enercon turbine in Switzerland (Tammelin et al. 1996).

c) Microwave heating: These systems are based on microwave technology, but have not been successfully implemented (Jasinski et al. 1998), considering the amount of energy it consumes.

d) Air barrier: The concept of protecting the blade surface by a layer of clean air has been conceived. The idea is to utilise air flow from inside of the blade and pushing it through the rows of small holes near the leading and trailing edges. This will generate a layer of clean air and, if necessary, heated air directly around the blade surface. This air flow would deflect the majority of water droplets and would melt the ice droplets that managed to strike the blade surface (Tammelin et al. 1996).

4.1.3 Problems Associated with Direct and Indirect Heating

The following points highlight the inherent disadvantages connected with these heating methods:

i) Leading edge heating elements are not effective in de-icing when turbines are at a standstill (e.g., during icing conditions combined with low-wind speeds). Further, the low damping of the natural frequency (edgewise vibrations) are added to the gravity loads as a result high stresses are formed on the main girder of many glass reinforced blades. This scenario leads to even higher stress in the wires or fibres of the heating elements. This is particularly true when the heating elements are made from carbon fibre whose modulus of elasticity is much higher when compared to the glass fibres. This means that the heated fibres end up bearing most portion of the loads resulting in cracks in the heating elements (Talhaug et al. 2005).

ii) The electrical heating elements with metallic nature like carbon fibre are prone to lightning strikes at an exposed site.

iii) Electrical fibres and laminate embedding may increase surface roughness. The airfoil contour must be free from any surface anomalies induced by these heating elements to avoid unnecessary disturbances of the laminar flow around the leading edge during ice-free conditions (Talhaug et al. 2005).

iv) For forced warm air thermal systems, the efficiency of the turbine goes down as shell structures become thicker and thermal resistance rises. In practice, very high temperatures are required inside the blades to keep the outer surfaces free of ice, even in mild conditions. Considering the maximum operating temperatures of thermoset composites, maintaining high temperatures inside the blade cavity to keep the blades free of ice would be a significant challenge (Maissan 2001).

v) In case one blade heater fails, a significant mass imbalance may be imposed on the rotor as each blade will experience different icing loads. One such situation occurred for Yukon Energy in 1996 (Maissan 2001). This loading problem can also occur when the run-back water on the blade from icing and blade heating can freeze after it passes the heated zone. The ice formed from run-back water has a higher density and may accumulate at positions where it can be aerodynamically harmful (Maissan 2001).

vi) Current heating technology consumes electricity and the break-even cost of such a heating systems depends on how much electricity production is lost due to icing and the price of electricity. Thermal anti-icing requires power in the range of 6 and 12 per cent of the output (Tammelin et al. 1996–1998). Therefore, when the financial benefits of a blade heating system are appraised it is imperative that the icing time, severity of icing, and potential wind resources are taken into consideration.

4.2 Passive Methods

Passive methods exploit the physical properties of the blade surface to remove or prevent ice, and are similar to active methods in their ability to act as de-icing or anti-icing systems.

4.2.1 Passive De-icing Systems

a) Blade flexing

A concept called as blade flexing has been proposed wherein designed blades are flexible enough to crack the ice loose from the surface. However, there is little published information on this subject. The disadvantage here is that thin layers of ice can adhere quite strongly to the blade surface and may not be brittle enough to crack loose from just the vibration and/or flexing. This can compromise the aerodynamic properties of the blade (Tammelin et al. 1996).

b) Electro-expulsive system

This system depends on very rapid, electromagnetically induced vibrations and is certified for use on Raytheon's Premier I business jet. However, there remains a lack of practical information on the usage of these systems for wind turbines (Makkonen et al. 2001).

Other passive systems such as active pitching of the blades by facing the blades into the sun are proposed to remove ice from turbine blades. Though these methods may work in light icing environments, their efficacy needs to be validated in heavy icing conditions. Moreover, such methods may damage the turbine and/or decrease the power production (Botta et al. 1998).

4.2.2 Passive Anti-icing Systems

In 1996, Yukon Energy introduced a black coloured coating called StaClean (Maissan 2001). StaClean is accredited by the manufacturer to be non-wetter and slicker than Teflon coatings with high impact and abrasion resistant. Subsequently these qualities were reported to be effective at reducing icing issues (Maissan 2006). Although, the black colour did not increase the blade surface temperatures during winter months (Weis and Maissan 2003), it in fact raised the temperatures of other sections of the blade coated with the manufacturer's original gelcoat.

It is a well-known fact that special coatings can reduce the shear forces between the ice and the blade's surface. The developments of an ideal surface coating for icing mitigation is still an ongoing research that eludes both aerospace and wind turbine industry. Critical to the comprehension and a potential resolution to these issues are specifically the composition and surface properties of the turbine blade. From this view point the following section addresses some of the new surface engineering techniques like slippery liquid-infused porous surfaces (SLIPS) and magnetic slippery extreme icephobic surfaces (MSEIS) that could be likely used in wind turbines in a holistic sense. It is been observed by the authors that the complexities involved in the practical application of these techniques on huge blade surfaces calls on expertise from materials science, aero-, thermo-, and structural-dynamics.

5. Water and Surface Interaction

There have been pioneering works that explain the nature of solid-liquid interactions, including the wetting and non-wetting scenarios. For instance, in 1805, Thomas Young described the equilibrium behaviour of a droplet on an ideal surface (Young 1805). In the early 1900s, the metallurgy community reported the variations in the contact angle of a droplet on a solid surface, which are critical to liquid adhesion and mobility which were described as 'hysteresis' (Rickard and Ralston 1917). But the mention of this phenomenon dates to Gibbs's work on the thermodynamic properties of surfaces that included a discussion of "the frictional resistance to a displacement of the [contact] line" (Gibbs 1878), and so-called contact angle hysteresis (CAH) continues to be investigated till date (Nosonovsky 2007, Krasovitski and Marmur 2005). Further understanding of non-ideal surfaces happened through the Wenzel (Wenzel 1936) and Cassie–Baxter equations. These theories established the facts that the surface can be stated as wetted or hydrophilic surface, if the contact angle' value of water is from 0° to 90°. On the contrast, a hydrophobic surface or the surface that remains non-wetted by water obtains the contact angle value larger than 90°. Surfaces that exhibit contact angle value > 150° are considered as super-hydrophobic where the water droplet will be almost spherical on the surface. When the contact angle value is smaller than 5° the surface is considered as super-hydrophilic, as illustrated in Fig. 1 (Ketelson et al. 2011). For almost any given material, the surface energy can be modified chemically or can be altered by engineering the surface roughness (Nosonovsky and Bhushan 2005). Together, all these theories have established the pre-requisite surface characteristics that are needed to engineer highly effective water-repellent materials. A timeline of several major advances in understanding the repellence phenomena (Michael et al. 2016) in Fig. 2.

6. Contact Angle and Wettability

Contact angle is generally used to measure the degree of hydrophobicity or hydrophilicity of the liquid, solid and gas phases (Chau 2009). Parameters that affect the formation of contact angle vary significantly (Kumar and Prabhu 2007), thus making it challenging to find reproducible contact angles in systems. To compensate this, many of contact angle equations have been constructed to demonstrate various states and scenario. With respect to this the foremost models are the Young's equation, Wenzel's equation and the Cassie-Baxter equation. While Young's equation describes

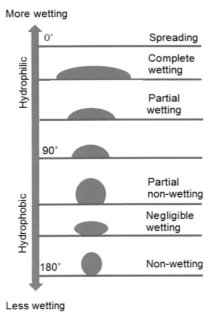

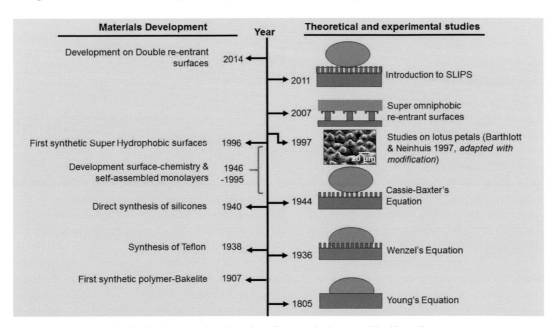

Fig. 1. Variations in the contact angle of a droplet on a solid surface resulting in hydrophobic and hydrophilic states.

Fig. 2. Timeline showing the major milestones in the area of liquid repellency.

wetting on a smooth surface, the Wenzel's and the Cassie-Baxter equations can be used to describe the wetting on rougher surfaces. In Fig. 3 the models are presented altogether—the surface roughness has been taken into consideration in Wenzel's and in Cassie-Baxter's equations.

Young reported the correlation between contact angle and surface tension on an ideal surface (i.e., chemically and topographically homogeneous). This theory states that there are three interfacial free energies affecting the contact angle of liquid: solid-liquid (SL), liquid-vapour (LV) and solid-vapour (SV), presented in Fig. 4.

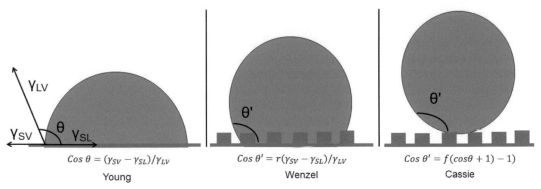

$$Cos\ \theta = (\gamma_{SV} - \gamma_{SL})/\gamma_{LV} \qquad Cos\ \theta' = r(\gamma_{SV} - \gamma_{SL})/\gamma_{LV} \qquad Cos\ \theta' = f(cos\theta + 1) - 1)$$

Young   Wenzel   Cassie

Fig. 3. The wetting models of Young, Wenzel and Cassie-Baxter (Nakajima 2011).

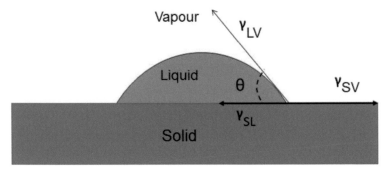

Fig. 4. Schematic presentation of a liquid droplet on a solid surface where the contact angle (θ) is due to three interphases (γ): solid-liquid (SL), liquid-vapour (LV) and solid-vapour (SV) (Della-Bona 2005).

As these interfacial tensions are balanced on tangential direction of non-deformable solid surface, Young's equation is constructed as in Eq. 2.

$$\gamma SL + \gamma LV\cos\theta_y = \gamma SV \qquad (2)$$

As the Young's equation describes the balance of surface forces, θ_y which is the Young's contact angle is balanced by the surface tension forces. The surface hydrophobicity increases with the decreasing surface free energy of the solid air interphase (γSV). Here the wettability of the surface is mainly determined by the surface chemical composition and does not account for the roughness (Chau et al. 2009, Kwok et al. 1998).

Wenzel studied the effects of surface roughness on the static contact angle and proposed that the surface geometry plays an equally important role along with the surface chemistry on the static contact angle. His studies indicated that the surface roughness made hydrophobic solid to behave even more hydrophobic. It was understood that the surface textures enhances the hydrophobic nature of the material. As the surface roughness increases, the effective surface area enlarges which makes the liquid less likely to spread on a rough hydrophobic substrate. These findings led to a dimensionless roughness factor which describes the ratio of the actual surface area over its apparent surface area (Wenzel 1936). Thus, according to the Eq. 3, the apparent contact angle is the related to the ideal contact angle, when the surface is rough and the droplet size is adequately large compared with the roughness scale. This equation also assumes that the liquid entirely fills the pores of the surface.

$$\cos\theta_w = r\cos\theta_y \qquad (3)$$

where θ_w is the Wenzel contact angle on a rough surface. θ_y is the Young's contact angle on a smooth surface and r is the average roughness ratio. For values of r > 1 indicate that the water contact angle on a smooth surface can be increased by texturing. If the surface is flat (r = 1), the Wenzel equation gives the Young's equation. The fundamental assumption of this model is that the liquid droplet is in complete contact with the solid surface which is called as the Wenzel state. In most cases, a droplet on hydrophobic surfaces with minor roughness would follow Wenzel's equation but it does not account for greater roughness and heterogeneous surfaces (i.e., with large surface area). Wenzel's equation can describe relations for homogeneous solid-liquid interphase. This was addressed by Cassie who modified the equation for contact angle changes, for two component surfaces (Eq. 4). The apparent contact angle related to the ideal contact angle on a heterogeneous surface can be expressed as (Cassie and Baxter 1944).

$$\cos \theta_c = f_1 \cos \theta_1 + f_2 \cos \theta_2 \tag{4}$$

where f_1 is the fractional area of the surface with θ_2, θ_c is the Cassie contact angle. The equation can be reduced into Cassie-Baxter equation, Eq. 5, for porous surfaces:

$$\cos \theta_c = f_1 \cos \theta_1 - f_2 \tag{5}$$

Here f_2 is the fraction of air space (open area), that makes $\cos\theta_2 = -1$, as $\theta = 180°$ as to a non-wetting state. In this state, air is trapped between the liquid droplet and the rough surface. Water in contact with the solid surface and air forms the so called fakir droplets which sit on the surface and on the trapped air. It is relevant to realise that on the surface, the Wenzel as well as the Cassie-Baxter states both can coexist in droplets (Quéré 2005). In Cassie-Baxter's proposal the uneven surface is composed of two fractions: fractional area f_1 with contact angle θ_1, and other area f_2 with θ_2, whereas $f_1 + f_2 = 1$.

The apparent contact angle in this situation can be described using the Cassie-Baxter equation, as given in Eq. 6:

$$\cos \theta_{rough} = \Phi_S \cos\theta_{flat} + \Phi_v \cos\theta_{LV} = \Phi\cos\theta_{flat} - (1 - \Phi_S) \tag{6}$$

Φ_S and Φ_v are the fractions of solid and air contacting the water ($\Phi_S + \Phi_v = 1$). Air under the droplet significantly enhances the apparent surface hydrophobicity because the contact angle of water θ_{LV} air is 180° (i.e., $\cos \theta_{LV} = -1$). In the above equation, monotonic decline of Φ_S results in an increase of θ_{rough} which finally turns to a superhydrophobic state. Thus, it is clear that two requirements are needed to gain superhydrophobicity—elevated contact angle value and minute contact angle hysteresis. To achieve this well-engineered surface chemistry with well fabricated surface roughness is required. Since, obtaining reproducible values for contact angles may be challenging as the measurements are vulnerable to many factors, the usage and reliability of the Young's, Wenzel's and Cassie-Baxter's equations has to also be taken into consideration before validating the results.

7. Factors Affecting Wettability

Super hydrophobic nature of the material surfaces can be defined by the sliding behaviour of the water droplet. This behaviour is governed by both surface tension and the gravity (Kim et al. 2008). On the sliding surface, the drop acquires an asymmetric shape as depicted in Fig. 5, where the contact angle on the sliding (advancing contact angle) becomes greater and the upper side (receding contact angle) gets reduced. The difference of the advancing and receding contact angle is defined as contact angle hysteresis. The so called hysteresis achieves its peak value as the liquid droplet starts to move down the sliding plane. It is well documented fact that a liquid droplet will roll off easily from the surface when the contact angle is greater than 150° and the hysteresis smaller than 5° (Quéré 2005). There are many physio-chemical factors that influence the hydrophobicity of the solid surfaces for, e.g., heterogeneity, surface roughness, shape and size of surface structures, e.g., particles (Chau 2009). The roughness of the surface is known to affect the hysteresis. However, the exact relationship of contact angle hysteresis and surface roughness may not be obvious and is difficult to be derived from the existing wetting models (Gao and McCarthy 2007, Kwon et al. 2010, McHale 2007). Additionally,

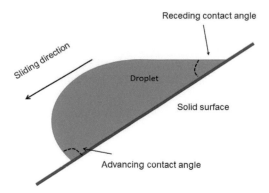

Fig. 5. Liquid droplet on a sliding plane exhibiting advancing and receding contact angles [Pan14].

the shape and size of the anomalies on the surface make every point in the surface unique. As a result, the apparent contact angle varies within the surface (Chau et al. 2009). However, studies have demonstrated the importance of topological influence and have shown that the contact angles cannot be foretold only by roughness factors (Spori et al. 2008, Starov et al. 2007, Veeramasuneni et al. 1997).

Roughness is influenced by the impurities and composition of different constituents on the material surface. Dissolution of functional groups and material anisotropy also has an influence on the contact angles of the droplets. Texturing and particle shapes on the surface are fundamental aspects that determine hydrophobicity (Ebert and Bhushan 2012, Khorasani et al. 2005, Patankar 2004, Ulusoy and Yekeler 2005). Hydrophobic surface can be made by numerous techniques as the final surface structure can have several shapes as seen in Fig. 6. For instance, Callies et al. used the silicon wafers to pattern micro-pillars to trap the air in the surface structure via conventional photolithography

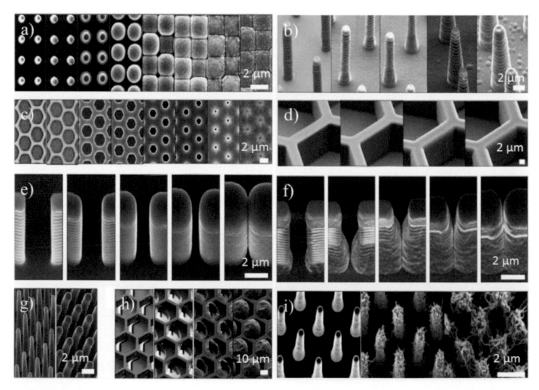

Fig. 6. Prospective textured micro- and nanostructures for hydrophobic surfaces. The surface texture resembles pillars, columns and honeycomb structure (Adapted with modifications from (Kim et al. 2012)).

techniques (Callies et al. 2005). It was shown that the degree of super hydrophobicity can be controlled by altering the density and height of the pillars. Other studies have shown the effects of the pillar shape and height and found effects of environmental resistance, effortless droplet rolling and pressure stability (Kim et al. 2012, Roach et al. 2008, Zheng et al. 2010). It was observed that randomly distributed particles on the material surface can also alter the wetting states. Surface roughness and wettability of milled industrial minerals namely, calcite, barite and quartz, showed that the degree of hydrophobicity increase with decreasing surface roughness. The particles which had rougher surfaces (0.55 to 4.49 µm) showed greater wettability (Ulusoy and Yekeler 2005).

There are several types of attractive forces that affect the solid surface and the liquid droplets. These forces can be electrostatic, van der Waals, chemical or capillary forces (Bhushan et al. 2009, Burton and Bhushan 2006). The surface tension of liquid minimises the surface area thereby, making the droplet spherical. The gravitational force affects only to larger drops than the capillary length of 2.7 mm thereby flattening the spherical droplet. When the droplet is smaller than this, the effect of gravity is insignificant and is often ignored (Kim et al. 2008).

8. Superhydrophobic Surfaces

In nature the essence of micro and/or nanostructure is in the heart of any material and has its own inspiring hydrophobic surfaces. There are many biological surfaces such as lotus leaves (Barthlott and Neinhuis 1997), anti-reflective eyes of moths (Genzer and Efimenko 2006), water strider's legs (Gao and Jiang 2004), shark skin (Bechert et al. 2000) and also duck feathers (Liu 2008) that are known for their superhydrophobicity. Great efforts have been made to unleash nature's secrets in exceptional and complex surface structures. Apart from the microstructure, there are also composite-like hydrophobic surfaces in nature such as different waxes and fibrils, papillae or microsetae (Koch et al. 2009). The epicuticular wax layer on biological surfaces has low surface energy which makes the superhydrophobicity possible (Tuteja et al. 2007). It is a well-known fact that any system tries to achieve a state which corresponds to a minimum energy, thus, liquid droplets and bubbles have a spherical shape on the surface (Nosonovsky and Bhushan 2008). It is stated that the hierarchical structure of the nature's own surface is responsible for the superhydrophobicity. The droplets of water sit effortlessly on the nanostructure as the bubbles of air fill the cavities beneath the droplets in the porous structure (Burton and Bhushan 2006). The trapped air inside the structure prevents the water droplet from entirely touching the surface, hence, superhydrophobicity is possible. The potential for anti-icing materials are abundant, promising and intriguing, albeit there are only a small number of actual products that are available in the market. The major limitation of utilising artificial super hydrophobic coatings in systems like wind turbine is that they suffer from poor durability against physical forces such as exposure to sunlight and abrasion by wind and dirt which can lead to permanent damage of the coating surfaces (Kim et al. 2008).

8.1 Superhydrophobic Polymers

Fluorinated polymers inherently possess low surface energy because of which they exhibit pronounced hydrophobicity. For instance, poly(tetrafluoroethylene), PTFE, is shown to have water contact angle between 112°–120° (McKeen 2006, Zhang et al. 2004) and are widely been used in various industries due to the polymer's extreme resistance towards elevated temperatures, corrosion, stress cracking and chemical reactions. It was observed that by stretching the PTFE more than 100 per cent of the original length, the surface roughening occurs because of which the fibrous microcrystals which have a great fraction of voids space in the surface gets exposed. This generates number of nanoscale features to gain superhydrophobicity. The results showed water contact angle improvement from 118° to 165° after stretching (Zhang 2004).

Superhydrophobicity can be achieved by roughening of surfaces of PTFE through plasma etching. Alternatively, fluorinated block polymers with microphase separation can be prepared. The water contact angle can be increased to around 170° by etching the surface with high-energy oxygen species generated by plasma (Morra et al. 1989, Shiu et al. 2004). Moreover, hydrocarbon polymers such as polypropylene, (PP), also seem to benefit in superhydrophobicity by the creation of surface roughness during plasma etching (Youngblood and McCarthy 1999). Whilst drying in humid conditions, a copolymer of methyl methacrylate and fluorinated acrylate monomers can develop hexagonally packed nanopores at the cast film surface (Yabu and Shimomura 2005). Likewise in poly(dimethylsiloxane), PDMS, when treated with laser ablation, the PDMS surfaces roughen up at microscale and can exhibit a contact angle of 160° with water sliding angle under 5° (Jin et al. 2005, Khorasani et al. 2005). Experiments by Lu et al. showed that porous surface could be fabricated onto low density polyethylene (LDPE) by adjusting nucleation rate of crystallisation and time. These studies showed that by controlling the crystallisation behaviour of the LDPE, water contact angle increased from 101° to around 173° (Lu et al. 2004).

8.2 Inorganic Hydrophobic Materials

Ceramics and metals are likely to have hydrophilic nature due to the large number of polar sites on the surface resulting in coordinative unsaturation. For example surface atoms on aluminium have deficiency of electrons (six electrons in their three sp2–hybrid orbitals). Thus, these atoms form a hydrogen bond with the interfacial water molecules to achieve a full octet. However in case of rare-earth oxides due to dissimilar electron structure, the incomplete 4f orbitals are prevented from interactions with the immediate water environment because of the full octet configuration of electrons in the 5s2p6 outer shells. This has resulted in the use of rare-earth oxides to accomplish superhydrophobic surfaces even in extreme conditions and rough wear treatments (Aytug et al. 2014, Azimi et al. 2013).

In this regard, cerium dioxide films, CeO_2, which are fairly inexpensive can be fabricated to have hierarchical structure for trapping air in the surface within micro- and nanoporous textures. Hence, the air-liquid interface is increased by the air: droplets cannot reach the shallows of the surface (Ishizaki and Saito 2010). The CeO_2 film was proven to maintain high water contact angle (152°) even after 21 days of exposure to sodium chloride aqueous solution. This showed that CeO_2 surface had good chemical stability and maintained durable in corrosive solution (Liang et al. 2013). High chemical stability CeO_2 films was also reported n pH ranging from 1 to 14 (Ishizaki and Saito 2010). Titanium dioxide, TiO_2, can also be used as a hydrophobic surface film when high performance for the coating is desired. TiO_2 layers are suitable for rapid decompositions of organic compounds in vapour and liquid states under UV-light. This photocatalytic activity can be further improved by enhancing the crystallinity of TiO_2, surface roughness and pore size alterations. The wettability in ZnO nanorods could be tuned from superhydrophocity to superhydrophilicity (i.e., the water contact angle shifted from 160° to near 0°) by treating with ultraviolet irradiation and storing samples in dark, respectively (Feng et al. 2004). In another study, it was concluded that high surface porousness and low surface free energy effect are crucial for surfaces hydrophobic nature (Hou et al. 2007).

It was demonstrated by Li et al. (Henna niemelä-anttonen 2015) that the monolithic materials microstructure and properties can be altered by changing the compositions and calcination temperatures of used precursors. This was shown using macroporous complex metal oxide monoliths, such as $MgAl_2O_4$, which were prepared by selective leaching of self-generated MgO sacrificial template from the sintered two phase composites. A significant improvement in water contact angle-increase from 8° to 150° was observed in this process. These studies demonstrate the fact that the reactive surface area can be achieved by fabricating large number of pores on a textured surface to gain hydrophobic materials with abundant applications ranging from one industry to another, for example many components in wind turbine industry can benefit these kinds of robust coatings.

8.3 Inorganic-Polymer Composite Materials

Many composite coatings made from inorganic particles and polymers show hydrophobicity. The main idea is to achieve the required porosity or texture with the inorganic particles as the polymer binds to these particles. Polymers with inherent hydrophobic characteristics can be applied with ease using spray coating and dip coating methods (Cao et al. 2009, Sellinger et al. 1998). The inorganic components for these coatings are generally metal oxides such as silica, titania and ceria particles which have been widely used (Bharathidasan et al. 2014, Farhadi et al. 2011). In these composites the inorganic particle size plays a crucial role on the hydrophobicity. The particle size in the composite structure needs to be optimal in order to prevent ice formation on the surface. Cao et al. in their studies showed that when the mixtures of polymer binder (acrylic polymer resin) and silica particles (20 nm–20 μm in diameter) were sprayed on Al substrate, the probability of icing increased with the increasing silica particle size (Cao et al. 2009).

There are studies that have reported robust polymer-inorganic composite superhydrophobic coatings. For instance (Zhou et al. 2012), used PDMS and fluorinated alkyl silane combinations to gain silica with fluorinated alkyl silane. a surface that was durable against acids, alkali, boiling water and abrasion. These composite materials could sustain 500 machines washes after applied on fabrics. Sellinger et al. constructed a nanocomposite material that mimicked natural structure of a balone shell. The polymerisation ensured locking of the nanotexture which bonded covalently the inorganic-organic interphase resulting in hard transparent coatings (Sellinger et al. 1998).

9. Slippery Liquid Infused Porous Surfaces (SLIPS)

Unlike superhydrophobic surfaces, one of the most potential ways to generate super liquid-repellent surfaces is a solid-liquid composite surface with pores filled with a liquid instead of air. The idea was that depositing a drop of immiscible liquid onto the impregnated surface would result in the liquid droplet to be non-sticking and will remain floating on the surface. Quéré had conceptualised these kinds of composite surfaces called as Slippery liquid infused porous surfaces (SLIPS) to have self-cleaning properties and also contemplated the characteristics of the ideal surface texture and its optimal size and shape (Quéré 2005). However the actual SLIPS were introduced by Wong et al. (Wong et al. 2011) who reported a bioinspired SLIPS with pressure-stable hydrophobicity. The gained surface had many advantages which included self-healing, optical transparency, dropwise condensation (Glavan et al. 2014, Lalia et al. 2013), repellency to ice and frost (Kim et al. 2012, Rykaczewski et al. 2013, Subramanyam et al. 2013), stain repellency (Epstein et al. 2012, Lafuma and Quéré 2011) and corrosion inhibition (Qiu et al. 2014, Wang et al. 2015). The illustration of the porous solid material infused with a lubricating liquid is shown in Fig. 7.

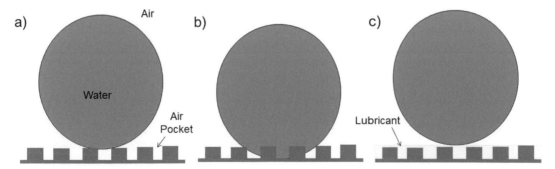

Fig. 7. (a) In Cassie-Baxter state, droplet sits on air cavities resulting in hydrophobicity. (b) Wenzel state has no air beneath the droplet, hence the surface in contact with water. (c) In Slippery liquid infused porous surface where a lubricating agent is impregnated to gain hydrophobic characteristics (Rutter 2011).

The essential criteria for SLIPS include adequate impregnation of the lubricating liquid onto the porous surface. It should be able to wet and stably adhere with the solid surface and finally the lubricating and impinging test liquids must be immiscible (Wong et al. 2011). The limitations of solid-air composite structure can be overcome by using SLIPS. The trapped air cannot always be an adequate defence towards low surface tension liquids—such as organic liquids and mixtures—especially when countered under pressure, for instance in windy situations. Moreover, the fragility of fabricated textures in solid-air structures remains as a challenge. When the porous substrate material is infused with lubricant oil, the gained solid-liquid structure repels the water droplet as seen in Fig. 7c. The lubricant is held at its place by the texture and features in the substrate, that is, the topology itself does not enable the water-repellency of the structure. While in the case of Cassic-Baxter state, the droplet of water lies on the air cavities whereas when compared to the Wenzel model, there are no air cavities under the droplet. In Fig. 7 (a, c), the surface can be stated as non-wetted.

9.1 Tuning Lubricant Impregnated Porous Surfaces

SLIPS have been tuned in many ways by using various lubricants and even wider spectrum of porous solids. Mostly these lubricants have been perfluoropolyethers (PFPE), and silicone oils. There have been some experiments using vegetable oils, such as olive and almond oils although not much commercial success have been attained (DuPont 2011). These kinds of SLIPS can be tailored in many ways based on the applications, i.e., from 3D printing to molding methods. The combination of the lubricant and the porous solid material depends heavily on the application and required properties. Table 1 shows the materials and methods used in creating SLIPS.

The suitability of the oils mentioned in Table 1 is definite as to the fact that they are colourless, non-flammable, non-reactive and have longer life. The polymer chain of perfluoropolyether is saturated with carbon (22 per cent by weight), fluorine (69 per cent by weight) and oxygen atoms (9 per cent by weight), as seen from the chemical structure in Fig. 8 (DuPont 2011).

Perfluorinated compounds have come under a lot of criticism as these compounds are persistent in nature and may be toxic towards organisms. The usage of perfluorinated compounds have been reduced worldwide. Hence, it is important to consider a suitable lubricating liquid used in de-icing applications ranging from any field of industries (Lindstrom et al. 2011, Zaggia and Ameduri 2012, Zushi et al. 2011). Likewise, silicone oils with different viscosities are used in making SLIPS. Silicone

Table 1. Reported substrate and lubricant pairs for SLIPS fabrication.

Sl. No.	Substrate	Lubricant	Fabrication Technique	Pore Size	Reference
1	Sterlitech Teflon membranes	Krytox oils	Replica moulding	$\geq$ 200 nm	(Wong et al. 2011)
2	Teflon membrane	Krytox 100, 103, 105	Replication	200 μm	(Daniel et al.2013)
3	Cellulose lauroyl ester-films	Perfluoropolyether	Spray coating	30–1060 nm	(Chen et al. 2014)
4	Al substrate, Poly(pyrrole)	Krytox 100	PPy electrodeposition	2 μm	(Kim et al. 2012)
5	Poly(vinylidene) fluoride-co-hexafluoropropyle	Polymer quartz oil, Krytox 1506	Electrospinning	2–4 μm	(Lalia et al. 2013)
6	Polyvinyl-4,4-dimethylazlactone, Polyethyleneimine	silicone oil, canola oil, coconut oil, olive oil	Layered assembly	Micro to nanoscale pores	(Manna and Lynn 2015)
7	Poly(vinylidene) fluoride-co-hexafluoropropyle/Butyl phthalate	Krytox 103	Non-solvent-induced phase separation	450–850 nm	(Okada and Shiratori 2014)
8	1H,1H,2H,2H-perfluorodecyltriethoxysilane on Al substrate	Perfluoropolyether (PFPE) Nascent™ FX-5200	Anodisation	< 100 nm	(Wang et al. 2015)

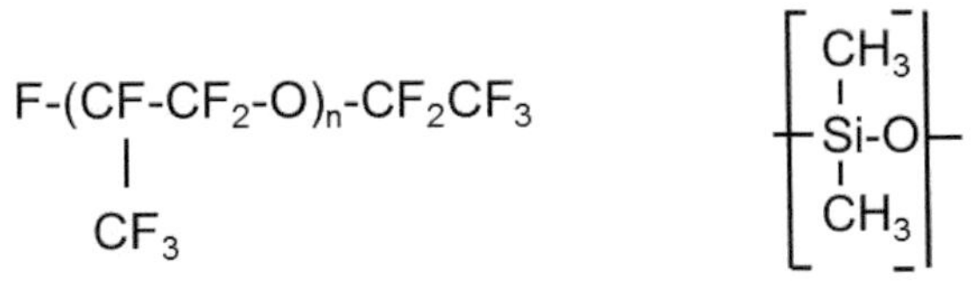

$$F\text{-}(CF\text{-}CF_2\text{-}O)_n\text{-}CF_2CF_3$$
$$|$$
$$CF_3$$

$$\left[\begin{matrix} CH_3 \\ | \\ -Si\text{-}O- \\ | \\ CH_3 \end{matrix}\right]$$

Perfluoropolyether, PFPE   Poly(dimethylsiloxane), PDMS

Fig. 8. Commonly used lubricants for SLIPS, Perfluorinated lubricants and silicone oils.

fluids work well in SLIPS since they are strongly hydrophobic, physiologically inert, lubricating by their nature and possess excellent release properties (Moretto et al. 2000). The SLIPS must have lubricant that is immiscible towards the repelled liquid—for example to water and also should have good spread ability and good affinity towards the porous matrix in comparison with water (Wang et al. 2015, Wong et al. 2011).

9.2 Multifunctional SLIPS

There have been several studies where superhydrophobic technology exhibiting anti-icing surfaces have shown promising results. However the functional life and cost of these coatings have been an issue (Kim et al. 2012). Wong et al. in his studies used commercially available materials like porous teflon membrane and low-surface-tension perfluorinated liquid as lubricant to achieve multifunctional SLIPS (Wong et al. 2011). Kim et al. described ice-repellent modified surfaces based on SLIPS-technology, as shown in Fig. 9. In the study, polypyrrole (PPy) was electrodeposited on Al substrates. The PPy with texture diameter of 2 μm was fluorinated and impregnated with the lubricating agent called Krytox 100. It was observed that the cooled droplets sled away from SLIPS-Al surface as compared to untreated Al where droplets froze before sliding away. This approach showed that SLIPS-coated metals surfaces suppressed ice accretion and had low ice adhesion strengths. The SLIPS-Al had

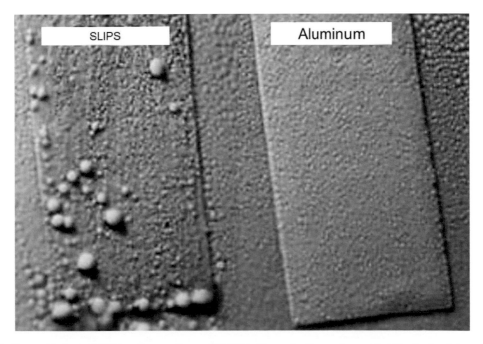

Fig. 9. Comparing the wettability and ice-repellency of SLIPS to Al samples. It was observed that if water droplets are frozen onto SLIPS, they tend to isolate—not to form continuous ice layer. Adapted with modifications (Kim et al. 2011).

ice adhesion strength of 15.6 ± 3.6 kPa as compared to untreated Al which showed values of 1360 ± 210 kPa at −10°C (Kim et al. 2012). Subramanyam et al. in their studies have shown with SEM imaging that ice-adhesion strength (measured using a force transductor) is dependent on the surface texture and decreases with increasing texture density. They used photolithography to gain square silicon micropost surfaces and coated with a lubricating octadecyltrichlorosilane (Subramanyam et al. 2013).

There has been debate whether the mechanisms of ice and water can be related while constructing superhydrophobic surfaces into icephobic. It is assumed that by lowering the surface energy and increasing the water contact angle, icephobicity can be achieved. But the mechanisms related to surface adhesion of water and ice differs because water cannot withstand shear stress but can support pressure or tensile compression. Thus, to understand superhydrophobic surfaces and their icephobic characteristics, it is essential to comprehend mechanical forces which affect ice and liquid droplets. Figure 10 shows the ice adhesion values for different material categories. It has been shown that ice adhesion increases as the receding contact angle decreases on smooth surfaces.

It is observed that if there are no proper sized pores, some superhydrophobic surfaces can strongly adhere to ice (Nosonovsky and Hejazi 2012). The assumption that measurement of hydrophobicity is a good indicator for materials icephobicity (Menini and Farzaneh 2009) is shown to be incorrect (Jung et al. 2011, Kulinich et al. 2011, Nosonovsky and Hejazi 2012). However studies by (Hejazi et al. 2013) showed that water and ice do have some features and properties in common which includes bouncing-off behaviour towards incoming droplets and interactions that are driven by Gibbs' surface energy minimisation. For instance, aluminum alloy 6061 was deposited with PTFE coating and showed 2.5 times decrease in ice adhesion. Water contact angle for PTFE-coatings were hydrophobic and was found to be around 130–140° (Menini and Farzaneh 2009). By studying the wetting behaviour, a correlation between low wetting hysteresis and ice adhesion strength was observed but no correlation between water contact angle and ice adhesion (Kulinich and Farzaneh 2009) was noted. It was noted that without SLIPS, i.e., with the same principle of air trapping into

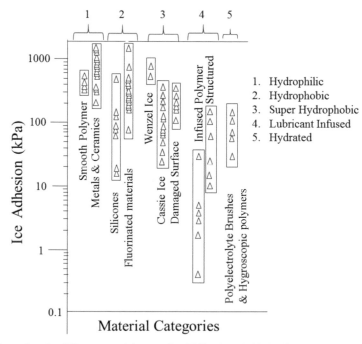

Fig. 10. Ice adhesion values for different material categories (Philseok et al. 2012, Liu et al. 2015, Vogel et al. 2013, Wang et al. 2013, Zhu et al. 2013, Urata et al. 2015, Wang et al. 2015, Petrenko and Peng 2003, Jellinek et al. 1978, Chen et al. 2014, Chernyy et al. 2014, Farhadi et al. 2011, Fu et al. 2014, Ge et al. 2013, Kulinich and Farzaneh 2009, Meuler et al. 2010, Momen et al. 2015, Sojoudi et al. 2015, Subramanyam et al. 2013, Susoff et al. 2013, Wang et al. 2014, Yin et al. 2015, Zou et al. 2011, Dou et al. 2014, Kulinich et al. 2011, Chen et al. 2013).

the porous surface excellent anti-icing properties could be achieved. Peng et al. (Peng et al. 2012) demonstrated this by dissolving poly(vinyldiene) fluoride (PVDF) into N,N-dimethylformamide and coating it to a wind turbine blade to create rough porous surfaces with pores ranging from 1 to 5 μm, into which the air could be trapped. These coated samples showed water contact angle of 156° and sliding angle of only 2° where the supercooled droplets rolled off and no ice crystals appeared to the surface of PVDF coating.

SLIPS have also presented similar properties to air trapped superhydrophobic surfaces for ice-repellence. However the lubricant depletion due to droplets can be seen as a major challenge in this technology. To avoid this scenario reservoir layer is ought to be developed beneath the porous membrane which should have a continuous and to some extent in-built permanent lubrication system. They should have the capability to shift volumes of lubricant onto the surface in a given intervals of time. This would be advantageous to maintain the continuous lubricant transfer throughout the operational life of the component (Wong et al. 20011). To get the lubricant releasing from the reservoir layer towards the top layer, surface energy gradient should be used to gain the movement. Since the pores are required to be under the capillary length of the lubricating fluids, gravity is expected to have no effect on the lubricant movement in the porous structure. This requires explicit understanding of the chemistry of the solids so that manufacturing of SLIPS can be done by cost-effective methods.

Based on SLIPS, many applications are being explored in a variety of industries. For example, enhancing condensation heat transfer (Anand et al. 2012, Xiao et al. 2013), omniphobic textiles (Shillingford et al. 2014), self-cleaning solar panels (Wong et al. 2011), biofouling prevention (Epstein et al. 2012), etc., are some applications where the engineered SLIPS could be used in the future. The problems of superhydrophobic and liquid infused surfaces which include wetting ring formation, cloaking, mechanical endurance of the micro- and nanoscale textures and lubricant depletion are being addressed by many researchers (Ludmila et al. 2013, Jianyong et al. 2014, Samaha et al. 2012, Yao et al. 2011).

Conclusion

Ice accretion poses significant challenges in infrastructures, marine applications, wind turbines, aerospace, power transmission, telecommunications and other industries. In this chapter, we have described on the ways in which ice prevention strategies have been employed using SLIPS. Although academically progress has been made, no single surface has shown the ability to rapidly shed impacting and condensing water droplets, suppress ice nucleation and reduce ice adhesion, in real time environments with required high durability and life. SLIPS can possess the key to become a next generation anti-icing solution wind turbine industries (Wong et al. 2011). Besides anti-icing, SLIPS have proven to have potential properties and characteristics in self-healing (Wong et al. 2011), anti-sticking (Richard and Quéré 2007), anti-fouling (Barthlott and Neinhuis 1997) and self-cleaning (Blossey and Scientifique 2003). These properties would enable easy removal of insect proteins and algae that adhere to the blade surface and impacting it aerodynamically thereby enhancing the energy and cost savings. Although anti-icing materials continue to be improved, each has limitations in some aspects. Authors believe that by understanding the successes and failures of each technology, it may be possible to design surfaces that incorporate features from multiple strategies to further improve versatility. Although, the integration of anti-icing materials with current wind turbine technologies has not seen extensive scope, it is important to develop materials that will meet the requirements for the eventual application.

References

Anand, S., A.T. Paxson, R. Dhiman, J.D. Smith and K.K. Varanasi. 2012. Enhanced condensation on lubricant-impregnated nanotextured surfaces. ACS Nano 6: 10122–10129.
Antikainen, P. and S. Peuranen. 2000. Ice Loads Case Study. Proc. BOREAS V Conference.

Aytug, T., M.P. Paranthaman, J.T. Simpson and D.F. Bogorin. 2014. Superhydrophobic Films and Methods for Making Superhydrophobic Films. U.S. Patent # 20,140,065,368 A1.

Azimi, G., R. Dhiman, H.M. Kwon, A.T. Paxson and K.K. Varanasi. 2013. Hydrophobicity of rare-earth oxide ceramics. Nature Materials 12: 315–320.

Barthlott, W. and C. Neinhuis. 1997. Characterization and distribution of water repellent, self-cleaning plant surfaces. Annals of Botany 79: 667–677.

Battisti, L., R. Fedrizzi, S. Dell'Anna and M. Rialti. 2005. Ice Risk Assessment for Wind Turbine Rotors Equipped with De-Icing Systems. BOREAS VII. FMI, Saariselkä, Findland. 11.

Bechert, D.W., M. Bruse and W. Hage, 2000. Experiments with three-dimensional riblets as an idealized model of shark skin. Experiments in Fluids 28: 403–412.

Bharathidasan, T., S.V. Kumar, M.S. Bobji, R.P.S. Chakradhar and B.J. Basu. 2014. Effect of wettability and surface roughness on ice-adhesion strength of hydrophilic, hydrophobic and superhydrophobic surfaces. Applied Surface Science 314: 241–250.

Bhushan, B., Y.C. Jung and K. Koch. 2009. Micro - nano- and hierarchical structures for superhydrophobicity, self-cleaning and low adhesion. Philosophical Transactions. Series A, Mathematical, Physical, and Engineering Sciences 367: 1631–1672.

Blossey, R. and C. Scientifique. 2003. Self-cleaning surfaces—virtual realities. Nature Materials 2: 301–306.

Botta, G., M. Cavaliere, S. Viani and S. Pospo. 1988. Effects of hostile terrains on wind turbine performances and loads: The acqua spruzza experience. Journal of Wind Engineering and Industrial Aerodynamics 74-76: 419–31.

Burton, Z. and B. Bhushan. 2006. Surface characterization and adhesion and friction properties of hydrophobic leaf surfaces. Ultramicroscopy 106: 709–719.

Callies, M., Y. Chen, F. Marty, A. Pépin and D. Quéré. 2005. Microfabricated textured surfaces for super-hydrophobicity investigations. Microelectronic Engineering 78-79: 100–105.

Cao, L., A.K. Jones, V.K. Sikka, J. Wu and D. Gao. 2009. Anti-Icing superhydrophobic coatings. Langmuir 25: 12444–12448.

Cassie, A.B.D. and S. Baxter. 1944. Wettability of porous surfaces. Transactions of the Faraday Society 40: 546–551.

Chau, T.T., W.J. Bruckard, P.T.L. Koh and A.V. Nguyen. 2009. A review of factors that affect contact angle and implications for flotation practice. Advances in Colloid and Interface Science 150: 106–115.

Chau, T.T. 2009. A review of techniques for measurement of contact angles and their applicability on mineral surfaces. Minerals Engineering 22: 213–219.

Chen, J., Dou Renmei, Cui Dapeng, Zhang Qiaolan, Zhang Yifan, Xu Fujian et al. 2013. Robust prototypical anti-icing coatings with a self-lubricating liquid water layer between ice and substrate. ACS Applied Materials & Interfaces 5: 4026–4030.

Chen, J., Z. Luo, Q. Fan, J. Lv and J. Wang. 2014. Anti-icecoating inspired by ice skating. Small 10: 4693–9.

Chernyy, Sergey, Järn Mikael, Shimizu Kyoko, Swerin Agne, Uttrup Steen, Pedersen et al. 2014. Superhydrophilic polyelectrolyte brush layers with imparted anti-icing properties: Effect of counter ions. ACS Applied Materials and Interface 6: 6487–6496.

Dalili, N., A. Edrisy and R. Carriveau. 2009. Review of surface engineering issues critical to wind turbine performance. Renewable and Sustainable Energy Reviews 13: 428–438.

Daniel, D., M.N. Mankin, R.A. Belisle, T.S. Wong and J. Aizenberg. 2013. Lubricant-infused micro/nano-structured surfaces with tunable dynamic omniphobicity at high temperatures. Applied Physics Letters 102: 231603 1–5.

Della-Bona, A. 2005. Characterizing ceramics and the interfacial adhesion to resin: II—the relationship of surface treatment, bond strength, interfacial toughness and fractography. Journal of Applied Oral Science 13: 101–109.

Dou, Renmei, Chen Jing, Zhang Yifan, Wang Xupeng, Dapeng, Cui, Song Yanlin et al. Anti-icing coating with an aqueous lubricating layer. 2014. ACS Applied Materials and Interfaces 6: 6998–7003.

DuPont. 2011. DuPont Performance Lubricants. Product overview.

Ebert, D. and B. Bhushan. 2012. Durable lotus-effect surfaces with hierarchical structure using micro- and nanosized hydrophobic silica particles. Journal of Colloid and Interface Science 368: 584–591.

Epstein, A.K., T.S. Wong, R.A. Belisle, F.M. Boggs and J. Aizenberg. 2012. Liquid-infused structured surfaces with exceptional anti-biofouling performance. Proceedings of the National Academy of Sciences of the USA 109: 13182–13187.

Farhadi, S., M. Farzaneh and S.A. Kulinich. 2011. Anti-icing performance of superhydrophobic surfaces. Applied Surface Science 257: 6264–6269.

Feng, X., L. Feng, M. Jin, J. Zhai, L. Jiang and D. Zhu. 2004. Reversible super-hydrophobicity to super-hydrophilicity transition of aligned ZnO nanorod films. Journal of the American Chemical Society 126: 62–63.

Fortin, G., J. Perron and A. Ilinca. 2005. Behaviour and Modeling of Cup Anemometers under Icing Conditions. IWAIS XI, Montréal, Canada.

Fu, Qitao, Wu Xinghua, Kumar Divya, Jeffrey W.C. Ho, Pushkar D. Kanhere, Narasimalu Srikanth et al. 2014. Development of sol-gel icephobic coatings: Effect of surface roughness and surface energy. ACS Applied Material Interfaces 6: 20685–20692.

Gao, X. and L. Jiang. 2004. Biophysics: Water-repellent legs of water striders. Nature 432.

Gao, L. and T.J. McCarthy. 2007. How Wenzel and Cassie were wrong. Langmuir 23: 3762–3765.

Ge, Liang, Guifu Ding, Hong Wang, Jinyuan Yao, Ping Cheng and Yan Wang. 2013. Anti-icing property of super hydrophobic Octadecyltrichlorosilane film and its ice adhesion strength. Journal of Nanomaterials 1–5.

Genzer, J. and K. Efimenko. 2006. Recent developments in superhydrophobic surfaces and their relevance to marine fouling: A review. Biofouling 22: 339–360.

Gibbs, J.W. 1878. On the equilibrium of heterogeneous substances. Transactions of the Connecticut Acadamy of Arts and Sciences 3: 343–524.

Gillenwater, D. 2008. Pertes de Puissance Associées aux Phénomènes Givrants sur une Éolienne Installée en Climat Nordique. Master Thesis, ÉTS, Montréal, Canada, 162.

Glavan, A.C., R.V. Martinez, A.B. Subramaniam, H.J. Yoon, R.M.D. Nunes, H. Lange et al. 2014. Omniphobic "R^F paper" produced by silanization of paper with fluoroalkyltrichlorosilanes. Advanced Functional Materials 24: 60–70.

Hejazi, V., K. Sobolev and M. Nosonovsky. 2013. From Superhydrophobicity to icephobicity: Forces and interaction analysis. Scientific Reports 3: 2194.

Henna niemelä-anttonen. 2015. Wettability and anti-icing properties of slippery Liquid infused porous surface. MS Thesis, Tampere University of Technology, Finland.

Holman, H.P. and T.D. Jarrell. 1923. The effects of waterproofing materials and outdoor exposure upon the tensile strength of cotton yarn. Indistrial and Engineering Chemistry 15: 236–240.

Homola, M.C. 2005. Wind energy in BSR: Impacts and causes of icing on wind turbines. For Interreg IIIB Project partners. Matthew Carl Homola, Narvik University College, Norway. 1–15.

Hou, X., F. Zhou, B. Yu and W. Liu. 2007. Superhydrophobic zinc oxide surface by differential etching and hydrophobic modification. Materials Science and Engineering A 452-453: 732–736.

Ishizaki, T. and N. Saito. 2010. Rapid formation of a superhydrophobic surface on a magnesium alloy coated with a cerium oxide film by a simple immersion process at room temperature and its chemical stability. Langmuir: The ACS Journal of Surfaces and Colloids 26: 9749–9755.

Jasinski, W.J., S.C. Noe, M.S. Selig and M.B. Bragg. 1997. Wind Turbine Performance Under Icing Conditions, Aerospace Sciences Meeting & Exhibit AIAA, Reno, USA.

Jasinski, W.J., S.C. Noe, M.S. Selig and M.B. Bragg. 1998. Wind turbine performance under icing conditions. Transactions of the ASME, Journal of Solar Energy and Engineering 120: 60–65.

Jellinek, H.H.G., H. Kachi, S. Kittaka, M. Lee and R. Yokota. 1978. Ice releasing block-copolymer coatings. Colloid and Polymer Science 256: 544–551.

Jianyong. Lv., Yanlin Song, Lei Jiang and Jianjun Wang. 2014. Bio-inspired strategies for anti-icing. ACS Nano 8: 3152–3169.

Jin, M., X. Feng, J. Xi, J. Zhai, K. Cho, L. Feng et al. 2005. Super-hydrophobic PDMS surface with ultra-low adhesive force. Macromolecular Rapid Communications 26: 1805–1809.

Jung, S., M. Dorrestijn, D. Raps, A. Das, C.M. Megaridis and D. Poulikakos. 2011. Are superhydrophobic surfaces best for icephobicity. Langmuir 27: 3059–3066.

Ketelson, H., S. Perry, G. Sawyer and J. Jean. 2011. Exploring the Science and Technology of Contact Lens Comfort.

Khorasani, M.T., H. Mirzadeh and Z. Kermani. 2005. Wettability of porous polydimethylsiloxane surface: Morphology study. Applied Surface Science 242: 339–345.

Kim, P., M.J. Kreder, J. Alvarenga and J. Aizenberg. 2008. Hierarchical or not? Effect of the length scale and hierarchy of the surface roughness on omniphobicity of lubricant-infused substrates. Nano Letters 13: 1793–1799.

Kim, P., M. Kreder and T.S. Wong. 2011. Slippery icephobic materials. Webpage.

Kim, P., W.E. Adorno-Martinez, M. Khan and J. Aizenberg. 2012. Enriching libraries of high-aspect-ratio micro or nanostructures by rapid, low-cost, benchtop nanofabrication. Nature Protocols 7: 311–327.

Koch, K., B. Bhushan and W. Barthlott. 2009. Multifunctional surface structures of plants: An inspiration for biomimetics. Progress in Materials Science 54: 137–178.

Krasovitski, B. and A. Marmur. 2005. Drops down the hill: Theoretical study of limiting contact angles and the hysteresis range on a tilted plate. Langmuir 21: 3881–3885.

Kulinich, S.A. and M. Farzaneh. 2009. How wetting hysteresis influences ice adhesion strength on superhydrophobic surfaces. Langmuir 25: 8854–8856.

Kulinich, S.A., S. Farhadi, K. Nose and X.W. Du. 2011. Superhydrophobic surfaces: Are they really icerepellent? Langmuir 27: 25–29.

Kumar, G. and K.N. Prabhu. 2007. Review of non-reactive and reactive wetting of liquids on surfaces. Advances in Colloid and Interface Science 133: 61–89.

Kwok, D.Y., C.N.C. Lam, A. Li, A. Leung, R. Wu, E. Mok et al. 1998. Measuring and interpreting contact angles: A complex issue. Colloids and Surfaces A: Physicochemical and Engineering Aspects 142: 219–235.

Kwon, Y., S. Choi, N. Anantharaju, J. Lee, M.V. Panchagnula and N.A. Patankar. 2010. Is the cassie-baxter formula relevant? Langmui 26: 17528–17531.

Laakso, T., E. Peltola, P. Antikainen and S. Peuranen. 2003. Comparison of Ice Sensors for Wind Turbines, BOREAS VI. FMI, Pyhätunturi, Finland.

Laakso, T., H. Holttinen, G. Ronsten, L. Tallhaug, R. Horbaty, I. Baring-Gould et al. 2003. State-of-the-Art of Wind Energy in Cold Climate 1–56.

Laakso, T. and E. Peltola. 2005. Review on blade heating technology and future prospects. BOREAS VII. FMI, Saariselkä, Finland.

Laforte, J., M. Allaire and C. Laforte. 2005. Demonstration of the feasibility of a new mechanical method of cable de-icing. Proc. 11th International Workshop on Atmospheric Icing of Structures, Montreal 347–352.

Lafuma, A. and D. Quéré. 2011. Slippery pre-suffused surfaces. EPL Europhysics Letters 96: 56001.

Lalia, B.S., S. Anand, K.K. Varanasi and R. Hashaikeh. 2013. Fog-harvesting potential of lubricant-impregnated electrospun nanomats. Langmuir 29: 13081–13088.

Liang, J., Y. Hu, Y. Fan and H. Chen. 2013. Formation of superhydrophobic cerium oxide surfaces on aluminum substrate and its corrosion resistance properties. Surface and Interface Analysis 45: 1211–1216.

Lindstrom, A.B., M.J. Strynar and E.L. Libelo. 2011. Polyfluorinated compounds: Past, present, and future. Environmental Science & Technology 45: 7954–61.

Liu, Y., X. Chen and J.H. Xin. 2008. Hydrophobic duck feathers and their simulation on textile substrates for water repellent treatment. Bioinspiration & Biomimetics 3: 46007.

Liu, Qi, Ying Yang, Meng Huang, Yuanxiang Zhou, Yingyan Liu and Xidong Liang. 2015. Durability of a lubricant-infused electrospray silicon rubber surface as an anti-icing coating. Applied Surface Science 346: 68–76.

Lu, X., C. Zhang and Y. Han. 2004. Low-density polyethylene superhydrophobic surface by control of its crystallization behavior. Macromolecular Rapid Communications 25: 1606–1610.

Ludmila B. Boinovich and Alexandre M. Emelyanenko 2013. Anti-icing potential of superhydrophobic coatings. Mendeleev Communications 23: 03–10.

Maissan, J.F. 2001. Wind power development in sub-arctic conditions with severe rime icing. Proc. Circumpolar Climate Change Summit and Exposition.

Maissan, J.F. 2006. Report on Wind Energy for Small Communities, Prepared for Inuit Tapiriit Kanatami 1–25.

Makkonen, L., T. Laakso, M. Marjaniemi and K.J. Finstad. 2001. Modeling and prevention of ice accretion on wind turbines. Wind Engineering 25: 3–21.

Makkonen, L. 1994. Ice and construction. Rilem report 13. 1st Edition. Chapman & Hall. London, England.

Manna, U. and D.M. Lynn. 2015. Fabrication of liquid-infused surfaces using reactive polymer multilayers: Principles for manipulating the behaviors and mobilities of aqueous fluids on slippery liquid interfaces. Advanced Materials 27: 3007–3012.

Mara, K., G. Kelk, D. Etkin, I. Burton and S. Kalhok. 1999. Glazed over: Canada copes with the ice storm of 1998. Environment Science and Policy for Sustainable Development 41: 6–11.

Mc. Burney, D. 1935. Coated fabrics in construction industry. Industiral and Engineering Chemistry 27: 1400–1403.

McHale, G. 2007. Cassie and Wenzel: Were they really so wrong? Langmuir 23: 8200–8205.

Mc Keen, L.W. 2006. Fluorinated Coatings and Finishes Handbook: The Definitive User's Guide. William Andrew.

Menini, R. and M. Farzaneh. 2009. Elaboration of Al_2O_3/PTFE icephobic coatings for protecting aluminum surfaces. Surface and Coatings Technology 203: 1941–1946.

Meuler, Adam J., Smith J. David, Varanasi Kripa K., Mabry Joseph M., McKinley Gareth H., Cohen Robert E. et al. 2010. Relationships between water wettability and ice adhesion. ACS Applied Materials and Interfaces 2: 3100–3110.

Michael, J., Kreder, Alvarenga Jack, Kim Philseok and Aizenberg Joanna. 2016. Design of anti-icing surfaces: Smooth, textured or slippery? Nature Reviews Materials Article number: 15003

Momen, G., R. Jafari and M. Farzaneh. 2015. Ice repellency behaviour of superhydrophobic surfaces: Effects of atmospheric icing conditions and surface roughness. Applied Surface Science 349: 211–218.

Moretto, H.-H., M. Schulze and G. Wagner. 2000. Silicones. Ullmann's Encyclopedia of Industrial Chemistry 1–38.

Morra, M., E. Occhiello and F. Garbassi. 1989. Contact angle hysteresis on oxygen plasma treated polypropylene surfaces. Journal of Colloid and Interface Science 132: 504–508.

Nakajima, A. 2011. Design of hydrophobic surfaces for liquid droplet control. NPG Asia Materials 3: 49–56.

National Transportation Safety Board. 1996. Aviation Accident Report AAR-96-01: In-flight Icing Encounter and Loss of Control Simmons Airlines, d.b.a. American Eagle Flight 4184 Avions de Transport Regional (ATR) Model 72–212, N401AM.

Nosonovsky, M. and B. Bhushan. 2005. Roughness optimization for biomimetic superhydrophobic surfaces. Microsystem Technologies 11: 535–549.

Nosonovsky, M. 2007. Model for solid–liquid and solid–solid friction of rough surfaces with adhesion hysteresis. The Journal of Chemical Physics 126: 224701.

Tadmor, R. 2004. Line energy and the relation between advancing, receding, and young contact angles. Langmuir 20: 7659–7664.

Nosonovsky, M. and B. Bhushan. 2008. Multiscale Dissipative Mechanisms and Hierarchical Surfaces: Friction, Superhydrophobicity, and Biomimetics. Springer Science & Business Media.

Nosonovsky, M. and V. Hejazi. 2012. Why superhydrophobic surfaces are not always icephobic. ACS Nano 6: 8488–8491.

Okada, I. and S. Shiratori. 2014. High-transparency, self-standable Gel-SLIPS fabricated by a facile nanoscale phase separation. ACS Applied Materials and Interfaces 6: 1502–1508.

Patankar, N.A. 2004. Mimicking the lotus effect: Influence of double roughness structures and slender pillars. Langmuir 20: 8209–8213.

Peng, C., S. Xing, Z. Yuan, J. Xiao, C. Wang and J. Zeng. 2012. Preparation and anti-icing of superhydrophobic PVDF coating on a wind turbine blade. Applied Surface Science 259: 764–768.

Petrenko, V.F. and S. Peng. 2003. Reduction of ice adhesion to metal by using self-assembling monolayers (SAMs). Canadian Journal of Physics 81: 387–393.

Philseok, Kim., Wong Tak-Sing, Alvarenga Jack, Kreder Michael J. and Adorno-Martinez Wilmer E. et al. 2012. Liquid-infused nanostructured surfaces with extreme anti-ice and anti-frost performance. ACS Nano 6: 6569–6577.

Qiu, R., Q. Zhang, P. Wang, L. Jiang, J. Hou, W. Guo et al. 2014. Fabrication of slippery liquid-infused porous surface based on carbon fiber with enhanced corrosion inhibition property. Colloids and Surfaces 453: 132–141.

Quéré, D. 2005. Non-sticking drops. Reports on Progress in Physics 68: 2495–2532.

Richard, D. and D. Quéré. 2007. Viscous drops rolling on a tilted non-wettable solid. Europhysics Letters 48: 286–291.

Rickard, T.A. and O.C. Ralston. 1917. Flotation. Mining and Scientific Press.

Roach, P., N.J. Shirtcliffe and M.I. Newton. 2008. Progress in superhydrophobic surface development. Soft Matter 4: 224–240.

Rutter, M.P. 2011. Plant offers slick strategy.

Rykaczewski, K., S. Anand, S.B. Subramanyam and K.K. Varanasi. 2013. Mechanism of frost formation on lubricant-impregnated surfaces. Langmuir 29: 5230–5238.

Samaha, M.A., H.V. Tafreshi and M. Gad-El-Hak. 2012. Influence of flow on longevity of superhydrophobic coatings. Langmuir 28: 9759–9766.

Seifert, H. 2003. Technical Requirements for Rotor Blades Operating in Cold Climate, BOREAS VI. FMI: 13.

Sellinger, A., P.M. Weiss, A. Nguyen, Y. Lu, R.A. Assink, W. Gong et al. 1998. Continuous self-assembly of organic-inorganic nanocomposite coatings that mimic nacre. Nature 394: 256–260.

Shillingford, C., N. MacCallum, T.S. Wong, P. Kim and J. Aizenberg. 2014. Fabrics coated with lubricated nanostructures display robust omniphobicity. Nanotechnology 25: 14–19.

Shiu, J.-Y., C.W. Kuo and P. Chen. 2004. Fabrication of tunable superhydrophobic surfaces. Proc. SPIE 5648. Smart Materials III, 325–332.

Sojoudi, H., G.H. McKinley and K.K. Gleason. 2015. Linker-free grafting of fluorinated polymeric cross-linked network bilayers for durable reduction of ice adhesion. Materials a Horizons 2: 91–99.

Spori, D.M., T. Drobek, S. Zürcher, M. Ochsner, C. Sprecher, A. Mühlebach et al. 2008. Beyond the lotus effect: Roughness influences on wetting over a wide surface-energy range. Langmuir 24: 5411–5417.

Starov, V.M., M.G. Velarde and C.J. Radke. 2007. Wetting and Spreading Dynamics. CRC Press.

Subramanyam, S.B., K. Rykaczewski and K.K. Varanasi. 2013. Ice adhesion on lubricant-impregnated textured surfaces. Langmuir 29: 13414–13418.

Susoff, M., K. Siegmann, C. Pfaffenroth and M. Hirayama. 2013. Evaluation of icephobic coatings-screening of different coatings and influence of roughness. Applied Surface Science 282: 870–879.

Talhaug, L., K. Vindteknik, G. Ronsten, R. Horbaty, I. Baring-Gould, A. Lacroix et al. 2005. Wind energy projects in cold climates. 1st edition Executive Committee of the International Energy Agency Program for Research and Development on Wind Energy Conversion Systems 1–36.

Tammelin, B., A. Heimo and M. Leroy. 2000. Wind Energy Production in Cold Climate (WECO). FMI: 41.

Tammelin, B. et al. 2005. Wind Turbines in Icing Environment: Improvement of Tools for Siting, Certification and Operation New Ice Tools: 127.

Tammelin, B., M. Cavaliere, H. Holttinen, C. Morgan, H. Seifert and K. Säntti. 1996–1998. Wind energy production in cold climate (WECO) (Photo courtesy of Kranz). 1–38.

Tuteja, A., W. Choi, M. Ma, J.M. Mabry, S.A. Mazzella, G.C. Rutledge et al. 2007. Designing superoleophobic surfaces. Science. New York 318: 1618–1622.

Ulusoy, U. and M. Yekeler. 2005. Correlation of the surface roughness of some industrial minerals with their wettability parameters. Chemical Engineering and Processing: Process Intensification 44: 555–563.

Urata, C., G.J. Dunderdale, M.W. England and A. Hozumi. 2015. Self-lubricating organogels (SLUGs) with exceptional syneresis-induced anti-sticking properties against viscous emulsions and ices. Journal of Material Chemistry A 3: 12626–12630.

Veeramasuneni, S., J. Drelich, J. Miller and G. Yamauchi. 1997. Hydrophobicity of ion-plated PTFE coatings. Progress in Organic Coatings 31: 265–270.

Vogel, N., R.A. Belisle, B. Hatton, T.S. Wong and J. Aizenberg. 2013. Transparency and damage tolerance of patternable omniphobic lubricated surfaces based on inverse colloidal monolayers. Nature Communications 4: 2176.

Wang, H., G. He and Q. Tian. 2012. Effects of nano-fluorocarbon coating on icing. Applied Surface Science 258: 7219–7224.

Wang, Y., J. Xue, Q. Wang, Q. Chen and J. Ding. 2013. Verification of icephobic/anti-icing properties of a superhydrophobic surface. ACS Applied Materials Interfaces 5: 3370–3381.

Wang, C., T. Fuller, W. Zhang and K.J. Wynne. 2014. Thickness dependence of ice removal stress for a polydimethylsiloxane nanocomposite: Sylgard 184. Langmuir 30: 12819–12826.

Wang, Y., Y. Xi, C. Jing, Z. He, J. Liu, Q. Li et al. 2015. Organogel as durable anti-icing coatings. Science China Materials 58: 559–565.

Weis, T.M. and J.F. Maissan. 2003. The effect of black blades on surface temperature for wind turbines.

Wenzel, R.N. 1936. Resistance of solid surfaces to wetting by water. Industrial and Engineering Chemistry 28: 988–994.

Wilson, P.W., W. Lu, H. Xu, P. Kim, M.J. Kreder, J. Alvarenga et al. 2013. Inhibition of ice nucleation by slippery liquid-infused porous surfaces (SLIPS). Physical Chemistry Chemical Physics 15: 581–585.

Wong, T.S., S.H. Kang, S.K.Y. Tang, E.J. Smythe, B.D. Hatton and A. Grinthal et al. 2011. Bioinspired self-repairing slippery surfaces with pressure-stable omniphobicity. Nature 477: 443–447.

Xiao, R., N. Miljkovic, R. Enright and E.N. Wang. 2013. Immersion condensation on oil-infused heterogeneous surfaces for enhanced heat transfer. Scientific Reports 3: 1988.

Yabu, H. and M. Shimomura. 2005. Single-step fabrication of transparent superhydrophobic porous polymer films. Chemistry of Materials 17: 5231–5234.

Yao, X., Y. Song and L. Jiang. 2011. Applications of bio-inspired special wettable surfaces. Advanced Materials 23: 719–734.

Yin, X., Z. Yue, D. Wang, Z. Liu, X. Pei et al. 2015. Integration of self-lubrication and nearinfrared photothermogenesis for excellent anti-icing/deicing performance. Advanced Functional Materials 25: 4237–4245.

Young, T. 1805. An essay on the cohesion of fluids. Philosophical Transactions 95: 65–87.

Youngblood, J.P. and T.J. McCarthy. 1999. Ultrahydrophobic polymer surfaces prepared by simultaneous ablation of polypropylene and sputtering of poly(tetrafluoroethylene) using radio frequency plasma. Macromolecules 32: 6800–6806.

Zaggia, A. and B. Ameduri. 2012. Recent advances on synthesis of potentially non-bioaccumulable fluorinated surfactants. Current Opinion in Colloid and Interface Science 17: 188–195.

Zhang, J., J. Li and Y. Han. 2004. Superhydrophobic PTFE surfaces by extension. Macromolecular Rapid Communications 25: 1105–1108.

Zheng, Q., C. Lv, P. Hao and J. Sheridan. 2010. Small is beautiful and dry. Science China Physics, Mechanics and Astronomy 53: 2245–2259.

Zhou, H., H. Wang, H. Niu, A. Gestos, X. Wang and T. Lin. 2012. Fluoroalkyl silane modified silicone rubber/nanoparticle composite: a super durable, robust superhydrophobic fabric coating. Advanced Materials 24: 2409–2412.

Zhu, L., J. Xue, Y. Wang, Q. Chen, J. Ding and Q. Wang. 2013. Ice-phobic coatings based on silicon-oil infused polydimethylsiloxane. ACS Applied Materials Interfaces 5: 4053–4062.

Zou, M., S. Beckford, R. Wei, C. Ellis, G. Hatton and M.A. Miller. 2011. Effects of surface roughness and energy on ice adhesion strength. Applied Surface Science 257: 3786–3792.

Zushi, Y., J.N. Hogarh and S. Masunaga. 2011. Progress and perspective of perfluorinated compound risk assessment and management in various countries and institutes. Clean Technologies and Environmental Policy 14: 9–20.

13

Towards a Universal Model of High Energy Density Capacitors

Francisco Javier Quintero Cortes,[1] *Andres Suarez*[2] and *Jonathan Phillips*[3,*]

1. Introduction

Over the last few decades, parallel efforts have been made to improve several distinctly different types of capacitors with the intent of developing/discovering capacitors with far higher energy density, and concomitantly power delivery capability, than the historic 'electrostatic' type of capacitors. Given the increased use of electric power, technological paradigms regarding capacitor electrical energy density are being re-examined. For example, the old paradigm was that capacitors, for any application, should be made with the material with the highest dielectric value. Yet 'supercapacitors', based on low dielectric constant materials, but high surface area electrodes, are clearly the commercial capacitor variety with the highest energy density. A second example regards materials with the highest dielectric constants. The old paradigm was that solids will always have the highest dielectric constants. Yet, the recent development of super dielectric materials (SDM), based on salt in solution, clearly shows that is not the case (Gandy et al. 2016, Cortes and Phillips 2015a,b, Fromille and Phillips 2014, Jenkins et al. 2016, Phillips 2016). Another paradigm under reconsideration is that batteries will always have higher energy density than capacitors.

Can capacitors be developed with higher energy density than the best batteries? If battery development had stalled over the past two decades, the above question could be answered in the positive for many uses, although with caveats. Capacitor research has clearly led to remarkable improvements in performance. Capacitor energy density has improved so dramatically that the best capacitor prototypes (ca. 450 J/cm^3) now appear to have surpassed classic battery types, including lead acid (ca. 360 J/cm^3), in terms of volumetric energy density (Divya and Østergaard 2009). But, battery technology has similar improvement and some commercial lithium ion batteries have almost an order of magnitude higher energy density than the best capacitor prototypes, and a couple of orders of magnitude higher than the best commercial capacitors.

There is a multitude of potential applications for the high energy density capacitors currently at the prototype state and under consideration for commercial development for which volume and weight are not primary cost drivers. Capacitors are ubiquitous in electric circuits for power rectification, filtering,

---

[1] School of Material Science and Engineering, Georgia Institute of Technology, Atlanta GA 30318.
[2] Department of Mechanical Engineering, Central Universidad de La Defensa, Marin, Spain.
[3] Energy Academic Group, Naval Postgraduate School, Monterey CA 93943.
* Corresponding author: jphillip@nps.edu

resonance tuning, data storage, etc. In many of these applications, capacitors with higher energy density will reduce weight, and/or volume, that is 'footprint'. But, the greatest need for improved capacitors is for systems in which very high power is needed in bursts or 'pulses'. Batteries are not suitable for pulsed power application, hence capacitors are a key technology, possibly irreplaceable, for enabling many pulsed power applications. High density power is needed for lasers, prototype fusion reactors, storing 'breaking energy' from electric trains, trucks and cars, for levelling power to the grid from 'green' sources with irregular/uncontrollable power delivery such as windmills and solar cells. Regarding the latter: the final economic barrier to photovoltaic solar cells and/or wind energy becoming the primary electric energy source for the grid is the availability of low cost electric energy storage. Battery cost is decreasing at such a pace that the net cost of solar/battery storage is predicted to be less than the net cost of gas fired turbines before 2020. Once this cost barrier is broken it is only a matter of time before solar will become the primary, not the back-up, source of electric power for the grid. Notably, there is no fundamental reason that volume and/or weight are significant cost drivers in grid storage application, thus if capacitors for which the cost per unit of energy storage is less than that of batteries, solar/capacitor systems could replace solar/battery systems in 'grid' application.

There are general advantages of capacitors, rather than batteries, as a source of electric power energy storage that impact uses and cost. One advantage relative to batteries is there are no chemical reactions taking place, hence capacitors can be cycled orders of magnitude more times than batteries without performance loss. Another abiding strength of energy storage capacitors is the ability to deliver and receive energy orders of magnitude faster than a battery of the same energy density. Since capacitors charge far more quickly than batteries, they are the preferred electric power source for systems requiring repeated bursts of power, and for the same reason are theoretically more convenient for applications such as consumer electronics, electric powered cars, etc. Capacitors are also less temperature sensitive than batteries making them a preferential source for power in some environments.

Another application of 'energy storage' style capacitors is for load levelling battery systems (Miller and Simon 2008, Dowgiallo and Hardin 1995, Cohen et al. 2014, Hemmati and Saboori 2016, Wong et al. 2016). Capacitors used in load leveling circuits can significantly improve battery lifetime, presently the limit on the lifetime of many systems such as satellites, by providing peak power. Batteries, which store energy chemically, are damaged by either delivering or receiving too much power. Capacitors can buffer battery based systems such that batteries never see power spikes. Thus, even though capacitors still lag batteries in energy density, they are an integral part of all high power systems, hence, higher energy density capacitors are needed to reduce the footprint, or improve the lifetime, of power systems in many applications. And perhaps someday capacitors will have sufficient energy density, coupled with fast charging rates and remarkable lifetimes that they will compete with batteries as the preferred electrical energy storage modality for many applications.

It must also be noted that there are caveats to facile comparison between batteries and capacitors on an energy density basis as a straight-up comparison misses 'energy quality' (Huggins 2010). Batteries and capacitors deliver energy, charge, etc., in very different fashions. In terms of energy delivery, batteries deliver almost constant voltage, 'high quality' electric energy, until the reactants that provide the chemical/electrical energy are nearly exhausted. Capacitors have 'lower quality' energy as they do not deliver energy at a constant voltage, but rather lose voltage exponentially during discharge. Hence, the two types of electrical energy storage device are not equivalent. Thus, even if capacitors reach the energy density of batteries, they will never be a simple 'drop in' substitute for batteries. That is, they can replace batteries, but only following appropriate changes in energy/power system design (Lu et al. 2007, Khaligh and Li 2010). In practice, this means to replace batteries with capacitors in many applications will require energy density > 2X that of a battery.

The objective of the present essay is limited to facilitating a fundamental/scientific understanding of the many forms of 'new' energy storage capacitors. This essay does not provide an economic survey of capacitor types, or a comparison of capacitors and batteries, although ultimately the use of capacitors in real world energy storage applications will depend on such an analysis. Clearly, a fundamental understanding should not be dismissed as irrelevant to economic development. Herein it is suggested development of economically attractive forms of the leading energy storage type capacitors, EDL Supercapacitors and New Paradigm (NP) Supercapacitors, requires a fundamental scientific understanding.

Uniquely, in this essay the science of all capacitors is explained on the basis of two simple postulates. It is shown this approach permits the development of a simple, 'universal', scientific framework for all capacitors. The first postulate is that all developments in capacitor technology can be explained using the very simple, and classic, equation for the capacitance of a parallel plate capacitor. A review of that equation (later sections) indicates, capacitance in practical devices can be changed only by either (i) increasing the electrode surface area, (ii) decreasing the distance between electrodes or (iii) increasing the dielectric constant. It is shown all capacitor research can be explained as an effort to achieve one of those three objectives.

The second postulate developed here is a simple model of the fundamental physics of dielectrics. In brief: polarisation of dipoles within dielectrics, opposite to the field on the electrodes, reduces the electric field created by charges on the electrodes everywhere in space. In turn, this allows more charges to be brought to the electrodes with less energy. In fact all successful increases in dielectric constant result from increases in either dipole density or dipole length. Finally, it should be noted that powders are often a key to capacitor performance. Specifically, in all cases increasing the electrode surface area involves the use of high surface area conductive powders, and in many cases high dielectric values are achieved by employing high surface area, highly porous, electrically insulating powders.

2. Capacitor Types

There are several approaches to categorising and discussing capacitors. One approach is to divide capacitors under study over the last couple of decades into two major categories:

  i) Electric Double Layer Capacitor (EDLC), herein labelled as EDL Supercapacitors, and
 ii) advanced dielectric based.

Each category can be further divided into sub-categories. Sub-classes of the former include (i) those based on high surface area carbon, (ii) those based specifically on graphene, clearly the object of the most research and resources, and more recently (iii) those based on high surface area conductive clays. Dielectric based capacitors are distinguished by the class of dielectric employed as all other parts of the capacitor are fundamentally the same.

Dielectric sub-classes include (i) ceramic, often improved forms of titanates, (ii) polymer based, (iii) mixed metal/insulating dielectrics, (iv) 'colossal dielectrics', and (v) superdielectric materials (SDM).

An alternative means to characterise capacitors, and the one preferred in this essay, is to place capacitors into categories based on the particular parameter improved/adjusted to maximise capacitance in the standard parallel plate capacitor equation (Resnick and Halliday 1978, Walker 2007):

$$C = \frac{\varepsilon \varepsilon_0 A}{t} \tag{1}$$

where ε is the dielectric constant, ε_0 is the permittivity of free space, A the electrode surface area, and t the distance between plates. Clearly, only A, ε and/or t can be modified.

2.1 Type I, Increasing Area

In the case of EDL Supercapacitors the parameter improved in the equation is the surface area of the electrode. The primary focus of research in this area is to find and employ the electrically conductive material with the highest surface area per unit mass, or unit volume (Conway 1999, Zhang and Zhao 2009, Wang et al. 2012). This is generally leading to the use of graphene in prototype capacitors of this type (Wang et al. 2009, Huang et al. 2012), but perhaps surprisingly, some alternatives, such as conductive clays (Ghidiu et al. 2014) and even conductive oxides, such as lead oxide (Zhang et al. 2014), are proving to be competitive. It is understood that the dielectric constant of the material in EDL Supercapacitors is fixed and of a relatively low value. It is also clear that the 'capacitance' is not sufficient to describe all observed behaviour. For example, for frequencies faster than 1 Hz virtually all EDL Supercapacitors discharge curves show a sharp drop in voltage at the initial moment of discharge, followed, by the 'expected' exponential decay in voltage. This behaviour, repeatedly observed, is not predicted or described by standard theory.

2.2　Type II, Increasing Dielectric Constant

There are four approaches to improving the dielectric constant. The first is to work with variants of the material with the 'best' dielectric constant, historically barium titanate (Reynolds et al. 2012, Kinoshita and Yamaji 1976, Arlt et al. 1985), other more complex titanates (Tong et al. 2013, Tang and Sodano 2013, Parizi et al. 2014), and more recently some polymer based materials (Prateek et al. 2016, Smith et al. 2014, Yu et al. 2016). The second approach is to make mixtures of barium titanate, or other 'standard' ceramic dielectrics, and metal particles (Pecharromán et al. 2004, Pecharromán et al. 2010, Saha 2004, Valant et al. 2006). The third approach is to create solid dielectric with dipoles based on defects including surface states. There is a great deal of evidence of the existence of colossal dielectric behaviour (Samara et al. 1990, Lunkenheimer et al. 2004, Yang et al. 2013, Lunkenheimer et al. 2009, Lunkenheimer et al. 2002), although there is no general agreement on the physics underlying the observation of colossal dielectric behaviour. The final approach is to employ newly discovered Superdielectric Materials, or SDM (Cortes and Phillips 2015a,b, Phillips 2016, Gandy et al. 2016, Jenkins et al. 2016, Fromille and Phillips 2014). One finding that complicates the selection of a simple means to organise a discussion of new types of capacitors is that some SDM materials do not have a constant dielectric value, and that the value of ε is not only a function of the material, but the thickness of the material as well (Cortes and Phillips 2015b).

2.3　Type III, Decreasing Electrode Distance

This approach is under development in conjunction with research into creating and handling smaller barium titanate and metal particles. It is a means to create high power density capacitors, primarily for 'on-chip' applications such as memory storage and micro-machine applications. Despite forming very thin layers, the energy density is not as high as 'predicted, in fact it is < 10 J/cm³ in all cases. It is clear that dielectric values do not scale. As the dielectric layer gets thinner there are clearly phenomenon that limit the ultimate dielectric value and hence energy density (see Capacitor Theory section).

Some capacitors do not easily fit any of the above characterisation schemes. For example, some dielectrics are not designed to maximise any of the parameters of Eq. (1), but rather energy, per Eq. (2), by having very high breakdown voltages:

$$\text{Energy} = 1/2 \ CV^2 \tag{2}$$

Specifically, there is research focused on discovery of materials with 'non exceptional' dielectric values, but very high breakdown voltages, herein labelled 'TYPE IV' (Tang and Sodano 2013, Kim et al. 2015, Parizi et al. 2014, Tong et al. 2013). Conceptually, such capacitors can have high energy density at very high voltages, even if the capacitance of the device is not high.

All of the capacitor types under study employ particles in some versions. For example, the high surface area conductive electrodes in EDL Supercapacitors are carbon powders, and all materials under consideration for 'next generation' EDL Supercapacitors, from graphene to conductive clay, are also particulates. Another example of powder materials are those employed in the most common high dielectric value based capacitors: Barium titanate and its derivatives. Generally the ceramics are employed in particle form, and it appears this will be the case in the future as well. Finally, regarding high energy density capacitors based on SDM, much of the work to date has been with powder SDM materials.

3. Capacitor Theory

To aid in understanding the new developments in capacitor technology a model, universal to all capacitors, containing some new elements, is presented. Also, the primary focus of the theory presentation is to provide a conceptual framework for understanding features of capacitors, rather than to present a mathematical prescription. The postulated theory of all capacitors under discussion herein is very straightforward. In particular, the theory is most clearly applied to 'electrostatic capacitors', capacitors composed of dielectric materials between two parallel conductive electrodes, as per Types II, III and IV, above. The analysis, presented below also pertains to Type I (EDLC), but the argument for that is more complex.

3.1 Geometric Dependence

Capacitance is a function of geometry, as discussed in more detail elsewhere (Resnick and Halliday 1978, Lerner 1996). All bodies can store free charge, and hence create a field, which leads to a capacitance. In terms of energy storage however, there is no argument regarding the preferred geometry: parallel electrodes yield the highest volumetric energy density. Thus, in this essay only the equation for parallel plate geometry, Eq. (1), is developed.

In a capacitor once opposite and equal charges are placed on the two electrodes of a parallel plate capacitor, a field, roughly a 'dipole' field at distances large compared to the dimensions of the capacitor, is produced in all space. As with all dipoles, and assuming the amount of charge on the plates is unchanging, the farther apart the two charge centers, the bigger the dipole and the stronger the field at all points in space. The character of the field determines the energetics of 'loading' the capacitor with additional charge. That is, the energy required to bring a charge from infinite distance to the electrode surface is equal to this simple equation:

$$V = \int E \cdot dl \tag{3}$$

where V is the voltage (scaler), E the electric field (vector) and l is a measure of distance along a path. Thus, the addition of more charge on the plates, increases the field everywhere, leading to an increase in voltage, that is energy required to bring the next charge up to the electrode, *ad nauseum.*

Clearly, as the electrodes get closer the dipole field is smaller at every point in space, thus the integral in Eq. (3), gets smaller. How can this be described 'physically'? That is, why is the field reduced as the two plates are drawn closer? Answer: As the plates are drawn closer the (+) field at any point in space created by the positive charges on one plate, is increasingly cancelled by the (−) field created by negative charges on the other plate, and vice versa. Indeed, the net field at any point is the vector sum of the fields from every charged species on the capacitor plate. As the plates get closer the (+/−) fields increasingly become closer in scaler magnitude at every point in space, yet still opposite in orientation, hence the fields get closer to cancelling at every point. In the limit of the two plates touching, the (+) and (−) fields exactly cancel, and Eq. (3) is net zero.

To further understand the fundamentals of parallel plate capacitors the definition of capacitance must be considered:

$$C = q/V \tag{4}$$

This is best done through a 'thought experiment'. Imagine, two parallel plates, well spaced. Move charge to create a (+) and a (−) plate, hence a voltage Vo is generated, as per Eq. (3). Now, imagine bringing the electrodes closer, without changing the amount of charge on the electrodes. As the electrodes are closer, the fields everywhere are now smaller, and the voltage (Eq. 3) is reduced! The voltage is lower, the charge is the same, and concomitantly, the capacitance, Eq. (4), is higher. This is consistent with Eq. (1) which indicates capacitance increases inversely with distance between electrodes. There is a message in this analysis: To increase capacitance find a means, while holding charge constant, to lower the fields created by charge on the capacitor everywhere. This will lead to a reduction in voltage between the electrodes, and a concomitant increase in capacitance.

3.2 Dielectric Impact on Field and Energy

The standard means to reduce the field everywhere, for given A and t values, is to place a polarisable media between the electrodes. As first noted by Faraday (Faraday 1832), and later confirmed by Maxwell (Maxwell 1891), the effect of putting a dielectric material in an electric field is to create dipoles, by separation of +/– charges already present within the dielectric material (Fig. 2). No net charge enters or leaves the dielectric.

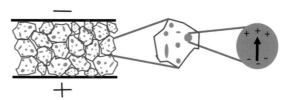

Fig. 1. Dipole Formation in a Powder Super Dielectric Material. Pictured is a representative cross section of a powder SDM, consisting of a high surface area electrically inert powder, such as alumina, in which the pores are filled with liquid containing dissolved ions, for example aqueous NaCl solutions. The ions in solution, induced by fields on the electrodes, move to form dipoles opposite in orientation to the electrode field.

The dipoles induced in the dielectric create fields oppositely polarised to the field created by the charges on the electrodes. Thus, the dipoles of the dielectric media will cancel, to some extent, the field created by the charges on the electrode everywhere. The field everywhere will be smaller, and consequently, the voltage difference (Eq. 3) between the electrodes will be smaller. Once again, per Eq. 4, lowering the voltage, while maintaining the charge, increases the capacitance.

In sum, the model consistently applied in this review is that all capacitive effects are due to electric polarisation leading to the formation of dipoles within a non-conducting material held between conducting electrodes, such as a solid dielectric or a non-conducting liquid containing mobile ions. In fact, this is the classic model of dielectrics (Kao 2004, Garnett 1904, Rysselberghe 1931, Debye 1929, Onsager 1936, Kirkwood 1939, Fröhlich and Maradudin 1959), and also may be considered a version of the ponderable media model (Maxwell 1891, Mansuripur et al. 2013). In brief the classic model is no net current results from exposure of a non-conducitve media to a field, but there is a net local charge separation as positive species migrate toward the anode and negative species migrate toward the cathode. As outlined elsewhere there are a variety of polarisation processes that occur including electronic polarisation, ionic polarisation, dipolar polarisation, interface polarisation, space charge polarisation (Kao 2004). Indeed, the particulars of the polarisation process are a function of the nature of the dielectric. For example, later it will be argued that in superdielectric materials the polarisation takes the form of the migration of dissolved ions, on the order of microns in some case, in a liquid media traversing the liquid, whereas in a ceramic dielectric, it is a short range separation of ions in a crystal. The separation that takes place in a solid $BaTiO_3$ type dielectric is less than the mean atomic separation distance, within a single atomic primitive cell. In this manuscript it is assumed for all types of dielectrics that there is a dominant mechanism of polarisation throughout the dielectric material.

It is important to reiterate the following critical features of the polarisation process as they impact capacitance. First, in parallel plate capacitors external field induced polarisation can be considered to result in the formation of electric dipoles. Second, the direction of the field produced by the dipoles is always opposite the applied field direction. Third, the induced dipoles reduce the net field within the dielectric relative to the field which would exist in the absence of a dielectric. Finally, for a parallel plate capacitor, the only type considered, the reduction in internal field, relative to the field within the dielectric that would exist in the absence of the dielectric, can be expressed very simply (Eq. 5). As per standard models, Gauss's Law can be applied and leads to the conclusion that the reduced field within a dielectric in a parallel plate capacitor is everywhere the same, as developed later.

$$E=Eo/\varepsilon \tag{5}$$

where E is the net field, Eo the applied field and ε the dielectric constant of the media. The more easily a material is polarised, the higher the value of ε. The above equation indicates that dielectrics that easily polarise, for which an isolated metal particle with very high permittivity is the most extreme example, will have high dielectric constants, and those that are difficult to polarise, for which vacuum is the most extreme example, will have low dielectric value.

All the above is well covered in the standard literature. Herein we add one explicit element that is only implicit elsewhere: *The field reduction created by the polarisation of the dielectric is not limited to the body of the dielectric, but rather occurs in all space.* The precise geometry of this reduction, even for a parallel plate capacitor, is complex. Clearly, the field outside the dielectric, unlike that within the dielectric of a parallel plate capacitor, is not a single value. The required reduction of the field at all points in space must be true if fields are 'conservative'. That is, if the field everywhere were not impacted by the dipoles within the dielectric, the energy required to bring a charge to the dielectric would be path dependent. To wit, an electron traversing the dielectric from the anode to the cathode would require less energy than one moved between the same endpoints, but on a path outside the dielectric media. Indeed, in the latter case the field, not influenced by the dielectric, would be higher at all points, making this a higher energy path (Eq. 3). This is a fundamental violation of Maxwell's electrodynamics, specifically path independence of voltage.

It might be argued that the conductive electrodes create a Faraday cage around the dielectric such that fields produced by the dielectric are 'cancelled' exterior to the dielectric. The counter argument is that the electrodes are not Faraday cages as they are not neutral, but rather contain excess charge. The fields produced by the polarised dielectric are organised to reduce the fields created by these excess charges throughout space. The net field in space can be considered a sum of fields: that resulting from the 'large dipole' field determined from co-consideration of charges on both electrodes, minus the field generated by the oppositely polarised dipoles of the dielectric material.

The above leads to a very simple basis for accepting/rejecting models intended to explain 'novel' dielectrics. Specifically, any model of dielectric behaviour that pertains solely to changes, field or other, within the dielectric media itself is not acceptable. The model must also result in a reduction of field outside the dielectric media such that fields remain conservative. In brief, this means any model of increased capacitance that does not include an element of electric polarisation within the dielectric, and a concomitant field reduction everywhere, is summarily rejected.

One final point will be made regarding the nature of the polarisation: The mechanism of polarisation impacts the time required for the polarisation process to take place (Kao 2004). As discussed later, the movement of charge in a ceramic type dielectric is over a sub-angstrom scale, hence the process is rapid, and the dielectric value is nearly constant with frequency until GHz in many cases. In contrast, in a super dielectric material the polarisation requires ions to travel distances of the order of microns, hence the dielectric constant decreases rapidly with frequency, as explained in more detail later.

Presenting the theory as per the above should not be necessary. It is simply a variation on dielectric theory as originally presented by Maxwell (Maxwell 1891) and Faraday (Faraday 1832), hence 'obvious'. Unfortunately, as discussed to some extent later in this essay, some more 'complex' models exist even in textbooks and these obscure the simplicity of capacitor theory.

3.3 Energy in Electrons

Surprisingly, the next logical consideration in understanding capacitor design is subject to significant misunderstanding: Where is the energy (Eq. 4) stored in a capacitor, per Eq. (2)? In many places, including textbooks (Resnick and Halliday 1978, Walker 2007), it is stated that the energy of a capacitor is 'stored in the electric field', yet this does not follow from the above analysis.

The argument that the 'all energy storage in the field' postulate is incorrect follows immediately from the argument that polarisation is responsible for all dielectric behaviour. It is argued that for capacitors of the same geometry, at the same voltage, the field at every point in space is identical, regardless of the presence/absence of a dielectric. Yet, in the presence of a dielectric the total energy of the capacitor can be dramatically higher than that of the same capacitor without a dielectric at the same voltage. Thus, a conundrum exists if the 'energy in field' postulate is accepted. There cannot be the same field yet dramatically different energies if the energy is stored in the field. There are two arguments used to support the above postulate. First Gauss's Law as it pertains to parallel plate capacitors (Kao 2004) shows that the field between the plates is independent of presence/absence of a dielectric. Specifically: in any closed shell containing no net charge, as per any non-conductive dielectric, there is no net field flux through the surfaces. Thus, in the symmetry existing between the electrodes of a parallel plate capacitor the field strength entering (positive flux) and exiting (negative flux) a rectangular body symmetric with the capacitor electrodes, must be identical, as discussed in more detail elsewhere (Resnick and Halliday 1978, Katz 2016). This can only be true if the electric field between the electrodes in a parallel plate capacitor is a single constant value independent of the presence/absence of a dielectric. From this it follows that the magnitude of the vector field in the dielectric is related only to the voltage. This fact is even used in the standard model of ionic transport in a slab of non-conductive material between two electrodes, and the applied field (V/cm) is the field employed in the models of ionic conductivity (Adamec and Calderwood 1977, Onsager 1934, Seitz 1987).

The second regards the field outside the dielectric. This is not a constant valued field, but it is argued that at a given capacitor voltage, and constant capacitor geometry, there is no basis for postulating a change in field. That is, we postulate a unique field: The field structure outside the region of the dielectric at a given capacitor voltage, and for a given capacitor geometry, is unique. The argument for the latter is based on voltage as a state property. In particular, an electron can be brought to the high voltage electrode via a path from infinity (0 volts), or from the low energy electrode through the dielectric, and the integrated value of the field along both paths must be the same (Eq. 3). By symmetry it is argued that the only solution that is consistent with this requirement for any dielectric is that for a given voltage between the electrodes the electric field is everywhere the same, independent of the presence/absence of a dielectric. The energy required to bring an electron to the electrode (Eq. 3) is independent of path.

In sum, at a given voltage the field in the dielectric, and outside the capacitor, is independent of the presence/absence of a dielectric. The field geometry is independent of all parameters, and the field magnitude at any point is only a function of the voltage difference between the electrodes. Yet the energy stored in the capacitor with a dielectric can literally be billions of times more (see *SDM* section) than that found in a capacitor of the same dimensions and at the same voltage, with no dielectric. Or, to put it another way: The fields in geometrically identical capacitors held at the same voltage, but in one case with a dielectric and in one case without a dielectric, are identical but vastly different energy is stored. Clearly, the energy is not stored in the field.

Where is the energy stored? The energy in a capacitor is stored in the individual electrons and positive species on the electrodes. For example, barium titanate has a dielectric constant of order 1,000. By Eq. (1) that means a capacitor with barium titanate dielectric has a capacitance 1,000 times greater than one absent barium titanate. By Eq. 4 this means there are 1,000 times more electrons 'stored' at a given voltage in the presence of barium titanate dielectric. To wit: The extra energy (Eq. 2) in the capacitor with the barium titanate dielectric is directly associated with 'extra electrons'.

For example, if, in the absence of barium titanate, there are 1,000 electrons with energy between 10 and 10.1 volts on the negative electrode, then there will be 1,000,000 electrons on the negative electrodes in the same voltage range in the presence of barium titanate. Each electron carries a specific energy (voltage), hence if there are more of them, there is proportionately more energy stored. An analogy: Imagine two dams of identical construction. The smaller damn has a small flat bottom lake behind it containing 1000 cubic meters of water. The second damn has a flat bottom lake of the same depth, but has 1,000, 000 cubic meters of water. Both lakes are emptied to generate electrical energy. The 1,000 times greater electric energy generated by emptying the larger lake arises not from some 'field', but simply from the fact there is more water to run through the turbines. Returning to the capacitor: If there is any stored 'field energy', it cannot be greater than the case in which no dielectric is present. Hence, only in the poorest capacitor is field energy possibly a consideration.

Another approach to demonstrating that the energy is in the electrons is to reflect on the fact, first pointed out by Maxwell (Maxwell 1891), that the energy in an assembly of charges, and the maximum energy that can be recovered, is equal to the amount of work required to assemble them from charges originally at 'infinity'. The amount of work is equal to

$$W = \int V \cdot I \, dt \qquad (6)$$

where W is the work required to assemble the charges, V the voltage on the capacitor, and I the current to the capacitor. There is no 'field' in this equation, but there is an accounting for the total charge added to the capacitor. As Maxwell shows, for a capacitor Eq. 6 yields Eq. 2 where V is the final voltage on the capacitor. For a given geometry, and voltage, a capacitor with a dielectric will clearly have a more charge/higher capacity, and concomitantly higher energy density than one without a dielectric. It is also important to note that experimental data in which voltage and current are both collected can be employed to determine the amount of energy entering a capacitor during charging, or the amount of energy leaving during discharge.

As noted earlier, the application of the model presented above is clear for electrostatic capacitors, but less apparent for EDLC. Yet, in fact, the model does apply to EDLC Supercapacitors, as is generally understood (Sverjensky 2001, Simon and Burke 2008). This is discussed later in the section on EDLC Supercapacitors.

3.4 Frequency Dependence Ceramics

Most mathematical models of capacitors presume that the capacitance is only a function of frequency (Debye 1912, 1929). These models are from a class of models of relaxation phenomenon, used to explain capacitance, inductance and even some magnetic behaviour (Dumesic and Topsøe 1977, Mørup and Topsøe 1976, Phillips et al. 1980). In the case of capacitance, the relaxation models relate the reduction with frequency to 'phase lag'. That is, the dipoles that create the dielectric value (ε) fail to follow the voltage applied to the electrodes as the frequency increases. The mathematical model assumes an in-phase component (ε') and an 'out of phase' component (ε''). That is, capacitors must be described as having a complex impedance. A net vector at angle ε between the two vectors, based on the relative magnitude of the in and out of phase components, is employed to analyse losses, particularly energy loss, associated with the phase lag. In fact, Debye suggested the analogy of frictional losses for an assembly of identical dipoles. That is, the dipoles are modeled as non-interacting objects 'swinging' in a viscous medium. The motion leads to frictional heating of the viscous media, hence requires energy. Thus, not all the energy added to a capacitor is available during discharge. In the perfect analogy the dipole lag follows a simple algebraic relationship as a function of frequency (Debye 1912, 1929). In real systems there is generally a more complex relationship between phase lag/energy loss and frequency.

One difficulty for modelling all capacitive behaviour using the standard relaxation model is the requirement that the absolute magnitude of the dipole vector does not change, only its direction relative to the applied field (angle δ). The model works to explain increasing loss with frequency, but nowhere in this well developed mathematical analysis is it anticipated that capacitance could also be a function of voltage. Yet, for all capacitors there is evidence of voltage dependent capacitance. The discussion of this effect must be divided into three parts; ceramic capacitors, EDL supercapacitors and NP Supercapacitors.

The effect for ceramic capacitors is noticed in two fashions, limiting computer speed, and limiting maximum energy density. First, in high speed electronics (always ceramic based) the discharge time for capacitance is critical in determining switching speeds. For example, the 'value' stored in a digital memory circuit cannot be determined until adequate time has elapsed for discharge or charge to occur. The discharge/charge rate thus limits the net rate of reading/writing memory. In this realm, it is understood that capacitance is a function of electric field gradient/voltage. Indeed, it is well documented that Class II dielectrics (ferroelectrics) show voltage dependent behaviour (Carter and Norton 2013, Niepce and Hugentobler 1991, Kaiser 1993). Hence, modelling the charge/discharge behaviour, critical for determining ultimate processing speed, is complex. Predictive 'fitting' models and equivalent circuits have been created to at least accurately predict behaviour (Drofenik et al. 2010, Heckel and Frey 2015). However; these models, basically complex equivalent circuits, do not clearly explain the underlying physical process that leads to the observed, and theoretically anticipated (Waser and Lohse 2001) voltage dependent capacitance.

One observation (Fig. 2) does suggest at least part of the physics of the observed voltage dependence in high speed switching capacitance: the capacitance of these ceramic capacitors decreases with increasing voltage. The same phenomenon is found for ceramic capacitors intended for energy storage (Reynolds et al. 2012). This suggests the same underlying physical basis exists in the two cases. To wit: As the voltage increases a phenomenon known as 'saturation' occurs in all types of ceramic dielectrics. In a saturated dielectric all of the dipoles are fully aligned once the 'saturation field', Es, is reached. As per Eq. 4, further increases in the voltage, concomitantly field, do not create additional dipole field from the dielectric, thus the effective capacitance appears to decrease above this voltage. It as if the voltage increases, but charge stored on the electrodes, barely changes. This will appear, as shown in Fig. 2, to lead to a decrease net capacitance above Es.

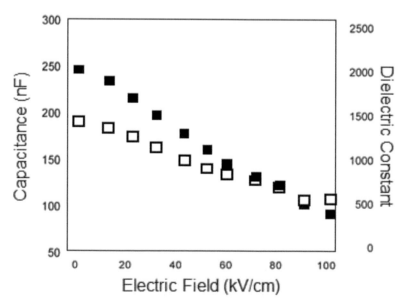

Fig. 2. Saturation of a thin film of BaTiO₃. A 500 nm thick layer of Barium Titanate was used as the dielectric in a capacitor and tested at various voltages. Measured capacitance (black squares) is nearly constant until the saturation field is reached, then decreases nearly linearly with increasing applied field. The apparent dielectric constant (empty squares) also decays even as saturation voltage is approached. Note the field never reaches the breakdown strength. Adapted from Reynolds et al. 2012.

3.5 Frequency Response Supercapacitors

The frequency response of both types of Supercapacitors is also associated with the physical process involved in switching dipole orientation as the field is reversed; however, this reversal is far more complex and slower in supercapacitors than in ceramic capacitors. For example, in super dielectric materials (SDM), as per Fig. 1, high dielectric values result from the separation of +/− salt ions dissolved in a liquid phase. This separation is induced by the fields produced in the dielectric by charges on the electrodes. As discussed elsewhere (Cortes and Phillips 2015a,b, Fromille and Phillips 2014), to create this separation takes time, as physical ions must travel distances, possibly of order microns. The analogy with motion in a viscous media suggested by Debye becomes real. The ions are literally travelling in a viscous media (e.g., water). At very low frequencies, the ions have time to reach maximum separation, dipoles lengths maximised, before discharge commences. This leads to capacitance independent of voltage, below the breakdown voltage. In contrast, as frequency increases, the dipoles do not have time to reach full length before discharge commences. Thus, the initial capacitance, as described later, is lower. That is, in contrast to the standard relaxation models, it is postulated that both the *magnitude* and the direction of the 'dielectric vector' changes. Yet, at the high frequencies, even as the discharge is occurring, the dipole lengths continue to increase as a field induced by charge on the electrodes is still leading to increased ion separation. This leads to the capacitance increasing even as charge is leaving the plates. The net result: At the beginning of discharge of an NP Supercapacitor the capacitance is low, but the capacitance increases even as the voltage across the NP Supercapacitor decreases.

The above description of NP Supercapacitors raises this question: Are dipoles forming in the far more commonly studied EDL Supercapacitors also? What is the physical nature of these dipoles, the kinetics of formation, and what does that suggest about the physical process that takes place as the frequency of the voltage applied to the electrodes increases? These questions are addressed later.

4. Measurement Method

Evaluation of data contained in reports of capacitive performance requires an understanding of the caveats of each approach to measuring engineering relevant performance characteristics. For proper circuit design the relevant parameters include some, or all, of the following: capacitance, capacitance as a function of frequency, energy density, energy density as a function of frequency, dielectric value, dielectric value as a function of frequency, equivalent circuit parameters for internal resistance and output resistance, power output, power output as a function of discharge time, and safe operational voltage range, and it is important to note the values reported are a function of the measurement technique employed.

There are four major approaches to measuring capacitance that are presently in use: (i) RC time constant, (ii) Impedance spectroscopy, (iii) cyclic voltammetry based on controlled linear increase/decrease in voltage with time, and (iv) constant current employing a galvanostat. Each one of these methods has caveats, and the choice of method can impact the reported measured values of parameters, as discussed below.

4.1 RC Time Constant

The first method is the historically employed method, and the one generally described in elementary text books (Resnick and Halliday 1978). It is reasonable to assume the dielectric values of a host of homogenous materials, listed in many tables (Dorf 2003), were determined long ago using this method. In the RC time constant method (Fig. 3) a capacitor is charged and discharged through a resistor (R) of known value and based on Eq. 7, where t is time, and V is the voltage as a function of time, the capacitance (C) determined from either the charge or discharge leg of the data:

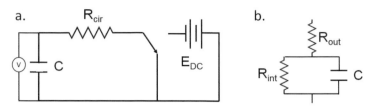

Fig. 3. RC Time Constant Circuit. a. Simple circuit used to charge/discharge capacitor (C) through external resistor (R). As shown the capacitor is discharging so power supply (E_{DC}) not in circuit, and voltage being measured with meter (V). b. The capacitor (C) is better described with an equivalent circuit, wherein there is a 'leak' current through an internal resistor (R_{int}) and output to the circuit flows through a resistor (R_{out}).

$$\frac{V}{V_0} = e^{-t/RC} \tag{7}$$

This method is still employed, particularly for assessing energy output from a novel capacitor over its full voltage operating range (Kim et al. 2015, Cortes and Phillips 2015a,b, Phillips 2016, Gandy et al. 2016, Jenkins et al. 2016, Fromille and Phillips 2014).

There are several easily identified difficulties with the method. First, the external resistor is invariably part of a 'voltage divider' circuit. A very simple version of an equivalent circuit for a 'capacitor' in series with a resistor is shown (Fig. 3), but far more complex versions can be found (Zubieta and Bonert 2000, Spyker and Nelms 2000). Assuming the simple equivalent circuit shown, 'R' in Eq. 7 is:

$$R_{net} = \frac{R_{cir} + R_{out}}{\dfrac{(R_{cir} + R_{out})}{R_{int}} + 1} \tag{8}$$

The net resistance, which should be used in Eq. 7 is R_{net}, and is approximately equal to the value selected for R_{cir} only if R_{out} is small relative to R_{cir}. Also R_{int} must be, relative to the other values, very large. If R_{int} is not particularly large relative to R_{cir}, then a significant fraction of the energy will 'leak' through the internal resistor and the capacitance determined will appear lower than the value in the equivalent circuit. If R_{int} is made to be relatively large, by picking a relatively small R_{cir} this may lead to R_{out} and R_{cir} being similar in magnitude. This will also distort the value of capacitance determined using Eq. 7. This discussion leads to one advantage of the RC time constant relative to the others: It is reasonably easy to determine the size of the leak resistor and the output resistor (Fig. 3) as described elsewhere (Gualous et al. 2003).

As capacitance is employed to determine dielectric constant this raises fundamental questions. Was proper care taken to assure that the dielectric values listed in many tables are accurate? For example, was proper adjustment made to eliminate the impact of the relative value of internal and circuit resistors? Another issue is the difficulty of using this method for determining capacitance and dielectric values as a function of frequency. In order to change discharge time, for example to make it short, the only parameter that can be changed is the value of R_{cir}, and as noted above, this will change the measured apparent capacitive value. This brings up another question. What was the frequency at which all of the tabulated dielectric values were determined? Are they intended to represent the static dielectric value?

4.2 Impedance Spectroscopy

The second method for measuring capacitance, impedance spectroscopy, is very seductive. It allows the user to collect data regarding dielectric values as a function of frequency over a very wide range of frequencies with relatively little effort. It is clearly the method most often employed in modern assessments of capacitors and novel dielectric materials (Kumari et al. 2017, Westerhoff et al. 2016, Oz et al. 2014). Indeed, it is the method employed in hand held multimeters which include a 'capacitance' setting. In essence; this method employs a low voltage signal (0 +/– 25 mV) to probe a capacitor. The voltage is restricted to this range as it is well understood in electronics that thermal noise, for all circuit elements, becomes a significant component of signal at higher voltages (Motchenbacher and Connelly 1993, Johnson 1927, Barsoukov and Ross Macdonald 2005, Ross Macdonald and Kenan 1987). At 300 K (ambient room temperature) above 25 mV a non-linear current develops. The non-linearity is particularly acute for p-n junction capacitance analysis (e.g., photovoltaic cells) where linearity requires operation below ~ 5 mV (Yadav et al. 2016). Therefore, the Z value becomes dependent on applied voltage and a key assumption of the method, the linearity of Ohm's Law ($Z = V/I$) fails.

Thus, in order to provide a thermal noise free signal only low voltage response is probed. The low voltage signal returned from the capacitor, after an 'instrument transfer function', is then deconvoluted (software), generally on the basis of an assumed 'electronic equivalent circuit' (EEC). Most often the data is employed to determine the capacitance as a function of frequency.

There are two fundamental concerns associated with employing this method to probe EDL Supercapacitors and NP Supercapacitors devices employed for energy storage/power delivery. First, the devices are only probed at very low voltages. As it is now thoroughly established that capacitance is not constant with voltage for the best energy storage type capacitors, even employing a voltage offset with IS (Barsoukov and Ross Macdonald 2005, Ross Macdonald and Kenan 1987), and clearly most energy is associated with the highest range- and not probed-portion of the operating voltage, there is no basis for projecting the values obtained with IS into an engineering expectation of the energy/power delivery associated with EDL or SDM supercapacitors. Indeed, an IS analysis will not even provide the user with a value of the maximum operating voltage.

There are several difficulties with the algorithms/circuit models employed in IS analysis which further reduce credibility. First, the equivalent circuit approach employed, particularly with commercial IS instruments using proprietary software, is not transparent. A review of the operating manual provided with most instruments reveals that the details of the equivalent circuit model employed are treated as proprietary information. That is, these manuals provide no information regarding the model elements in the equivalent circuits used to deconvolute the signal to provide dielectric values as a function of frequency, etc. There is the implication in basic equivalent circuit theory that capacitors are flexibly modelled as a series of transmission line (TL) elements (Conway 1999, Pean et al. 2016); however, in other cases it appears that TL or other models are improved by *ad hoc* addition of in phase/advance phase/lagging phase elements, also known as constant phase (CPE) elements, representing ideal inductors, capacitors and resistors (Macdonald 2006, Hirschorn et al. 2010). Second, IS devices do not use different assumed equivalent circuits for different types of capacitors. An accurate EEC is based on a clear mathematical analogy with the physical system (Randles 1947). It is not possible that the same equivalent circuit can represent each fundamentally different (e.g., SDM vs. EDLC) type of capacitor with equal accuracy (Macdonald 2006, Harrington and van den Driessche 2011). In brief, the inability of this method to measure energy or power over the entire proposed operating voltage range, and the uncertainty regarding the applicability of the built in algorithms particularly for supercapacitors, makes data collected using IS inappropriate for capacitors employed as energy storage or high power devices.

The IS method clearly has a range of applicability. It works very well with solid state dielectrics for which there is little lag between dipole orientation and driving voltage, hence very little 'energy loss', until frequencies of the MHz range are reached (Haertling 1999). Thus, it is an excellent technique for the analysis of the types of capacitors for any type of high frequency electronics. Outside the power supply, supercapacitors which have poor frequency response (below) will not be found in modern electronics.

4.3 Cyclic Voltammetry

The third method employed to determine capacitance is based on a constant increase in voltage (charging) followed by a constant decrease in voltage. Current is the measured, dependent, quantity. The capacitance is determined according to this mathematical model (Li et al. 2009):

$$Q = \frac{1}{2}(Q_{charge} + Q_{discharge}) = \frac{1}{2}\int_{t_1}^{t_2} i(t)dt \tag{9}$$

In practice, current is measured as a function of voltage, so Eq. (9) requires a change of independent variable to voltage:

$$v = \frac{v}{t} \Leftrightarrow dt = \frac{dV}{v} \tag{10}$$

where v is the rate of voltage increase (constant) expressed in volts/second. Thus, Eq. (9) becomes:

$$Q = \frac{1}{2v}\int_{V_1}^{V_2} i(V)dV \tag{11}$$

and this leads to the following expression for average capacitance:

$$C = \frac{\int_{V_1}^{V_2} i(V)dV}{2v(V_2 - V_1)} \tag{12}$$

over the selected voltage range. The only required data therefore is a record of current as a function of applied voltage. In fact, the quoted capacitance is generally an 'average' computed over the capacitor discharge (i.e., negative voltage range, most plotting conventions).

This method excels in two respects relative to the two discussed above. First, it can be employed over the entire operating range to provide a realistic assessment of energy storage potential, and second the actual stored charge is directly computed. Caveats include the fact that this is an 'average' capacitance over voltage. As discussed later, the capacitance of many high energy density capacitors is not voltage independent. In fact, voltage dependent capacitance is easy to spot in cyclic voltammetry. If the capacitance is independent of voltage, the measured current will be constant, and the typical I vs. V plot will be nearly a perfect rectangle. In a more typical case, absolute current increases as the absolute voltage value decreases during discharge. This indicates a voltage dependent capacitance, thus the capacitance determined (Eq. 12) is clearly an average. Another caveat is that capacitance is a function of frequency. Moreover; the dependence of capacitance on voltage increases with increasing frequency.

To determine frequency dependence, a set of experiments must be conducted at different values of the constant voltage rate, v. Although rarely done, the method proves effective (Li et al. 2009, Farma et al. 2013). Yet, many 'high power' capacitor based applications require power to be delivered in time frames of order 10 msec. A final caveat regards the presentation/interpretation of data. Often the maximum voltage is deliberately set below 1 volt (Li et al. 2009, Farma et al. 2013, Khomenko et al. 2005, Karthika et al. 2012, Ghidiu et al. 2014) and the 'result' is in F/g, a number which tells little about energy storage in the absence of a full voltage scan. Also, generally, the data is determined at a single, very low, rate of voltage increase, corresponding to frequency far less than 1 Hz (Khomenko et al. 2005, Sani and Shanono 2013), or even a set of frequencies that are all below 1 Hz (Li et al. 2009, Ghaffari et al. 2013, Stoller et al. 2008, Ghidiu et al. 2014). In these very slow charge/discharge studies a 'best' result for capacitance is reported, as capacitance rolls off with frequency (below).

4.4 Galvanostat Method

The fourth method employed to determine capacitance is the constant current method, generally conducted using a programmable galvanostat. The basis of the method is derived from taking the derivative of Eq. 4, and the assumption that capacitance is a constant value at all voltages. To wit:

$$\frac{d(C)}{dt} = \frac{d(\frac{q}{V})}{dt} = 0 \tag{13}$$

This reduces to:

$$\frac{-q}{V^2}\frac{dV}{dt} + \frac{1}{V}\frac{dq}{dt} = 0 \tag{14}$$

and in the constant current ($dq/dt = I$) case this reduces to:

$$\frac{1}{\frac{dV}{dt}} = C \tag{15}$$

Thus, if the capacitance is not a function of voltage, then the application of constant current should yield a curve of voltage vs. time that is a straight line. The slope of this line should have the same magnitude, but opposite sign, for charging and discharging.

As shown, once the discharge time is reduced, which in a galvanostat is accomplished by increasing the value of the constant current, the charge and discharge curves are not linear (Fig. 4). This is most easily interpreted as demonstrating that the capacitance changes as a function not only

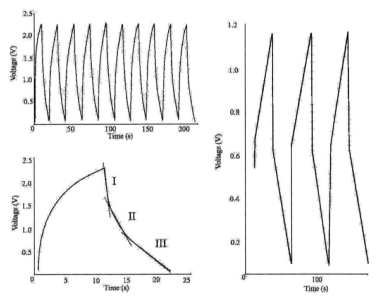

Fig. 4. Voltage Dependent Capacitance Discharge of Supercapacitors. TOP LEFT: Constant Current Charge/Discharge data collected for NP Supercapacitor made from fumed silica saturated with a 30 wt per cent NaCl aqueous solution (5). BOTTOM LEFT: An expansion of one of the TOP LEFT cycles shows that the discharge voltage is not of constant slope. This indicates capacitance is a function of voltage, even at relatively long discharge times. As shown the discharge is modelled as consisting of three different near constant capacitance regions. RIGHT: Using the identical protocol to that used to collect the data on the left, and the same instrument (commercial galvanastat, BioLogics 300), the constant current charge/discharge cycles of a commercial 2 F EDLC supercapacitor are found to be extremely non-linear even for discharge times of approximately 10 seconds.

of frequency, but also of voltage. Of course, over any very small voltage range, as per example the ~ 25 mV test employed using IS, the capacitance is nearly constant.

Two sets of questions arise from the finding that constant current analysis applied to supercapacitors does not lead to the behaviour 'expected' for ideal capacitors. The first set of questions: How does one 'label' such capacitors? Are the concepts of dielectric constant and even capacitance meaningful? The second set of questions regard the physical processes that lead to the voltage dependent capacitance.

The first set of questions is complex, and there is no general agreement on an answer; although there are approaches that can serve as guides. For example, in studies of NP Supercapacitors discussions of capacitance and dielectric value are generally restricted to low voltage, e.g., less than one volt (Cortes and Phillips 2015a, Phillips 2016, Jenkins et al. 2016, Fromille and Phillips 2014). In essence, conventional analysis/labelling is restricted very formally to a very restricted voltage range, a range which it is shown experimentally that capacitance is constant. Greater emphasis is placed on total energy and power density, values well measured by the constant current method, over the full operational voltage range as a function of discharge time (Christen and Carlen 2000). This is reasonable as it is anticipated that use of these capacitors will be restricted to energy/power applications.

As described in the theory section, and below in the sections on each type of capacitor, the origin of non-conformity with the accepted standard behaviour of 'constant capacitance with voltage' has many origins. For example, in SDM dipoles form via the physical separation of ions in a viscous media. Analysis of response of ions in this environment is complex and not presently available. At a minimum a model would include momentum, viscosity as a function of velocity relative to the media, coulombic force as a function of the immediate voltage, etc.

4.5 Data Reporting

In addition to questions regarding data collection methods there are questions regarding data reporting. For example, it was argued recently that reporting the energy density of graphene based supercapacitors on a mass basis, a standard approach, is not particularly informative for the many applications for which volume energy density is far more important. It was pointed out that once the low density of graphene is accounted for, as well as packaging, on a volumetric basis even the best graphene based capacitors are not significantly better than commercial carbon based supercapacitors. It was also argued that simply reporting results on a Ragone chart is not sufficient, particularly given the rapid degradation of many prototypes. Moreover, other factors such as self-discharge rate, output resistance, etc., are rarely reported. All of these reporting failures make comparisons difficult (Jonscher 1999, Murali et al. 2013).

In sum, 'the measurement method is the message'. The tool used to measure capacitance has a major impact on the reported results. For example, an IS measurement produces a reliable capacitance value of frequency for small voltage swings, and, if limitations are properly understood, it works well for ceramic capacitors. However; IS inherently fails to provide reliable information regarding potential energy and power delivery capability of supercapacitors. In contrast, a constant current analysis provides reliable information regarding energy and power characteristics over a broad range of frequencies, but cannot provide single metric values as a function of frequency. That is, it does not provide 'capacity' and 'dielectric constant' values as a function of frequency for energy storage type capacitors.

5. Principal Categories of Proposed Energy Storage Capacitors

In this section we review the principle details of three types of 'energy storage' capacitors, solid dielectrics or 'ceramic' dielectrics, EDLC Supercapacitors, and NP Supercapacitors. Each of these is further broken down into sub-categories.

Some common themes and concerns are found for capacitors of all types. An example of a theme is the clear understanding that the formation of large dipoles, where large implies a charge separation greater than a single atom, within the dielectric is the key to high dielectric values, hence high energy density, in all capacitor types. An example of a common concern is the magnitude of the gap between 'prototype' and 'commercial potential', a gap clearly a function of the type of dielectric. Some prototypes, for example high energy density ceramic capacitors, are built on such a small scale, employing such complex fabrication techniques, that it seems unlikely they will ever be built on a sufficient scale to compete with carbon based EDL Supercapacitors for application to energy storage or power delivery. The expense associated with substituting graphene for carbon also appears to limit the future wide use of graphene based EDL supercapacitors. Activated carbon is likely to remain the electrode material of choice for EDL Supercapacitors.

There are also some significant issues which are clearly not thoroughly addressed in the scientific literature. What are the values of the internal resistance and output resistances of the capacitors studied? How long will they hold charge? What is the frequency dependence of the capacitance of EDL Supercapacitors? What are the actual shapes of charge/discharge curves (e.g., voltage vs. time) for all types of capacitors over the full voltage operating range? What is the full voltage operating range?

5.1 Ceramic Dielectrics

5.1.1 Barium Titanate

The most commonly employed and most studied high dielectric material is barium titanate because it is the homogeneous solid with the highest dielectric value. This reflects the large natural dipole value of barium titanate, in fact all solids in the class of 'ferroelectrics' have relatively high dielectric values due to the formation of natural dipoles in the crystal structure. The dielectric constant is a function of many parameters including frequency, firing protocol, and grain size (Bologna Alles et al. 2005). The general consensus is that solid dielectrics with the lowest defect density, and largest grain size will have the highest effective dielectric constant. For this reason significant effort is employed devising protocols to reduce defects and increase grain size (Tan et al. 2015).

Many studies show the factors influencing the behaviour of barium titanate. For example, in Fig. 2 it is shown to be a function of field strength. Once 'saturation' is reached the energy density and effective dielectric decrease linearly with increasing applied voltage. As shown in Fig. 5, the dielectric constant of $BaTiO_3$ is also clearly a function of temperature, particle size, and frequency.

As shown the dielectric value can be as high as ~ 10,000 over a narrow temperature range, although in practice, capacitors are generally employed near room temperature where the dielectric constant of a well prepared, low defect density $BaTiO_3$ is close to 2,000.

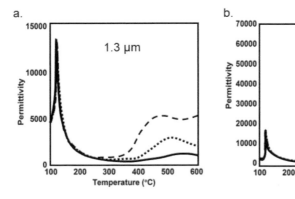

Fig. 5. Parameters Impacting $BaTiO_3$ Behaviour. Dielectric constant at different frequencies as a function of temperature for conventionally sintered $BaTiO_3$ with two different grain sizes 1.3 µm (a) and 32.0 µm (b). Dashed lines correspond to 1 kHz, dotted lines to 10 kHz, and solid lines to 100 kHz. Adapted from (Tan et al. 2015).

The temperature dependence reflects the mechanism of dipole formation in barium titanate. Specifically, in an applied field a dipole forms due to relative motion of the Ti^{4+} ion in the lattice. The abrupt change in the measured dielectric constant correlates to a structural change, as described elsewhere (Mishra et al. 2014). Above the Tc (~ 130C) the crystal has a cubic structure, and below it has a tetragonal structure. In both structures there is a net separation between the + and − charge centers in the unit cell, the length of which is a function of temperature. This separation creates dipoles, with temperature dependent length, that tends to orient along the field lines imposed on the dielectric within a parallel plate capacitor.

Generally pure $BaTiO_3$ is not employed commercially because its behaviour is not optimal at normal deployed temperatures, and its behaviour is too strong a function of temperature to insure reliable circuit performance. Thus, additives that shift the Curie temperature closer to ambient such as $SrTiO_3$ and $BaZrO_3$ are employed. It is also possible to broaden the temperature range of maximum dielectric value by appropriate processing strategy or the addition of other species such as ZrO_3. It is reported that dielectric values as high as 20,000 can be achieved near ambient temperatures for properly prepared barium titanate based dielectrics (Levinson 1987, Niepce and Hugentobler 1991).

It is notable that there are no commercial high energy density $BaTiO_3$ based capacitors. Indeed, there is no solid evidence that the actual energy density exceeds ca. 2 J/cm^3. This reflects the impact of both saturation and breakdown voltage on very thin dielectric layers required to maximise energy density, as discussed elsewhere (Reynolds et al. 2012). It is particularly important to note the severe limitations on maximum energy density created by 'saturation' on barium titanate, and all other ceramic dielectrics. The true effect of saturation means dielectric values measured at 'normal' ceramic thickness (ca. > 50 micron) cannot be extrapolated to very fine dielectric thickness (< 50 micron) of the type required for high energy density. Thus, the recently published potential energy density of anti-ferroelectrics (Chauhan et al. 2015), less than 5 J/cm^3, is likely high. Those values are apparently computed solely on the basis of breakdown voltage and dielectric values measured far below the breakdown voltage. Yet, it is clear the dielectric value decreases dramatically well before breakdown is reached due to saturation (Fig. 2).

5.1.2 Alternative Ceramics

Given the failure of $BaTiO_3$ based capacitors to achieve high energy density, alternative approaches to creating high energy density storage using ceramic dielectrics have been explored. Indeed, there would be significant advantages to a true high energy density ceramic based capacitor including high energy density even at high frequency, which would reduce charging time, and great durability. Recently dielectric materials have been designed to create high energy density ceramic capacitors based on various combinations of high dielectric values and high breakdown voltages. The results are intriguing as high energy density capacitors have been created, order of 50 J/cm^3, and the energy is only, apparently, a weak function of operating frequency. In contrast, for supercapacitors energy density is a strong function of frequency.

This low dielectric value/high voltage approach is based on the fundamental equations of energy density in capacitors. Per Eq. 2, the energy density of a capacitor is proportional to voltage square and linearly proportional to capacitance, which is the basis for focusing the attention on increasing the operating voltage of capacitors. In principle, combining Eqs. 1 and 2, the energy density of such a capacitor would be the dielectric constant times the permittivity of free space times the breakdown voltage (voltage per unit distance) square, the dielectric thickness is out of the equation. Combining this with Eq. (1) leads to this relation for parallel plate capacitors (Burn and Smyth 1972, Ortega et al. 2012, Li et al. 2013):

$$E = 0.5 \; \epsilon\epsilon_0 (B_d)^2 d^{-2} \tag{16}$$

where B_d is the thickness corrected breakdown voltage, and E is the energy density. Note, although the standard term employed is breakdown 'voltage' the controlling parameter is really breakdown electric field. The field value increases inverse to thickness, thus the applied voltage at which breakdown occurs is inverse to dielectric thickness.

Using this correlation (Eq. 15) for barium titanate, assuming a dielectric constant of 1000 and a breakdown voltage of 4 MV m^{-1} (Branwood et al. 1962), the energy density would be 0.1 J cm^3, which is three orders of magnitude lower than EDL Supercapacitors. In contrast, a material with a dielectric constant of 10 and a breakdown voltage of 400 MV m^{-1} would have an energy density of 14 J cm^3, which is still not very high, but illustrates the impact of having a higher breakdown voltage over a higher dielectric constant.

One method employed in this effort has been to focus on improving breakdown voltage by improved material processing techniques. Given that breakdown strength is a function of defects, grain size, porosity and other factors (Wu et al. 2007, Young et al. 2007, Huang et al. 2010), methods to increase grain size and reduce defects in high dielectric constant ceramics by various processing improvements have been made. For example, there are many studies in which low percentages (ca. < 10 per cent) of 'glass' (silica containing compounds) are added to high dielectric constant materials (e.g., strontium barium niobate (Qu et al. 2002, Lenzo et al. 1967, Ewbank et al. 1987)) in order to reduce the number of defects and aid sintering (Chen et al. 2016a,b, Wang et al. 2014).

The net result of this work (Chen et al. 2016b) theoretically increased the energy density to the order of 5 J/cm^3, a value far less than that reported for all types of supercapacitors. Moreover; the reported energy densities are only theoretical as the actual energy density at the breakdown voltage was never measured. Indeed, these reported values are very likely high as the impact of saturation was not included in the computation.

A particularly successful fabrication was that of a complex perovskite structure, $Pb_{0.91} La_{0.09} (Ti_{0.65} Zr_{0.35})O_3$, described as a relaxor ferroelectric thin film, by Hao et al. This had a reported energy density of order 30 J/cm^3 (Hao et al. 2012). Yet there are issues regarding this report including both the measurements employed to determine the energy density and the complexity and scale of the fabrication method. Regarding the first: It is difficult to understand the method employed by Hao et al. to make the energy density measurements, and no raw data for independent verification is available. Regarding the second: In order to create the perovskite structure, nine reagents were employed in a five step, 26 + hour process. The final capacitors generated were 1 micron thick covered by circular metal electrodes 0.20 mm in diameter. The process and scale are appropriate to micro-machine ('on chip') circuit energy/power requirements, but difficult to scale for commercial energy and power requirements.

Others have worked on the synthesis of novel materials that have either, or both, a large dielectric constant and a large breakdown voltage. The ultimate material has not been found, and trends in evaluated materials suggest it may not exist. Most of the materials with the highest breakdown voltages are polymers, and they tend to have very low dielectric constants. At the same time, the materials with the largest dielectric constants have relatively low breakdown voltages as per barium titanate. Another difficulty with this approach is it fails to account for two phenomena; (i) dielectric constant is impacted by both voltage and thickness, and (ii) in all known systems 'saturation' (Reynolds et al. 2012) effectively caps the ultimate energy density at voltages considerably lower than the breakdown voltage.

5.1.3 High Breakdown Voltage

There have been many efforts to capitalise directly on high breakdown voltages (ca. > 3 MV/cm), but low dielectric value materials to create high energy density capacitors (Larcher and Tarascon 2015). For example, one report regards dielectric composites composed primarily of nanowires of $Ba_x Sr_{1-x} TiO_3$, and suggests these can be configured into capacitors with energy density of order 20 J/cm^3 (Tang and Sodano 2013). However, the dielectric constant of these materials is very low < 20 in all cases, and the predicted high energy density is based on high breakdown voltages. Hence, real capacitors based on these materials would have to be extremely thin, and the stability tenuous. For example, a 10 micron thick dielectric layer of this material operated at 30 V (breakdown of air for this thickness) would have an energy density of about 10^{-3} J/cm^3. To reach the predicted value of ~ 20 J/cm^3 a 10 micron thick dielectric layer would have to be operated at 500 V. This is close to, but below, the reported breakdown voltage of the material; however, given the breakdown voltage of air, the operational parameters are clearly problematic. Others report that a sol gel can achieve similar or higher energy density (ca. 40 J/cm^3), but once again the analysis is based on materials with very low dielectric constants (< 25), in very thin layers, operated at voltages far above the breakdown of air (Kim et al. 2015). There are also extrapolated energy density values for ceramic capacitors greater than 20 J/cm^3, but again these extrapolations require the assumed application of voltages far above the breakdown voltage for air for very thin dielectric layers (Parizi et al. 2014, Tong et al. 2013).

Another example of a complex synthesis process, on a very small scale that yielded high energy density values is that of Smith et al. (Smith et al. 2014). They report a high energy density, 30 J/cm^3, for a polymeric film with low dielectric value (~ 75 J/cm^3) and a measured high breakdown voltage (> 250 MV/m). In this case the film, Poly(vinylidene fluoride-trifluoroethylene-chlorotrifluoroethylene) Terpolymer (neat), synthesis took more than 16 hours. The final capacitor consisted of a 4 microns thick polymer layer with sputtered metal film, 1 mm in diameter, electrodes. Once again, scaling to a practical device appears daunting and the voltage required to achieve high energy density is well above the breakdown voltage of air.

Another frequently encountered issue regards evaluation of the veracity of the energy density measurements. For example, Smith et al. 2014 report employing 'charge-discharge' measurements without providing any detail as to the system employed.

To create dielectric materials closer to the ideal of combining high dielectric constant and high breakdown voltage complex fabrications have been explored with the intent of creating a structure which combines the high breakdown voltage value of one component with the high dielectric value of a second. For example, one fabrication consisted of high dielectric value BaTiO$_3$ in low volume fraction (< 10 per cent) mixed into a high breakdown voltage, but low breakdown polymer (PVDFHEP) (Tong et al. 2013). The reported maximum energy density still never surpassed 4.7 J/cm^3 but some key information such as the precise voltages employed, and testing frequency are not given. In other cases, higher energy density is reported, but the values are extrapolated from dielectric values obtained using the IS technique (Parizi et al. 2014).

5.1.4 Composite dielectrics

In Table 1 the energy densities of composites with the highest reported energy densities, without critique of the method employed to determine these values, are given. The polymer-ceramic composite with the highest energy density in the table is few-nanometer thin polymer bilayer, one of the layers is a ferroelectric polymer, and the other one is a charge-blocking polymer: the first one provides the dielectric properties and the second one provides the high breakdown voltage (800 MV m^{-1}) (Kim et al. 2015). In this case, energy densities, exceeding 40 J/cm^3 were measured with a variation on the RC time constant method. Other composites involve decorating a polymer with ceramics of the family of barium titanate, as shown in Fig. 6 (Zhang et al. 2016, Puli et al. 2016). Some pure ceramic or polymer materials have been engineered to increase both dielectric constant and breakdown voltage, but the energy density that electrostatic capacitors based on this kind of dielectrics is not greater than 30 J cm^{-3} (Hao et al. 2012, Smith et al. 2014).

In sum, there are reliable reports that dielectric materials can be engineered to have either high breakdown voltages, high dielectric values, or, to a limited extent, a combination of both. The highest energy densities reported, more than 40 J/cm^3 compete well with commercial supercapacitors, but not with prototype supercapacitors (below). This type of capacitor also has the advantage of a low rate of 'roll off' of energy density with increasing frequency than all types of supercapacitors.

Table 1. High energy density dielectric capacitors: dielectric capacitors with energy density above 20 J/cm³.

Dielectric Material	Energy Density (J/cm³)	Dielectric Constant	Reference
Silica sol-gel coated by an organic acid monolayer	40	22	(Kim et al. 2015)
Polymer loaded with $BaTiO_3$ @ TiO_2	31.2	35	(Zhang et al. 2016)
$Pb_{0.91}La_{0.09}(Ti_{0.65}Zr_{0.35})O_3$	28.7	360–1200	(Hao et al. 2012)
Polymer	27	77	(Smith et al. 2014)
Core-shell ceramic-polymer composite	22.5	56	(Puli et al. 2016)

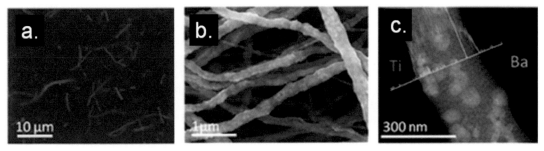

Fig. 6. Designed ceramic-polymer composites. (a) SEM image of the polymer-ceramic composite decorated with the nanofibers (b) SEM image of TiO_2 nanofibers decorated with barium titanate nanoparticles, (c) TEM image of same. From (Zhang et al. 2016).

There are notable practical limitations for electrostatic capacitors based on designed high dielectric materials which must be overcome before capacitors employing these dielectrics can be fielded. These include the complexity of the synthesis of the dielectric materials, the concomitant cost, and the technology employed to create ultra small scale thickness dimension. Also, the field strength required to produce the high energy density is often far higher (e.g., 100 X) than the breakdown voltage of air at the same length scale. Additionally, some of the measurements methods employed are difficult to understand leading to some uncertainty regarding claimed energy densities. Indeed, in some cases the maximum energy density claims appear to be extrapolated from low voltage measurements and do not include the potential impact of 'saturation' on net dielectric values at actual operating voltages (Reynolds et al. 2012).

5.1.5 Added Metal Particles

Another means to improve the dielectric value of ceramics is via the addition of metal particles. Potentially, this could lead to higher energy density capacitors, although the potential for this is not the focus of the literature on the topic. Generally, the focus is on fundamental principles leading to the observed changes in permittivity and conductivity.

There is a remarkably long history of study of the changes in properties of mixing metal particles (high conductivity and dielectric) into low conductivity/low dielectric media. In all studies, conducted over more than one hundred years, it is clear the net dielectric value increases, in some cases by an order of magnitude, with the volumetric fraction of metal particles in a ceramic matrix material. The first discussion of the subject of dielectrics composed of layers of materials of different dielectric value is found in Maxwell (Maxwell 1891). Experimental studies of the phenomenon and further development can be found in many places in the early 20th century (Wagner 1913, 1914).

One of the remarkable early advances was the development of an algebraic model, known as the Maxwell-Garnett model (Garnett 1904), that accurately predicts electrical properties of composites consisting of low concentrations (ca. < 20 vol %) of high conductivity/dielectric value particles in a low conductivity matrix. The model is limited to volumetric loadings of less than about 20 vol per cent because of the phenomenon known as percolation (Gerenrot et al. 2003). Indeed, above 25 per cent weight loading metal particle addition will actually form a continuous path between electrodes, leading to a short circuit. This model continues to be developed to permit it to be used for the addition of several particle types, with particles of different sizes and shapes (Koledintseva et al. 2006).

Presently there are at least three branches of research regarding the impact of adding metal particles, below the percolation limit, to low dielectric value matrix materials. One community is focused on the addition of metal particles to polymeric materials (Dang et al. 2004, Psarras et al. 2003, Tsangaris et al. 1991, Tsangaris et al. 1996).

The modelling of the 'polymer' community is essentially based on the Maxwell-Garnett approach, and it is clear the data, generally showing less than an order of magnitude increase in dielectric constant, is well explained. It is also clear that the identity of the metal, the size and shape of the particles, have a relatively minor impact on the measured properties.

In addition to the literature on metal particle addition to polymeric materials, there is a parallel literature on the impact of metal addition to standard ceramic dielectrics such as barium titanate. Experiments clearly show metal particle addition to standard dielectrics, such a barium titanate, does increase the dielectric constant, and by as much as an order of magnitude (Tsangaris et al. 1991). As with the polymer studies, the dielectric value increases with the amount of metal added below percolation. The hypotheses presented in the literature to explain the impact of metal particle addition on ceramic mixtures do not cite earlier work, and are mutually exclusive. One class of explanations is that the increased dielectric values observed result from the peculiar properties of nanoscale metal particles (Saha 2004, Lewis 2005, Ravindran et al. 2006). This is not consistent with the observation that 50 micron scale large metal particles produce the same effect. A second model is based on the observation that the dielectric increases with metal loading right up to the percolation limit (Efros 2011). This suggested to some workers that some properties of mixtures, including dielectric value, are sensitive to percolation and move toward infinite value at that limit. This concept was developed mathematically (Bergman and Imry 1977, Efros and Shklovskii 1976, Efros 2011).

There are both unanswered experimental and conceptual questions regarding the impact of metal particles added to ceramics. One problem with the experimental results is that they are based on Impedance Spectroscopy. As noted earlier in this essay, that technique, which is limited to voltages of less than 25 mV, is not suited to a determination of energy density. There is insufficient data to determine the effective energy density of ceramic dielectrics containing metal particles. The conceptual issue is that the postulated mechanisms of increased capacitance appear fundamentally flawed. Specifically, any acceptable model must include a mechanism for reducing the field everywhere in space, not just within the dielectric. No aspect of any model based on perturbation results in changes in any property, including field strength, outside the bounds of the dielectric.

Herein an alternative model, consistent with the general hypothesis presented herein regarding the relationship between dipoles formed in the dielectric and dielectric value, is postulated: Metal particles, with near infinite permittivity, form dipoles once voltage is applied to the electrodes. Therefore, below the percolation limit the metal particles will increase net dipole moment within the dielectric, and reduce the field everywhere. As the permittivity of metal is effectively higher than any dielectric, the field reduction, both within the dielectric and in the rest of space, due to the metal particles is greater than the non-conductive dielectric it displaces. Thus, metal loaded composite dielectric will have a net higher dielectric constant than the metal free material. The effect of increasing metal loading will be to increase this effect, until the percolation limit is reached. At the percolation limit, a short will form and the capacitor will no longer function as such.

5.1.6 Colossal Dielectric Materials

Another proposed means to create high energy ceramic based capacitors is to employ 'colossal dielectric' materials (Lunkenheimer et al. 2009). There are a number of ceramic based materials which have IS measured dielectric values in the 10^5 range. Among the reported colossal dielectric materials are 'transition metal oxides' such as $CaCu_3Ti_4O_{12}$ and related systems as well as lanthanum series oxides such as $La_{2-x}Sr_xNiO_4$ (Lunkenheimer et al. 1992, Lunkenheimer et al. 1995) and other complex oxides containing lanthanum series and metal species (Lunkenheimer et al. 1992).

The high dielectric values measured with IS suggest these materials could lead to a breakthrough in capacitor energy density, yet significant stumbling blocks remain. First, it has not been established that the materials will not suffer breakdown at relatively low electric field strength, hence voltage. Second, it has not been established that these materials do not reach saturation at relatively low electric field strength (Hu et al. 2013). As noted previously, IS measurements can lead to exaggerated expectations regarding performance at thicknesses and electric field strengths commensurate with high energy density.

There is also disagreement with the mechanism that leads to the high dielectric values observed at low voltage. The following have been suggested: hopping of localised charge carriers (Long 1982, Jonscher 1999, Elliott 1987, Mott and Davis 1979), barrier layer (BLC) effects (Wang and Zhang 2006, Li et al. 2006, Zhang et al. 2005, Zang et al. 2005, Fiorenza et al. 2008), surface state dipole formation, and charge density waves. In all cases, the models require, consistent with the fundamental theory discussed above, that dipoles form that are larger than those found in traditional high dielectric materials such as barium titanate.

The hopping mechanism is based on the concept of pinning electrons in defects several atomic lengths from the positively charged clusters ('metal-oxygen polyhedra') within the structure such that they form dipoles considerably longer in length than those found in ferroelectrics such as barium titanate (Hu et al. 2013, Li et al. 2006, Whangbo and Subramanian 2006, Wuttig et al. 2007). The highest dielectric values reported for those promoting this mechanism, $\sim 10^6$, only exist at temperatures above ~ 600 K (Hu et al. 2013). At room temperature, the dielectric values are always $< 10^5$, hence are low compared to those reported for superdielectric material. The measured dependence on temperature is consistent with the requirement that defects sites in the lattice are at an elevated energy. Thus, the population of charged species in the defect sites should increase, possibly exponentially, with temperature leading to a concomitant increase in dielectric value with temperature.

The surface state formation concept is based on polarisation across a very thin insulating layer (Lunkenheimer et al. 2009). This concept is confusing as atomic scale polarisation is assumed to occur in electric double layers, but the theoretical dielectric values derived are less than 30 in all cases (see Table 2). Charge-density wave (CDW) are variations in the electronic charge density is a periodic function of position that may be incommensurate with the crystal lattice. In theory this leads to the formation of net dipoles between the layers and concomitantly high dielectric values (Whangbo and Subramanian 2006, Dumas et al. 1983, Grüner 1988, Fleming et al. 1986, Cava et al.

Table 2. Capacitance of various EDLC systems. From Pandolfo and Hollenkamp 2006.

Material	Capacitance ($\mu F\ cm^{-2}$)	Surface Area (m^2g^{-1})
Carbon Fiber Cloth	6.9	1630
Activated Carbon	19	1200
Carbon Aerogel	23	650
Graphite Cloth	10.7	630
Carbon Black	10	230
Graphite Powder	35	4

1986). It is notable that this behaviour is generally observed only at very high frequency (Wuttig et al. 2007, Zhang et al. 2005).

In sum, there are many reports of dielectric constants as high as 10^6 recorded at very low voltages (ca. < 100 mV) in the literature for some complex metal and metal/lanthanide oxide structures. There is a large number of competing explanations for the observed high dielectric values postulated, but none is universally accepted. Indeed, as noted by others, it is very hard to experimentally verify a particular model (Hu et al. 2013). None of these 'novel' materials has been shown appropriate for use in energy storage applications. There is nearly complete reliance on the use of the IS technique to determine the dielectric properties, a method singularly unsuited for predictions of energy density.

5.2 Electric Double Layer Capacitors

A remarkable quantity of research is devoted to EDL Supercapacitors, including many reviews cited hundreds and even thousands of times (Simon and Gogotsi 2008, Wang et al. 2012, Zhang and Zhao 2009, Zhi et al. 2013). Thus, the focus here is not to provide a complete review, but rather to demonstrate this type of capacitor fits the pattern established in the Introduction.

First is a review of the theory of EDL Supercapacitors including the consistency with the fundamentals outlined above: (i) the primary means to increase the capacitance of this class is to increase the surface area of the electrode and (ii) the formation of dipoles in the double layer is the source of dielectric properties. Second, there is a discussion of activated carbon based EDL Supercapacitors, the only commercial variety, with an emphasis on the divergence of real material behaviour from the ideal theoretical behaviour. Finally, the use of graphene as the ultimate electrode material is discussed, and associated caveats reviewed.

5.2.1 Theory of EDL Supercapacitors

In this brief review we outline the model that clearly explain this salient fact regarding EDLC: Both the capacitance and energy density increases linearly with the volume (or mass) of the capacitor. In fact, the units employed to characterise supercapacitors, Farads/gm or Farads/cm^3 reflect this reality. At first this appears to be somewhat of a puzzle: Why doesn't the capacity and energy density decrease as the spacing between 'electrodes' increases, a natural result of increasing the volume, as per Eq. 1?

The answer to the above question is a first order model: *An EDL Supercapacitor is essentially an electrolytic cell*. Like an electrolytic cell, and clearly distinct from a ceramic capacitor, an EDL Supercapacitor contains two phases, a liquid electrolyte phase with mobile solubilised ions, and a solid electrode (generally carbon) phase upon which a polarised 'electric double layer' forms from species in the liquid phase upon the application of a voltage to the electrodes. Moreover; the presence of the electrolyte phase generally limits the maximum operating voltage, hence energy density (Eq. 2) to the liquid breakdown, generally of order 1.5 volts (Qu 2009).

In an EDL Supercapacitor, the carbon powder which fills most of the volume, is not 'dielectric', but rather solid 'electrode'. The end pieces, generally made of metal, that connect the capacitor to a circuit are simply 'collector plates'. Each bit of electrically conductive carbon associated with one electrode, all of which must be in electrical contact with the other bits and the collector plate, is all at the same potential, hence is one large electrode. The surface area of the carbon thus is the value of 'A', electrode surface area, in Eq. 1. This has several important implications, including the fact that the capacitance increases linearly with the amount of carbon, and hence linearly with the volume or mass of carbon. Also, it suggests, but later shown to be a questionable assumption, that the higher the surface area of the carbon per unit volume, or mass, the higher the capacitance per volume or mass.

The need for the ionically conductive electrolyte found in all. EDLC is also consistent with the 'electrolytic cell' model. To wit: the EDLC can be considered to have an internal voltage divider. The full voltage 'drop' is divided into three segments; the drop across the anode, the drop across the cathode and the drop across the media between. A similar model is applied to electrolytic cells. To maximise energy and charge on the electrodes, it is critical that virtually all the voltage 'drop' is across the double layers at the electrodes. A conductive electrolyte phase, virtually no electric resistance, insures this is the case.

Another component of all EDL Supercapacitors, an insulating separator, is also consistent with this model. The separator prevents the two carbon electrodes from directly touching, hence creating a short circuit. In a classic electrolytic cell with hard, physically separated electrodes there is no need for a separator. The electrodes are simply designed not to touch. A supercapacitor packed with loose powder is a short circuit waiting to happen.

The employment of a divider to prevent a short circuit raises the question, why is there no short circuit via the conductive electrolyte? The answer is that the conductive electrolyte is never in direct contact with the electrodes. The full circuit must be considered a linear set of elements: (i) anode to anode double layer, (ii) anode double layer to electrolyte, (iii) electrolyte to cathode double layer, (iv) cathode double layer to cathode, as illustrated in Fig. 7. In that scheme neither anode nor cathode is ever in direct electric contact with the electrolyte.

Other features of supercapacitors are easy to understand using this conceptual framework as well. For example, as with an electrolysis cell, there is a strict voltage limit for capacitive behaviour. Above the upper voltage limit, chemical processes begin at the electrodes and capacitive effects are negligible (Bard and Faulkner 2000). An elementary, and only intended for pedagogical purposes, model of the voltage limit can be computed assuming the 'double layer' consists simply of one atomic layer of polarised water molecules. The empirical 'break down' voltage of water, that is the voltage gradient at which water becomes conductive due to ionisation, is 70 MV/m. Next, assume, per usual, the size of water molecule 0.275 nm. Thus, a single layer of water should break down at a voltage of 0.02 V. In fact, aqueous supercapacitors are generally reliable to about 2.3 volts, although they age much more slowly if operated below 2.0 volts. This relatively high voltage clearly shows that the electric double layer is more complex than a single molecular layer of water, and can be a heterogeneous structure incorporating, for example dissolved ions from the electrolyte solution. The nature/structure of the double layer started with a model proposed by Helmholtz more than one hundred and fifty years ago (Helmholtz 1853) and modelling continues (Bockris et al. 1963, Conway 1991).

In this first order model, all of the 'conductive surface' in either electrode of an EDLC is at the same voltage. Next, the 'double layer' on each of these surfaces is identical. Thus, the 'voltage drop' from the solution phase to the electrode surface through the double layer is the same. Hence, each small surface area is accurately described by Eq. 1. In brief, each 'fragment' of carbon (electrode) and its associated double layer ('dielectric') are the components of a parallel plate capacitor.

The frequency response of EDL Supercapacitors is very sluggish compared to ceramic capacitors. Indeed, even in many commercial supercapacitors this sluggish behaviour can impact the effective dielectric value and energy density even at very low frequencies, ca. 1 Hz (Fig. 8). This sluggish

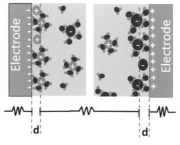

Fig. 7. Dipole Formation in EDL Supercapacitors. Double layer formed at the surface of the conducting electrodes. Inside the double layer, solvent molecules are trapped between the electrode and the ions. The double layer is formed on both electrodes, forming a capacitor on each side, with a distance d, corresponding to the thickness of the double layer. These capacitors are separated by a membrane that prevents electrical contact and represents a resistance to ionic movement. Adapted from Biener et al. 2011.

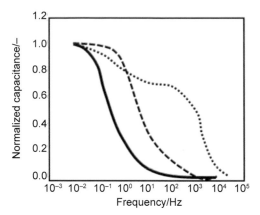

Fig. 8. Capacitance of EDL Supercapacitors vs. Frequency. Capacitance data was normalised for a given frequency of a capacitor using activated carbon fiber cloth (solid line), disordered multiwalled carbon nanotubes (dashed line), and vertically aligned multiwalled carbon nanotubes (dotted line) as electrode materials. Note the capacitance drops sharply, for all three types of EDL Supercapacitors, at frequencies orders of magnitude lower than the frequencies at which capacitance drops sharply (ca. Mhz) for ceramic capacitors. In this respect, the first words of the title of the original paper must be noted: 'Excellent frequency response', modified from Honda et al. 2007.

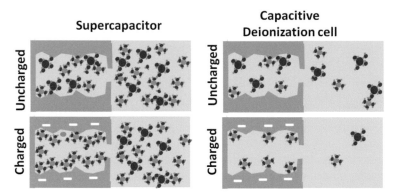

Fig. 9. Ionic Migration in EDL Supercapacitors. Upon polarisation, ions travel from the bulk of the liquid to the surface of the electrode (in gray), traveling also through the tortuous pores. When the concentration of ions in solution is large (left), the bulk remains almost unchanged, but when the concentration is low, the liquid can be depleted of ions (right). Both scenarios involve large ionic displacement through the pores of the electrodes. Physical motion of ions, as per batteries, requires time. As frequency increases, the double layer is not able to completely form during the time alloted, thus capacitance is reduced (Fig. 8) with increasing frequency.

behaviour reflects the need for the double layer, composed of mobile liquid phase species, to form/ dissolve/re-form during each voltage cycle (Fig. 7).

One final consideration is required to fully develop the first order model of EDL Supercapacitors as simply a huge system of parallel plate capacitors. Specifically, the values normally associated with parallel plate capacitors are not available: Neither the value of ε the dielectric constant of the double layer material, nor the thickness of the double layer ('t' of Eq. (1)), are measured. That is, for EDLC, the parameters ε, ε_o and t of Eq. (1) are lumped together, as they are difficult/impossible to measure independently. This lumped value is referred to as a capacitance/unit surface area (Table 2). Indeed, for aqueous electrolyte/carbon EDLC the typical nominal value given is 20 microF/cm² (Conway 1999, Simon and Burke 2008), although in practice this varies with carbon employed (Table 2).

To summarise the 'electrolysis cell' analogy: An EDL supercapacitor is a form of electrolysis cell (e.g., two metal foils, connected to anode/cathode of a power supply, both placed in an electrically conductive liquid media), in which the metal foils have been replaced by high surface area conductive carbon. Indeed, it has always been known that the amount of charge stored on the electrodes in such a cell, at voltages below the chemical reaction threshold, is proportional to the surface area of the electrodes, and very weakly a function of their separation distance (Bard and Faulkner 2000). As in classical electrolysis theory, the ability of the electrode to store charge is associated with the 'polarisation' of a thin layer of liquid adjacent to the electrode, the so-called electric double layer. The precise structure of this EDLC is somewhat of a controversy, some indicating it is one molecular layer thick, others suggesting a multi-layer, more complex structure. The history of the supercapacitor, found elsewhere (Pandolfo and Hollenkamp 2006) is consistent with early development following this conceptual framework.

Table 3. High energy density supercapacitors: EDL Supercapacitors with energy density greater than 200 J/g or 200 J/cm^3.

Electrode Material	Specific Energy (J/g)	Energy Density (J/cm³)	Reference
Carbon Nanofibers	886	443*	(Kim et al. 2016a)
Cobalt hydroxide nanowires (voltage range –0.1 V to 0.3 V)	792	–	(Mahmood et al. 2015)
Reduced graphene oxide (ionic liquids)	653	326*	(Sahu et al. 2015)
Hydrous ruthenium oxide and multiwalled carbon nanotubes	486	–	(Chaitra et al. 2016)
Graphene	457	324	(Xu et al. 2014)
Cobalt hydroxide nanorods decorating 3D carbon foam	450	–	(Patil et al. 2016)
Gold	353	–	(Kim et al. 2016b)
Graphene, ionic liquid electrolyte	308	154*	(Liu et al. 2010)
Vanadiom oxide and polypyrrole	295	–	(Bai et al. 2014)
Iron-decorated carbon	269	–	(Guan et al. 2015)
Graphene	252	126*	(Zhu et al. 2011)
Activated carbon from rice bran	252	115	(Hou et al. 2014)
Carbon nanofibers and silica based ionic liquid	220	110*	(Lawrence et al. 2016)
$MnCo_2O_4$	217	–	(Hui et al. 2016)
Reduced graphene oxide	211	106*	(Mohanapriya et al. 2016)

*Volumetric densities are calculated assuming a density of 0.5 g cm^{-3} and only considering active material (Murali et al. 2013). All values of energy density are either taken from the original paper or calculated using the highest operational voltage shown therein.

5.2.2 Activated Carbon Based Supercapacitors

The first order 'electrolytic cell' model outlined above only provides a general, loose, framework for conceptual understanding of EDL Supercapacitors. Real systems prove to be more complex, thus the first order model does not provide a recipe for predicting or improving performance. All commercial supercapacitors are currently based on the use of activated carbon. These capacitors do not behave as predicted by the model in several significant fashions.

An example of the failure of the simple model regards the difference between the double layer prediction of the impact of pore size and its measured impact. There is a general rule that for a pore to contribute to the double layer surface area, the pore must be 2 times the diameter of the molecules that form the double layer. According to the general understanding of the double layer, a small pore cannot accommodate the formation of a double layer, hence cannot polarise to allow charge storage. Yet, it has been demonstrated that micropores in carbons, that is pores too small to accommodate a true double layer, do contribute significantly to the capacitance of activated carbons, the most commonly employed 'electrode material' found in EDL Supercapacitors, as shown in Fig. 10.

Another feature that plays a surprising and significant role in determination of net capacitance, and internal resistance/leak rate is the surface chemistry of the carbon powder (Menéndez et al. 1996, 1997). The effect of oxygen and other surface groups on these parameters is generally to reduce

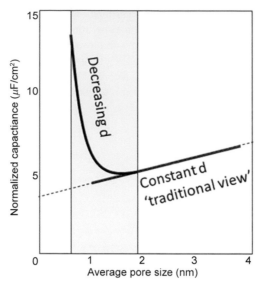

Fig. 10. Impact of Activated Carbon Pore Structure—The normalised capacitance for a number of activated carbons, all generated using different techniques, graphed. The 'traditional view' is that micropores do not participate in double layer formation, hence charge storage. The work by Chimola et al. 2006 is inconsistent with the standard view as it indicates that very small micropores (in the order of the size of the solvated ions) in some activated carbons contribute significantly to capacitance. Modified from Chmiola et al. 2006.

capacitance and increase leak rate (Yoshida et al. 1990, Morimoto et al. 1996). Another departure of real systems from the model is the finding that the measured capacitance, as a function of surface area, diverges from the predicted trend. The prediction arising from the first order model is that the capacitance on a normalised basis, generally taken as capacitance/g, should increase linearly with measured surface area/gram. As shown in Fig. 11 for activated carbons there is a flattening of the curve as the surface area increases, and the capacitive values at high surface area are far below those predicted based on a linear extrapolation.

A final notable empirical fact inconsistent with models: The measured value of F/cm² for all carbons is generally significantly less than that, ~ 20 F/cm² the value predicted by theory. Another notable failure of the theory is the finding that carbon nanotubes, on a F/gm basis perform far less well than anticipated (Talapatra et al. 2006, Pandolfo and Hollenkamp 2006) leading to a very low F/cm² value.

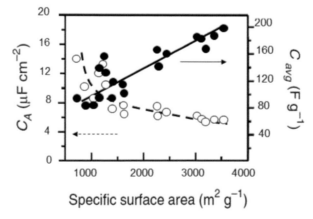

Fig. 11. Electric Double Layer Capacitance Variation with Specific Surface Area for Porous Carbon Materials. The values were obtained from averaging over the entire voltage range, and the lines are only qualitative indicators of the general trends (Modified from: Ji et al. 2014).

Another issue with activated carbon based EDL Supercapacitors regards the method of reporting. Generally the energy density, capacitance and other values are reported relative to mass (Table 3). Yet, in many applications the effective deployment limit is available volume, and relative to this parameter these materials do not perform as well. Indeed, as reported in several places the usual densities are of the order of 0.5 gm/cm³ (Burke 2007, Murali et al. 2013). Attempts to increase volumetric capacitance and energy density by compression methods have simply reduced the volumetric capacitance significantly (Yang et al. 2013).

In building practical, high energy density EDL Supercapacitors, there are variety of factors considered in selecting an activated carbon powder to serve as the electrode: high electrical conductivity, specific surface area, high accessibility of the surface area, and stability including corrosion resistance. These can serve as guideline, but in the final analysis, knowledge of these parameters is insufficient to predict behaviour. Hence, the actual selection of activated carbons must in the final analysis be done on an empirical basis.

Another practical consideration is operation as a function of temperature. Clearly, operation near engines will often occur at elevated temperatures, and operations in some environments will occur where the ambient temperature is quite low. As shown in Fig. 12 the capacitance increases with temperature (Liu 2010). Similar trends are reported elsewhere (Shulka et al. 2000). This is not only significant operational information, it is also demonstrates consistency with the model proposed here: In EDL Supercapacitors the frequency response is limited due to the need for a double layer to form via the migration over many atomic lengths, through the liquid phase, of molecular and ionic species. As the temperature rises species move more quickly and viscosity drops, allowing dipoles to form more quickly, consistent with the data presented. The activation energy implied by the collected data (ca. 20 kcal/g mole) is significantly higher than that required for diffusion (ca. 2 kcal/g mole). Are chemical bonds broken in this process? This is clearly a question for further research.

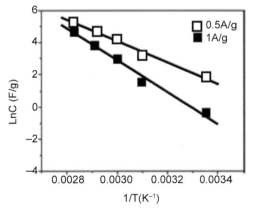

Fig. 12. Temperature Effects. There is an Arrhenius relationship between measured capacitance for a given charge/discharge current and temperature. This is consistent with the suggestion that double layer formation requires significant motion of ionic species. Modified from Liu 2010.

5.2.3 Graphene Based EDL Supercapacitors

Despite the uncertain relationship between surface area and net capacitance, much research is still predicated on simply increasing the electrode surface area. This is leading research away from activated carbon. The material generally believed to be the 'ultimate' in terms of electrode surface area is graphene (Dato et al. 2008, Luhrs and Phillips 2014, Tatarova et al. 2014, Tatarova et al. 2013).

Theoretical graphene is single layer (sp² hybridised) graphite, which has a high theoretical surface area (> 2600 m²/gm), is highly conductive, and is chemically stable. Real 'bulk' graphene powder generally consists of up to ~ 10 planes of stacked graphite basal planes. The theoretical maximum capacitance level for single layer graphene is 550 F/gm (El-Kady et al. 2012). Combined with a purported safe operating voltage in aqueous systems of 2.3 V, and assuming a density of 0.5 g/cm^3, this is consistent with a maximum energy density of ~ 750 J/cm^3. It has also been demonstrated that graphene supercapacitors permit very rapid charge/discharge (Shao et al. 2015, Bonaccorso et al. 2015).

A major barrier to achieving predicted maximum theoretical capacitance with graphene is that graphene prefers, thermodynamically, to reform into graphite by re-stacking processes (Fang et al. 2012, Yang et al. 2011). The restacking process invariably leads to performance degradation (Zhao et al. 2015).

Various methods have been employed to reduce re-stacking including creating composites of graphene-polymer, or graphene transition metal oxide and introducing heterogeneous materials between the planes of graphene, an example is presented in Fig. 13 (Shao et al. 2015, Xu et al. 2015, Meng et al. 2013, Song et al. 2015). All of these methods include complex synthesis protocols, and generally they degrade sharply over time. More complex structures, based on syntheses in which graphene is combined with other high porosity/high surface area carbon based structures to create layered structures (Zhuang et al. 2013, Zhuang et al. 2014, Yang et al. 2015, Li et al. 2014) consisting of alternating layers of graphene and molecular scale microporous carbons. Indeed, the high porosity of the microporous layer suggests they can serve as supercapacitor electrodes, independent of associated graphene layers.

The multi-layered materials as synthesised, consist of particles/sheets from the nano to the micron in dimension, are generally wrinkled on the nanoscale. The precise structure can be manipulated including distance between layers, although not precisely and homogeneously, via control of variables such as the ratio of graphene to the polymer component. The capacitive performance of some of these materials has been studied, and values as high as nearly 450 F/gm reported (Zhuang et al. 2014).

A final issue regarding reports on graphene as well as activated carbon based EDL Supercapacitors is the viability of energy density data. There are two major caveats to the methods employed to report capacitance. First, data for a full charge/discharge analysis is either not collected at all, or only over a very limited voltage range. Without this full cycle data it is not possible to obtain even a good

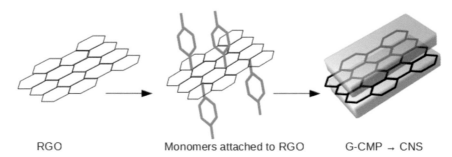

RGO   Monomers attached to RGO   G-CMP → CNS

Fig. 13. Designer Graphene. Structures are designed to maximised surface area, prevent 'fusion' of layers, and permit electrolyte penetration to all surfaces. RGO: Reduced graphene oxide, G-CMP: Graphene-based conjugated microporous polymers, CNS: porous carbon nanosheet. From: Yuan et al. 2016.

estimate of energy released during discharge. For example (Zhuang et al. 2014) it is clear 'effective' capacitance is a function of voltage. That is, the slope of the V vs. I curve in the discharge leg, which is inversely proportional to capacitance for a constant current process (Eq. 4), clearly reduces as the voltage increases. Indeed, for capacitance independent of voltage the use of constant current charge/ discharge cycles will yield a saw tooth pattern, as shown in Fig. 5, but it generally does not. In fact, the capacitances reported often represent the highest values that is those recorded at only a few tens of millivolts. In virtually all cases effective capacitance decreases with increasing voltage. For this reason, a direct computation of energy from Eq. 2 is impossible, and it is necessary to collect data for voltage and current as a function of time over the entire voltage range, and determine energy by integration of experimental data using Eq. 5.

The second difficulty with most data is that it is only collected at one discharge rate, or roughly one frequency. For example, in Fig. 14, the discharge times are clearly of the order of thousands of seconds, or at frequencies less than 10^{-3} Hz. As there is capacitive 'roll off' for all capacitor types, data collected at ca. 10^{-3} Hz is only valid for a limited range of capacitor applications. In particular, total energy released will fall with increased frequency of operation. For many applications, from engine start up to laser firing, the capacitors performance at higher frequency, ca. 10^3 Hz, is far more important.

For engineering purposes data must be available, based on full voltage cycle analysis, over a broad frequency range. Recently, there is evidence of an increasing awareness of this issue. As shown in Fig. 14, data was collected over a range of frequencies for variety of polymer/graphene layer composites by changing the current but over a limited voltage (1.0 volts) and frequency range (< 0.1 Hz). And, it is clear capacitance, hence net energy density, does fall with increased frequency. The key question: Why does energy density fall as frequency increases? The answer: Dielectric constant, hence capacitance is proportional to dipole length and dipole density. Similarly, a drop in energy density indicates a drop in capacitance, and concomitantly a drop in either dipole length or density. One of the key points of the present manuscript is that as the frequency of applied voltage increases, dipole length, effectively becomes shorter. In the case of ceramic capacitors for which travel lengths are sub-angstrom, this effect is minimal until MHz frequencies. For supercapacitors in which ions must travel nanometers if not microns, the onset of incomplete dipole formation is observed at low frequencies. This is attributed to the inability of the ions, as period is shortened, to reach the position that maximises the dipole length. Hence, the theory predicts for EDL Supercapacitors that even at relatively low frequency, the dipoles are shorter, possibly less dense, leading to a lower net dielectric and effectively a lower capacitance.

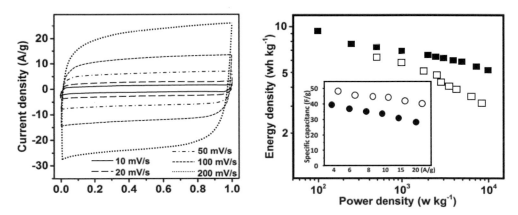

Fig. 14. Graphene Supercapacitor Frequency Dependence. LEFT—To control discharge time with either voltammetry the voltage increase rate is changed. For 'rapid' discharge (200 mV/s or 5 sec discharge period) the capacitance is a function of voltage, whereas for slow discharge (10 mV/sec or 100 second discharge period) it is independent of voltage. RIGHT—As the discharge period is shortened, the total energy (TE) of discharge is reduced; however, the period (T) is also reduced. Typical for supercpacitors, the ratio of TE/T, used to compute power, increases as TE gets shorter. This results in the type of 'Ragone curve' (Energy density vs. Power Density) shown. Inset—Typical for all supercapacitors the shorter the discharge time, the lower the capacitance. In this study the capacitance dropped by more than half when the discharge time was cut from 100 seconds to 10 seconds. From Yuan et al. 2016.

5.3 Super Dielectric Materials

Super Dielectric Materials (SDM) were invented based on a single aspect of the theory of dielectrics developed above. Specifically, the larger the dipoles of a dielectric, the more effectively the dielectric material cancels the dipole field of charges on the electrodes. Thus, SDM were invented/designed to create dielectric materials with the longest dipoles ever generated with little sacrifice to dipole density. Indeed, it is generally presumed that ferroelectrics, such as barium titanate, which form the largest dipoles of all crystalline solids when polarised by an electric field, should be the best solid dielectrics. The explanation for the high dielectric values observed in colossal dielectric materials is that these materials have 'long' dipoles, in fact longer than those found in barium titanate. Thus, SDM can be considered simply a further, and significant, step in the historical development of dielectric materials with dipoles of increasing length, and concomitantly higher dielectric values.

SDM dielectrics are designed to deliberately create a new form of 'dielectric' material, one with 'very, very long' dipoles, that is dipoles of the order hundreds of Å in length. This was not accomplished through the discovery of a new solid capable of polarising to create very long dipoles, but rather by postulating that in aqueous solutions containing dissolved salt, an applied field causes the +/– salt ions to move in opposite directions. This creates dipoles equal in length to the average ionic separation. The length of the polarised 'unit of water' determines the dipole length.

Three different types of SDM have been described in the literature: Powder SDM (P-SDM), Tube SDM (T-SDM) and Fabric SDM (F-SDM). In all cases the high dielectric values are attributed to the long dipoles that form in the liquid phase due to ion separation in response to an applied field. In all cases the ionic solutions, primarily aqueous to date, are the same. The difference between these types is the inert phase which holds the salt solutions. P-SDM are created by saturating high surface area porous refractory oxides such as alumina, silica and fumed silica with salt solutions, T-SDM are based on filling the linear pores in anaodised titania with salt solutions, and F-SDM are based on filling fabrics such as woven nylon with salt solutions.

Table 4. Super Dielectric Materials: dielectric materials with dielectric constants greater than 10^5 at low frequencies and room temperature.

Super Dielectric Material	Highest Dielectric Constant	Energy Density (J/cm³)	Reference
Nylon fabric saturated with aqueous NaCl	1.2×10^{11}	0.7	(Phillips 2016)
Fumed silica powder with aqueous NaCl	1×10^{11}	1	(Jenkins et al. 2016)
TiO$_2$ nanotubes with aqueous NaCl	7×10^9	390	(Gandy et al. 2016)
Porous alumina powder with aqueous NaCl	1×10^{11}	< 1	(Cortes and Phillips 2015a)
TiO$_2$ nanotubes with aqueous NaNO$_3$	7×10^9	230	(Cortes and Phillips 2015b)
Porous alumina powder with boric acid	5×10^9	< 1	(Fromille and Phillips 2014)
Illuminated perovskite	1×10^7	N/A	(Juarez-Perez et al. 2014)
Wet perovskite	1×10^7	N/A	(Almond and Bowen 2015)
Li/Na ion solid electrolytes	1×10^7, 1×10^9 at 170°C	N/A	(Braga et al. 2016)

5.3.1 Powder SDM

The specific material employed as the first invented/discovered SDM was a high porosity powder, a form of alumina, filled with a solution of DI water saturated with boric acid. The particular alumina employed had a measured average pore size of 500 Å. Enough 'acid water', that is water containing dissolved boric acid, was added to the alumina to create a state of incipient wetness, a concept employed in the synthesis of heterogenous catalysts. That is, enough 'acid water' was added to the alumina powder to fill all the pores, without any excess/free water. In practice, this creates a material with the mechanical consistency of slightly wet paste. This form of SDM is now called Powder-SDM (P-SDM).

In water, boric acid dissolves to create (+) and (−) ions. In the field created between the two electrodes of a parallel plate capacitor the (+) ions will move toward the negative electrode and the (−) ions toward the positive electrode. This will create a long electric dipole. In the first study it was assumed that the 'unit of solution' over which a dipole forms is the same as the size of the average pore, about 500 Å in length.

There are several empirical aspects of the behaviour of SDM noted in the first study, conducted only at very long discharge periods using the RC time constant method, that bear review as they imply limitations to the use of SDM based capacitors for high energy density storage and power delivery. First, it was found there was an ultimate limit to the voltage of ~ 1.4 volts. It was argued that above this voltage there is a breakdown ('electrolysis') of water that creates a short circuit. Similar phenomena are believed to limit the ultimate voltage of aqueous based supercapacitors, as discussed earlier. Second, above 0.8 volts there is a sharp drop in capacitance. Again, a similar phenomenon is observed for supercapacitors, although, the creation of different regions of capacitance is not always observed until higher frequencies. Third, the dielectric constant of the original P-SDM material was found to be a true constant of the material. That is, the measured net capacitance was inversely related to the dielectric layer thickness over a wide thickness range. This is only possible for a material with a constant dielectric value. Finally, all capacitance values were based on the R-C time constant method, and the discharge times were from hundreds to thousands of seconds. Later studies established that the dielectric constant values 'roll-off' with frequency, suggesting the recorded dielectric values are upper limits.

The results of the second study of P-SDM were consistent with the general SDM theory that any solution containing dissolved ions in a liquid phase will act as an SDM. In the second study the same alumina employed in the first study was used, but NaCl (salt) was employed to create the aqueous solution, and in three different concentrations. Moreover; in the second study a model that allowed a quantitative test was developed. To wit: Dipole strength is a linear function of two parameters, dipole length and the charge of the dipole. In this first quantitative model of SDM behaviour, it was assumed that the relative dielectric constant of two materials is simply the ratio of the net dipole strength:

$$\text{Dielectric constant ratio} = \frac{\text{Dipole length Material 1} \times \text{Mobile Charge Density Material 1}}{\text{Dipole length Material 2} \times \text{Mobile Charge Density Material 2}} \quad (17)$$

This equation was employed to predict the dielectric constant ration between the aqueous salt solution SDM (Material 1) and barium titanate (Material 2) with a presumed dielectric constant of 1,000. The mobile charge density of a solid, for example $BaTiO_3$, is equal the density of the ions within the solid which 'move' to form dipoles. For a liquid, the mobile charge density is assumed to be equal the number of ions dissolved in the solution. A simple model of the number of dissolved ions present in the 'salt water' of the first dielectric materials indicated 7.5×10^5 ions in the typical acid water 'drop/pore', whereas there are about 10^6 barium titanate primitive cells, hence dipoles, in the same volume. It was also assumed that the volume taken up by the solid component of the P-SDM, that is the alumina lattice, was about equal to the 'dead space' in an imperfect packing of the particles used in standard barium titanate dielectric. In sum, this analysis indicates, perhaps surprisingly, that the 'mobile charge density' of the pH neutral salt solution employed was roughly half the mobile charge density of barium titanate. Thus, the anticipated increase in dielectric constant was no more than 1,000 times greater than barium titanate, which has a measured dielectric constant of ~ 2×10^3. Hence, the authors anticipated dielectric constant for the first P-SDM: 2×10^6. The dielectric constant measured in the second SDM study was far greater, between 3×10^7 and 2×10^9, below approximately 1 volt, falling sharply above that voltage. That is, the measured increase in dielectric constant was at least 1000 times greater than that predicted by the model. The authors concluded that the use of materials, specifically aqueous solutions containing dissolved ions, creates materials with super dielectric constants, as anticipated, but that the model underestimates the observed increases in dielectric constant relative to that of ferroelectric solids. The observed dielectric values of SDM suggest the SDM model as originally postulated is incomplete.

It was also found that the salt and the 'matrix' material impact the ultimate dielectric constant. Indeed, P-SDMs with dielectric constants as high as 60 billion (6.0×10^{10}) were created in P-SDM consisting of aqueous sodium chloride solutions (salt water) in the same alumina used in the first study. The results also suggest that the theory needs some refinement particularly regarding the impact of ion concentration and 'pore size' on the ultimate dielectric constant. For example, the highest dielectric constants were not for aqueous solutions that were approximately 90 per cent 'saturated' with NaCl, but rather those that were closer to 50 per cent saturated.

In a third study of P-SDM the effect of matrix material was studied. In place of porous alumina, fumed silica was employed. The measured dielectric constants for very slow discharges (RC time constant method), and a distance between plates of just over 2 mm, were about a factor of two higher than those observed below 1.0 volts using the alumina material, approximately 1.1×10^{11}, at high concentrations of NaCl (ca. 30 weight per cent) dissolved in DI water.

The high dielectric values observed for fumed silica saturated with aqueous NaCl solutions was a 'puzzle' given the expectation of small pore size. Given an average particle size in fumed silica of ~ 5 nm, and that pores in a packed bed should be of size order the particle size, this raises the question of dipole size, and subsequently the anticipated value of the dielectric constant as per Eq. 5. According to theory, a 'pore' with a dimension of 5 nm filled with 'salt water' will not display significant SDM like behaviour. However; it is well known that the particles in wet fumed silica do not simply lie like a 'box of rocks', but rather are attached via hydrogen bonding allowing aggregates of primary particles to form agglomerates such that pores the size of the agglomerates are expected. The effective pore length, that is the length over which salt ions can diffuse in a reasonable time, is difficult to accurately model. To a first approximation it was assumed the diffusion length equals the total distance between electrodes.

To explore this first order model, the impact of total dielectric thickness on the dielectric constant was studied and it was found, unlike the case for the alumina based SDM that the dielectric constant decreased steadily with decreasing distance between electrodes. This was interpreted to be consistent with the model for fumed silica based SDM: effective pore length proportional to distance between electrodes. Hence, as the distance between electrodes is reduced, the average dipole length is reduced with the net effect of lower dielectric value.

Another avenue of study initiated in this third PSDM study was the impact of frequency on the effective capacitance of SDM. For this work the measurement protocol was also changed from 'RC time constant' to 'constant current' using a commercial galvanostat (Bio-Logic 300). This study was undertaken because the theory of SDM suggests a relatively slow development of dipoles and concomitantly high dielectric values. Indeed, SDM are based on field driven convection of ions inside the liquid filling the pores of the electrically insulating matrix material. Given sufficient time this leads to the creation of 'super' dipoles.

It takes time for the ions to move and form the multi-micron length dipoles. Ions initially randomly dispersed in a liquid must travel the order of the thickness of the dielectric layer to form super dipoles. Similarly, once formed, it takes time for the dipoles to 'dissolve' allowing electrons on the electrodes to release. This model of multi-micron ionic travel under the influence of an applied field implies a time dependence to the capacitance of SDM. Qualitatively, the model clearly indicates that if the applied field is switched too quickly from one polarity to the opposite, the net ionic separation will be reduced. The ions that form the dipole will simply not have time to travel to their equilibrium, fully polarised, positions. This in turn implies smaller effective dipoles at higher frequency. This should lead to a smaller measured dielectric and capacitance values with increasing frequency (Fig. 15).

Studies clearly showed that the dielectric constant and capacitance decreased as the discharge period decreased. It was found that the dielectric constant for a discharge time of 1 second was roughly an order of magnitude less than that measured for a discharge time of ten seconds. However; a true general rule was not found as other factors such as dielectric thickness and salt concentration were found to impact the rate of capacitive 'roll off' with decreasing discharge time. In sum, the study did suggest that for SDM the capacitance roll off is significant even for relatively 'long' (e.g., 1 second) discharge times; although the net 'power' delivered actually increased with decreasing discharge time. This reflected the fact power is a ratio of energy released to the discharge time. For all SDM, both values decrease as the discharge current increases/period decreases, but at different rates. In the case of SDM most work to date shows a trend: The rate of net energy decrease is less than the rate of decrease in the discharge period. Thus, 'power' increases as the period of discharge decreases. That is, the power density increases with increasing frequency.

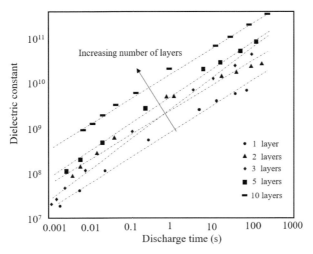

Fig. 15. Dielectric constant as a function of discharge time in fiber SDM capacitors. Capacitors were created from layers of nylon fabric filled with aqueous solutions of 30 wt per cent NaCl. From Phillips 2016.

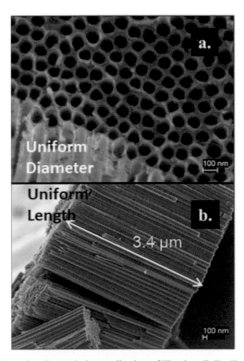

Fig. 16. SEM image of Titania Nanotubes Formed via Anodisation of Titanium Foils. Top (a) and lateral (b) SEM views of titania nanotube arrays. It can be seen that tubes are open at the top and are highly organised and oriented. Modified from Cortes and Phillips 2015b.

5.3.2 Tube-SDM

The success of the SDM theory as applied to P-SDM, suggested that matrix structures other than packed particles might work as well. In particular, the dielectric properties of anodic titania films filled with various aqueous salt solutions were studied.

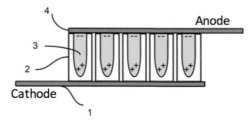

Fig. 17. Working Principle of T-SDM Based NP Supercapacitors. Cross section representation includes the electrodes, labeled 1 and 4. Electrode 1 is the parent titanium foil, still attached to the tubes of amorphous titania formed via anodistion. Electrode 4 is generally a form of graphite. The direction of the voltage depicted here is the only polarisation under which s superdielectric behaviour is observed. Ions migrate, as shown, to oppose the applied field forming dipoles approaching the size of the tubes. Modified from Cortes and Phillips 2015b.

As shown in Fig. 16 upon anodisation an oxide film consisting of a closely packed array of 'tubes' of TiO_2, orthogonal to the original surface, form (Cortes et al. 2016). These tubes are hollow and opened at the top, with an internal diameter generally of the order 100 nm across and as deep as the oxide film. According to the SDM theory, once filled with an aqueous salt solution these structures should be excellent dielectrics, as (giant) dipoles of the same length, generally many microns, as the pores should form (Fig. 17).

As anticipated, the dielectric constants for these T-SDM materials were extremely high, but a function of the pore length. In the first study aqueous sodium nitrate solutions were used to fill the pores of anodised titania of various lengths. Even for pores as short as ~ 3 micron the dielectric constant was greater than 10^7 for long discharge times (ca. 100 seconds) even at 2 volts. The thicker the oxide layer, the larger the dielectric constant. For 18 micron pores the dielectric constant was consistently $> 10^8$, again even at 2 volts.

It was argued the model for P-SDM (above) applies to T-SDM with some modifications. In brief:

$$\text{Dielectric constant} \propto \text{Dipole Length} \cdot \text{Dipole Charge} \cdot \text{Dipole Density} \qquad (18)$$

In the 'powder' version it is assumed the dipole length is proportional to the average pore size in the powder medium. In P-SDM the pore/dipole length, is independent of the thickness of the dielectric layer. In contrast, for TSDM the dipole lengths are proportional to the tube lengths. The dipole lengths are a linear function of the 'thickness' of the dielectric. Hence, for a given salt concentration (dipole density), the dielectric constants observed should be proportional to the length of the tubes.

$$\text{Dielectric constant} \propto \text{Tube Length} \cdot \text{Dipole Charge} \cdot \text{Dipole Density} \qquad (19)$$

Moreover, in the TSDM the dipole density will be proportional to the total number of salt molecules. In turn, the number of salt molecules will be proportional to the product of tube volume and the salt concentration. As the only parameter of the tubes that varies from sample to sample is the length, this leads to this version of the model:

$$\text{Dielectric constant} \propto \text{Tube Length} \cdot \text{Dipole Charge} \cdot (\text{Tube Length} \cdot \text{free salt concentration}) \qquad (20)$$

This is expressed as:

$$\varepsilon \propto t^2 \cdot S \qquad (21)$$

where t is the tube length and S is the salt concentration. A test of this model based on employing different tube lengths showed excellent quantitative agreement between model and data (Cortes and Phillips 2015b).

There were several findings of this study that suggested Novel Paradigm Supercapacitors (NP Supercapacitors), that is capacitors that employ SDM dielectrics, might lead to successful commercialisation. First, the energy density achieved, $\sim$ 230 J/cm^3 of dielectric, was far better than any commercial supercapacitor, and competitive with the best prototype supercapacitors (Table 4). Second, the voltage achieved before a sharp drop in capacitance, > 2.0 volts, was significantly higher than that achieved with P-SDM. For capacitors, the higher the voltage, the better the 'quality'/ usability of the stored energy.

There is no clear understanding of the difference in maximum voltage between P- SDM (< 1.2 V) and T-SDM ($\sim$ 2 V). It was suggested that the TiO$_2$/underlying titanium interface at the bottom of the pores in the anodised structure (Fig. 17) prevents extra charge carriers, produced via the electrolysis of the liquid phase, from completing a circuit. Thus 'Short circuit'/low dielectric conditions require the breakdown of the interface to produce some type of charge carrier. It has been reported that the interface formed through fluoride mediated anodisation of titanium is a Schottky contact, with a breakdown voltage of about 2.5 V (Lai et al. 2005). This hypothesis is also supported by the fact that these capacitors only act as such in one polarity, and act as resistors if connected in the other direction.

One additional study of T-SDM based on anodised titania showed that the ultimate energy density is a function of the salt employed. Indeed, using NaCl as the salt rather than sodium nitrate increased the energy density to $\sim$ 390 J/cm^3 (Gandy et al. 2016), but only for discharge times of several hundred seconds. In all other respects the study showed the same agreements and disagreements with the theory as the first T-SDM study. The fundamental findings in both are (i) SDMs exist as predicted by theory, (ii) the energy density of T-SDM is independent of tube length, all other parameters unchanged, and (iii) remarkably high energy density can be achieved. The fundamental difference with theory is the impact of salt concentration. Indeed, dielectric constant, capacitance, and energy density were not found to be a linear function of ion density as predicted by theory.

The power density of T-SDM requires direct measurement at different frequencies, besides the 'slow' discharges easily studied with RC circuits. Initially, it might be expected that this kind of SDM will have relatively long charge/discharge times (tens to hundreds of seconds) because the ions have to travel through several microns of liquid upon polarisation. However, a look at the diffusivity constant of ionic salts in water can provide a different perspective. Take NaCl for example, its diffusivity coefficient is in the order of 10^{-4} cm^2/s (Fell and Hutchison 1971), which means that even without an external field, just by random diffusion, the ions can travel a net distance in the order of 100 μm/s. In other words, if the pore length is below 10 μm, the ions would take less than 0.1s to move from one end to the other. This would be facilitated by the highly organised, nanometer-sized, straight pores in the T-SDM, and this is assuming the electric field would not enhance the rate at which ions move. This promising feature, naturally, requires empirical evidence.

5.3.3 Fabric-SDM

In order to demonstrate the true generality of SDM theory and to demonstrate a potential 'easy' route to the creation of high energy density NPS, a study of fabric based superdielectric materials (F-SDM) was conducted. Specifically, a commercial nylon fabric was saturated with an 90 per cent NaCl saturated water solution and tested in for capacitors consisting of between 1 and 10 layers, for dielectric constant, power, energy density. Also, a thorough testing of these parameters as a function of the discharge time, and number of layers, was conducted. It was found (Fig. 15) that below 1 volt the capacitance followed this simple relationship for all capacitors, including those with 1 layer and those with 10 layers.

Capacitance = $C_{100}*(100/DT)^{-0.55}$

In the equation C_{100} is the capacitance measured at a discharge time of 100 seconds, and DT is the discharge time in seconds. The same equation also represents the dielectric value for all the capacitors tested with the obvious substitution of the dielectric constant at 100 seconds (D_{100}) for the capacitance at 100 seconds. It is notable that the dielectric constant and the capacitance increased with the number of layers, although the energy density gradually decreased as the number of layers increased.

Finally, it should be noted, these low surface area material (< 1 m^2/gm) had extraordinary dielectric values, as anticipated based on the SDM theory.

Conclusion

The intention of this chapter is to show the connections between apparently disparate efforts to improve the energy/power density of capacitors. What does research on EDL Supercapacitors, NP Supercapacitors, ceramic capacitors, metal loaded ceramic capacitors, etc., have in common? First, all are designed to address the same challenges. How can the footprint of high energy density capacitors be reduced to allow them to be more broadly used in applications from satellites to electric automobiles? Can these more robust, 'simpler' devices even replace batteries? A second common link: In all the fields of capacitor research, progress to increase energy density is significant. In fact the best capacitors, both commercial and prototype, are EDL Supercapacitors, with prototypes having measured volumetric energy density greater than 500 J/cm^3, but only at very long discharge times, ca. > 30 seconds. Given the success of these capacitors it is not surprising that overwhelmingly most investment, both commercial and research, is focused there. Yet, prototype studies, precursors to the 'next generation' of commercial products, show there is strong performance competition from other types of capacitors. Indeed, the best prototype NP Supercapacitors, a newly invented category for which little investment has been made, and have measured energy density of ~ 400 J/cm^3, also at long discharge times. Also there is evidence that capacitors based on improved solid state ceramic capacitors will soon have energy densities approaching 50 J/cm^3. Ultimately, ceramic dielectrics may be preferred as the energy density is not a significant function of frequency until MHz range is reached. In contrast, all forms of supercapacitors can show significant loss in dielectric value and energy density even at ca. 1 Hz.

The best prototype high surface area carbon powder electrode supercapacitors now have higher volumetric energy density than the best batteries of about a generation ago, e.g., Pb acid. However; batteries are a 'moving target' and the energy density of the modern commercial Li-ion battery (ca. 2400 J/cm^3) is nearly an order of magnitude better than these modern capacitors. Thus, at present it seems unlikely that capacitors will ever achieve the energy density of batteries, hence 'high energy density' capacitors, even as they continue to improve, will fill select functions, such as providing high power bursts for battery powered systems as a means to extend battery life, or in domains for which high power and/or rapid recharge are more important than energy density.

Another theme linking all research in the area is that there is only one engineering/science narrative in the field. To wit: All capacitors competing in energy storage/high power space can be described by the equation (Eq. 1), for parallel plate capacitors. Most research in the field is focused on modifying one of the three parameters in that equation: distance between layers, dielectric constant or electrode surface area. There is one exception: A small segment of the total activity in capacitor development is devoted to increasing energy density by finding dielectrics with high breakdown voltage, although at present this is strictly at the pre-prototype stage.

As noted, most research, and virtually all commercial development, for high power/energy density capacitors is focused on increasing the surface area ('A' in Eq. 1) of the electrode material, that is EDL Supercapacitors. These capacitors are based on replacing metal electrodes with powders composed of high surface area carbon, and in some cases high surface area electrically conductive clays. Presently, the focus in this area is on further increasing the electrode surface area, through the use of graphene, which is believed to be the ultimate high surface area conductive material. Yet, there is empirical evidence no simple correlation exists between surface area/gram, and capacitance/gram. Indeed, all evidence indicates there is a falloff in this value with increasing carbon surface area. This suggests that the ultimate energy density achieved via this approach will be less than the 'theoretical maximum' of ~ 750 J/cm^3. Indeed, it is likely to be a far lower value as that projected value assumes that graphene can be tightly packed. Experimental results indicate tight packing of graphene results in the formation of actual graphite and the concomitant loss of surface area.

There is less, but still significant, academic research focused on improving a second parameter in Eq. 1, the dielectric constant (ε), as a means to increase energy density. In all cases this is linked to a third major 'connecting' theme of this chapter: Increasing ε always involves increasing the length of dipoles that form in the dielectric upon exposure to a field. This is true for novel dielectrics as different as NP Supercapacitors, and colossal dielectrics. Although the models for these two classes of dielectrics, as well as other types such as metal loaded ceramic capacitors, are superficially different, in fact they are fundamentally the same. For example for Colossal dielectrics, the precise mechanism is in dispute, but all proposed models are based on the fundamental principle that the dielectric value is a function of the length and concentration of dipoles, which form in solid dielectrics upon exposure to an electric field. All models of colossal dielectrics postulate the formation of dipoles longer than those found in ferroelectrics such as barium titanate. Similarly, the remarkable dielectric values found for SDM are attributed to the long dipoles that can form via the separation of ions in a liquid solution induced by the application of an electric field. It is interesting to note that many desalination processes are based on ion migration to anode/cathode from 'salt water' in a system very similar to the EDL supercapacitors. The success of this technology demonstrates the reality of charge separation and concomitant dipole formation upon exposure of an ionic solution to an electric field.

For energy storage activated carbon powder based supercapacitors are currently unchallenged commercially. Significant effort is underway to create a 'second generation' of carbon powder super capacitors based on graphene replacing activated carbon. Indeed, prototypes show energy density of order 500 J/cm^3, but the expense of employing graphene is significant. It is also clear that EDL Supercapacitors always show significant roll-off in capacitance at frequencies orders of magnitude lower than that found in ceramic based capacitors. This is linked to the time required for ions to travel, certainly nanometers, possibly farther, and organise an electric double layer.

Are there alternatives to high surface area electrically conductive powder based supercapacitors? Very recently, a new class of materials, called Superdielectric materials, composed of salt solutions confined by a solid matrix, have been shown to have dielectric constants as much as 5 orders of magnitude higher than even colossal dielectric materials. Prototypes with energy density of about 400 J/cm^3, rivalling the best graphene based supercapacitors, have been built and tested. The theory of these materials is that in the electric field created by voltage applied to the electrodes of a parallel plate capacitor, cations and anions created by dissolved salts will move in opposite directions through the liquid solution in which they are dissolved, based on polarity. This potentially could result in the formation of dipoles that can be millimeters in length, although details of the 'dipole structures' are not presently known.

There are several 'classes' of SDM including high surface area powders saturated with salt water, anodised titania saturated with salt water, fabrics saturated with salt water, and additional classes are in development. Capacitors employing these dielectric materials, NP Supercapacitors, have been specifically tested for energy density. The highest 'static' energy density (discharge time order of 100 seconds) is of order 450 J/cm^3. Moreover, the cost of the key materials in some forms of NP Supercapacitors appears to be minimal.

The frequency response of some NP supercapacitors has been thoroughly characterised. Simple algebraic expressions for the 'roll off' with frequency empirically determined. Not surprisingly, capacitors based on this type of dielectric are 'slow'. In theory this is because as the polarity on the electrodes is switched, ions in the liquid respond by travelling toward the opposite electrode. This can require travel though a viscous liquid of several microns in some cases. If the frequency of the applied field is too short for full re-formation of oppositely polarised dipoles, the effective dipole length will be shortened, and concomitantly the net dielectric value reduced. In contrast, very little data regarding the frequency response of EDL Supercapacitors is available. The data that is available shows slow response, such that even at ca. 1 Hz the capacitance is already dropping. This data suggests, that like NP Supercapacitors, EDL Supercapacitors require ion travel. That is, the slow response is consistent with double layer formation requiring significant ion travel through the electrolyte.

Throughout this chapter, the different approaches to capacitive energy storage were discussed and compared in terms of one fundamental equation, and their various working principles were outlined. As a final note, one last comparison will be made. This time in terms of performance updating the Ragone chart (Fig. 18) to include some of the recent advances in NP Supercapacitors. The results are that all NPSupercapacitors employing aqueous salt solutions in anodised titania is restricted to a rectangular region on the Ragone chart.

It is reasonable to argue that all the NP Supercapacitors outperform EDL Supercapacitors. The best EDL supercapacitors produce results that lie within a 'blob' shaped region, whereas the measured T-SDM lie along a line for any particular NP Supercapacitor. There is some variation on the line position of the NP Supercapacitor lines as a function of salt employed, but all lines fall within the rectangle shown. The primary finding: all NPSupercapacitors, based on directly measured values, perform 'above' the EDL Supercapacitor region.

Future research can focus on overcoming factors that limit the ultimate energy density of present generation capacitors. For example, one limit to the energy density of both types of supercapacitors is 'breakdown' of the ion containing liquid phase. Even the best commercial EDL supercapacitors cannot be operated at above ~ 2.7 volts. The best NP Supercapacitors are reported to operate no higher than 2.3 volts. Electrolytic solutions with higher breakdown voltages could potentially improve the energy density of both types.

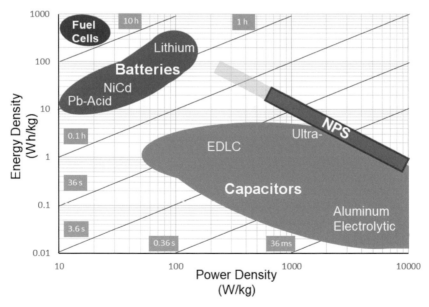

Fig. 18. NP Supercapacitor Performance Plotted On a Ragone Chart. The dark NPS 'rectangle' is based on directly fitted data collected over a range of discharge times for ~ 8 micron long anodised titania tubes, filled with three different salts in aqueous solutions. The data were collected using the constant current method operated with a wide range of currents. This enabled determination of Energy Density and Power Density over more than four orders of magnitude of discharge times (ca. 50 seconds to .005 seconds). The light section of the NPS rectangle is an extrapolation of the fitted curves. EDL supercapacitor, fuel cell and battery regions are from the open literature.

References

Adamec, V. and J.H. Calderwood. 1977. Electric-field-enhanced conductivity in dielectrics. J. Phys. D: Appl. Phys. 10(6): L79.

Almond, D.P. and C.R. Bowen. 2015. An explanation of the photoinduced giant dielectric constant of lead halide perovskite solar cells. J. Phys. Chem. Lett. 6(9): 1736–1740.

Arlt, G., D. Hennings and G. de With. 1985. Dielectric properties of fine-grained barium titanate ceramics. J. Appl. Phys. 58(4): 1619–1625.

Bai, M.-H., L.-J. Bian, Y. Song and X.-X. Liu. 2014. Electrochemical codeposition of vanadium oxide and polypyrrole for high-performance supercapacitor with high working voltage. ACS Appl. Mater. Interfaces 6(15): 12656–12664.

Bard, A.J. and L.R. Faulkner. 2000. Electrochemical Methods: Fundamentals and Applications, 2nd Edition, Wiley Global Education.

Barsoukov, E. and J. Ross Macdonald. 2005. Impedance Spectroscopy: Theory, Experiment, and Applications, John Wiley and Sons.

Bergman, D.J. and Y. Imry. 1977. Critical behavior of the complex dielectric constant near the percolation threshold of a heterogeneous material. Phys. Rev. Lett. 39(19): 1222–1225.

Biener, J., M. Stadermann, M. Suss, M.A. Worsley, M.M. Biener, K.A. Rose et al. 2011. Advanced carbon aerogels for energy applications. Energy Environ. Sci. 4(3): 656.

Bockris, J.O.M., M.A.V. Devanathan and K. Muller. 1963. On the structure of charged interfaces. Philos. T. Roy. Soc. A. 274(1356): 55–79.

Bologna Alles, A., R. Vanalstine and W. Schulze. 2005. Dielectric properties and aging of fast-fired barium titanate. Lat. Am. Appl. Res. 35(1): 29–35.

Bonaccorso, F., L. Colombo, G. Yu, M. Stoller, V. Tozzini, A.C. Ferrari et al. 2015. 2D Materials. Graphene, related two-dimensional crystals, and hybrid systems for energy conversion and storage. Science 347(6217): 1246501.

Braga, M.H., M. Helena Braga, J.A. Ferreira, A.J. Murchison and J.B. Goodenough. 2016. Electric dipoles and ionic conductivity in a Na glass electrolyte. J. Electrochem. Soc. 164(2): A207–A213.

Branwood, A., J.D. Hurd and R.H. Tredgold. 1962. Dielectric breakdown in barium titanate. Brit. J. Appl. Phys. 13(10): 528–528.

Burke, A. 2007. R&D considerations for the performance and application of electrochemical capacitors. Electrochim. Acta 53(3): 1083–1091.

Burn, I. and D.M. Smyth. 1972. Energy storage in ceramic dielectrics. J. Mater. Sci 7(3): 339–343.

Carter, B. and M.G. Norton. 2013. Ceramic Materials: Science and Engineering, Springer Science and Business Media.

Cava, R.J., P. Littlewood, R.M. Fleming, R.G. Dunn and E.A. Rietman. 1986. Low-frequency dielectric response of the charge-density wave in (TaSe 4) 2 I. Phsy. Rev. B 33(4): 2439–2443.

Chaitra, K., P. Sivaraman, R.T. Vinny, U.M. Bhatta, N. Nagaraju and N. Kathyayini. 2016. High energy density performance of hydrothermally produced hydrous ruthenium oxide/multiwalled carbon nanotubes composite: Design of an asymmetric supercapacitor with excellent cycle life. J. Mater. Chem. A 25(4): 627–635.

Chauhan, A., S. Patel, R. Vaish and C.R. Bowen. 2015. Anti-ferroelectric ceramics for high energy density capacitors. Materials (Basel) 8: 8009–8031.

Chen, G.H., Z.C. Li, T. Yang and Y. Yang. 2016a. Enhanced energy storage properties of strontium barium niobate ceramics by glass addition. J. Mater. Sci-Mater. El. 27(12): 12820–12825.

Chen, G.H., J. Zheng, Z.C. Li, J.W. Xu, Q.N. Li, C.R. Zhou et al. 2016b. Microstructures and dielectric properties of $Sr_{0.6}Ba_{0.4}Nb_2O_6$ ceramics with BaCu (B2O5) addition for energy storage. J. Mater. Sci-Mater. El. 27(3): 2645–2651.

Chmiola, J., G. Yushin, Y. Gogotsi, C. Portel, P. Simon and P.L. Taberna. 2006. Anomalous increase in carbon capacitance at pore sizes less than 1 nanometer. Science 313(5794): 1760–1763.

Christen, T. and M.W. Carlen. 2000. Theory of Ragone plots. J. Power Sources 91(2): 210–216.

Cohen, I.J., J.P. Kelley, D.A. Wetz and J. Heinzel. 2014. Evaluation of a hybrid energy storage module for pulsed power applications. IEEE transactions on plasma science. IEEE Plasma Sci. Soc. 42(10): 2948–2955.

Conway, B.E. 1991. Transition from "Supercapacitor" to "Battery" behavior in electrochemical energy storage. J. Electrochem. Soc. 138(6): 1539.

Conway, B.E. 1999. Electrochemical Supercapacitors: Scientific Fundamentals and Technological Applications, Springer Science and Business Media.

Cortes, F.J.Q., P.J. Arias-Monje, J. Phillips and H.R. Zea. 2016. Empirical kinetics for the growth of titania nanotube arrays by potentiostatic anodization in ethylene glycol. Mater. Design. 96: 80–89.

Cortes, F.J.Q. and J. Phillips. 2015a. Novel materials with effective super dielectric constants for energy storage. J. Electron Mater. 44(5): 1367–1376.

Cortes, F.J.Q. and J. Phillips. 2015b. Tube-super dielectric materials: Electrostatic capacitors with energy density greater than 200 J·cm^{-3}. Materials 8(9): 6208–6227.

Dang, Z.-M., Y.-H. Zhang and S.-C. Tjong. 2004. Dependence of dielectric behavior on the physical property of fillers in the polymer-matrix composites. Synthetic Met. 146(1): 79–84.

Dato, A., V. Radmilovic, Z. Lee, J. Phillips and M. Frenklach. 2008. Substrate-free gas-phase synthesis of graphene sheets. Nano Lett. 8(7): 2012–2016.

Debye, P. 1912. Einige resultate einer kinetischen theorie der isolatoren. Phisik. Zeits 13: 97–100.

Debye, P.J.W. 1929. Polar molecules, Chemical Catalog Company, Incorporated.

Divya, K.C. and J. Østergaard. 2009. Battery energy storage technology for power systems—An overview. Electr. Pow. Sys. Res. 79(4): 511–520.

Dorf, R. ed. 2003. CRC Handbook of Engineering Tables, CRC Press, NY.

Dowgiallo, E.J. and J.E. Hardin. 1995. Perspective on ultracapacitors for electric vehicles. IEEE Aero. El. Sys. Mag. 10(8): 26–31.

Drofenik, U., A. Musing and J.W. Kolar. 2010. Voltage-dependent capacitors in power electronic multi-domain simulations. In The 2010 International Power Electronics Conference - ECCE ASIA -, 643–650. IEEE.

Dumas, J., C. Schlenker, J. Marcus and R. Buder. 1983. Nonlinear conductivity and noise in the quasi one-dimensional blue bronze $K_{0.30} Mo O_3$. Phys. Rev. Lett. 50(10): 757–760.

Dumesic, J.A. and H. Topsøe. 1977. Mössbauer spectroscopy applications to heterogeneous catalysis. Adv. Catal. 121–246.

Efros, A.L. 2011. High volumetric capacitance near the insulator-metal percolation transition. Phys. Rev. B. 84(15). Available at: https://link.aps.org/doi/10.1103/PhysRevB.84.155134.

Efros, A.L. and B.I. Shklovskii. 1976. Critical behaviour of conductivity and dielectric constant near the metal-non-metal transition threshold. Phys. Satus Solidi (b) 76(2): 475–485.

El-Kady, M.F., V. Strong, S. Dubin and B.K. Richard. 2012. Laser scribing of high-performance and flexible graphene-based electrochemical capacitors. Science 335(6074): 1326–1330.

Elliott, S.R. 1987. A.c. conduction in amorphous chalcogenide and pnictide semiconductors. Adv. Phys. 36(2): 135–217.

Ewbank, M.D., R.R. Neurgaonkar, W.K. Cory and J. Feinberg. 1987. Photorefractive properties of strontium-barium niobate. J. Appl. Phys. 62(2): 374–380.

Fang, Y., B. Luo, Y. Jia, X. Li, B. Wang, Q. Song et al. 2012. Renewing functionalized graphene as electrodes for high-performance supercapacitors. Adv. Mater. 24(47): 6348–6355.

Faraday, M. 1832. Experimental researches in electricity. Philos. T. Roy. Soc. 122(0): 125–162.

Farma, R., M. Deraman, Awitdrus, I.A. Talib, R. Omar, J.G. Manjunatha et al. 2013. Physical and electrochemical properties of supercapacitor electrodes derived from carbon nanotube and biomass carbon. Int. J. Electrochem. Sc. 8(1): 257–273.

Fell, C.J.D. and H.P. Hutchison. 1971. Diffusion coefficients for sodium and potassium chlorides in water at elevated temperatures. J. Chem. Eng. Data 16: 427–429.

Fiorenza, P., R. Lo Nigro, C. Bongiorno, V. Raineri, M.C. Ferarrelli, D.C. Sinclair et al. 2008. Localized Electrical Characterization of the Giant Permittivity Effect in $CaCu_3Ti4O_{12}$ Ceramics. Appl. Phys. Lett. 92(18): 182907.

Fleming, R.M., R.J. Cava, L.F. Schneemeyer, E.A. Rietman and R.G. Dunn. 1986. Low-temperature divergence of the charge-density-wave viscosity in K 0.30 MoO3, (TaSe 4) 2 I, and TaS 3. Phys. Rev. B 33(8): 5450–5455.

Fröhlich, H. and A. Maradudin. 1959. Theory of dielectrics. Phys. Today 12(2): 40–42.

Fromille, S. and J. Phillips. 2014. Super dielectric materials. Materials 7(12): 8197–8212.

Gandy, J., F.J.Q. Cortes and J. Phillips. 2016. Testing the tube super-dielectric material hypothesis: Increased energy density using NaCl. J. Electron Mater. 45(11): 5499–5506.

Garnett, J.C.M. 1904. Colours in metal glasses and in metallic films. Philosophical Transactions of the Royal Society A: Mathematical, Philos. T. Roy. Soc. A 203(359-371): 385–420.

Gerenrot, D., L. Berlyand and J. Phillips. 2003. Random network model for heat transfer in high contrast composite materials. IEEE Trans. Adv. Pack. 26(4): 410–416.

Ghaffari, M., Y. Zhou, H. Xu, M. Lin, T.Y. Kim, R.S. Ruoff et al. 2013. High-volumetric performance aligned nano-porous microwave exfoliated graphite oxide-based electrochemical capacitors. Adv. Mater. 25(35): 4879–4885.

Ghidiu, M., M.R. Lukatskaya, M.-Q. Zhao, Y. Gogotsi and M.W. Barsoum. 2014. Conductive two-dimensional titanium carbide 'Clay' with high volumetric capacitance. Nature 516(7529): 78–81.

Grüner, G. 1988. The dynamics of charge-density waves. Rev. Mod. Phys. 60(4): 1129–1181.

Gualous, H., D. Bouquain, A. Berthon and J.M. Kauffmann. 2003. Experimental study of supercapacitor serial resistance and capacitance variations with temperature. J. Power Sources 123(1): 86–93.

Guan, C., J. Liu, Y. Wang, L. Mao, Z. Fan, Z. Shen et al. 2015. Iron oxide-decorated carbon for supercapacitor anodes with ultrahigh energy density and outstanding cycling stability. ACS Nano 9(5): 5198–5207.

Haertling, G.H. 1999. Ferroelectric ceramics: History and technology. J. Am. Ceram. Soc. 82(4): 797–818.

Hao, X., Y. Wang, J. Yang, S. An and J. Xu. 2012. High energy-storage performance in Pb 0.91 La 0.09 $(Ti_{0.65} Zr_{0.35})O_3$ relaxor ferroelectric thin films. Jrn. J. Appl. Phys. 112(11): 114111.

Harrington, D.A. and P. van den Driessche. 2011. Mechanism and equivalent circuits in electrochemical impedance spectroscopy. Electrochim Acta 56(23): 8005–8013.

Heckel, T. and L. Frey. 2015. A novel charge based SPICE model for nonlinear device capacitances. In 2015 IEEE 3rd Workshop on Wide Bandgap Power Devices and Applications (WiPDA). 2015 IEEE 3rd Workshop on Wide Bandgap Power Devices and Applications (WiPDA). IEEE 141–146.

Helmholtz, H. 1853. Ueber einige Gesetze der Vertheilung elektrischer Ströme in körperlichen Leitern mit Anwendung auf die thierisch-elektrischen Versuche. Ann. Phys. (Berl.) 165(6): 211–233.

Hemmati, R. and H. Saboori. 2016. Emergence of hybrid energy storage systems in renewable energy and transport applications—A review. Renew. Sust. Energ. Rev. 65: 11–23.

Hirschorn, B., M.E. Orazem, B. Tribollet, V. Vivier, I. Frateur and M. Musiani. 2010. Constant-phase-element behavior caused by resistivity distributions in films. Renew. Sust. Energ. Rev. 157(12): C458.

Honda, Y., M. Takeshige, H. Shiozaki, T. Kitamura and M. Ishikawa. 2007. Excellent frequency response of vertically aligned MWCNT electrode for EDLC. Electrochemistry 75(8): 586–588.

Hou, J., C. Cao, X. Ma, F. Idrees, B. Xu, X. Hao et al. 2014. From rice bran to high energy density supercapacitors: A new route to control porous structure of 3D carbon. Sci. Rep. UK 4: 7260.

Huang, J., Y. Zhang, T. Ma, H. Li and L. Zhang. 2010. Correlation between dielectric breakdown strength and interface polarization in barium strontium titanate glass ceramics. Applied physics letters 96, no. 4. Proceedings of the 15th IEEE Int. Ferro. 042902.

Huang, Y., J. Liang and Y. Chen. 2012. An overview of the applications of graphene-based materials in supercapacitors. Small 8(12): 1805–1834.

Huggins, R. 2010. Energy Storage. Springer Science & Business Media.

Hui, K.N., K.S. Hui, Z. Tang, V.V. Jadhav and Q.X. Xia. 2016. Hierarchical chestnut-like $MnCo_2O_4$ nanoneedles grown on nickel foam as binder-free electrode for high energy density asymmetric supercapacitors. J. Power Sources 330: 195–203.

Hu, W., Y. Liu, R.L. Withers, T.J. Frankcombe, L. Norén, A. Snashall et al. 2013. Electron-pinned defect-dipoles for high-performance colossal permittivity materials. Nat. Mater. 12(9): 821–826.

Jenkins, N., C. Petty and J. Phillips. 2016. Investigation of fumed silica/aqueous NaCl superdielectric material. Materials 9(2): 118.

Ji, H., X. Zhao, Z. Qiao, J. Jung, Y. Zhu, Y. Lu et al. 2014. Capacitance of carbon-based electrical double-layer capacitors. Nat. Commun. 5: 3317.

Johnson, J.B. 1927. Thermal agitation of electricity in conductors. Nature 119(2984): 50–51.

Jonscher, A.K. 1999. Dielectric relaxation in solids. Journal of Physics D: Appl. Phys. 32(14): R57–R70.

Juarez-Perez, E.J., R.S. Sanchez, L. Badia, G. Garcia-Belmonte, Y.S. Kang, I. Mora-Sero et al. 2014. Photoinduced giant dielectric constant in lead halide perovskite solar cells. J. Phys. Chem. Lett. 5(13): 2390–2394.

Kaiser, C.J. 1993. The Capacitor Handbook, Springer Science and Business Media.

Kao, K.-C. 2004. Dielectric Phenomena in Solids: With Emphasis on Physical Concepts of Electronic Processes, Academic Press.

Karthika, P., N. Rajalakshmi and K.S. Dhathathreyan. 2012. Functionalized exfoliated graphene oxide as supercapacitor electrodes. Soft Nanoscience Letters 02(04): 59–66.

Katz, D.M. 2016. Physics for Scientists and Engineers: Foundations and Connections, Extended Version with Modern, Cengage Learning.

Khaligh, A. and Z. Li. 2010. Battery, ultracapacitor, fuel cell, and hybrid energy storage systems for electric, hybrid electric, fuel cell, and plug-in hybrid electric vehicles: State of the Art. IEEE T. Veh. Technol. 59(6): 2806–2814.

Khomenko, V., E. Frackowiak and F. Béguin. 2005. Determination of the specific capacitance of conducting polymer/nanotubes composite electrodes using different cell configurations. Electrochim. Acta 50(12): 2499–2506.

Kim, C.H., J.-H. Wee, Y.A. Kim, K.S. Yang and C.-M. Yang. 2016a. Tailoring the pore structure of carbon nanofibers for achieving ultrahigh-energy-density supercapacitors using ionic liquids as electrolytes. J. Mater. Chem. A 4(13): 4763–4770.

Kim, S.-I., S.-W. Kim, K. Jung, J.-B. Kim and J.-H. Jang. 2016b. Ideal nanoporous gold based supercapacitors with theoretical capacitance and high energy/power density. Nano Energy 24: 17–24.

Kim, Y., M. Katchaperumal, V.W. Chen, Y. Park, C. Fuentes-Hernandez, M.-J. Pan et al. 2015. Bilayer structure with ultrahigh energy/power density using hybrid sol-gel dielectric and charge-blocking monolayer. Adv. Energy Mater. 5(9): 1500767.

Kinoshita, K. and A. Yamaji. 1976. Grain-size effects on dielectric properties in barium titanate ceramics. J. Appl. Phys. 47(1): 371–373.

Kirkwood, J.G. 1939. The dielectric polarization of polar liquids. J. Chem. Phys. 7(10): 911–919.

Koledintseva, M.Y., R.E. DuBroff and R.W. Schwartz. 2006. A maxwell garnett model for dielectric mixtures containing conducting particles at optical frequencies. Pr. Electromagn. 63: 223–242.

Kumari, N., V. Kumar, S.K. Singh, S. Khasa and M.S. Dahiya. 2017. Synthesis modified structural and dielectric properties of semiconducting zinc ferrospinels. Physica E 86: 168–174.

Lai, Y.K., L. Sun, C. Chen, C.G. Nie, J. Zuo and C.J. Lin. 2005. Optical and electrical characterization of TiO_2 nanotube arrays on titanium substrate. Appl. Surf. Sci. 252: 1101–1106.

Larcher, D. and J.-M. Tarascon. 2015. Towards greener and more sustainable batteries for electrical-energy storage. Nat. Chem. 7(1): 19–29.

Lawrence, D.W., C. Tran, A.T. Mallajoysula, S.K. Doorn, A. Mohite, G. Gupta et al. 2016. High-energy density nanofiber-based solid-state supercapacitors. J. Mater. Chem. 4(1): 160–166.

Lenzo, P.V., E.G. Spencer and A.A. Ballman. 1967. Electro-optic coefficients of ferroelectric strontium barium niobate. Appl. Phys. Lett. 11(1): 23–24.

Lerner, L.S. 1996. Physics for Scientists and Engineers, Jones and Bartlett Publishers.

Levinson, 1987. Electronic Ceramics: Properties: Devices, and Applications, CRC Press.

Lewis, T.J. 2005. Interfaces: nanometric dielectrics. J. Power Sources 38(2): 202–212.

Li, H., J. Wang, Q. Chu, Z. Wang, F. Zhang and S. Wang. 2009. Theoretical and experimental specific capacitance of polyaniline in sulfuric acid. J. Phys. D. Appl. Phys. 190(2): 578–586.

Li, L., X. Yu, H. Cai, Q. Liao, Y. Han and Z. Gao. 2013. Preparation and dielectric properties of $BaCu(B_2O_5)$-Doped $SrTiO_3$-based ceramics for energy storage. Mat. Sci. Eng. B 178(20): 1509–1514.

Li, M., A. Feteira, D.C. Sinclair and A.R. West. 2006. Influence of Mn doping on the semiconducting properties of $CaCu_3Ti_4O_{12}$ ceramics. Appl. Phys. Lett. 88(23): 232903.

Li, X.J., W. Xing, J. Zhou, G.Q. Wang, S.P. Zhuo, Z.F. Yan et al. 2014. Excellent capacitive performance of a three-dimensional hierarchical porous graphene/carbon composite with a superhigh surface area. Chemistry 20(41): 13314–13320.

Liu, C., Z. Yu, D. Neff, A. Zhamu and B.Z. Jang. 2010. Graphene-based supercapacitor with an ultrahigh energy density. Nano Lett. 10(12): 4863–4868.

Long, A.R. 1982. Frequency-dependent loss in amorphous semiconductors. Adv. Phys. 31(5): 553–637.

Luhrs, C.C. and J. Phillips. 2014. Reductive/Expansion Synthesis of Graphene. US Patent 8,894,886.

Lunkenheimer, P., V. Bobnar, A.V. Pronin, A.I. Ritus, A.A. Volkov and A. Loidl. 2002. Origin of apparent colossal dielectric constants. Phys. Rev. B. 66(5). http://dx.doi.org/10.1103/physrevb.66.052105.

Lunkenheimer, P., R. Fichtl, S.G. Ebbinghaus and A. Loidl. 2004. Nonintrinsic origin of the colossal dielectric constants in$CaCu_3Ti_4O_{12}$. Phys. Rev. B. 70(17). http://dx.doi.org/10.1103/physrevb.70.172102.

Lunkenheimer, P., G. Knebel, A. Pimenov, G.A. Emel'chenko and A. Loidl. 1995. Dc and Ac Conductivity of La2CuO4+δ. Zeitschrift Für Phys. Rev. B. 99(1): 507–516.

Lunkenheimer, P., S. Krohns, S. Riegg, S.G. Ebbinghaus, A. Reller and A. Loidl. 2009. Colossal dielectric constants in transition-metal oxides. Eur. Phys. J-Spec. Top. 180(1): 61–89.

Lunkenheimer, P., M. Resch, A. Loidl and Y. Hidaka. 1992. Ac Conductivity in La_2CuO_4. Phys. Rev. Lett. 69(3): 498–501.

Lu, Y., H.L. Hess and D.B. Edwards. 2007. Adaptive Control of an Ultracapacitor Energy Storage System for Hybrid Electric Vehicles. In 2007 IEEE International Electric Machines & Drives Conference. http://dx.doi.org/10.1109/iemdc.2007.383565.

Macdonald, D.D. 2006. Reflections on the history of electrochemical impedance spectroscopy. Electrochim Acta 51(8-9): 1376–1388.

Mahmood, N., M. Tahir, A. Mahmood, J. Zhu, C. Cao and Y. Hou. 2015. Chlorine-doped carbonated cobalt hydroxide for supercapacitors with enormously high pseudocapacitive performance and energy density. Nano Energy 11: 267–276.

Mansuripur, M., A.R. Zakharian and E.M. Wright. 2013. Electromagnetic-force distribution inside matter. Physical Review. A 88, no. 2. http://dx.doi.org/10.1103/physreva.88.023826.

Maxwell, 'A Treatise on Electricity and Magnetism, Vol I', 3rd Edition Clarendon Press 1891.

Menéndez, J.A., J. Phillips, B. Xia and L.R. Radovic. 1996. On the modification and characterization of chemical surface properties of activated carbon: In the search of carbons with stable basic properties. Langmuir: Langmuir 12(18): 4404–4410.

Menéndez, J.A., B. Xia, J. Phillips and L.R. Radovic. 1997. On the modification and characterization of chemical surface properties of activated carbon: Microcalorimetric, electrochemical, and thermal desorption probes. Langmuir: Langmuir 13(13): 3414–3421.

Meng, Y., K. Wang, Y. Zhang and Z. Wei. 2013. Hierarchical porous graphene/polyaniline composite film with superior rate performance for flexible supercapacitors. Adv. Mater. 25(48): 6985–6990.

Miller, J.R. and P. Simon. 2008. Materials Science. Electrochemical capacitors for energy management. Science 321(5889): 651–652.

Mishra, A., N. Mishra, S. Bisen and K.M. Jarabana. 2014. Frequency and temperature dependent dielectric studies of $BaTi_{0.96}Fe_{0.04}O_3$. J. Phys. Conf. Ser. 534(24): 012011.

Mohanapriya, K., G. Ghosh and N. Jha. 2016. Solar light reduced graphene as high energy density supercapacitor and capacitive deionization electrode. Electrochim. Acta 209: 719–729.

Morimoto, T., K. Hiratsuka, Y. Sanada and K. Kurihara. 1996. Electric double-layer capacitor using organic electrolyte. J. Power Sources 60(2): 239–247.

Mørup, S. and H. Topsøe. 1976. Mössbauer studies of thermal excitations in magnetically ordered microcrystals. J. Phys. D. Appl. Phys. 11(1): 63–66.

Motchenbacher, C.D. and J.A. Connelly. 1993. Low Noise Electronic System Design. J. Wiley & Sons.

Mott, N.F. and E.A. Davis. 1979. Electronic Processes in Non-Crystalline Materials. Oxford: Clarendon Press.

Murali, S., N. Quarles, L.L. Zhang, J.R. Potts, Z. Tan, Y. Lu et al. 2013. Volumetric capacitance of compressed activated microwave-expanded graphite oxide (a-MEGO) electrodes. Nano Energy 2(5): 764–768.

Niepce, J.C. and D. Hugentobler. 1991. Capacitors. In Concise Encyclopedia of Advanced Ceramic Materials. Elsevier, pp. 53–57.

Onsager, L. 1934. Deviations from ohm's law in weak electrolytes. J. Chem. Phys. 2(9): 599–615.

Onsager, L. 1936. Electric moments of molecules in liquids. J. Am. Chem. Soc. 58(8): 1486–1493.

Ortega, N., A. Kumar, J.F. Scott, D.B. Chrisey, M. Tomazawa, S. Kumari et al. 2012. Relaxor-ferroelectric superlattices: High energy density capacitors. J. Phys. Condens. Matter 24(44): 445901.

Oz, A., S. Hershkovitz and Y. Tsur. 2014. Electrochemical impedance spectroscopy of supercapacitors: A novel analysis approach using evolutionary programming. In, 162 7: 76–80. AIP Conf. Proc. AIP Publishing LLC.

Pandolfo, A.G. and A.F. Hollenkamp. 2006. Carbon properties and their role in supercapacitors. J. Power Sources 157(1): 11–27.

Parizi, S.S., A. Mellinger and G. Caruntu. 2014. Ferroelectric barium titanate nanocubes as capacitive building blocks for energy storage applications. ACS Appl. Mater. Interfaces 6(20): 17506–17517.

Patil, U.M., R.V. Ghorpade, M.S. Nam, A.C. Nalawade, S. Lee, H. Han et al. 2016. PolyHIPE derived freestanding 3D carbon foam for cobalt hydroxide nanorods based high performance supercapacitor. Sci. Rep. 6: 35490.

Pazde-Araujo, C., R. Ramesh and G.W. Taylor. 2001. Science and Technology of Integrated Ferroelectrics: Selected Papers from Eleven Years of the Proceedings of the International Symposium of Integrated Ferroelectronics. CRC Press.

Pean, C., B. Rotenberg, P. Simon and M. Salanne. 2016. Multi-scale modelling of supercapacitors: From molecular simulations to a transmission line model. J. Power Sources 326: 680–685.

Pecharromán, C., F. Esteban-Betegón, J.F. Bartolomé, G. Richter and J.S. Moya. 2004. Theoretical model of hardening in zirconia–nickel nanoparticle composites. Nano Lett. 4(4): 747–751.

Pecharromán, C., F. Esteban-Betegón and R. Jiménez. 2010. Electric field enhancement and conduction mechanisms in Ni/ $BaTiO_3$ percolative composites. Ferroelectrics 400(1): 81–88.

Phillips, J. 2016. Novel superdielectric materials: Aqueous salt solution saturated fabric. Materials 9(11): 918.

Phillips, J., B. Clausen and J.A. Dumesic. 1980. Iron pentacarbonyl decomposition over grafoil. Production of small metallic iron particles. J. Phys. Chem. 84(14): 1814–1822.

Prateek, V.K. Thakur and R.K. Gupta. 2016. Recent progress on ferroelectric polymer-based nanocomposites for high energy density capacitors: Synthesis, dielectric properties, and future aspects. Chem. Rev. 116(7): 4260–4317.

Psarras, G.C., E. Manolakaki and G.M. Tsangaris. 2003. Dielectric dispersion and Ac conductivity in—Iron particles loaded—polymer composites. Compos. Part A Appl. Sci. Manuf. 34(12): 1187–1198.

Puli, V.S., M. Ejaz, R. Elupula, M. Kothakonda, S. Adireddy, R.S. Katiyar et al. 2016. Core-shell like structured barium zirconium titanate-barium calcium titanate–poly(methyl Methacrylate) nanocomposites for dielectric energy storage capacitors. Polymer 105: 35–42.

Qu, D. 2009. Mechanistic studies for the limitation of carbon supercapacitor voltage. J. Appl. Electrochem. 39(6): 867–871.

Qu, Y.Q., A.D. Li, Q.Y. Shao, Y.F. Tang, D. Wu, C.L. Mak et al. 2002. Structure and electrical properties of strontium barium niobate ceramics. Mater. Res. Bull. 37(3): 503–513.

Randles, J.E.B. 1947. Kinetics of rapid electrode reactions. Discuss. Faraday Soc., 1, p.11.

Ravindran, R., K. Gangopadhyay, S. Gangopadhyay, N. Mehta and N. Biswas. 2006. Permittivity enhancement of aluminum oxide thin films with the addition of silver nanoparticles. Appl. Phys. Lett. 89(26): 263511.

Resnick, R. and D. Halliday. 1978. Physics. Wiley.

Reynolds, G.J., M. Kratzer, M. Dubs, H. Felzer and R. Mamazza. 2012. Electrical properties of thin-film capacitors fabricated using high temperature sputtered modified barium titanate. Materials 5(12): 644–660.

Ross Macdonald, J. and W.R. Kenan. 1987. Impedance Spectroscopy: Emphasizing Solid Materials and Systems. Wiley-Interscience.

Rysselberghe, P.V. 1931. Remarks concerning the clausius-mossotti law. The Journal of Physical Chemistry 36(4): 1152–1155.

Saha, S.K. 2004. Observation of giant dielectric constant in an assembly of ultrafine Ag particles. Phys. Rev. B 69(12). http://dx.doi.org/10.1103/physrevb.69.125416.

Sahu, V., S. Shekhar, R.K. Sharma and G. Singh. 2015. Ultrahigh performance supercapacitor from lacey reduced graphene oxide nanoribbons. ACS Appl. Mater. Interfaces 7(5): 3110–3116.

Samara, G.A., W.F. Hammetter and E.L. Venturini. 1990. Temperature and frequency dependences of the dielectric properties of $YBa_2\,Cu_3\,O_6 + X$ ($X \cong 0$). Phys. Rev. B 41(13): 8974–8980.

Sani U.S and I.H. Shanono. 2013. A study on carbon electrode supercapacitors. International Journal of Engineering Research and Technology 2: 2597.

Seitz, F. 1987. The Modern Theory of Solids. Dover Publications.

Shao, Y., M.F. El-Kady, L.J. Wang, Q. Zhang, Y. Li, H. Wang et al. 2015. Graphene-based materials for flexible supercapacitors. Chem. Soc. Rev. 44(11): 3639–3665.

Shukla, A., S. Sampath and K. Vijayamohanan. 2000. Electrochemical supercapacitors: Energy storage beyond batteries. Curr. Sci. 79(12): 1656–1661. Retrieved from http://www.jstor.org/stable/24104124.

Simon, P. and A.F. Burke. 2008. Nanostructured Carbons: Double-Layer Capacitance and More. Electrochem. Soc. Interface.

Simon, P. and Y. Gogotsi. 2008. Materials for electrochemical capacitors. Nat. Mater. 7(11): 845–854.

Smith, O.L., Y. Kim, M. Kathaperumal, M.R. Gadinski, M.-J. Pan, Q. Wang et al. 2014. Enhanced permittivity and energy density in neat poly(vinylidene fluoride-trifluoroethylene-chlorotrifluoroethylene) terpolymer films through control of morphology. ACS Appl. Mater. Interfaces 6(12): 9584–9589.

Song, Y., T.-Y. Liu, X.-X. Xu, D.-Y. Feng, Y. Li and X.-X. Liu. 2015. Pushing the cycling stability limit of polypyrrole for supercapacitors. Adv. Funct. Mater. 25(29): 4626–4632.

Spyker, R.L. and R.M. Nelms. 2000. Classical equivalent circuit parameters for a double-layer capacitor. IEEE Trans. Aerosp. Electron. Syst. 36(3): 829–836.

Stoller, M.D., S. Park, Y. Zhu, J. An and R.S. Ruoff. 2008. Graphene-based ultracapacitors. Nano Lett. 8(10): 3498–3502.

Sverjensky, D.A. 2001. Interpretation and prediction of triple-layer model capacitances and the structure of the oxide-electrolyte-water interface. Geochim. Cosmochim. Acta 65(21): 3643–3655.

Talapatra, S., S. Kar, S.K. Pal, R. Vajtai, L. Ci, P. Victor et al. 2006. Direct growth of aligned carbon nanotubes on bulk metals. Nat. Nanotechnol. 1(2): 112–116.

Tang, H. and H.A. Sodano. 2013. Ultra high energy density nanocomposite capacitors with fast discharge using $Ba0.2Sr_{0.8}TiO_3$ nanowires. Nano Lett. 13(4): 1373–1379.

Tan, Y., J. Zhang, Y. Wu, C. Wang, V. Koval, B. Shi et al. 2015. Unfolding grain size effects in barium titanate ferroelectric ceramics. Sci. Rep. 5: 9953.

Tatarova, E., A. Dias, J. Henriques, A.M. Botelho do Rego, A.M. Ferraria et al. 2014. Microwave plasmas applied for the synthesis of free standing graphene sheets. J. Phys. D: Appl. Phys. 47(38): 385501.

Tatarova, E., J. Henriques, C.C. Luhrs, A. Dias, J. Phillips, M.V. Abrashev et al. 2013. Microwave plasma based single step method for free standing graphene synthesis at atmospheric conditions. Appl. Phys. Lett. 103(13): 134101.

Tong, S., B. Ma, M. Narayanan, S. Liu, R. Koritala, U. Balachandran et al. 2013. Lead lanthanum zirconate titanate ceramic thin films for energy storage. ACS Appl. Mater. Interfaces 5(4): 1474–1480.

Tsangaris, G.M., N. Kouloumbi and S. Kyvelidis. 1996. Interfacial relaxation phenomena in particulate composites of epoxy resin with copper or iron particles. Mater. Chem. Phys. 44(3): 245–250.

Tsangaris, G.M., G.C. Psarras and A.J. Kontopoulos. 1991. Dielectric permittivity and loss of an aluminum-filled epoxy resin. J. Non-Cryst. Solids 131-133: 1164–1168.

Valant, M., A. Dakskobler, M. Ambrozic and T. Kosmac. 2006. Giant permittivity phenomena in layered $BaTiO_3$–Ni composites. J. Eur. Ceram. Soc. 26(6): 891–896.

Wagner, K.W. 1914. Erklärung der dielektrischen Nachwirkungsvorgänge auf Grund Maxwellscher Vorstellungen. Arch. Elektrotech. 2(9): 371–387.

Wagner, K.W. 1913. Zur Theorie der unvollkommenen Dielektrika. Ann. Phys. 345(5): 817–855.

Walker, J.S. 2007. Physics, Pearson Prentice Hall.

Wang, C.C. and L.W. Zhang. 2006. Surface-Layer Effect in $CaCu_3Ti_4O_{12}$. Appl. Phys. Lett. 88(4): 042906.

Wang, G., L. Zhang and J. Zhang. 2012. A review of electrode materials for electrochemical supercapacitors. Chem. Soc. Rev. 41(2): 797–828.

Wang, Y., Z. Shi, Y. Huang, Y. Ma, C. Wang, M. Chen and Y. Chen. 2009. Supercapacitor devices based on graphene materials. J. Phys. Chem. C 113(30): 13103–13107.

Wang, Z., H.J. Li, L.L. Zhang and Y.P. Pu. 2014. Effects of BaO–B_2O_3–SiO_2 glass additive on dielectric properties of $Ba(Fe0.5Nb0.5)O3$ ceramics. Mater. Res. Bull. 53: 28–31.

Westerhoff, U., K. Kurbach, F. Lienesch and M. Kurrat. 2016. Analysis of lithium-ion battery models based on electrochemical impedance spectroscopy. Energy Technol. 4(12): 1620–1630.

Whangbo, M.-H. and M.A. Subramanian. 2006. Structural model of planar defects in $CaCu_3 Ti_4 O_{12}$ exhibiting a giant dielectric constant. Chem. Mater. 18(14): 3257–3260.

Wong, D.N., D.A. Wetz, J.M. Heinzel and A.N. Mansour. 2016. Characterizing rapid capacity fade and impedance evolution in high rate pulsed discharged lithium iron phosphate cells for complex, high power loads. J. Power Sources 328: 81–90.

Wuttig, M., D. Lüsebrink, D. Wamwangi, W. Wełnic, M. Gillessen and R. Dronskowski. 2007. The role of vacancies and local distortions in the design of new phase-change materials. Nat. Mater. 6(2): 122–128.

Wu, Z.H., M.H. Cao, Z.Y. Shen, H.T. Yu, Z.H. Yao, D.B. Luo et al. 2007. Effect of glass additive on microstructure and dielectric properties of $SrTiO_3$ ceramics. Ferroelectrics 356(1): 95–101.

Xu, Y., Z. Lin, X. Zhong, X. Huang, N.O. Weiss, Y. Huang et al. 2014. Holey graphene frameworks for highly efficient capacitive energy storage. Nat. Commun. 5. http://dx.doi.org/10.1038/ncomms5554.

Xu, Y., G. Shi and X. Duan. 2015. Self-Assembled three-dimensional graphene macrostructures: Synthesis and applications in supercapacitors. Acc. Chem. Res. 48(6): 1666–1675.

Yadav, P., K. Pandey, V. Bhatt, M. Kumar and J. Kim. 2016. Critical aspects of impedance spectroscopy *in silicon* solar cell characterization: A review. Renewable Sustainable Energy Rev. http://linkinghub.elsevier.com/retrieve/pii/S1364032116309509.

Yang, X., C. Cheng, Y. Wang, L. Qiu and D. Li. 2013. Liquid-mediated dense integration of graphene materials for compact capacitive energy storage. Science 341(6145): 534–537.

Yang, X., X. Zhuang, X. Huang, J. Jiang, H. Tian, D. Wu et al. 2015. Nitrogen-enriched hierarchically porous carbon materials fabricated by graphene aerogel templated schiff-base chemistry for high performance electrochemical capacitors. Polym. Chem. 6(7): 1088–1095.

Yang, X., J. Zhu, L. Qiu and D. Li. 2011. Bioinspired effective prevention of restacking in multilayered graphene films: Towards the next generation of high-performance supercapacitors. Adv. Mater. 23(25): 2833–2838.

Yang, Y., X. Wang and B. Liu. 2013. $CaCu_3Ti_4O_{12}$ ceramics from different methods: Microstructure and dielectric. J. Mater. Sci.—Mater. Electron. 25(1): 146–151.

Yoshida, A., I. Tanahashi and A. Nishino. 1990. Effect of concentration of surface acidic functional groups on electric double-layer properties of activated carbon fibers. Carbon 28(5): 611–615.

Young, A., G. Hilmas, S.C. Zhang and R.W. Schwartz. 2007. Effect of liquid-phase sintering on the breakdown strength of barium titanate. J. Am. Ceram. Soc. 90(5): 1504–1510.

Yuan, K., T. Hu, Y. Xu, R. Graf, G. Brunklaus, M. Forster et al. 2016. Engineering the morphology of carbon materials: 2D porous carbon nanosheets for high-performance supercapacitors. J. Mater. Chem. 3(5): 822–828.

Yu, S., F. Qin and G. Wang. 2016. Improving the dielectric properties of poly(vinylidene fluoride) composites by using poly(vinyl pyrrolidone)-encapsulated polyaniline nanorods. J. Phys. D: Appl. Phys. 4(7): 1504–1510.

Zang, G., J. Zhang, P. Zheng, J. Wang and C. Wang. 2005. Grain boundary effect on the dielectric properties of $CaCu_3 Ti_4 O_{12}$ ceramics. J. Phys. D: Appl. Phys. 38(11): 1824–1827.

Zhang, J.L., P. Zheng, C.L. Wang, M.L. Zhao, J.C. Li and J.F. Wang. 2005. Dielectric dispersion of $CaCu_3Ti_4O_{12}$ ceramics at high temperatures. Appl. Phys. Lett. 87(14): 142901.

Zhang, L.L. and X.S. Zhao. 2009. Carbon-based materials as supercapacitor electrodes. Chem. Soc. Rev. 38(9): 2520.

Zhang, W., H. Lin, H. Kong, H. Lu, Z. Yang and T. Liu. 2014. High energy density PbO2/activated carbon asymmetric electrochemical capacitor based on lead dioxide electrode with three-dimensional porous titanium substrate. Int. J. Hydrogen Energy 39(30): 17153–17161.

Zhang, X., Y. Shen, B. Xu, Q. Zhang, L. Gu, J. Jiang, J. Ma, Y. Lin, and C.-W. Nan. 2016. Giant energy density and improved discharge efficiency of solution-processed polymer nanocomposites for dielectric energy storage. Adv. Mater. 28(10): 2055–2061.

Zhao, J., H. Lai, Z. Lyu, Y. Jiang, K. Xie, X. Wang et al. 2015. Hydrophilic hierarchical nitrogen-doped carbon nanocages for ultrahigh supercapacitive performance. Adv. Mater. 27(23): 3541–3545.

Zhi, M., C. Xiang, J. Li, M. Li and N. Wu. 2013. Nanostructured carbon–metal oxide composite electrodes for supercapacitors: A review. Nanoscale 5(1): 72–88.

Zhuang, X., F. Zhang, D. Wu and X. Feng. 2014. Graphene coupled schiff-base porous polymers: Towards nitrogen-enriched porous carbon nanosheets with ultrahigh electrochemical capacity. Adv. Mater. 26(19): 3081–3086.

Zhuang, X., F. Zhang, D. Wu, N. Forler, H. Liang, M. Wagner et al. 2013. Two-dimensional sandwich-type, graphene-based conjugated microporous polymers. Angew. Chem. Int. Ed. 52(37): 9668–9672.

Zhu, Y., S. Murali, M.D. Stoller, K.J. Ganesh, W. Cai, P.J. Ferreira et al. 2011. Carbon-based supercapacitors produced by activation of graphene. Science 332(6037): 1537–1541.

Zubieta, L. and R. Bonert. 2000. Characterization of double-layer capacitors for power electronics applications. IEEE Trans. Ind. Appl. 36(1): 199–205.

Index